BOTANICAL RESEARCH AND PRACTICES

THE FLORA OF NARA DESERT, PAKISTAN

RAHMATULLAH QURESHI

Nova Science Publishers, Inc.
New York

For permission to use material from this book please contact us:
Telephone 631-231-7269; Fax 631-231-8175
Web Site: http://www.novapublishers.com

Library of Congress Cataloging-in-Publication Data

The flora of Nara Desert, Pakistan / editor: Rahmatullah Qureshi.
p. cm.
Includes index.
ISBN 978-1-62081-638-7 (hardcover)
1. Plants--Thar Desert (India and Pakistan) 2. Desert plants--Thar Desert (India and Pakistan) I. Qureshi, Rahmatullah.
QK358.5.F48 2012
581.7'54--dc23 2012013591

Published by Nova Science Publishers, Inc. † New York

Contents

List of Tables		vii
List of Figures		ix
List of Plates		xi
Foreword		xiii
Preface		xv
Chapter I	Introduction	1
Chapter II	Geology, Topography and Soils	3
Chapter III	Climate	9
Chapter IV	Plants and Vegetation Types	13
Chapter V	Previous Work and Present Study	19
Chapter VI	Materials and Methods and Plan of the Flora	21
Chapter VII	Results	25
Figures		285
References		309
Index		313

List of Tables

Table 1. Shows the Mean of Monthly Maximum and Highest Temperature Recorded in the Year 1998-2000 **10**

Table 2. Alphabetically Family-wise Arranged List of Plants along with Local Name, Life Form and Life Span **25-33**

Table 3. Contribution of Groups/Families in Terms of Genera and Species of the Flora of Nara Desert **33-34**

Table 4. Life Form of the Flora of Nara Desert, Pakistan **35**

List of Figures

Figure 1. Blepharis scindica (Acanthaceae) [A]; Limeum indicum (Aizoaceae) [B]; Trianthema portulacastrum (Aizoaceae) [C]; Achyranthus aspera (Amaranthaceae) [D] **285**

Figure 2. Aerva javanica var. bovei (Amaranthaceae) [A]; A. javanica var. javanica (Amaranthaceae) [B]; Amaranthus viridis (Amaranthaceae) [C]; Celosia argentea (Amaranthaceae) [D] **286**

Figure 3. Digera muricata (Amaranthaceae) [A]; Aristolochia bracteolata (Aristolochiaceae) [B]; Calotropis procera (Asclepiadaceae) [C]; Leptadenia pyrotechnica (Asclepiadaceae) [D] **287**

Figure 4. Oxystelma esculentum (Asclepiadaceae) [A]; Amberboa ramosa (Asteraceae) [B]; Carthamus oxycantha (Asteraceae) [C]; Eclipta prostrata (Asteraceae) [D] **288**

Figure 5. Iphiona grantioides (Asteraceae) [A]; Launaea procumbens (Asteraceae) [B]; Pluchea lanceolata (Asteraceae) [C]; Sonchus oleraceus (Asteraceae) [D] **289**

Figure 6. Xanthium indicum (Asteraceae) [A]; Heliotropium crispum (Boraginceae) [B]; Heliotropium subulatum (Boraginceae) [C]; Trichodesma indicum (Boraginceae) [D] **290**

Figure 7. Senna italica (Caesalpiniaceae) [A]; Cadaba farinosa (Capparidaceae) [B]; Capparis decidua (Capparidaceae) [C]; C. spinosa (Capparidaceae) [D] **291**

Figure 8. Cleome brachycarpa (Capparidaceae) [A]; C. scaposa (Capparidaceae) [B]; C. viscosa (Capparidaceae) [C]; Gynandropsis gynandra (Capparidaceae) [D] **292**

Figure 9. Vaccaria hispanica (Caryophyllaceae) [A]; Haloxylon salicornicum (Chenopodiaceae) [B]; H. stocksii (Chenopodiaceae) [C]; Chenopodium album (Chenopodiaceae) [D] **293**

Figure 10. Chenopodium murale (Chenopodiaceae) [A]; Salsola imbricata (Chenopodiaceae) [B]; Saueda fruticosa (Chenopodiaceae) [C]; Convolvulus arvensis (Convolvulaceae) [D] **294**

Figure 11. Convolvulus prostratus (Convolvulaceae) [A]; Cressa cretica (Convolvulaceae) [B]; Ipomoea aquatica (Convolvulaceae) [C]; Citrullus colocynthis (Cucurbitaceae) [D] **295**

Figure 12. Cucumis melo var. agrestis (Cucurbitaceae) [A]; Mukia maderaspatana (Cucurbitaceae) [B]; Bergia aestivosa (Elatinaceae) [C]; Alhagi maurorum (Fabaceae) [D] **296**

Figure 13. Crotalaria burhia (Fabaceae) [A]; Crotalaria medicagnea (Fabaceae) [B]; Indigofera argentea (Fabaceae) [C]; I. cordifolia (Fabaceae) [D] **297**

Figure 14. Indigofera hochstetteri (Fabaceae) [A]; Lathyrus aphaca (Fabaceae) [B]; Medicago polymorpha (Fabaceae) [C]; Melilotus alba (Fabaceae) [D] **298**

Figure 15. Melilotus indica (Fabaceae) [A]; Tephrosia falciformis (Fabaceae) [B]; T. uniflora var. petrosa (Fabaceae) [C]; T. villosa (Fabaceae) [D] **299**

Figure 16. Vicia sativa (Fabaceae) [A]; Glinus lotoides (Molluginaceae) [B]; Mollugo cerviana (Molluginaceae) [C]; M. nudicaulis (Molluginaceae) [D] **300**

Figure 17. Boerhavia procumbens (Nyctaginaceae) [A]; Commicarpus boisseiri (Nyctaginaceae) [B]; Rumex dentatus (Polygonaceae) [C]; Anagallis arvensis var. coerulea (Primulaceae) [D] **301**

Figure 18. Zizyphus nummularia (Rhamnaceae) [A]; Schweinfurthia papilionacea (Scrophulariaceae) [B]; Physalis minima (Solanaceae) [C]; Withania somnifera (Solanaceae) [D] **302**

Figure 19. Corchorus aestuans (Tiliaceae) [A]; C. tridens (Tiliaceae) [B]; Phyla nodiflora (Verbenaceae) [C]; Tribulus terrestris (Zygophyllaceae) [D] **303**

Figure 20. Aeluropus lagopoides (Poaceae) [A]; Avena fatua (Poaceae) [B]; Brachiaria eruciformis (Poaceae) [C]; Cenchrus bifloris (Poaceae) [D] **304**

Figure 21. Cenchrus ciliaris (Poaceae) [A]; Cymbopogon jwarancusa (Poaceae) [B]; Cynodon dactylon (Poaceae) [C]; Dactyloctenium aegyptium (Poaceae) [D] **305**

Figure 22. Desmostachya bipinnata (Poaceae) [A]; Lasiurus scindicus (Poaceae) [B]; Ochthochloa compressa (Poaceae) [C]; Phalaris minor (Poaceae) [D] **306**

Figure 23. Phragmites australis (Poaceae) [A]; Polypogon fugax (Poaceae) [B]; Setaria pumila (Poaceae) [C]; Stipagrostis plumosa (Poaceae) [D] **307**

List of Plates

Plate 1. Crest habitats with xerophytic perennial shrubs **4**

Plate 2. Slope habitats occupied by some perennial plant species **4-5**

Plate 3a. Flat habitats giving a look of mini forest of *Prosopis cineraria* **5**

Plate 3b. Cactus like groove of *Euphorbia caducifolia* in Chotiari wetlands island **6**

Plate 3c. Wild datepalm groove in Nara Desert **6**

Plate 4. Saline/sodic lands (*Kharror*) pioneered by halophytes **8**

Plate 5. Wetlands represented by hydrophytes **16**

Plate 6. Hilly tracts showing lithophytes in the Rohri Hills **17**

Foreword

The flora of the Nara Desert is a comprehensive account of the geomorphology, climate and the flora that this region contains. More than 250 species are described with their distribution, habitat details, phenology, and local names. Reference to other floras and relevant synonymy is also given. The book is illustrated with maps and excellent photographs.

I am particularly pleased to be asked to write this Foreword, because I am passionate about Floras, especially desert Floras. Regional Floras, as this is one, are particularly useful as they concentrate on a smaller area, and where more information on species, their local uses, common names, and conservation status can be placed. A good regional Flora puts together information about a small area in greater detail – which can be well used in its own right or for a bigger country wide flora. This is one such work and I am glad to be able to see it completed.

Deserts are harsh ecosystems where plants, limited by the availability of water and good soil, have adapted themselves in various ways to live and reproduce. They are excellent examples and research tools for life in arid environments from where we have, and are still learning, on how to live under high temperatures, excessively saline soils and very little available water. With the global climate change towards warming which would result in aridity in many areas, these desert plants form an invaluable tool for us to study and research on ways for plants to live under arid conditions, and to use the research for better use of arid environments and cultivation in a sustainable way.

As a Flora writer and editor myself I know how much time and effort it takes to gather data and produce a book. I congratulate Rahmatullah Qureshi on his excellent work and commend this Flora to all who work on arid environments.

Shahina A. Ghazanfar
Head of Temperate Regional Team
Herbarium, Library, Art & Archives Directorate
Royal Botanic Gardens Kew,
Richmond, Surrey TW9 3AB, UK.

Preface

The complete survey of the flora of Nara desert was delineated. This area being hostile and harsh climate has neglected for the comprehensive and scientific study point of view. The present work is thus a contribution of more than 10 years exploration and collection of plant specimens from the whole study area.

This is the first ever effort to present the correct arrangement of our plants by following the system of Cronquest (1983) with slight alteration. The present work includes 59 families, 167 genera and 266 species including 3 ferns, 1 gymnosperm and 64 monocots. Detailed descriptions of the plants along with dichotomous keys to the families, genera and species have been provided with complete references of the taxa, voucher specimens from own collections. Besides, local names, localities and world distribution have also been provided along with field notes.

The illustrations of the representative species incorporated in the books are original and based on the specimens cited in the text.

These diagrams will provide insight to make it more popular amongst the young botanists of the country, foreigners, foresters, range managers and agriculturist. While conducting my field expedition in the northern part of the desert, I would like to express my thanks to Sardar Mohammad Bux Khan Maher, who provided base camp facilities at Dilkusha Bungalow at Sardar Garh (Ghotki) and to LASMO oil Pakistan for providing accommodation in their Camp site during plant expedition. Thanks go to Indus For All Programme (IFAP), a subsidiary organization of World Wide Fund for Nature (WWF), Pakistan for providing opportunity to work on the vegetation of Chotiari wetlands complex located in the southern extreme of Nara Desert called *Achhro Thar* (white desert).

I am very much indebted to my supervisor Prof. Dr. Raza Bhatti for his continuous help, support and encouragement. Without his personal relentless efforts, it was almost impossible to complete this project.

Thanks go to Dr. Muqarrab Shah, PMNH, Islamabad his support. I would express my thanks to Mr. Shakeel Ahmed Khaskheli for making illustration of desert plants. I thankfully acknowledge Pakistan Science Foundation, Islamabad for providing funds for my research project entitled, "Floristic Study of Arid Zone (Desert- Nara, Region) Sindh, Pakistan, No. SALU/ENVR (45)".

Thanks goes to my PhD scholar Ms Humaira Shaheen for cross checking and some edit work on the manuscript. I am also thankful to my parents and wife for their loyal support.

I am also indebted The publishers M/S Nova Science Publishers, N.Y. and especially Ms. Naydya Gordrich, Manager for publishing a technical Book, which is first of its kind on one of the deserts of Pakistan, since the establishment of Pakistan.

Dr. Rahmatullah Qureshi
Associate Professor of Botany,
Pir Mehr Ali Shah Arid Agriculture University,
Murree Road, Rawalpindi
Pakistan

Chapter I

Introduction

1. The Study Area

Sindh is the southeastern Province of Pakistan, bounded on the east by the Indian border of Rajasthan, in the south by the Ran of Kutch, the Arabian Sea, in the west by the arid rocky mountains of Baluchistan and in the North by irrigated plains of Punjab.

About 88,000 sq. km. area of Sindh has been regarded as Arid Zone that is nearly 60% of total geographical area of Sindh province (Anonymous, 1992). Administratively, Sindh Arid Zone Development Authority (SAZDA) has divided the Arid Zone of Sindh Province into three regions as follows:

(i) Kohistan (the western side of the Indus Valley),
(ii) Thar (the eastern area of Sindh Province).

Thar Desert further subdivided into the **Nara region** in the North and **Thar region** in the South. The **"Nara Desert"** lies between 26° to 28° North and 68° to 70° East. The altitude is between 50 and 115 meters above mean sea level (amsl). It spreads over from Taluka Ubaro, Daharki, Mirpur Mathelo & Khan Pur Mahar of Ghotki District, Rohri and Saleh Pat taluka of Sukkur District, Nara Taluka of Khairpur district; Taluka Sanghar and Khipro of district Sanghar. Eastern side boundaries of all these talukas have been marked by Rajasthan, India (Jodhpur & Jaisalmer). The study area is semiarid in nature.

The northern part lies between sandy arid plains in the west and semi-arid transitional plain in the east. The topography is marked with hills along steep slopes and extensive alluvium plain covered with aeolian sand deposits. Gullying is also a common feature in submontane region. North of the fertile soil up to the border is the plain area of interior drainage. This plain is full of sand hills, sand scapes and several low depressions in which salt and soda are deposited on drying.

About 80% of the area is a gently undulating sandy plain with frequent sand dunes, the number and size of which increases from West to East. The general surface of the desert is a wild maze of sand dunes and sand ridges, occasionally rising 500 feet above the general surface. The Desert soils include rolling to hilly sandy soils and clayey flood plain soils, where the soils are formed of deep sands. They are moderately calcareous and largely aeoline. In these places the wind blown material is mixed with old alluvium.

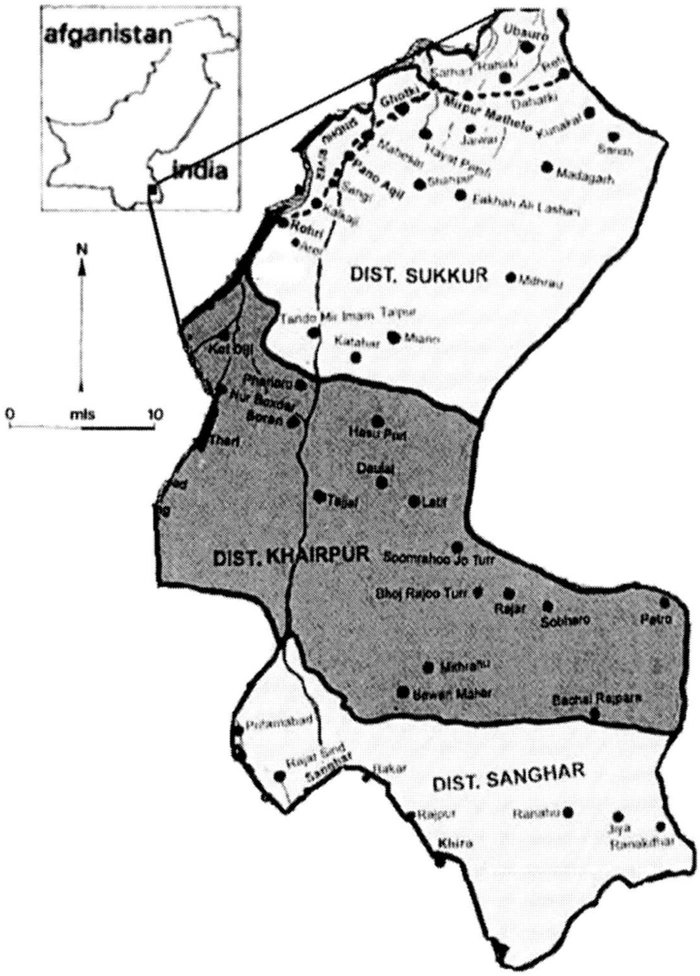

The Nara canal forms the eastern margins of the eastern valley section. This is covered with blown sand and crossed by infinite meander scars and narrow depressions locally known as "*Dhoros*". Some of *Dhoros* (dry water channels) are roughly 300-600 feet wide and 2-10 meter long. These dry water channels are without apparent continuation at either end, and seem to merge into the neighboring sand hills. Apparently these scars are fragments of old drainage systems, now disrupted by river shifts or sand encroachments. In addition to dry channels, some 400 dhands occupy the valley. Some of these dhands (shallow lakes) are as much as 3 miles long and one mile broad. These dhoros and numerous strips of meander flood plain indicate that in the recent past the River Indus flowed across most of the eastern valley section.

Chapter II

Geology, Topography and Soils

1. Geology

Nara desert forms a part of Indo-Genetic plain formed by the gradual filling up of the trough lying between the foothills of Himalayas and central core of the subcontinent. The alluvial soils brought by the river Indus lies over the tertiary shale and limestone. These rocks belong to Paleozonic, Mezozonic and Tertiary geological time scale (Anonymous, 2002).

2. Topography

The Nara Desert is rolling in desert landscape to nearly level in cultivated areas along the Nara canal. Soils of the study area were formed in aeolian sands originally deposited in Pleistocene age, but reworked by wind until recently to take the form of sand dunes. A minor extent of soils has been formed in a narrow alluvial plain of Nara canal, which has its origin as a small inundation river.

The topography of an area under investigation is distinctly marked with sandy hills, steep slopes and vast low laying areas locally known as *Patt*. The accumulation of sand in huge mass in the form of a hill is known as sand dune/ridges. Dune has crest; the topmost portion, swale (flank); the middle portion and foot; the base of the dune (Plates 1-3). The area between two successive dunes is called interdunal valley.

The dunes are of two major types either stabilized or unstabilized one. The unstabilized dune is a movable heaps of sand, not really fixed on the ground but can be moved with the wind direction. It lacks or with scanty vegetation, whereas, the stabilized dunes are characterized by some mixed population of plant species. The area under investigation has stabilized dunes with low lying flat area; the valleys.

The typical features responsible for such dunes are the wind, blows steadily against an obstacle, which results in the collection of sand on the windward side, forming a long slope of sand. Whereas it's other side, the leeward side, the dune drops in a sharp curving cliff.

Plate 1. Crest habitats with xerophytic perennial shrubs.

Plate 2. (Continued).

Plate 2. Slope habitats occupied by some perennial plant species.

Plate 3a. Flat habitats giving a look of mini forest of *Prosopis cineraria*.

Plate 3b. Cactus like groove of *Euphorbia caducifolia* in Chotiari wetlands island.

Plate 3c. Wild datepalm groove in Nara Desert.

Hummocky dunes are also common features of this area. These are formed when sand gets trapped around a perennial trees or shrubs growing over the top of the sand dunes (crest) or its sides (slopes). It has been observed that the majority of the older bushes are growing on the hummocks, whereas, the younger ones found without the hummock.

Panhwar (1964) demonstrated that the Thar in Hyderabad and Khairpur Division (at present Hyderabad, Mirpurkhas, Sukkur) is formed by sand, silt, salts and fine sea shells blowing from Runn of Kuchh, extended up to Rohri in the recent geological times. In Tharparkar districts due to high wind velocities the sand hills which are some times 300 feet high, lies south west to north east along the direction of wind. He further explained that the southern Khairpur where wind velocity decreases the direction of the sand hills changes slightly and becomes south- south-west to north-north-east. In northern Khairpur and Sukkur districts (presently includes Ghotki district), the sand hills lie south-north. The sand hills of Khairpur are not as high as those of Thar. Further he explained that there are *Dhandh* (lakes/wetlands) which are formed by rainwater sweeping down the sand mounds. Regarding the dhands in the close proximity of Nara Canal, he described them to get recharged continuously due the seepage from the canal. The rain water percolates down the sand where it is held by impervious clay which lies under the sand. The water so trapped or perched out in the form of spring, are locally called "*SIM*". All lakes in the desert area are formed by this process called as *Sim*. The salt contents in these lakes depend upon the salts dissolved by rain water and appearing as *Sim*. Water in most the dhands is brackish as these salts keep on accumulating in them year after year unless these salts are removed by human hands when water evaporates leaving cakes of salt behind. In some cases water from wells in the neighborhood is available at higher water table then in the dhands and is potable. This is actually the "*SIM*" water, which instead of seeping into lakes seeps into the wells.

The dhands are shallow, but some of them extend about a mile in length. At one time, these lakes yielded soda ash. Dhands close to Nara are even now always full of water due to seepage from the canal but this water no longer so brackish has instead become abode of crocodiles and water loving reeds and canes like '*Sar'*. The lakes away from the canal are highly saline. The range of alkalinity of the lakes of Khaipur is higher as compared with the Thar lakes. There is always a dispute about the causes of this phenomenon. It is believed that the salt of the lakes whether in Khairpur or Sanghar district are basically of similar ingredients but in the Khairpur lakes due to present of some bacteria or organic material the sodium salts are converted into soda ash.

3. Soils

The soil of the desert is sandy silty and moderately calcareous. The entire project area consists of level and gentle slopping plains broken by sand dunes. Brown sandy soil, desert soils and gravelly foot slopes constitute the major soil types in the area. The color of the sand dunes is generally grey, greyish-white or yellowish-grey.

The desert margin mostly consists of flat ground having saline/sodic soils commonly known as "*Kharror*" (Plate 4). Baqri et al. (1992) carried out investigation of soil and geology of Cholistan Desert.

They reported a large number of shells of Pelecypodes and Gastropodes from the river beds revealing the presence of large water bodies at the time of immemorial which got dried due to extreme climate change. The presence of Pelecypodes and Gastropodes from the flat area/valleys at various places of project area is probably due to the occurrence of water in the time immemorial.

Plate 4. Saline/sodic lands (*Kharror*) pioneered by halophytes.

Chapter III

Climate

Climate, topography and edaphic factors play a significant role in determining the vegetation (the total aggregation of plants of an area). Most of the areas of Nara Desert classified as an arid tract, characterized by extreme temperature (a long dry summer), severe drought supplemented by high wind velocity, low relative humidity, high evaporation rate and too scanty rainfall. The southern part (Khipro, Sanghar) is relatively semi-arid in nature which supports somewhat dense vegetation.

1. Temperature

There is no meteorological station within the study area but four stations Sukkur, Padidan, Nawab Shah & Moen Jo Daro are located within about 100 km radius of the project area. The information on temperature for the years 1998, 1999 & 2000 collected from the above weather stations. The mean of monthly maximum and the highest temperature for the period of 3 years is provided in the Table-2. This table shows that the hottest months are May-July, where temperature ranges from 47.6-51°C. The lowest temperature 26.9-28.1°C is recorded in the month of January.

Such environment has great influence on the natural vegetation/flora of the area. The herbaceous communities and ephemeral species may not emerge every year. These species are well adapted to survive the drought periods and respond quickly in the case of rain. After emerging with the first shower, these species grow instantly, produce seeds and complete their life cycle in a span of 7 to 8 weeks. Their seeds remain dormant throughout the dry period.

2. Rainfall

The whole desert continues to get heated till onset of monsoon by the middle of January. The monsoon currents receive from the Bay of Bengal and Arabian Sea. More than 90% rainfall is received from June to September. The month of July is the rainiest receiving 90 to 95 % rainfall. The average annual rainfall varies from 88 mm to 135 mm.

Table 1. Shows the Mean of Monthly Maximum & Highest Temperature Recorded in the year 1998-2000

Month	**Sukkur**		**Padidan**		**Nawab Shah**		**Moen Jo Daro**	
	Min. °C	**Max °C**	**Min. °C**	**Max. °C**	**Min. °C**	**Max.° C**	**Min. °C**	**Max. °C**
January	26.9	31	28.1	39	28.9	33.5	27.1	29
February	31	38	31.7	38	32.7	37	31	35
March	37.2	42.2	38.3	43	39.3	43	37.3	40.4
April	43.4	47	44.3	48	44.1	48	43.8	45.3
May	46.7	49	47.9	51	47.4	49	47.1	48.5
June	47.3	50	48.4	51	47.4	51	47.4	48.9
July	44.7	47	45.4	48	44.3	48	45.9	47.6
August	42	47	42.9	47	41.8	47	42.3	46
September	40.9	43.3	42	44.4	41.4	44.5	40.4	43.6
October	39.4	44	40.7	45	40.7	43	40.4	49.5
November	34.7	38	35.7	38.3	36.3	39	34.4	36.7
December	28.2	33	29.2	33	30	35	28.4	29.6

Five to ten percent of the annual rainfall occurs in the form of local showers during winter season and thunder storms particularly during June-July.

There is a scarcity of water in the Nara desert. Water supplies are scanty which is limiting the agriculture practice. The ground water resources are limited and are met at a depth of 50-300 feet from the surface. The only source of water for human beings and livestock is from dugout/natural ponds, in which the water is stored during monsoon. The livelihood of people are largely dependent on their livestock i.e. sheep, goat and camels.

3. Water Resources

3.1. Surface Water

Nara canal, its tributaries and natural surface water impoundments (locally called as dhands) are the source of surface water. This canal and its system is one of the Sindh's 14 main canal systems. In terms of Canal Command Area (CCA), the Nara command is the largest. The canal originates from Sukkur Barrage at Sukkur and was built in 1932. Where there is Nara Water Command there is agriculture. This canal is perennial except for a closure period in mid of January. The highest discharges in the Nara canal are recorded in May, June and July. The Nara Canal Command Area (NCCA) supplies ample fresh water in this area, where the locals use it for drinking, domestic and irrigate agriculture fields as well. Water table in the agriculture lands is normally 3-5 m. During field surveys 40 dhands were visited. These dhands within the desert area are used by the locals and livestock for domestic purposes.

3.2. Subsoil/Ground Water Resources

There is a scarcity of water in the area under investigation resulting in limited agriculture practice. The depth of water table near the Nara canal is found to be 2.5m to 5m and gradually increases to 10 to 18m in eastern and western desert margins. Groundwater flow in the area is perpendicular to Nara canal in southwest direction. The source of water for human beings and livestock is from dugout/natural ponds, in which the water is stored during monsoon. The source of "*dhands*" is seepage water from Nara canal and surfacing of shallow ground water. Some of the major *Dhands* in the area are Berwari Dhand at Tajjal, Nangpir Dhand near Goth Ghulam Muhammad, and Samni dhand north of Methewaro Goth. Dhands within the desert are used by locals and livestock for domestic purposes, although not all of them are drinkable. Dhands along the Nara canal contain sweet water.

4. Relative Humidity

The minimum relative humidity is recorded in the summer season, particularly in April and may when it ranges between 35 to 60 per cent in the morning hours and 10 to 30 per cent in the afternoon. The maximum humidity is noted during the rainy season, particularly mid

July to mid August (90%). During the winter season, particularly mid December to February, the relative humidity varies from 50-60% in morning and 25 to 35 per cent in the afternoon (Anonymous, 2002).

5. Wind Velocity

During hot and rainy seasons, the winds blow between southwest and west. Dust storms are very commonly observed during summer season particularly in the northern arid areas. Maximum dust storm occurs during May and June in the study area. Winds are very light and variable during post-monsoon period and winters i.e. October to February.

Chapter IV

Plants and Vegetation Types

Plant Types

The vegetation in this region is sparse consisting mainly of stunted, thorny or prickly shrubs and perennial herbs capable of drought resistance. Trees are few and scattered. The ephemerals come up during the rainy season, complete their life cycle before the advent of summer and the bulk of the area is once more transformed into open sandy plain, desolate and barren. Four types of plant groups observed in this area which are described as below on the basis of their characters and their performance:

True Xerophytes

Succulence
Covered with trichomes.
Reduce the size and length of transpiring parts (i.e. leaves), leaflessness etc. includes *Haloxylon stocksii, Arthrocnemum indicum, Saueda fruticosa, Salsola imbricata, Aerva* spp., *Glinus lotoides*, etc.

Semi-xerophytes

Plants which grow in arid climate in the presence of sufficient amount of precipitation. *Acacia nilotica, Calotropis procera, Heliotropium europium, H. currasavicum* and *Withania somnifera* are the members of this group. These plants are mostly growing on the periphery of the desert.

Pseudo-xerophytes

Ephemerals which complete their life cycle within 6-8 weeks before the advent of summer includes *Neurada procumbense, Mollugo* spp., *Gisekia pharnacoides,*

Cleome brachycarpa and *Boerhavia procumbense* and grasses like *Eragrostis* spp., *Stipagrostis plumosa* and *Aristida* spp. are the member of this category.

Hydrophytes

Plants which live in water or in very moist habitat are members of this category. The members of this category are Bulboschoenus glaucus, Cyperus rotundus, C. laevigatus, Fimbristylis dichotoma, Phragmites karka, Potamogeton perfoliatus, Salvinia molesta, Schoenoplectus littoralis and Typha spp.

Vegetation

Based on rainfall ranges, apparently this desert has two seasons. During the moonsonic showers, most of the plants grow vigorously. This is the best growing period and maximum area is found occupied with all types of plants especially ephemerals. From middle of September to middle of June, most of the area gets dried and the annual flora disappeared. During the dormant period, only perennial plants mainatain the vegetation cover especially on dunal and sloppy areas of the desert by shedding their leaves and drying up their aerial parts. Due to trapping down maximum rainfall water in the subsoil, the perennial plants in the valleys and plains are capable to continue their growth. Since the climate is more or less homogenous, the vegetation seems to edaphic controlled. The vegetation may be:

Sand Dunes and Interdunal Areas (Plates 1-3)

This is the most common habit of this region. Sand dunes of different types. They may be stabilised, partially stabilised or unstabilised and barren. The common trees and shrubs found on stabilised and unstabilised sand dunes are: *Acacia senegal, Prosopis cineraria, Salvadora oleoides, Tamarix aphylla* and *Calligonum polygonoides.* The other common plant species of sand dunes include *Aerva javanica, Leptadenia pyrotechnica, Crotalaria burhia, Tephrosia* spp., *Indigofera* spp., *Tribulus longipetalus,* etc. The common grasses are: *Aristida* spp., *Cenchrus ciliaris, C. biflorus, Cymbopogon jwarancusa, Dactyloctenium aegypticum, Eragrostis minor, Lasiurus sindicus, Panicum turgidum* and *Stipagrostis plumosa.*

The inter-dunal gaps support more luxuriant vegetation due to greater availability of moisture. The common trees and shrubs are: *Acacia jacqimontii, A. senegal, Prosopis cineraria, Salvadora oleoides, Tamarix aphylla, Calotropis procera, Capparis decidua, Ziziphus nummularia* etc. In addition, almost all the plants found on sand dunes grow more profusely in the interdunal area.

At the foot of the isolated hills and rock outcrops, the windward sides of which hold up the movement of sand, blockage sand dunes are often formed. On these dunes, a mixture of characteristic elements of sand dunes and rock outcrops like *Acacia Senegal, Calligonum polygonoides* are found.

The un-stabilized sand dunes are least covered with plants. However, *Aerva javanica, Crotalaria burhia, Dipterygium glaucum, Leptadenia pyrotechnica*, etc. are some of the pioneer species to colonize these dunes. The very crests of such sand dunes are often colonized by varying frequency of *Lasiurus sindicus, Panicum turgidum* and *Citrulus colocynthis*, etc.

Sandy and Hummocky Plains (Plate 3a,b,c)

Sandy and hummocky plains constitute the major part of the desert and plants growing in such habitats form some of the characteristic associations of the region. The common trees and shrubs are *Acacia senegal, Calligonum polygonoides, Calotropis procera, Capparis decidua, Dipterygium glaucum, Euphorbia caducifolia, Prosopis cineraria, Salvadora oleoides, Tamarix aphylla* and *Ziziphus nummularia.* In addition, under-shrubs and herbs like *Aerva javanica, Arnebia hispidissima, Boerhaavia procumbens, Convolvulus prostratus, Crotalaria burhia, Heliotropium crispum, H. strigosum, Indigofera oblongifolia, I. cordifolia, Leptadenia pyrotechnica, Tephrosia uniflora,* etc. are also common. The most common creepers are *Citrulus colocynthis, Cucumis melo* var. *agrestis, C. prophetarum* and *Mukia maderaspatana.* Most of these plants possess well developed root system and occur in open clump formations with plenty of vacant spaces between them which are occupied by several ephemerals and grasses like the species of *Aristida* spp., *Cenchrus* spp. and *Lasiurus sindicus.*

Saline Tracts (Plate 4)

There are large tracts spread throughout the desert. The common plants of this habitat are Aeluropus lagopoides, Cressa cretica, Haloxylon stocksii, Salsola baryosma, Suaeda fruticosa, Tamarix indica, Zygophyllum simplex, etc.

Aquatic Habitat (Plate 5)

The area maintains considerable number of fresh water, salt lakes and tanks with rocky substratum. Chotiari is the largest fresh water reservoir that lies on western flanks of *Achro Thar* (white sandy desert), the last boundary of Nara Desert towards Sanghar. The Reservoir occupies an area of about 18,000 hectares and has water storage capacity of 0.75 Million Acre Feet (MAF) flooding an area of approximately 160 km2. Chotiari reservoir is created in a natural depression that exists along the left bank of the Nara canal. The depression area is encircled by sand hills towards north, east and south-east, while towards the west and south lay the Nara canal. All these aquatic habitats support a reasonable number of aquatic species like *Bulboschoenus glaucus, Cyperus rotundus, C. laevigatus, Eclipta prostrata, Fimbristylis dichotoma, Phragmites karka, Potamogeton perfoliatus, Salvinia molesta, Schoenoplectus littoralis* and *Typha* spp.

Gravelly/Rocky Plains (Plate 6)

Gravel, formed by the action of wind, covers fairly large area of the region. Some of the common plants of this habitats are: *Cleome scaposa, C. bracycarpa, Fagonia indica, Indigofera* spp., *Heliotropium crispum, Iphiona grantioides, Leptadenia pyrotechnica, Blepharis sindica*, etc. certain plants of gravel are prostrate and star like with the branches remaining appressed to the ground e.g. *Euphorbia clarkeana, E. granulata, Indigofera cordifolia, I. hochstetteri, Mollugo cerviana, M. nudicaulis, Tribulus terrestris,* etc. The common trees or shrubs of this habitat are *Calotropis procera, Capparis decidua, Prosopis cineraria, Salvadora oleoides, Zizyphus nummularia,* etc. At certain spots, the gravel plains maintain grass-legume association. The common grasses are *Aristida* spp., *Cenchrus ciliaris, C. biflorus, Cymbopogon jwarancusa, Dactyloctenium aegypticum, Eragrostis minor, Lasiurus sindicus* and *Stipagrostis plumosa.*

The gravel plains are marked by deep valleys or shallow depressions, particularly near the foot of hills, the typical rock plants which become prominent along with some ramblers in such habitats are *Ephedra ciliata, Convolvulus prostrates*, *Kikxia ramisosissima,* etc. are found. The hill-slopes, however, maintain better vegetation due to the accumulation of sand and better water holding capacity. The common plants of these habitats are *Acacia senegal, Capparis decidua, Grewia tenax, Maerua crassifolia* and *Ziziphus nummularia.*

Plate 5. Wetlands represented by hydrophytes.

Plate 6. Hilly tracts showing lithophytes in the Rohri Hills.

The vegetation is comparatively denser at the foothills and nearby areas. *Prosopis cineraria* and *Tamarix aphylla* were main forest forming trees in such habitats. Other plant species include *Blepharis sindica, Cappris decidua, Salvadora oleoides, Schweinfurthia papilionacea, Senna italica, Tephrosia* spp., *Aristida adscensionis,* etc.

Endemic and Rare Taxa

The desertic zone due to its characteristic topography, geology, edaphic and climatic factors maintains a peculiar type of vegetation, not found elsewhere in other areas. Therefore, the recorded taxa are native to this area. Plant species such as *Aerva javanica, Alysicarpus monilifer, Calligonum polygonoides, Capparis deciduas, Cenchrus biflorus, C. prieurii, Citrullus colocynthis, Convolvulus prostratus, Crotolaria burhia, Ephedra ciliata, Gisekia pharnacioides, Gynandropsis gynandra, Lasiurus sindicus, Leptadenia pyrotechnica, Prosopis cineraria, Rhynchosia schimperi, Salvadora oleoides, Sueda fructicosa, Tamarix aphylla, Tecomella undulata, Tephrosia falciformis* and *Withania coagulans* are endemic to desert habitats. Some of the species such as *Anticharis glandulosa, A. linearis, Schweinfurthia papilionacea, Maerua crassifoloia, Boerhavia rubicunda* are also endemic to Rohri Hills.

Most of plant species possess high economic value and serve the natives for fulfilling their daily life requirement especially forage and medicinal values. The most notable medicinal plants are *Convolvulus prostratus, Cressa cretica, Leptadenia pyrotechnica, Salvadora* spp., *Tephrosia purpurea, Solanum surattense, Capparis deidua* and *Citrullus colocynthis*. Whereas, *Prosopis cineraria, Zizyphus nummularia, Capparis deciduas* and *Acacia nilotica* are some renowned plant species having high fodder values.

Chapter V

Previous Work and Present Study

The scrutiny of literature revealed that a considerable work has been done on different parts of the country; however no work was previously reported as a whole on the flora of Nara Desert. Chaudhy & Chuttar (1966) carried out preliminary work and enlisted 122 plants from the Thar Desert, covering the areas of Tharparkar district. In the Nara Desert, floristic work was started in 1998 with the financial support of Pakistan Science Foundation, Islamabad. Upto 2001, only 150 plant species belonging to 110 genera and 42 families were documented by the authors (Bhatti *et al.*, 2001; Qureshi, 2004) covering Northern parts of this desert. Later on, the work was continued to explore Southern part covering *Acchro Thar* (Qureshi, 2008) to produce detailed flora of Nara Desert. The presented work reports thus a comprehensive floristic account of the NaraDesert based on exploration of the whole area with regular intervals during 1998-2009.

Chapter VI

Materials and Methods and Plan of the Flora

The presented piece of work is based on comprehensive explorations, thorough study of the plants of the Nara desert by the author spread over 10 years (1998-2007). For this purpose, detailed field studies have been undertaken to collect plant specimens at regular intervals.

A wide range of plant specimens have been collected to examine the possible variations in the taxa. The specimens were processed by conventional method for drying and making herbarium sheets. The herbarium sheets made in duplicates/triplicate and quadruplicates. The field data then, transferred to herbarium labels.

The determinate and un-determinate specimens were also sent to Natural History Museum, Islamabad. Later on, the same material was checked and correlated with own determinations.

Thus one set of each specimen has been deposited in the aforementioned herbaria. This practice has an advantage that the specimen will be more accessible to scientific research or if a disaster (e.g. Fire or Flood) be falls much scientific data can be saved in the recipient herbaria.

During plant collection, relevant data such as collector's name, date of collection, observation on height, habit, habitat, abundance, ecological features, flowering & fruiting period, flower color etc. has been recorded.

Local names, life form and life for species have been recorded and provide in Table 2.

Deposition of Plant Specimens

The identified species have been deposited at the Shah Abdul Latif University Botanical Garden & Herbarium (SALUBGH) at Shah Abdul Latif University, Khairpur for record.

Plan of Flora

The present work mainly deals with angiosperms. It represents a total of 59 families, 167 genera and 266 species including 3 ferns, 1 gymnosperm and 64 monocots with 92 illustrations drawn from the collected specimens. The families are arranged according to Cronquist system of classification (1981) except in few cases in which Hooker's classification (1862-83) is followed for splitting of the families.

The family and generic names are provided with the names of authors along with updated descriptions by various workers. The genera and subordinate taxa in the text are given in sequence as their number appeared in keys under each family.

Keys

Dichotomous keys, based on macroscopic characters are provided for the identification of the rank of family to infra-specific taxa.

Nomenclature

Naming of taxa is based according to the ICBN. The source of valid combination is that of Flora of Pakistan (Nasir & Ali 1970-1989; Ali & Nasir 1989-1991; Ali & Qaiser, 1993-2007) which was mostly consulted in correct specific name and their systematic affiliation.

Wherever, the species is not included in aforesaid work, other regional floras or recent literatures (Cooke, 1903-1906; Jafri, 1966; Matthew, 1981-83; Batanouny, 1981; Boulos, 1991; Shetty & Singh, 1987 & 1991; Bhandari, 1987) have been cited.

Descriptions

Most of the descriptions are based on author's own observations. In case of inadequate material, the descriptions have been borrowed from recent authentic literature. The local names of the taxa are collected from the study area are included in Table 1 wherever possible.

Flowering and Fruiting Period

The phonological data given in this work is based on author's own observations which are recorded during plant specimen collection from the field areas.

Specimens Examined

Some selected specimens out of the material examined by the author have been cited to indicate the distribution of the species in the project area.

Illustration and Photographs

Botanical Illustrations and photographs are an important supplement to the herbarium collection by providing information on habit and floral characters. Ninety two illustrations depicting habits of plants are provided in this work.

Chapter VII

Results

The present work includes 266 species distributed across 167 genera and 59 families, including 3 ferns, 1 gymnosperm and 64 monocots. There were four monocotyledon families and amongst them, Poaceae family contributed largest number of taxa (28 genera, 47 spp.), followed by Cyperaceae (4 genera and 14 spp.). Out of 51 dicotyledonus families, Fabaceae contributed highest taxa (12 genera and 26 spp.), followed by Asteraceae (10 genera, 13 spp.), Boraginaceae (4 genera, 9 spp.), Capparidaceae (6 genera, 9 spp.), Euphorbiaceae (4 genera, 8 spp.) and Amaranthaceae, Chenopodiaceae, Solanaceae (5 genera, 7 spp. each) and Convolvulaceae (4 genera and 5 spp.). The contribution of families is provided in Table 3.

There were 11 life forms/habits of the flora recorded from the Nara Desert. Table 4 depicted the life form of the existing vegetation that reveals that most of the area was dominated by herbs (117 spp.), followed by grasses (47 spp.), shrubs (34 spp.), subshrubs (20 spp.), sedges (14 spp.) and trees (13 spp.).

Table 2. Alphabetically family-wise arranged list of plants along with local name, life form and life span

S. No.	Family Name/Plant Species	Local Name	Life-Form/Habit	Life-Span
	PTERIDOPHYTES			
	Adiantaceae			
1	*Adiantum capillus-veneris* L.	Pursha	Fern	Perennial
	Equisetatceae			
2	*Equisetum arvense* L.	----	Fern	Perennial
	GYMNOSPERM			
	Ephederaceae			
3	*Ephedra ciliata* Fisch. & Mey. ex C.A.Meyer.	Santh	Climbing Shrub	Perennial
	Marsileaceae			
4	*Marsilea minuta* L.	Lusani	Fern	Perennial
	MONOCOTYLEDONS			
	Cyperaceae			
5	*Bolboschoenus affinis* (Roth.) Drobov	Kaluro	Sedge	Perennial
6	*B. glaucus* (Lam.) S.G. Smith	Kaluro	Sedge	Perennial

Table 2. (Continued)

S. No.	Family Name/Plant Species	Local Name	Life-Form/Habit	Life-Span
7	*Cyperus arenarius* Retz.	Moniah Gaah	Sedge	Perennial
8	*C. articulatus* L.	Kal	Sedge	Perennial
9	*C. bulbosus* Vahl	----	Sedge	Perennial
10	*C. conglomeratus* Rottb.	Moniah Gaah	Sedge	Perennial
11	*C. difformis* L.	----	Sedge	Perennial
12	*C. iria* L.	----	Sedge	Perennial
13	*C. laevigatus* L.	----	Sedge	Perennial
14	*C. longus* L.	----	Sedge	Perennial
15	*C. rotundus* L.	Kabah	Sedge	Perennial
16	*Fimbristylis bisumbellata* (Forssk.) Bubani	Kal	Sedge	Perennial
17	*F. cymosa* R. Br.	Kal	Sedge	Perennial
18	*Schoenoplectus litoralis* (Schrad.) Palla	Kaluro	Sedge	Perennial
	Liliaceae			
19	*Asphodelus tenuifolius* Cavan	Basri	Herb	Annual
	Poaceae			
20	*Aeluropus lagopoides* (L.) Trin. ex. Thw	Kalar Gaah	Grass	Perennial
21	*Aristida adscensionis* L.	Lumb Gaah	Grass	Annual
22	*A. hystricula* Edgew.	Lumb Gaah	Grass	Annual
23	*A. funiculata* Trin. & Pupr.	Lumb Gaah	Grass	Annual
24	*A. mutabilis* Trin. & Rupr.	Lumb Gaah	Grass	Annual
25	*Avena fatua* L.	Jangli Jai	Grass	Annual
26	*Brachiaria eruciformis* (J. E. Smith) Griseb.	Sawri	Grass	Annual
27	*B. ramosa* (L.) Stapf	Sawri	Grass	Annual
28	*B. ovalis* (L.) Stapf	Sawri	Grass	Annual
29	*Cenchrus biflorus* Roxb.	Mohabbat buti	Grass	Annual
30	*C. ciliaris* L.	Bhurt	Grass	Annual
32	*C. pennisetiformis* Hochst. & Steud. ex Steud.	Bhurt	Grass	Annual
33	*C. prieurii* (Kunth) Maire	Bhurt	Grass	Annual
35	*C. setigerus* Vahl.	Bhurt	Grass	Perennial
31	*Chloris barbata* Sw.	Gaah	Grass	Annual
34	*Cymbopogon jwarancusa* (Jones) Schult.	Kattan	Grass	Perennial
36	*Cynodon dactylon* (L.) Pers.	Chhabar	Grass	Perennial
37	*Dactyloctenium aegyptium* (L.) P. Beauv.	Gandheer Gaah	Grass	Annual
38	*D. aristatum* Link	Gandheer Gaah	Grass	Annual
39	*D. scindicum* Boiss.	Gandheer Gaah	Grass	Annual
40	*Desmostachya bippinata* (L.) Stapf.	Drabh	Grass	Perennial
41	*Dichanthium annulatum* (Forsskal) Stapf.	Dinohi	Grass	Perennial
42	*Digitaria sanguinalis* (L.) scop.	Karash	Grass	Perennial
43	*Dinebra retroflexa* (Vahl.) Panzer.	Gaah	Grass	Annual
44	*Echinochloa colona* (L.) Link	Sanwak	Grass	Annual
45	*Eragrostis barrelieri* Dav.	Gaah	Grass	Annual
46	*E. cilianensis* (All.) Lut. ex F.T. Hubbard	Gaah	Grass	Annual
47	*E. ciliaris* (L.) R.Br.	Gaah	Grass	Annual
48	*E. minor* Host.	Makhni Gaah	Grass	Annual

S. No.	Family Name/Plant Species	Local Name	Life-Form/Habit	Life-Span
49	*E. tenella* (L.) P. Beauv. ex Roem. & Schult.	Gaah	Grass	Annual
50	*Imperata cylindrica* (Linn) Raeuschel.	Drabh	Grass	Annual
51	*Lasiurus sindicus* Henr.	Booro	Grass	Annual
52	*Leptothrium senegalense* (Kunth) W.D. Clayton	Gaah	Grass	Annual
53	*Ochthochloa compressa* (Forssk..) Hilu.	Mandhano	Grass	Annual
54	*Panicum turgidum* Forsk.	Sewan	Grass	Perennial
55	*Paspalidium geminatum* (Forssk.) Stapf in Prain	Gaah	Grass	Perennial
56	*Paspalum paspalodes* (Michx.) Scribner	Gaah	Grass	Perennial
57	*P. scrobiculatum* L.	Gaah	Grass	Perennial
58	*Phalaris minor* Retz.	Banak	Grass	Perennial
59	*Phragmites australis* (Cav.) Trin. ex Steud.	Nar/Naro	Grass	Annual
60	*P. karka* (Retz.) Trin. ex Steud.	Nar/Naro	Grass	Perennial
61	*Polypogon fugax* Nees ex Steud.,	Nor puchh	Grass	Perennial
62	*Saccharum bengalense* Retz.	Booro	Grass	Perennial
63	*S. griffithii* Munro ex Boiss.	Booro	Grass	Perennial
64	*S. spontaneum* L.	Kanh	Grass	Perennial
65	*Setaria pumila* (Pouir.) Roem. & Schult.	Sawri	Grass	Annual
66	*Stipagrostis plumosa* (L.) Munro ex T. Anders.	Lumb Gaah	Grass	Annual
	Typhaceae			
67	*Typha domingensis* Pers.	Pan	Shrub	Perennial
68	*T. elephantina* Roxb.	Deer/Pan	Shrub	Perennial
	DICOTYLEDONS			
	Acanthaceae			
69	*Barleria acanthoides* Vahl	----		
70	*Blepharis sindica* Stocks ex. Anders	Utangan	Herb	Annual
	Aizoaceae			
71	*Gisekia pharancoides* L.	Buto	Herb	Annual
72	*Limeum indicum* Stocks ex T. Anderson.	Dhoor Chhapri	Herb	Annual
73	*Sesuvium sesuvioides* (Fenzl) Verdc.	Lonak	Herb	Annual
74	*Trianthema portulacastrum* L.	Waho	Herb	Annual
75	*T. triquettra* Rottl. and Willd.	Wasanh/Waho	Herb	Annual
76	*Zelya petandra* (L.) Jeffrey	Wasanh/Waho	Herb	Annual
	Amaranthaceae			
77	*Achyranthes aspera* L.	Ubat Kandri	Herb	Annual
78	*Aerva javanica* var. *bovei* Webb.	Booh	Sub-shrub	Perennial
79	*A. javanica* var. *javanica* (Burm.f.) Juss. ex. Journ. A. Schultes.	Booh	Sub-shrub	Perennial
80	*Amaranthus graecizans* subsp. *thllungianus* (Nevski) Gusev.	Mariro	Herb	Annual
81	*A. virdis* L.	Mariro	Herb	Annual
82	*Celosia argentea* L.	Boobak		
83	*Digera muricata* (L.) Mart.	Lulur	Herb	Annual
	Arecaceae			
84	*Phoenix sylvestris* (L.) Roxb.	Jhangli khaji	Tree	Perennial
	Aristolochiaceae			
85	*Aristolochia bracteolata* Lamk.	Zeerawund	Herb	Annual
	Asclepiadaceae			

Table 2. (Continued)

S. No.	Family Name/Plant Species	Local Name	Life-Form/Habit	Life-Span
86	*Calotropis procera* (Willd.) R. Br.	Ak	Shrub	Perennial
87	*Leptadenia pyrotechnica* (Forssk.) Decne.	Khip	Shrub	Perennial
88	*Oxystelma esculentum* (L.f.) R. Br.	Phuli	Climbing Shrub	Perennial
89	*Pentatropis nivalis* (J.F.Gmel.) Field & J.R.I.Wood	Wan Wehri	Climber	Perennial
	Asteraceae			
90	*Amberboa ramosa* (Roxb.) Jafri.	----	Herb	Annual
91	*Carthamus oxycantha* Bieb.	Pohli	Bushy subshrub	Annual
92	*Cirsium arvense* (L.) Scop.	Leh	Herb	Annual
93	*Echinops echinatus* DC.	Kanderi Bhattar	Bushy subshrub	Annual
94	*Eclipta prostrata* (L.) L.	Bhangrio	Herb	Annual
95	*Iphiona grantioides* (Boiss.) Anderb.	Gandraf	Herb or subshrub	Annual
96	*Launaea procumbens* (Roxb) Ramayya & Rajagopal.	Bhattar	Herb	Biennial
97	*L. resedifolia* (L.) Kuntze	Bhattar	Herb	Biennial
98	*Pluchea lanceolata* (DC.) Oliv. & Hiern.	Phaar Buti	Shrub	Perennial
99	*P. wallichiana* DC.	----	Herb	Annual
100	*Sonchus asper* (L.) Hill.	Bhattar	Herb	Annual
101	*S. oleraceus* L.	Bhattar	Herb	Annual
102	*Xanthium strumarium* L.	Bhurt	Shrub	Annual
	Bignoniaceae			
103	*Tecomella undulata* (Sm.) Seem.	Lohiro	Shrub/Small tree	Perennial
	Boraginaceae			
104	*Arnebia hispidissima* (Sieber) DC.	Harnub	Herb	Annual
105	*Cordia gharaf* (Forssk.) Ehren. ex Asch.	Lesuro	Small tree	Perennial
106	*C. myxa* L.	Lesuro	Small tree	Perennial
107	*Heliotropium crispum* Desf.	Kharsan	Herb	Annual
108	*H. curassavicum* L.	Kharsan	Herb	Annual
109	*H. europeum* L.	Uth Charo	Herb	Annual
110	*H. strigosum* Willd.	Kharsan	Herb	Annual
111	*H. subulatum* (Hochst. ex DC.) Vatke.	Kharsan	Herb	Annual
112	*Trichodesma indicum* (L.) R.Br.	Gaaozaban	Hrerb	Annual
	Burseraceae			
113	*Commiphora stocksiana* (Engl.) Engl.	Bhayi	Shrub	Perennial
114	*C. wightii* (Arn.) Bhandari	Gugur	Shrub	Perennial
	Brassicaceae			
115	*Farsetia hamiltonii* Royale.	Lathio	Herb	Annual
116	*F. jacquemontii* Hook. f. & Thoms.	Lathio	Herb	Annual
117	*Coronopus didymus* (L.) Smith	Jhangli halon	Herb	Annual
	Caesalpiniaceae			
118	*Caesalpinia bonduc* (L.) Roxb.	Pahar Wal	Climbing Shrub	Perennial

S. No.	Family Name/Plant Species	Local Name	Life-Form/Habit	Life-Span
119	*Senna holosericea* (Fresen.) Greuter	Sindhi senna	Herb	Annual
120	*S. italica* Mill.	Ghora Wal	Herb	Annual
	Capparidaceae			
121	*Cadaba farinosa* Forssk.	Khabri Wal	Shrub	Perennial
122	*Capparis decidua* (Forssk.) Edgew.	Kirar	Shrub	Perennial
123	*C. spinosa* L.	Golaro	Sub-shrub	Perennial
124	*Cleome brachycarpa* Vahl.	Dhanaar Khathoori	Herb	Annual
125	*C. scaposa* D.C.	Khathoori	Herb	Annual
126	*C. viscosa* L.	Kinni Buti	Herb	Annual
127	*Dipterygium glaucum* Decne.	Phair	Bushy subshrub	Perennial
128	*Gynandropsis gynandra* (L.) Briq.	Kinro	Herb	Annual
129	*Maerua crassifolia* Forssk.	Khabar	Shrub	Perennial
	Caryophyllaceae			
130	*Vaccaria hispanica* (Miller) Rauschert	Taklo	Herb	Annual
131	*Spergula arvensis* L.	Lani	Herb	Annual
	Chenopodiaceae			
132	*Arthrocnemum indicum* (Willd.) Moq.	Lano	Shrub	Perennial
133	*Chenopodium album* L.	Chil	Herb	Annual
134	*C. murale* L.	Chil	Herb	Annual
135	*Haloxylon salicornicum* (Moq.) Bunge ex Boiss.	Lano	Shrub	Perennial
136	*H. stocksii* (Boiss.) Benth. & Hook.	Sacho Lano	Shrub	Perennial
137	*Salsola imbricata* Forssk.	Lano	Sub-shrub	Perennial
138	*Saueda fruticosa* (L.) Forssk.	Lani	Sub-shrub	Perennial
	Convolvulaceae			
139	*Cressa cretica* L.	Oin	Herb	Annual
140	*Convolvulus arvensis* L.	Naro	Climbing herb	Annual
141	*C. glomeratus* Choisy.	Kirhanj	Prostrate herb	Annual
142	*C. prostratus* Forssk.	Kirhanj	Herb	Annual
143	*Ipomoea aquatica* Forssk.	Naro	Herb	Annual
144	*I. carnea* (Martius ex Choisy) D. F. Austin	Akri	Shrub	Perennial
145	*Seddera latifolia* Hochst. et Steud.	----	Shrub	Perennial
	Cucurbitaceae			
146	*Citrullus colocynthis* (L.) Schrad.	Trooh	Prostrate subshrub	Perennial
147	*Ctenolapis cerasiformis* (Stocks.) Hooker.f.	----	Prostrate herb	Annual
148	*Cucumis melo* var. *agrestis* Naudin.	Mitero	Prostrate herb	Annual
149	*C. prophaterum* L.	Kando Mitero	Prostrate herb	Annual
150	*Luffa echinata* Roxb.	Jhangli toori	Climber	Annual
151	*Mukia maderaspatana* (L.) M. J. Roem.	Chibhri Wal	Prostrate herb	Annual
	Cuscutaceae			

Table 2. (Continued)

S. No.	Family Name/Plant Species	Local Name	Life-Form/Habit	Life-Span
152	*Cuscuta chinensis* Lam.	Bepari	Stem parasite	Annual
	Elatinaceae			
153	*Bergia aestivosa* Wight & Arn.	----	Sub-shrub	Perennial
	Euphorbiaceae			
154	*Chrozophora plicata* (Vahl) A. Juss. ex Spreng.	Nil Kanthi	Sub-shrub	Annual
155	*Euphorbia caducifolia* Haines	Thor	Shrub	Perennial
156	*E. clarkeana* Hook. f.	Kher Wal	Herb	Annual
157	*E. hirta* L.	Kheer Wal	Herb	Annual
158	*E. thymifolia* L.	Kheera Wal	Herb	Annual
159	*Phyllanthus maderaspatensis* L.	Hazar dani	Herb	Annual
160	*P. reticulatus* Pair.	Kamho	Shrub	Perennial
161	*Ricinus communis* L.	Heran buti	Shrub	Perennial
	Fabaceae			
162	*Alhagi maurorum* Medic.	Kandero	Sub-shrub	Perennial
163	*Alysicarpus monilifer* (L.) DC.	----	Herb	Annual
164	*A. scariosus* Grah. ex Thwaites	----	Herb	Annual
165	*Argyrolobium roseum* ssp. *ornithopodioides* (Jaub. & Spach) Jafri & Ali	----	Herb	Annual
166	*Crotolaria burhia* Ham. ex Benth.	Chagg	Sub-shrub	Biennial
167	*C. medicaginea* var. *medicaginea* Lamk.		Herb	Annual
168	*Indigofera argentea* Burm.f.	Lathio	Herb	Annual
169	*I. cordifolia* Heyne ex Roth	Lathio	Herb	Annual
170	*I. hochstetteri* Baker	Bheekar	Herb	Annual
171	*I. linifolia* (L.) Retz.	Vekar	Herb	Annual
172	*I. oblongifolia* Forssk.	Jhil	Sub-shrub	Perennial
173	*I. sessiliflora* DC.	Vekar	Herb	Annual
174	*Lathyrus aphaca* L.	Jhangli matri	Herb	Annual
175	*Medicago polymorpha* L.	Mena	Herb	Annual
176	*Melilotus alba* Desr.	Sinjhi	Herb	Annual
177	*M. indica* (L.) All	Sinjhi	Herb	Annual
178	*Rhynchosia minima* (L.) DC.	Wan Wehri	Climber	Annual
179	*R. pulverulenta* Stocks	Wan vehri	Climber	Biennial
180	*R. schimperii* Hochst. ex Boiss.	----	Sub-shrub	Annual
181	*Sesbania bispinosa* (Jacq.) W. F. Wight	Jantur	Sub-shrub	Annual
182	*Tephrosia falciformis* Romaswami.	Drebarr	Herb	Annual
183	*T. purpurea* (L.) Pers.	Andhri/Siringh	Sub-shrub	Annual
184	*T. strigosa* (Dalz.) Sant. & Mahesh.	----	Herb	Annual
185	*T. uniflora* var. *petrosa* (Blatter & Hallberg) Gillett & Ali	Andhri/Siringh	Sub-shrub	Annual
186	*T. villosa* (L.) Pers.	Andhri/Siringh	Sub-shrub	Annual
187	*Vicia sativa* L.	Matri	Herb	Annual
	Frankeniaceae			
188	*Frankenia pulveulenta* L.	----	Herb	Annual

S. No.	Family Name/Plant Species	Local Name	Life-Form/Habit	Life-Span
	Fumariaceae			
189	*Fumaria indica* (Hausskn.) H.N. Pugsley	Doonhi Buti/Shahtaro	Herb	Annual
	Geraniaceae			
190	*Monsonia heliotropoides* (Cav.) Boiss.	----	Herb	Annual
	Hydrocharitaceae			
191	*Hydrilla verticillata* (L. f.) Royale	Machhi Gaah	Herb	Annual
	Lamiaceae			
192	*Leucaus aspera* (Willd.) Link.	Goomi Buti	Herb	Annual
193	*Salvia santolinifolia* Boiss.	----	Bushy	Perennial
	Malvaceae			
194	*Abutilon fruticosum* Guill. & Perr.	Pat Teer	Sub-shrub	Perennial
195	*A. indicum* (L.) Sweet	Pat Teer	Shrub	Perennial
196	*A. muticum* (Del.ex DC.) Sweet	Pat Teer	Shrub	Perennial
197	*Hibiscus lobatus* (J.A Murray) O. Ketze.	----	Herb	Annual
198	*Malvastrum coromendelianum* (L.) Garcke	----	Sub-shrub	Perennial
199	*Pavonia glechomaefolia* S. Abedin	----	Herb	Annual
	Menispermaceae			
200	*Cocculus hirsutus* (L.) Diels	Fareed Buti	Climber	Perennial
201	*C. pendulus* (J.R. & Forst) Diels	----	Liana	Perennial
	Mimosaceae			
202	*Acacia jacquemontii* Benth.	Banwar	Shrub	Perennial
203	*A. nilotica* ssp. *cupressiformis* (J.L. Stewart) Ali & Faruqi	Sindhi Babur	Tree	Perennial
204	*A. nilotica* subsp. *indica* (Benth.) Brenan	Sindhi Babur	Tree	Perennial
205	*A. senegal* (L.) Willd.	Khaunr	Tree	Perennial
206	*Prosopis cineraria* (L.) Druce	Kandi	Tree	Perennial
207	*P. juliflora* (Swartz) DC.	Devi	Shrub	Perennial
	Molluginaceae			
208	*Glinus lotoides* L.	Kotak	Herb	Annual
209	*Mollugo cerviana* (L.)Ser.	Hazar Dani	Herb	Annual
210	*M. nudicaulis* Lamk.	Hazar Dani	Herb	Annual
211	*M. pentaphylla* L.	Hazar Dani	Herb	Annual
	Moraceae			
212	*Ficus religiosa* L.	Pipul	Tree	Perennial
213	*F. bengalensis* L.	Barr	Tree	Perennial
	Moringaceae			
214	*Moringa oleifera* Lamk.	Suhandaro/Suhanjro	Tree	Perennial
	Neuradaceae			
215	*Neurada procumbens* L.	Chhipri	Herb	
	Nyctaginaceae			
216	*Boerhavia diffusa* L.	Dakhri	Herb	Annual
217	*B. procumbens* Banks ex Rxb.	Dakhri	Herb	Annual
218	*B. rubicunda* Steud.	----	Herb	Annual
219	*Commicarpus boisseiri* (Heimer) Cufod.	Dakhri	Herb	Annual
	Orobanchaceae			
220	*Cistanche tubulosa* (Schrenk) Hook. f.	Bhunjro	Root parasite	Annual

Table 2. (Continued)

S. No.	Family Name/Plant Species	Local Name	Life-Form/Habit	Life-Span
	Oxalidaceae			
221	*Oxalis corniculata* L.	Khatti buti	Herb	Annual
	Polygalaceae			
222	*Polygala erioptera* DC.	Hiran Buti	Herb	Annual
223	*P. irregularis* Boiss.	Hiran Buti	Herb	Annual
	Polgonaceae			
224	*Calligonum polygonoides* L.	Phog	Shrub	Perennial
225	*Persicaria barbata* (L.) Hara	----	Herb	Perennial
226	*P. glabra* (Willd.) M. Gómes	----	Herb	Perennial
227	*Polygonum plebejum* R.Br.	Kheer Wal	Herb	Annual
228	*Rumex dentatus* subsp. *klotzschianus* (Meisn.) Rech. f.	Jhangli palak	Herb	Annual
	Potamogetonaceae			
229	*Potamogeton perfoliatus* L.	Machhi Gaah	Herb	Annual
	Primulaceae			
230	*Anagallis arvensis* var. *coerulea* (L.) Gouan	Bili buti	Herb	Annual
	Portulacaceae			
231	*Portulaca oleracea* L	Lonak	Herb	Annual
	Rhamnaceae			
232	*Zizyphus nummularia* (Burm. F.) Wt. & Arn.	Jhangoori Ber	Shrub	Perennial
	Salicaceae			
233	*Populus euphratica* Olivier.	Bahan	Tree	Perennial
	Salvadoraceae			
234	*Salvadora oleoides* Decne.	Jar/Peroon	Shrub	Perennial
235	*S. persica* L.	Khabbar	Shrub	Perennial
	Scrophulariaceae			
236	*Anticharis glandulosa* Aschers.	Gaamesh	Herb	Annual
237	*A. linearis* (Benth.) Hochst. ex Aschers.	----	Herb	Annual
238	*Bacopa monnieri* (L.) Wettestein in Engl. & Prantl.	Kuari lonak	Herb	Biennial
239	*Kickxia ramosissima* (Wall.) Janchen	----	Herb	Annual
240	*Schweinfurthia papilionacea* (Burm. f.) Boiss.	Akri/Paneer Wal	Sub-shrub	Annual
	Solanaceae			
241	*Lycium edgeworthii* Dunal.	----	Sub-shrub	Perennial
242	*Datura fastuosa* L.	Charyo Dhatooro	Herb	Biennial
243	*Physalis minima* L.	Khat mithro	Herb	Annual
244	*Solanum surattense* Burm.f.	Kanderi Wal	Sub-shrub	Annual
245	*S. nigrum* var. *nigrum* L.	Kanwal Buti	Herb	Annual
246	*Withania coagulans* (Stocks) Dunal	Paneer	Shrub	Perennial
247	*W. somnifera* (L.) Dunal	Akri	Shrub	Perennial
	Sterculiaceae			
248	*Melhania denhamii* R. Br.	----	Herb	Annual
	Tamaricaceae			

S. No.	Family Name/Plant Species	Local Name	Life-Form/Habit	Life-Span
249	*Tamarix aphylla* (L.) Karst.	Lawo	Tree	Perennial
250	*T. dioica* Roxb. ex Roch	Lai	Shrub	Perennial
251	*T. indic*a willd.	Lai	Shrub	Perennial
252	*T. pakistanica* Qaiser	Lai	Shrub	Perennial
253	*T. passernioides* Del ex. Desv var. *macrocarpa* Ehrenb	Lai	Shrub	Perennial
	Tiliaceae			
254	*Corchorus aestuans* L.	Datehri	Herb	Annual
255	*C. depressus* (L.) Stocks.	Mundheri	Herb	Annual
256	*C. tridens* L.	Datehri	Herb	Annual
257	*C. trilocularis* L.	Datehri	Herb	Annual
258	*Grewia tenax* (Forssk.) Aschers & Schweinf.	----	Shrub	Perennial
	Verbenaceae			
259	*Phyla nodiflora* (L.) Greene.	Bukkan	Herb	Annual
	Zygophyllaceae			
260	*Fagonia bruguieri* DC. var. *rechingeri* Had	Dramaaho	Bushy subshrub	Annual
261	*F. indica* var. *schweinfurthii* Hadidi	Dramaaho	Bushy subshrub	Annual
262	*Seetzania lanata* (Willd.) Bullock.	----	Herb	Annual
263	*Tribulus longipetalus* Viv.	Bakhro/Bhurt	Herb	Annual
264	*T. ochroleucus* (Maire) Ozenda & Quezel	Bakhro/Bhurt	Herb	Annual
265	*T. terrestris* L.	Bakhro/Bhurt	Herb	Annual
266	*Zygophyllum simplex* L.	Jand Laani	Herb	Annual

Table 3. Contribution of groups/families in terms of genera and species of the flora of Nara Desert

S. No.	Group/plant family	No. of Genera	No. of Species
	PTERIDOPHYTES		
1	Adiantaceae	1	1
2	Equisetatceae	1	1
3	Marsileaceae	1	1
	GYMNOSPERM		
1	Ephederaceae	1	1
	MONOCOTYLEDONS		
1	Poaceae	28	47
2	Cyperaceae	4	14
3	Typhaceae	1	2
4	Liliaceae	1	1
	DICOTYLEDONS		
1	Fabaceae	12	26
2	Asteraceae	10	13
3	Boraginaceae	4	9
4	Capparidaceae	6	9
5	Euphorbiaceae	4	8
6	Amaranthaceae	5	7
7	Chenopodiaceae	5	7
8	Convolvulaceae	4	7

Table 3. (Continued)

S. No.	Group/plant family	No. of Genera	No. of Species
9	Solanaceae	5	7
10	Zygophyllaceae	4	7
11	Aizoaceae	5	6
12	Cucurbitaceae	5	6
13	Malvaceae	4	6
14	Mimosaceae	2	6
15	Polgonaceae	4	5
16	Scrophulariaceae	4	5
17	Tamaricaceae	1	5
18	Tiliaceae	2	5
19	Asclepiadaceae	4	4
20	Molluginaceae	2	4
21	Nyctaginaceae	2	4
22	Brassicaceae	2	3
23	Caesalpiniaceae	2	3
24	Acanthaceae	2	2
25	Burseraceae	1	2
26	Caryophyllaceae	2	2
27	Lamiaceae	2	2
28	Menispermaceae	1	2
29	Moraceae	1	2
30	Polygalaceae	1	2
31	Salvadoraceae	1	2
32	Arecaceae	1	1
33	Aristolochiaceae	1	1
34	Bignoniaceae	1	1
35	Cuscutaceae	1	1
36	Elatinaceae	1	1
37	Frankeniaceae	1	1
38	Fumariaceae	1	1
39	Geraniaceae	1	1
40	Hydrocharitaceae	1	1
41	Moringaceae	1	1
42	Neuradaceae	1	1
43	Orobanchaceae	1	1
44	Oxalidaceae	1	1
45	Potamogetonaceae	1	1
46	Primulaceae	1	1
47	Portulacaceae	1	1
48	Rhamnaceae	1	1
49	Salicaceae	1	1
50	Sterculiaceae	1	1
51	Verbenaceae	1	1
		167	266

Table 4. Life form of the flora of Nara Desert, Pakistan

S. No.	Life form	No. of spp.	Percentage
1	Hrerbs	117	43.98
2	Grasses	47	17.67
3	Shrubs	34	12.78
4	Subshrubs	20	7.52
5	Sedges	14	5.26
6	Trees	13	4.89
7	Climbers	9	3.38
8	Bushs	6	2.26
9	Ferns	3	1.13
10	Parasites	2	0.75
11	Liana	1	0.38
		266	100.00

Keys to Families

1a. Vascular plants reproducing by spores.. I. Pteridophytes
1b. Vascular plants reproducing by seed.. II. Gymnospermae
2a. Stems jointed; leaves reduced to a sheath at the nodes........................ 1. Equisetaceae
2b. Stems not jointed; leaves (fronds) not reduced to a sheath:
3a. Plant amphibious or aquatic; stipe green and herbaceous fan-shaped...............3. Marsileaceae
3b. Plant shade loving, terrestrial; stipe dark and lustrous, not fan-shaped..............2. Adiantaceae
4a. Flowers arranged in cones. Ovules naked, borne on the surface
of megasporophyll..4. Ehphederaceae
4b. Flowers not arranged in cones. Ovules enclosed in ovary
formed by the stigma bearing carpels:
5a. Embryo with 2 cotyledons; leaf venation usually reticulate; floral parts
4 or 5-merous or in multiples of these numbers,
very rarely in three or sixes...III. Magnoliopsida
5b. Embryo with 1 cotyledon; leaf venation usually parallel;
floral parts usually in three or sixes.......................................IV. Liliopsida
6a. Flowers with both calyx and corolla:
6b. Flowers with one floral envelop (calyx):
7a. Petals free:
7b. Petals united:
8a. Gynoecium apocarpous:
8b. Gynoecium syncarpous or carpel 1:
9a. Flowers unisexual, 3-merous, climbers, leaves simple.......................6. Menispermaceae
9b. Flowers bisexual, 5-merous; various formed, leaves compound:
10a. Flowers hypogynous:
10b. Flowers perigynous:
11a. Placentation parietal or ovules along the ventral suture:
11b. Placentation axile, basal or apical:
12a. Ovary borne on a long gynophores..25. Capparidaceae

12b. Ovary not borne on gynophores:
13a. Trees or shrub:
14a. Leaves decompounds; seeds winged..27. Moringaceae
14b. Leaves simple, small and narrow; seeds with tuft of hairs.......................21. Tamaricaceae
13b. Herbs:
15a. Flower zygomorphic; leaves pinnately divided..7. Fumariaceae
15b. Flower actinomorphic; leaves not pinnately dissected:
16a. Flower cruciform; stamens tetradynamous...26. Brassicaceae
16b. Flowers not cruciform; stamens not tetradynamous:
17a. Leaves alternate, stipulate; sepals 4; petals 4; ovary on gynophores........25. Capparidaceae
17b. Leaves opposite; exstipulate; sepals 4-6-lobed; stamens 4-6;
ovary simple..22. Frankeniaceae
18a. Stamens more than twice the petals:
18b. Stamens as many or twice as many as the petals:
19a. Sepals imbricate in bud; leaves exstipulate:
20a. Ovary deeply 2-3-lobed...38. Zygophyllaceae
20b. Ovary not in lobular form:
19b. Sepals valvate leaves stipulate:
21a. Stamens monadelphous, united at the base with the petals; anthers 1-locular..20. Malvaceae
21b. Stamens free or connate at the base, free from the petals;
anthers 2-locular..18. Tiliaceae
22a. Placentation basal or free central, bearing 2-many ovules; ovary 1-locular:
22b. Placentation axile, ovary 2-many-chambered:
23a. Sepals 4-5; stamens hypogynous...15. Caryophyllaceae
23b. Sepals 2, connate into a tube; stamens perigynous............................13. Portulacaceae
24a. Flowers without hypogynous disk, all bisexual:
24b. Flowers with hypogynous disk, unisexual or bisexual:
25a. Sepals valvate in bud; stamens united into a tube;
anthers alternating with staminodes..19. Sterculiaceae
25b. Sepals imbricate in the bud; stamens free with staminodes:
26a. Sepals 5; the inner larger and petaloid; petals 3; stamens 8....................36. Polygalaceae
26b. Not combination of above characters:
27a. Stipules absent; leaves usually alternate:
28a. Leaves digitate with 3 obcordate leaflets;
stamens 10, united at the base...39. Oxalidaceae
28b. Leaves simple; stamens less 10, free:
27b. Stipules present; leaves opposite or whorled:
29a. Aquatic herbs; leaves entire; flowers minute, in dense
axillary fascicles; fruit a capsule...17. Elatinaceae
29b. Terrestrial herbs with segmented or compound leaves;
flowers prominent, solitary to subumbellate:
30a. Receptacle elongated and tapering into a point, the 5, 1-seeded carpels
coalescent round it, the style remaining as long twisted
beak on each segment of the fruit..40. Geraniaceae
30b. Receptacle not much elongated, the ovary tipped by a single style;
fruit a schizocarpi...38. Zygophyllaceae

31a. Leaves exstipulate; ovary 2-5-celled..37. Burseraceae
31b. Leaves stipulate; ovary 2-3-celled:
32a. Stamens alternate with the petals; ovary 3-celled; fruit a capsule.............34. Euphorbiaceae
32b. Stamens opposite the petals; ovary 2-4-celled; fruit a drup or berry..........35. Rhamnaceae
33a. Fruit a legume:
33b. Fruit various but not legume:
34a. Flowers actinomorphic...30. Mimosaceae
34b. Flowers zygomorphic:
35a. Flowers papilionaceous; stamens diadelphous...32. Fabaceae
35b. Flowers not papilionaceous; stamens not diadelphous.........................31. Caesalpiniaceae
36a. Calyx adnate to the ovary: 36b. Calyx free from the ovary:
37a. Petals present:
37b. Petals absent:
38a. Calyx-tube dilated at length, spiny and forming with ripe carpels
as orbicular disk; plants tomentose...29. Neuradaceae
38b. Characters not as above:
39a. Leaves often fleshy; stamens many..10. Aizoaceae
39b. Leaves glaucous or tomentose; stamens definite, not more than 10........14. Molluginaceae
40a. Ovary superior:
40b. Ovary inferior:
41a. Corolla irregular:
42a. Stamens as many as the corolla lobes:
43a. Stamens opposite the corolla lobes:
44a. Calyx ebracteolate; style 1; fruit capsule...28. Primulaceae
44b. Calyx bracteolate; styles various; fruit various:
43b. Stamens alternating with corolla lobes:
45a. Carpel 2, free, united by their styles or stigma:
46a. Pollen in waxy masses; anthers adnate to
the compound stigma...41. Asclepiadaceae
46b. Pollen not in waxy masses; anthers free from stigma:
45b. Carpel united to form a single ovary:
47a. Ovary deeply 4-lobed; style gynobasic; fruit formed of 2 or
4 nutlets; flowers in helicoids cymes...45. Boraginaceae
47b. Ovary 1-2-locular; style short; fruit a berry or drupe;
flowers in dense axillary fascicles or panicles......................................33. Salvadoraceae
48a. Fruit of 2-4 nutlets or drupaceous:
49a. Leaves opposite; corolla ± zygomorphic...46. Verbenaceae
49b. Leaves alternate; corolla actinomorphic:
48b. Fruit a few to many-seeded capsule or berry:
50a. Sepals free or united only at the base, the free portion longer than the cup;
styles more than 1 (if 1 then branched); ovules 1-2 in each cell.........43. Convolvulaceae
50b. Sepals united to form a cup or tube, the tips alone free;
style 1; ovules numerous...42. Solanaceae
41b. Stamens fewer than the corolla lobes:
41b. Corolla irregular:
51a. Stamens 4, didynamous, mostly shrubs:

51b. Stamens 2, herbs...48. Scrophulariaceae
52a. Corolla convolute in bud; capsule elastically or explosively dehiscent,
throwing the discoid seeds from their retinacula.................................50. Acanthaceae
52b. Corolla imbricate in bud; fruit of 2 or more bony nutlets Or drupaceous....46. Verbenaceae
53a. Stamens 5:
53b. Stamens 4 or less:
54a. Calyx 5-partite; some of the filaments wooly; capsule 2-valved..........48. Scrophulariaceae
54b. Calyx tubular-urceolate, closely surrounding the ovary;
filaments not wooly; capsule not circumscisile...................................42. Solanaceae
55a. Ovule 1 in each of the 1-4 locules:
55b. Ovules 2-many in each locule or in a 1-locular ovary:
56a. Style gynobasic; fruit achene like nutlets..47. Lamiaceae
56b. Style terminal; fruit a drupe or berry...46. Verbenaceae
57a. Ovary 1-locular; root parasites on higher plants; placentae parietal....... 49. Orobanchaceae
57b. Ovary 2-locular; stem parasites; placentae axile.......................................44. Cuscutaceae
58a. Corolla lobes rolled up lengthwise in bud; capsule springing open elastically,
often borne on a stalk...50. Acanthaceae
58b. Corolla lobes not rolled up lengthwise in bud; capsule not springing
open elastically, not stalked:
59a. Seeds winged; shrubs or trees...51. Bignoniaceae
59b. Seeds not winged; mostly herbs..48. Scrophulariaceae
60a. Tendril bearing herbs; flowers in axillary cyme;
fruit a berry...23. Cucurbitaceae
60b. Herbs, shrubs or climbers without tendrils; flowers crowded in heads;
fruit a sessile achene..52. Asteraceae
61a. Flowers bisexual:
61b. Flowers unisexual or polygamous:
62a. Ovary superior:
62b. Ovary inferior...5. Aristolochiaceae
63a. Leaves with sheathing ochrear stipules..16. Polygonaceae
63b. Leaves exstipulate:
64a. Perianth absent; fruit a capsule; seeds hairy.......................................24. Salicaceae
64b. Perianth present; fruits various; seeds not hairy:
65a. Perianth tubular, petaloid; fruits glandular...9. Nyctaginaceae
65b. Perianth not tubular, bract like; fruits not glandular:
66a. Flowers bracteate; the bracts and sepals mostly scarious....................12. Amaranthaceae
66b. Flowers ebracteate or if bracteate the bracts not scarious.................11. Chenopodiaceae
67a. Trees with milky sap; flowers enclosed in a hollow receptacle
which becomes the frui...8. Moraceae
67b. Herbs, shrubs or trees; flowers not so arrange:
68a. Ovary 3-celled with 1 or 2 ovules in each cell...................................34. Euphorbiaceae
68b. Ovary 1-celled...12. Amaranthaceae
69a. Terrestrial, trees or shrubs. Stem with leaf-scars. Leaves pinnately or
palmately compound. Inflorescence a spadix.
Bracts hard, stony and spathaceous...55. Arecaceae
69b. Aquatic or marshland herbs. Stem without leaf-scars. Leaves simple,

entire. Inflorescence not spadix. Bracts not hard, stony and not spathaceous:
70a. Perianth absent. Flowers in the axil of dry, chaffy, usually imbricated bracts (glumes or scales) arranged in spikelets:
70b. Perianth present. Flowers not in the axil of dry, chaffy, usually imbricated bracts (glumes or scales):
71a. Flowers in the axils of glumes or between two glumes united in spikelets. Spikelets arranged variously:
72a. Leaf sheath splitting (open). Fruit a caryopsis......................................57. Poaceae
72b. Leaf-sheath not splitting (closed). Fruit an achene....................................56. Cyperaceae
71b. Flowers not enclosed or subtended by glumes, but surrounded by thread-like or spathulate forked scales, arranged in spikes...................58. Typhaceae
73a. Terrestrial or marshy plants; perianth usually petaloid...............................59. Liliaceae
73b. Aquatic, submerged plants; perianth indistinct:
74a. Flowers unisexual, enclosed in spathaceous or opposite bracts; carpel 2-15...53. Hydrocharitaceae
74b. Flowers bisexual, ebracteate; carpel 4...54. Potamogetonaceae

Systematic Account

I. Pteridophytes

1. Equisetaceae

Herbs with subterranean rhizomes; stems hollow, grooved, jointed, often with whorls of branches at the nodes; leaves reduced to a toothed sheath at each node; sporangia in compact terminal cone-like heads on the main stems and sometimes on the branches.

Equisetum

L., Sp. Pl., ed. 1, 1096 (1753); Boulos in Fl. Egypt 1: 2 (1999).
Description of the genus is that of the family.

Equisetum arvense L., Sp. Pl. 1061 (1753); Stewart in Nasir & Ali., Ann. Cat. Vasc. Pl. W.Pak. & Kashm., 2 (1972).

Aerial stems dimorphic; vegetative stems green, branched, 2-35 cm. Branches in regular whorls, ascending, solid; ridges 3-4; valleys channeled, the 1st internode of each branch longer than subtending stem sheath; sheath teeth attenuate. Fertile stems brown, lacking stomates, unbranched, shorter than vegetative stems, with larger sheaths, fleshy, ephemeral.
Cones: maturing in early spring.
Herbarium specimens examined: NARA DESERT: *Rahmatullah Qureshi*, 1360 (SALUBGH).
Field remarks: Rarely collected from Nara river banks.
Distribution: C. Asia, Iran, Afghanistan and Pakistan (Nasir & Yasin, 1987).

2. Adiantaceae

Rhizomes with opaque scales; stipes dark and lustrous; sori usually submarginal, often covered by the reflexed leaf-margin; spores tetrahedral or globose.

About 1150 species belonging to 56 genera are reported cosmopolitan especially distributed in tropics (Boulos, 1999).

Adiantum

L., Sp. Pl., ed. 1, 1096 (1753); Boulos in Fl. Egypt 1: 3 (1999).

Shade loving plants; rhizomes with very narrow brown scales; fronds spaced or tufted, 1-4-pinnate; pinnae with flabellate venation, toothed or lobed; spines brownish to black, glabrous, lustrous; sori on the inner surface of the reflexed margin of pinnules.

About 150 species mostly distributed as cosmopolitan especially in tropical America (Boulos, 1999).

Adiantum capillus-veneris L., Sp. Pl., ed. 1, 1096 (1753); Stewart in Nasir & Ali., Ann. Cat. Vasc. Pl. W.Pak. & Kashm., 4 (1972); Boulos in Fl. Egypt 1: 3 (1999).

Rhizome scales linear, brownish; fronds spaced; stipe blackish or brownish, glabrous; lamina bright reen, 1-2(-3)-pinnate, ca. 10-30 x 2-10 cm in size; pinnules glabrous, shallowly to deeply lobed; soral flaps 2-4 x 0.8-1.8 mm.

Herbarium specimens examined: NARA DESERT: *Rahmatullah Qureshi*, 1341 (SALUBGH).

Field remarks: A rare species of moist habitat.

Distribution: C. Asia, Iran, Afghanistan and Pakistan (Nasir & Yasin, 1987).

3. Marsileaceae

Description mainly as for *Marsilea* (65 species); also *Pilucaria*, frond filiform (6 species); *Renellidium*, frond 2 lobes (1 species), neither occurring in Pakistan.

Marsilea

L., Sp. Pl., ed. 1, 1096 (1753); Boulos in Fl. Egypt 1: 7 (1999).

Amphibious or aquatic ferns; rhizomes creeping, rooting at the nodes; frond with 4 pinnae on long stipes; pinnae subsessile, fan-shaped with cuneate base; sporocapous hard, on short pedicel.

About 65 species distributed in tropics and temperate regions (Boulos, 1999).

Marsilea minuta L., Mant. Alt. 308 (1771); in Nasir & Ali., Ann. Cat. Vasc. Pl. W.Pak. & Kashm., 21 (1972); Boulos in Fl. Egypt 1: 8 (1999).

Adults plants almost glabrous; fronds with long stipes, up to 12 cm; pinnae 1-2.5 cm; sporocarps in groups of 2-4, each on a pedicel at least twice as long as the sporocarp; sporocarp ellipsoid; apex rounded to truncate, with 2-teeths at the base.

Herbarium specimens examined: NARA DESERT: *Rahmatullah Qureshi,* 1342 (SALUBGH).
Field remarks: A very rare species mostly found in brackish water channels.

II. Gymnospermae

4. Ephedraceae

Dioecious shrubs. Wood without resin ducts. Leaves opposite, reduced. Male flowers in more or less ovoid spikes with usually 2 connate bracts and 1-3 sporangia. Female flowers 2, enclosed by a membranous 2-partite or -fid sheath, surrounded at base by numerous imbricate bracts. Fruit usually red, berry-like. Seeds 2.

One genus and 42 species distributed in warm temperate America, Asia and Europe. About nine species are reported from Pakistan (Nasir & Nasir, 1987).

Ephedra

L., Sp. Pl. 1041 (1753); Gen. Pl. 462 (1754); E. Nasir, M.A. Siddiqi & Z. Ali, Gymnosperms of W. Pak. 10 (1969); E. Nasir & Yasin J. Nasir in Nasir & Ali, Fl. Pak. 186: 27-34 (1987).

Erect, decumbent or straggling rarely climbing shrubs. Branches green, opposite or whorled, terete, striate, commonly articulated at nodes, often bear narrow leafy blades. Leaves reduced, mostly deciduous, opposite or whorled. Flowers unisexual, oftenly dioecious. Male strobili 2-4 per node, subglobose to oblong, short; each strobilus with 2-8 opposite pairs of bracts, of these the lower 1 or 2 pairs sterile and the rest bearing one flower each. Perianth of 2 opposite compressed membranous sepals. Stamens 1-8; anthers on a central column, 2-celled, opening by a short lid at the apex. Female strobili similarly 2-4 at each node; each strobilus with many pairs of sterile bracts and a terminal ovule and single integument prolonged into a styliform tube (tubillus), enclosed in a bag. Seeds 1 or 2.
A single genus with its 42 species distributed in warm temperate America, Asia and Europe. In Pakistan, ca. 9 species are recorded (Nasir & Nasir, 1987).

Ephedra ciliata Fisch. & Mey. ex C.A.Meyer, Monogr. Gatt. Eph. 100 (1846); Stewart in Nasir & Ali., Ann. Cat. Vasc. Pl. W.Pak. & Kashm., 22 (1972); E. Nasir & Yasin J. Nasir in Fl. Pak., 178-186: 28 (1987).

Syn: *E. foliata* Boiss., Fl. Or. 5: 716 (1884); *E. foliata* var. *ciliata* Stapf in Akad. Wiss. Wien, Math-Naturwiss. Kl. Denkschr, 56: 49 (1889).

A climbing shrub with many horsetail-like hanging branches, up to 2.5 m high. Branches fascicled, slender, green or glaucesent, smooth, striate; internodes ca. 1.5-10 cm. Leaves 2-4, shortly connate at the base, ca. 1-2 x 0.1-2 cm in size, narrowly linear, setaceous. Male strobili 1 or 2-3 together per node, ovate, obtuse, tetragonal; peduncle ca. 2-4 cm; flowers 6-12 ; bracts connate, rotund, apex obtuse, ca. 1-2 mm in size. Perianth obovate, exceeding the bracts, ca. 2-3 mm long. Staminal column equalling the perianth; stamens 3-4.

Female strobili 2-3-flowered, ovate, on long peduncle or in short fascicles, with 3 pairs of bracts. Flowers usually in pairs, equalling the bracts. Integument at first oblong, obtusely 3-gonous; tubillus ca. 1.5 mm, exserted, straight. Berry globose, ca. 6-9 mm across, subhyaline, milky white. Seeds 2, ca. 6-9 mm long, oblong-globose, brownish-black.
Flowering period: February-March.
Herbarium specimens examined: NARA DESERT: *Rahmatullah Qureshi*, 655, 822, 859, 1044, 1322 (SALUBGH).
Field remarks: A relatively rare species mostly recorded associated with *Salvadora oleoides* and *Capparis decidua* in hilly areas.
Distribution: C. Asia, Iran, Afghanistan and Pakistan (Nasir & Yasin, 1987).

III. Magnoliopsida

5. Aristolochiaceae

Vines or erect, low or prostrate herbs. Leaves simple, alternate, entire or 3-5-lobed, petiolate, exstipulate. Inflorescence solitary axillary, racemose or in clusters. Flowers bisexual, often large, usually lurid, zygomorphic or actinomorphic, epigynous, bracteate or not. Perianth simple, petaloid, brownish or purple, more or less adnate below the ovary, regular or irregular, tubular, mouth oblique, 3-lobed or with a unilateral limb, sometimes an inner whorl of 3 teeth present, lobes valvate. Stamens 6, free or adnate in 1 or 2 series to a fleshy stylar column (gynostegium); filament short and thick when present; anthers 4-celled, erect with parallel cells, extrose, dehiscing dorsally. Ovary inferior, partly or completely partitioned into 4-6-cells; ovules numerous on parietal in each locule, connivent placenta, anatropous; style short and columnar; stigma as many as the carpel, lobed. Fruit septicidal capsule, dehiscing basally. Seeds numerous, variable in form, albumen copious, fleshy, embryo minute.

A tropical family with 6 genera and about 450 species with representation in Pakistan by the following genera and 3 species (Qaiser, 1977).

Aristolochia

L., Sp. Pl. 960 (1753); Gen. Pl. ed. 5: 410 (1754); Boiss., Fl. Or. 4: 1047 (1879); Benth. & Hook. f., Gen. Pl. 2: 123 (1880); Clarke in Hook. f., Fl. Brit. Ind. 574 (1885); Cooke, Fl. Bomb. Pres. 3: 15 (1906); Qaiser in Nasir & Ali, Fl. Pak. 110: 1 (1977).

Vines. Leaves simple, alternate, cordiform-reniform or oblong, chartaceous, margin entire; petiole dilated at the base. Inflorescence solitary, axillary, terminal or leaf-opposed raceme. Perianth zygomorphic or actinomorphic, coloured; tube inflated at the base enclosing the stylar column, then contracted, hairy within; limb oblique, (1)2 or (-3)-lipped. Stamens (5-) 6 or multiples, adhering to stylar column; anthers oblong-linear, sessile. Ovary 5-6-celled, oblong, ribbed; placentation parietal or axile; ovules anatropous, 2-seriate; style divided into 3, 5 or 6 linear or obtuse lobes; stigmatic lobes as many as style lobes, connivent or spreading. Capsule 6-nerved. Seeds compressed; testa smooth, warty or winged.

A large genus of about 400 species mostly distributed in tropics. Represented in Pakistan by 2 native and 1 cultivated species (Qaiser, 1977).

Aristolochia bracteolata Lamk. Encyl. I:258 (1783); Stewart in Nasir & Ali., Ann. Cat. Vasc. Pl. W. Pak. & Kashm., 200 (1972); Qaiser in Nasir & Ali., Fl. Pak., 110:26 (1977). **Figure 3B** Syn: *A. bracteata* Retz., Obs. Bot. 5:29 (1789).

Perennial prostrate herb with an unpleasant smell. Stem up to 50 cm long, branched, striate, pubescent having glandular hairs. Leaves ca. 4 cm across, reniform or broadly ovate, usually obtuse, base cordate to broadly sagittate with a wide shallow sinus, margin wavy, glaucous beneath, finely reticulately veined, 5-nerved, glabrous to sparsely glandular hairs above; petiole up to 3 cm long. Flowers solitary, axillary, bracteate, ca. 5 cm long; pedicel 5-10 mm long. Bracts large, orbicular, 1-1.5 cm long. Perianth greenish; tube ca. 2 cm long, cylindric with a trumpet-shaped mouth, glabrous without, hirsute within, base inflated which is subglobose in shape; mouth ca. 0.8 cm across; limb entire, 1-lipped; lip ca. 3 cm, linear, rolled back, apex emarginate, margin revolute, finely reticulately veined, dark purple in colour. Stamens 6, sessile, included in the dilated part of the perianth-tube; anthers ca. 0.6 mm, united with a hexa-lobed gynostegium, lobes ovate, acute. Ovary inferior, ca. 8 x 0.4 mm; style ca. 2, columnar, thick, 6-lobed; stigmatic lobes connivent, acute. Capsule 2-3 cm long, 12 ribbed, oblong-ellipsoid, glabrous. Seeds 5-7 mm long, flat; testa thin, covered with round glands on one side and dark coloured.

Flowering period: June-September.

Herbarium specimens examined: NARA DESERT: *Rahmatullah Qureshi*, 1251 (SALUBGH).

Field remarks: This species was rarely found from Achhro Thar from calcarious and gravelly soils.

Distribution: Tropical Africa, Arabia, Sri Lanka, India (Deccan Gujrat) and Pakistan (Qaiser, 1977).

6. Menispermaceae

Stragglers, twiners or lianas, rarely shrubs or small trees. Leaves alternate, simple (rarely trifoliate), usually palminerved, sometimes peltate, exstipulate. Inflorescence mostly axillary or on old wood and leafless branchlets, in racemes or panicles, rarely umbels. Flowers dioecious, mostly 3-merous; bracts sometimes foliaceous. Sepals usually 3+3, free or sometimes connate, imbricate, inner whorl sometimes valvate, often unequal. Petals usually 3+3, mostly free, sometimes enclosing stamens. Male: Stamens usually 6 (4 or 8), rarely α, antipetalous; filaments free or variously connate or fused into a column with peltate apex; anthers extrose, 2-celled or falsely 4-celled, longitudinally or occasionally transversely dehiscent; pistillodes present or 0. Female: Staminode 6 or 0; Carpels 3 or 6, rarely 1, usually free; ovules solitary, ventral, usually amphitropous; styles short usually recurved, or 0; stigma often capitate-discoid, entire or lobed. Fruit drupaceous, often compressed; style-scar subterminal or sub-basal; endocarp usually ornamented. Seeds curved, globose or horseshoe shaped; albumen even or ruminate or 0; cotyledons flat or semiterete, foliaceous or fleshy, appressed or spreading.

A family with ca. 65 genera and nearly 350 species, mostly tropical and subtropical. Represented in Pakistan by 4 genera and 5 species (Siddiqi, 1974).

Cocculus

DC., Syst. 1: 515 (1817), nom, cons.; Boiss., Fl. Or. 4: 1201 (1879); Jafri in Fl. Kar. 127 (1966); Siddiqi in Fl. Pak. 74: 7 (1974).
Menispermum L., Gen. Pl. ed. 5. 518 (1754); *Cebatha* Forssk., Fl. Aegypt. Arab. 171 (1775); *Leaeba* Forssk., l. c. 172; *Holopeira* Miers, l. c. 52 (1851).

Climbing or sarmentose shrubs. Leaves multicostate. Flowers dioecious or polygamous, in axillary, usually short panicles, minute, greenish and 3-merous. Sepals 6, in 2-series, the inner larger. Petals 6, shorter than sepals, usually auricled, concave. Male flowers: Stamens 6, embraced by the petals, filaments free; anthers subgobose, cells bursting transversely. Female flower: Staminode 6 or 0. Ovaries 3-6; ovule solitary; styles usually cylindric. Drupe mostly 3, laterally compressed; endocarp horseshoe shaped, dorsally keeled and tuberculate, side excavate, scar of the style basal. Seeds horseshoe shaped; albumen scanty; cotyledons linear, flat, appressed.

A genus of 35 species, distributed throughout the tropical and subtropical countries of the world. Represent by 2 species in Pakistan (Siddiqi, 1974).

Key to secies:
1a. Male flowers in axillary panicles.. **1. C. hisutus**
1b. Male flowers sessile or shortly pedicellate in axillary clusters.................. **2. C. pendulus**

1. Cocculus hirsutus (L.) Diels in Engler, Pflanzenr. IV. 94: 236 (1910); Stewart in Ali and Nasir, Ann. Cat. Vasc. Pl. W. Pak. & Kashm., 282 (1972); Siddiqi in Nasir & Ali., Fl. Pak., 74: 8-9 (1974).

Syn. *Menispermum hirsutum* L., Sp. Pl.341 (1753); *Cocculus villosus* DC. Syst. 1:525 (1817); *Cebatha hirsuta* (L.) O. Ktze., Rev. Gen. 1:9 (1891).

A climbing under shrub, up to 5 m high. Stem fairly branched, woody at the base, herbaceous above; branchlets striate, softly and densely villous, green. Leaves alternate, ca. 1-4.5 x 0.5-2.5 cm in size, ovate-oblong, lanceolate-cordate, base obtuse-truncate, apex retuse or mucronate, margin ciliate, chartaceous, 3-nerved, soft villous on both surfaces; petiole ca. 0.5-1 cm long, densely villous. Male inflorescence: in small axillary cymose panicles, ca. 3-6 cm long; peduncle ca. 1-2.5 cm. Bracts minute, subulate, hairy. Sepals 3+3, free; outer ones obovate; inner ones oblanceolate, hairy above, the inner the larger. Petals 6, greenish, membranous, obovate, apex emarginate, auricled at the base, embracing the stamens, smaller than the petals of female flowers, glabrous. Stamens 6, free, enclosed by petals; anthers vertically dehiscing. Female inflorescence: in axillary fascicles, 2-3 together, rarely racemes. Sepals 3+3. Petals thick and fleshy, divided at the base into 2 triangular lobes with swollen bases, claw hairy. Ovary 3, smooth; style terete; stigmas thick and reflexed. Drupe ca. 0.5 cm across, laterally compressed; style-scar basal, smooth, reddish purple; endocarp transversely rigid.
Flowering period: February-March.

Herbarium specimens examined: NARA DESERT: Rohri Hills, *Rahmatullah Qureshi*, 517, 618, 635, 656, 658, 663, 1046 (SALUBGH).
Field remarks: A common species recorded from hilly areas.
Distribution: India, Pakistan and tropical Africa (Siddiqi, 1974).

2. Cocculus pendulus (J.R. & Forst) Diels in Engler, Pflanzenr. IV. 94: 237 (1910); Stewart in Ali and Nasir, Ann. Cat. Vasc. Pl. W. Pak. & Kashm., 282 (1972).

Syn: *Cebatha pendula* (J.R. & G. Forst.) O. Ktze.; *Cocculus leaeba* (Del.) DC.; *Epibatrium pendulum* J.R. & G. Forst.; *Menispermum leaeba* Del.

Scandent shrubs with long slender, puberulous branchlets. Stem up to 15 cm in diameter, branches c. 5-6 m long. Leaves 1.6-5 cm long, 0.5-2 cm broad; oblong-lanceolate, ovate or trapezoid, base truncate, cuneate, rounded or trilobed-hastate, apex obtuse, mucronate or emarginate, generally glabrous or slightly puberulous on both sides, basal nerves 3-5. Flowers minute. Male flowers in axillary panicles 0.5-2 cm long, with peduncles up to 1.5 cm long, pedicels 0.5 mm or absent; sepals ovate-elliptic, fleshy or membranous, thickened at the base, the outer 3,1-1.5 mm long, 0.4-0.7 mm broad, sparsely puberulous, the inner 3 larger, puberulous to glabrous; petals ovate-obovate, 0.8-2 mm long, 0.5-1 mm broad; stamens 0.8-1.5 mm long. Female flowers: 1-2 on short axillary peduncles, 0.7-1.3 cm long; carpels c. 1 mm long. Drupe reddish when fresh but turning black when dried, reniform, compressed, 4-7 mm long, 4-5 mm broad; endocarp ribbed on the lateral faces and without a prominent crest, not perforated in the centre.
Flowering Period: Almost throughout the year.
Herbarium specimens examined: NARA DESERT: Adam Sultan, *Rahmatullah Qureshi*, 1329 (SALUBGH).
Field remarks: A rare lian recorded from Hilly tract.
Distribution: Tropical and subtropical India, Pakistan and Africa (Siddiqi, 1974).

7. Fumariaceae

Annual or perennial herbs, mostly with brittle stem and watery sap, erect to sprawling or straggling, often glabrous. Leaves usually alternate, pinnatisect. Inflorescence a terminal or leaf-opposed raceme or spike, rarely 1(or 2-3)-flowered. Flowers small, bisexual, usually irregular, hypogynous, pedicellate and bracteate. Sepals 2, small, scale-like, caducous. Petals 2+2, in two dissimilar whorls, imbricate; outer anterior petal longer and larger and flat, one or both spurred or succate, ± connivent, posterior gibbous at base; inner ones small, erect, spathulate, apically coherent over anthers. Stamens 3+3, opposite to outer petals; central anther of bundle 2-celled, other 1-celled. Ovary superior, 1-celled; ovules 1 or 2 on parietal placentae; style filiform; stigma obtuse or lobed. Fruit a 2-valved capsule or nutlet, indehiscent, 1-seeded. Seeds with minute embryo and copious endosperm.

A family of about 14 genera and nearly 400 species, chiefly of the north temperate zone and S.E. Africa; represented in Pakistan by 2 genera and ca. 30 species, mostly in the hilly regions (Jafri, 1974).

Fumaria

L., Sp. Pl. 699 (1753); Gen. Pl. ed. 5: 314 (1754); Boiss., Fl. Or. 1: 132 (1867); Hook. f., Fl. Brit. Ind. 1: 127 (1872); Jafri in Fl. Pak. 73: 39 (1974).

Annual (rarely perennating), often scandent herb, glabrous or subglabrous, glaucous. Leaves cauline, alternate, irregularly 2-4 pinnatisect with many usually very narrow segments. Inflorescence leaf-opposed a short raceme, bracteolate; bracts often short, linear, rarely as long as or slightly longer than pedicel. Flowers small, irregular, white, pink or purplish, spur very short. Sepals 2 or obsolate. Petals 4, erect or connivent; the 2 outer dissimilar, the lower flat or concave, the upper gibbous or spurred at the base; the 2 inner clawed, keeled, tips free or coherent. Stamens 6, diadelphous, with a basal spur enclosed in the spur of the petal. Ovary 1-celled, usually suborbicular with 1 (-2) ovules on each placentae; style short, filiform; stigma entire or shortly 2-lobed. Fruit 1-seeded, indehiscent, subglobose nut, rugulose or smooth when dry, with 2 apical pits. Seeds usually large, suborbicular, brownish.

A genus of ca. 50 species, chiefly of the temperate and cooler regions of the Old World and often as weeds; only 2 species are recorded for our area (Jafri, 1974).

Fumaria indica (Hausskn.) H.N. Pugsley in Journ. L. Soc. Bot. 44: 313 (1919); Stewart in Ali and Nasir, Ann. Cat. Vasc. Pl. W. Pak. & Kashm., 293 (1972); Jafri in Nasir & Ali., Fl. Pak., 73: 39-40 (1974).

Syn: *F. vaillantii* Loisel var. *indica* Hausskn in Flora, 56: 443 (1873); *F. parviflora* auct. Non Lam.: Burkill, List. Fl. Pl. Baluch. (reprint ed.). 7 (1956); *F. parviflora* ssp. *Vaillantii* (Lois.) Hook. F. & Thomas., Fl. Ind. 1: 238 (1995).

An erect or diffuse leafy herb up to 45 cm long; branchlets grooved, puberulous. Leaves 2-3 pinnatisect, ca. 2.5-6 x 1.5-2.5 cm; petiole ca. 0.5-3 cm long, clasping the stem, upper one sessile. Pinnae 5-8, membranous, linear-oblanceolate, flat, apex apiculate, margin entire,; petiolule ca. 0.2-0.3 cm. Inflorescence of about 15 to 20-flowered, ca. 2-4.5 cm, leaf-opposed racemes. Flowers alternately arranged, pink with purple tips. Bracts ca. 1.5 x 0.5 mm in size, ovate, apex acuminate, margin dentate, glabrous, hyaline, base greenish, lateral nerves obscure. Sepals 2, ovate. Petals 2+2, elliptic-oblong, ca. 3-3.5 x 0.5-0.6 mm in size, membranous

Flowering period: March-June.

Herbarium specimens examined: NARA DESERT: Chundko, *Rahmatullah Qureshi*, 654, 683, 834 (SALUBGH).

Field remarks: A common weed of wheat crop and waste places.

Distribution: Pakistan, India, Afghanistan & C. Asia; introduced elsewhere (Jafri, 1974).

8. Moraceae

Armed or unarmed, monoecious or dioecious, shrubs, climbers or trees, rarely herbs (*Dorstenia*), usually lacticiferous. Leaves alternate or opposite, simple, blade entire, dentate or sometimes lobed, penninerved or 3- or 5-nerved from the base, lateral nerves with or without intercostals, often with cystoliths, margin entire or variously toothed, petiolate;

stipules often caduous. Flowers unisexual, racemose, aggregated in globose or elongate receptacles, either exposed on a head or clusters or enclosed within, forming syconia (*Ficus*). Tapals 4(5), free and imbricate or valvate, or often connate enclosing the pistils in female flower, usually persistent, rarely accrescent. Stamens 1-4(5), inflexed in bud or straight (*Artocarpus, Ficus*), antitepalous, exserted on included; anthers 2-celled, dehiscence longitudinal. Pistillode present or 0. Ovary superior, sometimes embedded in fleshy receptacle, 1-locular; ovule 1 on lateral or pendulous placenta; style simple or forked, often excentric. Fruits indehiscent, always enveloped by persistent Perianth when young, later often exposed and aggregated in one of the ways: i. drupes/achenes-discrete (simple); ii. Anthocarps-enclosed in syncarps (multiple); seed 1.

A family of ca. 40 genera and ca. 1400 species, distributed in tropical, subtropical and a few in temperate regions. It is represented in Pakistan by 5 genera and 32 species, of which 16 are native (Ghafoor, 1985).

Ficus

L., Sp. Pl. 1059 (1753); Gen. Pl. ed. 5. 482 (1754); Hutch., Gen. Fl. Pl. 2: 175 (1967); Ghafoor in Nasir & Ali, Fl. Pak., 171: 2 (1985).

Trees, at times erect or climbing shrubs, often epiphytic when young, with milky sap. Leaves alternate and spiral/distichous or decussate (both alternate and decussate in *F. exasperata*), blade entire, dentate or lobed, 3- or 5-nerved from the base, lateral nerves pinnate, usually arching, margin entire, rarely crenate or dentate, apex acuminate, sometimes acute or obtuse, petiolate; stipules single or paired, deciduous leaving annular sacrs. Figs (syconia) cauliflorus and/or o branchlets, axillary, solitary, paired or clustered, sessile or stalked, generally globose, inner wall lined by minute flowers, with or without internal bristles, openinig by apical ostiole, lined by numerous, sterile, imbricate bracts (orifice); outer bracts 3-7, persistent or deciduous. Flowers unisexual, of 4 different kinds: (i) Male-ostiolar or disperse, with 1-3 stamens; anthers to celled, dehiscence longitudinal; (ii) Female-usually disperse, with 1-celled ovay; style excentric; stigma 1 or 2; (iii) Gall flower abundant and disperse, with sterile, often pedicellate ovary, enclosing an insect; (iv) Neuter flowers a few or 0, with tepals only; sex organs 0. Tepals 2-7, free or connate, glabrous or hairy. Figs either monoecious with male, femal and gall flowers in the same syconium r dioecious with male and gall flowers in the gall plants and female with or without neuter flowers in the seed plants. Mature syconium smooth or ribbed, with or without sclerotic cells, rarely splitting through the orifice. Achenes hardly enlarged, with or white or brown, thinly pulpy; seeds often compressed; testa membranous; albumen copious or 0.

Nearly 1000 species, distributed in tropics and subtropics, especially Indomalesia and Polynesia. It represented in Pakistan by 24 species, of which 11 are native (Ghafoor, 1985).

Key to species:

1a. Leaves glabrous, ovate-rounded, apex acuminate, long tailed.....................**1. F. religiosa**

1b. Leaves tomentose, coriaceous, orbicular-ovate to elliptic, apex obtuse.........**F. bengalensis**

1. Ficus religiosa L., Sp. Pl. 1059 (1753); Stewart in Ali and Nasir, Ann. Cat. Vasc. Pl. W. Pak. & Kashm., 192 (1972); Ghafoor in Nasir & Ali, Fl. Pak., 171: 33-34 (1985).

Syn: *Urostigma religiosa* Gasp. Ric. Caprif. 82. t. 7. ff. 1-5 (1845).

A large glabrous tree up to 15 m high. Leaves ca. 5-15 x 5-12 cm, alternate, ovate or ovate-suborbicular, margin entire or undulate, apex caudate-acuminate, tail 3-5 cm long, base truncate or cordate, coriaceous, shining; lateral nerves 6-8 pairs, 3-5-nerved from the base, arching near margin, flattened above, raised below, intercostals transverse, zigzag in the middle, glabrous above, puberulous below; petiole up to 4 cm long, glandular at apex below; stipules deciduous, leaving annular scars. Receptacles sessile, axillary, 1 or 2, ca. 10 x 8 mm in size, smooth, puberulous without, depressed-globose, reddish-purple when ripe; bracts 4-5, copular, ca. 6 mm, shortly connate, obtuse, persistent. Tapals 3-5, shortly connate, ca. 2 mm long, brownish, glabrous. Male flowers very few, sessile, situated near the mouth; stamen 1. Fertile female flowers sessile or pedicellate; ovary abovoid-glabrous, ca. 1.5-1 mm in size, dark brown on stylar side; style ca. 2 mm long, erect or curved, tapering. Gall flowers as female, sessile. Achense globose-ellipsoid, ca. 2 x 1.5 mm in size, dark brown.
Flowering & Fruiting period: April-June.
Herbarium specimens examined: NARA DESERT: Chundko, *Rahmatullah Qureshi*, 1331 (SALUBGH).
Field remarks: A rare tree found in settled areas near moist places.
Distribution: Pakistan, India, Bangladesh, Ceylon, China, Burma and Thailand; introduced and cultivated in S.E. Asia, Middle East, North Africa (Egypt, Libya), U.S.A. and elsewhere (Ghafoor, 1985).

2. Ficus bengalensis L., Sp. Pl. 1059 (1753); Stewart in Ali and Nasir, Ann. Cat. Vasc. Pl. W. Pak. & Kashm., 192 (1972); Ghafoor in Nasir & Ali, Fl. Pak., 171: 33-34 (1985).
Syn: *Urostigma bengalense* Gaspar, Nov. Gen. Fic. 7 (1844).

A large evergreen tree with spreading branches having downward directed aerial roots. Leaves ca. 10-12 x 8-10 cm in size, orbicular-ovate to elliptic, margin entire, apex obtuse, tomentose-coriaceous, base cordate; stipules sheathing, leaving annular scars on branchlets. Receptacles ca. 1.2-1.3 cm in diam., axillary, sessile, globose, puberulent, red when ripe.
Flowering & Fruiting Period: June-September.
Herbarium specimens examined: NARA DESERT: Chundko, *Rahmatullah Qureshi*, 1330 (SALUBGH).
Field remarks: A tree found in settled areas.
Distribution: India, sub-Himalayan tract, Bengal and hill-slopes of S. India (Bhandari, 1978).

9. Nyctaginaceae

Unarmed, erect or trailing herbs or armed scandent shrubs or trees. Leaves opposite and usually unequal or alternate, simple, penninerved, margin entire, undulate or repand, petiolate, estipulate. Flowers actinomorphic, 5-merous, bisexual or unisexual by reduction, sessile or stalked, aggregated in umbellate, corymbose or thyrsoid clusters, subtending by 1-3 bracts, free or forming an involucre; bracteoles are present or 0. Perianth-lobes 5 or 10, salver-form or funnel-shaped, constricted near the middle or below, basal part sepaloid, ribbed or with wart-like appendages, articulate with the pedicel or not, persistent (enveloping the fruit) and becoming an anthocarp; upper portion petaloid, limb even or 5- 10-lobed, plicate or valvate, deciduous. Stamens 1-10 or more, shortly connate around ovary, exserted or included; filaments capillary, as long as, or longer than perianth-tube; anthers didynamous, dehiscence longitudinal, inflexed, dorsifixed. Ovary superior, stalked or sessile, sometimes seated on a disc, 1-locular; ovule solitary, basal; style elongate; stigma peltate, lobed or fimbriate, rarely 2-fid. Anthocarp indehiscent, ribbed, often with knob-like or warted appendages, hardened in fruit; epicarp thin. Seed erect; testa thin, smooth; albumen mealy; embryo straight with convolute cotyledons or incurved; radical inferior.

A family of 30 genera and about 290 species distributed in tropical and subtropical parts of both the hemispheres, especially America. In Pakistan it is represented by 5 genera and 10-11 species (Nasir, 1977).

Key to genera:

1a. Anthocarp 5-ribbed, 3 mm long, without wart like appendages.
Flowering pedicel less than 0.5 cm long...**1. Boerhavia**

1b. Anthocarp 10-ribbed, > 7 mm long, with wart-like appendages.
Flowering pedicel more than (1) 1.5 cm long.................................**2. Commicarpus**

1. Boerhavia

L., Sp. Pl. 3: (1753); Gen. Pl. ed. 5: 119 (1754); Boiss., Fl. Or. 4: 1044 (1879); Benth. & Hook. f., Gen. Pl. 4: 708 (1885); Jafri in Fl. Kar. 110 (1966); Hutch., Gen. Flow. Pl. 2: 268 (1967); Nasir in Fl. Pak. 115: 3 (1977).

Erect or diffuse, glabrescent to puberulous herbs. Branchlets swollen at base; indumentum of sessile, cystolith-like, appressed hairs. Leaves in unequal pairs, nerves ca. 4 pairs, often glaucous below. Panicle of umbellate, capitate clusters, terminal or in the axils of bracts; pedicel articulate; bracts (basal) and bracteoles (apical) 1 each, lanceolate, caduous. Perianth funnel shaped, constricted half way down, basal part 5-ribbed, with(out) glandular hairs, upper part expanding. Stamens 1-3, as long as perianth tube. Ovary stalked; style as long as filaments; stigma peltate. Anthocarp club shaped or obconic, 5-ribbed, glabrous or glandular, swelling in water, slimy.

A genus of about 40 species distributed to tropical to subtropical, S. Europe and S. United States. In Pakistan it is represented by 4-5 species (Nasir, 1977).

Key to secies:

1a. Flowers in axillary unbranched cyme.. **1. B. procumbens**

1b. Flowers in terminal, lax, leafless cymose panicles and occasionally
simple axillary cyme:

2a. Plants erect. Flowers solitary, rarely paired;
fruits not glandular along ridges...**2. B. rubicunda**

2b. plants prostrate or decumbent. Flowers 2-7 (-10) per cluster in terminal umbels;
fruits densely glandular throughout. ....**3. B. diffusa**

1. Boerhavia procumbens Banks ex Roxb., Fl. Ind. 1: 148 (1820); Nasir in Nasir & Ali., Fl. Pak., 115: 4-6 (1977). **Figure 17A**

Syn: *B. diffusa* auct. Mult., non L.; *B. coccinea* sensu R.R. Stewart, in Ann. Cat. Vasc. Pl. W. Pak. & Kashm, 233 (1972).

A perennial diffuse, straggling, puberulous herb up to 70 cm; roots elongated, fusiform tuberous tap root. Stem woody beneath, divericately branched, slender, cylindric, thickened at nodes, minutely pubescent or subglabrous; branches often reddish coloured. Leaves simple, opposite, ovate or oblong, obtuse to nearly suborbicular, apex obtuse to acute, margin sinuate-repand, in unequal pairs, ca. 2-4 x 2-3.5 cm in size, green glossy above, silvery white beneath, often pink coloured; petiole ca. 3.5 cm long, slender. Panicle ca. 3-6 cm, axillary and terminal, 3-10-flowered. Bracts and bracteoles ca. 1.5-2.5 mm long, ovate, apex acute-acuminate, margin membranous, puberulous. Flowers ca. 3-4.5 mm long, pink; pedicel ca. 0.3 mm. Perianth-tube urceolate or funnel-form, constricted about the middle, ca. 4 mm; lower part prominently 5-ribbed; upper part deep pink, ca. 4 mm across, deciduous. Stamens 2-3, slightly exserted; filaments 3 mm, slender, rosy, arising in the axil of a scale; anthers ca. 0.5 mm. Ovary ca. 0.6 mm, ovoid; style ca. 3 mm long, slender ending in an obtuse peltate stigma. Anthocarp club-shaped, ca. 3 mm, glandular-hairy, apex rounded, 1 seeded.

Flowering period: September-August

Herbarium specimens examined: NARA DESERT: *Rahmatullah Qureshi*, 18, 20, 32, 53, 93, 243, 714, 944, 1080, 1131, 1205, 1216 (SALUBGH).

Field remarks: A very common species found in all mircohabitats.

Distribution: S. Asia, India and Pakistan (Nasir, 1977).

2. Boerhavia rubicunda Steud. Nom. ed. 2. 1: 213 (1840), *nom. nud.*; Nasir in Nasir & Ali, Fl. Pak., 115: 7-9 (1977).

Syn: *B. elegans* Choisy in DC. Prodr. 13 (2): 453 (1846); Boiss, l. c. 1049, include. var. *stenophylla*.

An erect, puberulous to glabrous herb or subshrub up to 4 ft high, arising from woody root-stock. Stem annual, woody below, much branched from the base, glabrous above, thinly pubescence near the base, terete. Leaves ca. 1-4.5 x 0.5-1 cm in size, linear-oblong or oblong-lanceolate, apex obtuse and usually mucronulate, base gradually narrowed and rounded, margin entire, rugose and canescent beneath (white when dry); petioles ca. 2-4 mm long. Inflorescence large, lax, di or trichomously branched cymes above the leaves, up to 20 cm long; branches slender, glabrous in the axil of 1-4 mm long; bracts lanceolate, apex acute, ca. 2-3 mm long; bracteoles ca. 1 mm long, lanceolate, acute, deciduous. Flowers pale pink to pinkish, ca. 2 mm across; pedicels ca. 1.5-3 cm long, very slender, glabrous. Perianth: upper portion short, campanulate, the basal portion developing into fruit. Stamens 2, dum-bell shaped. Anthocarps ca. 2-3 mm long, narrow to oblong-clavate, 5-ribbed, glabrous, minutely hairy between ribs. Stigma capitate.

Flowering period: July-August

Herbarium specimens examined: NARA DESERT: *Rahmatullah Qureshi*, 1324 (SALUBGH).

Field remarks: A very rare species found in Rohri Hills.

Distribution: Africa, S. Arabia, India and Pakistan (Sindh and Baluchistan) (Nasir, 1977).

3. **Boerhavia diffusa** L., Sp. Pl., ed. 1, 3 (1753); Nasir in Nasir & Ali., Fl. Pak., 115: 7-9 (1977).

Syn: *B. repens* L., Sp. Pl. 3 (1753); *B. repens* var. *diffusa* (L.) Hook. F. in FBI. 4: 709 (1885).

A perennial herb up 65 cm long. Stem prostrate, decumbent or ascending, divaricately branched, cylindric, swollen at the nodes, ± viscid-puberulent, rarely purplish. Leaves opposite or subopposite, two of a node unequal, the larger 4-6 cm long, smaller 2-4 mm long, broadly ovate or suborbicualr, base obtuse to rounded or subcordate, margin undulate to sinuate, green and glabrous above, paler with minute, whitish scales beneath; petioles ca. 1-4 cm long, slender, puberulent, grooved above. Flowers in pedunculate, glomerulate clusters arranged in slender, long stacked, axillary or terminal corymbs or rarely panicles. Perianth ca. 3 mm long, the ovarial part of the tube ca. 0.7-1 mm long, minutely viscid-glandular, contracted above the ovary; limb pink, campanulate, ca. 1.7-2 x 3.5 mm, with 5 narrow densely viscid, vertical bands outside. Stamens 2-3, slightly exserted; filament pink, the lobes yellow. Stigma white, capitates. Fruit ca. 3-3.5 x 1-1.5 mm in size, obovoid or subellipsoid, rounded above, slightly cuneate below, broadly and bluntly 5-ribbed, very glandular throughout.

Flowering & Fruiting Period: Almost round the year.

Herbarium specimens examined: NARA DESERT: Chotiari Dam, Sanghar, *Rahmatullah Qureshi*, 1332 (SALUBGH).

Field remarks: A rare species found in Chotiari resorvior (Achhro Thar).

Distribution: India, sub-Himalayan tract, Bengal and hill-slopes of S. India (Bhandari, 1978).

2. Commicarpus

Standley, Contrib. U.S. Nat. herb. 12: 373 (1909); Hutch., Gen. Flow. Pl. 2: 268 (1967); Nasir in Fl. Pak. 115: 9 (1977).

Perennial suffruticose herbs or scrambling shrubs. Leaves opposite, entire, often sinuate-repand. Flowers bisexual. Involucral bracts present. Perianth-tube campanulate or infundibuliform; lobes 5. Stamens 2-5; filaments unequal, connate at the base. Style filiform; stigma peltate. Anthocarp clavate, 10-ribbed. Viscid and glandular with large wart-like glands.

A genus of 16 species distributed in tropical and subtropical regions. Represented in Pakistan by 4 species (Nasir, 1977).

Commicarpus boisseiri (Heimer) Cufod., Bull. Jard. Bot. Bruxelles 23: 79 (1953); Nasir in Nasir & Ali., Fl. Pak., 115:1-12 (1977). **Figure 17B**

Syn. *Boerhavia plumbaginea* Cav. var. *viscosa* Boiss., Fl. Or. 4: 1044 (1879); *B. boisseiri* heimerl apud Vierhapper in Denks. Akad. Wiss. Wien Math. Nat. Kl. 71: 346 (1907); *B. verticillata* auct. non Poir.; Hook. f., Fl. Brit. Ind. 4: 710 (1885); *Commicarpus stellatus* sensus F.G. Bhopal & M.N. Choudari, l.c. 75.

A scandent perennial herb up to 100 cm long. Stem woody at the base, herbaceous above, branched, terete, greenish, glabrous, ± fleshy; gradually increasing from the base, nodes swollen; internode ca. 2-7.5 cm long. Leaves opposite, unequal, ca. 1-2.5 cm across,

ovate to subcordate-suborbiculate, base obtuse, apex retuse or cuspidate, margin repand or sinuate, fleshy, glabrous above, glaucescent below; petiole ca. 0.3-1.5 cm long. Inflorescence an axillary and superposed umbel, ca. 2-4 cm long; peduncle ca. 1-3.5 cm long. Bracts 4, at verticel (Involucral bracts), lanceolate, hyaline, glandular hairy, ca. 2-3 mm long. Flowers verticellate, 3-5-flowered per node, ca. 12 mm long, campanulate; pedicel ca. 0.5-1.5, slender. Perianth pink-violet, funnel-shaped, 5-lobed; tube ca. 6 mm, lower part ca. 4 mm, 10-ribbed, raphids present; lobes ca. 2-3 x 3-4 mm in size, obcordate, apex deeply retuse, glandular-pubescent on above. Stamens 3, sub-exserted; filament ca. 7-8 mm long, slender; anthers bilobed, dorsifixed, oblong-suborbicular globose, ca. 1 x 1.7 mm in size, pinkish. Ovary ca. 1 x 0.8 in size, ovoid-subglobose, shortly stalked; style ca. 11-12 mm long, ± glandular; stigma capitate, ca. 1 mm across. Anthocarp ca. 10 x 4 mm in size, turbinate, nodding, slimy, greenish, apex rounded, with a ring of 5-large wart-like glands around the apex, sticky.
Flowering period: September-October.
Herbarium specimens examined: NARA DESERT: *Rahmatullah Qureshi*, 411 (SALUBGH).
Field remarks: A rare species of gravelly soils.
Distribution: India, S. Iran and Pakistan (Nasir, 1977).

10. Aizoaceae

Annual or perennial diffuse herbs or small shrubs. Leaves decussate to spuriously whorled, simple, those of a pair usually very unequal, membranous or subsucculent; petioles sometimes dilated, sheathing and amplexicaul; stipules scarious, fugacious, rarely 0. Inflorescence leaf-opposed, axillary and/or terminal, of umbellate cymes or clusters; flowers sometimes solitary, enclosed in petiolar sheath. Flowers 5-merous, bisexual, actinomorphic; bracteoles scarious. Calyx-tube obconical, sometimes adnate to petiolar sheath; sepals 5, imbricate in bud, persistent, rarely subsucculent, petaloid, margin thin. Petals more commonly absent or numerous and linear. Stamens (3)5 to α, the outer ones petaloid staminodes, hypogynous or inserted on the top of calyx-tube; filaments basally connate or free; anthers oblong, 2-locular, dehiscing longitudinally; staminodes occasional. Ovary superior, 1-5 celled, usually united, rarely free; ovules α on axile or basal placentae; styles as many as cells, longitudinally stigmatose. Capsule obconical or terete, usually enclosed by calyx-tube, loculicidal or circumscissile, sometimes indehiscent. Seed(s) 1 or α per cell (sub)reniform, compressed, concentrically ribbed or granular.

A family of 128 genera and ca. 1170 species, many of which are cultivated. S. and tropical Africa, S. America, W, Indies, Mediterranean and tropical Asia. It is represented in our country by 8 genera and 11 species (Nasir, 1975).

Key to genera:

1a. Style 1..**1. Trianthema**
1b. Styles 2-5:
2a. Raphides very distinct in leaves and sepals. Carpels ....**2. Gisekia**
2b. Raphides 0. Carpels united:
3a. Plants usually viscid-glandular hairy. Ovary 2-seeded.
Fruit a schizocarp..**3. Limeum**
3b. Plants papilose to glabrous. Ovary many seeded. Fruit a capsule:

4a. Flowers glomerulate; lid of the capsule splitting into 2 valves;
Style 2; seeds 4..**4. Zaleya**
4b. Flowers solitary; lid of the capsule remaining in one piece; styles 2-5;
seeds more than 4...**5. Sesuvium**

1. Trianthema

L., Sp. Pl. 223 (1753); Gen. Pl. ed. 5. 105 (1754); Boiss., Fl. Or. 2: 766 (1872); C.B. Clarke in Fl. Brit. Ind. 2: 660 (1879); Jafri in Fl. Kar. 115 (1966); Y. Nasir in Nasir and Ali, Fl. Pak., 41: 10 (1973).

Popularia Forsk., Fl. Aegypt, Arab. 69 (1775); *Portulacastrum* Juss. Ex Medic., Phil. Bot. 1: 99 (1789).

Diffusely branched, prostrate herbs with glabrous or papilose branchlets. Leaves decussate, opposite, those of a pair very unequal, linear-orbicular, obovate or oblong, subsucculent, entire; petioles dilated and sheathing at the base, amplexicaul, leaf-bases often connate in pairs;stipules often inserted on the sheaths. Flowers axillary, clustered or solitary, sessile or pedicellate, enclosed in the sheath or not. Calyx-tube obconical; lobes 5, coloured within, ridged in the middle, apex mucronate. Stamens 5-15(20), alternating with sepals; filaments filiform, inserted on the calyx. Ovary obconical-terete, 1- or 2-celled, apex truncate; ovules 2 or a few, on basal placentae; styles short, 1 or 2, longitudinally stigmatose. Capsule circumscissile, with a truncate operculum, sometimes beaked; upper part with 1 or 2 seed(s), basla part membranous, (1-)2- 9-seeded; seeds subreniform, concentrically ribbed.

A genus of 20 species, tropical to subtropical, Africa, America, Asia and Australia. In Pakistan, 2 species are recorded (Nasir, 1973).

Key to species:

1a. Flowers in clusters. Stamens 5.. **1. T. triquetra**
1b. Flowers solitary. Stamens 10...**2. T. portulacastrum**

1. Trianthema triquetra Rottl. & Willd. In ges. Naturfr. Berl., Schr. 4: 181 (1803); Stewart in Ali and Nasir, Ann. Cat. Vasc. Pl. W. Pak. & Kashm., 235 (1972); Townsed in Nasir and Ali, Fl. Pak., 41: 6 (1973).

Syn: *T. crystallina* (*auct. plur*) Roxb. Fl. Ind. 2: 444 (1832) & *sensu* FBI. *non.* Forsk.; Vhla, Symb. Bot. 1: 32 (1790); *T. sedifolia* Visiani, Pl. Aeg. Et Nubia, 19. t. 3. f. 1 (1836).

A succulent, prostrate, annual herb up to 30 cm; branchlets many, radiating in all directions, terete, reddish and papilose, especially the younger parts. Leaves ca. 4-15 x 1.5-4 mm in size, linear or linear-ellipticbase sub-acute, apex obtuse, margin entire, greyish-green, glabrous, succulent; petiole up to 3 mm long, dilated at base into a semi-amplexicaul sheath, the two adjacent sheaths free with each other; each sheath bearing at base a pair of minute, dentiform stipules, one on each side; stipules triangular. Flowers sessile, solitary or in clusters of 2-5, often crowded on short branchlets, the latter free from the petiolar sheaths, ca. 2 mm across. Perianth longitudinally ribbed throughout, 5-cleft to half way down; tube ca. 1.5 mm long, obconical; lobes 0.5 mm, obliquely erect or preading, recurved, ovate-triangular, apex obtuse with a dorsal, subapical, minute mucro. Stamens 5, alternating with the perianth;

filaments ca. 0.5 mm long; anthers purplish. Ovary ca. 1 mm long, obconical, turbinate, depressed above centrally around the excentric style; style 1, ca. 0.5 mm long, persistent. Pyxidium ca. 2 mm long, lid circularca. 1.5 mm broad, flat, depressed in the centre; amphora 1-seeded. Seeds ca. 1 mm, orbicular, concave-convex, brownish black, faintly ribbed.
Flowering period: July-November.
Herbarium specimens examined: NARA DESERT, *Rahmatullah Qureshi*, 878, 956, 995, 1128 (SALUBGH).
Field remarks: A common species found in saline habitats.
Distribution: Africa, Arabia, Indian & Pakistan (Nasir, 1973).

2. Trianthema portulacastrum L., Sp. Pl. 223 (1753); Jafri in Nasir and Ali, Fl. Pak., 41: 5-6 (1973). **Figure 1C**

Syn: *T. monogyna* L., Mant. 69 (1767); *T. obcordata* Roxb., Hort. Beng. 34 (1814).

Annual up to 60 cm long. Stem sparsely pubescent. Leaves sub-orbiculate to obovate, ca.4x3 cm in size, obtuse or acute; petiole ca.1.5 cm long, papillose to glabrous, base dilated, clasping with 2 lateral stipular appendages. Flower axillary solitary, sessile, bisexual, incomplete, regular, having two bracts, hairy outside, unicellular whitish in colour, sparsely hairy, ca. 2x3 mm in size. Extra tabular cover represented by 5-blunt projection at distal end; lower one tube closely sheathed by the leaf base; 15-16 mm wide. Sepals 5, 8-9 mm long, oblong, entire, apiculate, shortly aristate, twisted, imbricate. Stamens 10-8, unequal, basifixed, anthers pinkish, ca. 1x0.8 mm in size; filament 2-7 mm long. Ovary cup-shaped ca. 5x4 mm in size; style ca. 6 mm long having crystal unicellular appendages on its dorsal side; stigma globose covered by crystal like unicellular appendages or trichomes. Capsule ca. 6 mm long, comprises of Pyxidium at lower end and lid ca. 4 mm long. Seeds 7-11 in pyxidium, 1-3 in lid; reniform, rough with appendages, epidermal surface aculeate, ca. 2 mm broad.
Flowering Period: May- October
Herbarium specimens examined: NARA DESERT: a kharif crop weed, *Rahmatullah Qureshi*, 1333 (SALUBGH).
Field remarks: A common weed of Kharif season and wasteland.
Distribution: Africa, Arabia, India & Pakistan (Nasir, 1973).

2. Gisekia

L., Mant. 2: 554 (1771); C.B. Clarke in Fl. Brit. Ind. 2: 664 (1879); Jafri in Fl. Kar., 119 (1966); Y. Nasir in Nasir and Ali, Fl. Pak., 41: 10 (1973).

An annual diffuse branched herb. Leaves opposite or falsely whorled, subfleshy, linear or spathulate, abounding in raphides; stipules 0. Inflorescence in lax or dense axillary sessile fascicles or shortly peduncled umbellate cymes. Flowers bisexual or polygamous, 5-merous. Sepals 5, equal, herbaceous or with membranous margins. Petals 0. Stamens 5-15, hypogynous; filaments dilated at the base; anthers oblong. Carpels usually 5, distinct, sessile on a small torus, 1-celled; ovule solitary, basal; styles as many as the carpel. Fruit of 5 free membranous compressed papilose indehiscent 1-seeded carpels. Seeds vertical, compressed, subreniform; embryo annular.

A genus of 5 species distributed in Africa, Arabia, India and W. Pakistan. It represents 1 species in Pakistan (Nasir, 1973).

Gisekia pharancoides L. Mantis. Alt. 2: 562 (1771); Stewart in Nasir & Ali., Ann. Cat. Vasc. Pl. W.Pak. & Kashm., 234 (1972); Y. Nasir in Nasir and Ali, Fl. Pak., 41: 10 (1973).

Syn: *G. molluginoides* Wight, Calcutta J. Nat. Hist. 7: 162 (1847).

A procumbent or diffuse herb upto 40 cm high. Stem much branched, herbaceous, terete, prostrate or ascending, glabrous, pinkish. Leaves ca. 1-3.5 x 0.5-1 cm in size, pairs unequal, oblong or obovate-lanceolate, apex obtuse to apiculate, base cuneate, margin entire, fleshy, glaucous; petiole ca. 4 mm, not dilated at the base. Cymes ca. 1 cm, axillary, subumbellate; 10-flowered. Flowers ca. 5 mm across, greenish white; pedicel ca. 4 mm, slender. Bracts ovate, apex acute, margin entire, ca. 0.5 mm. Sepals ca. 2 mm, elliptic-oblong, apex obtuse, margin membranous, hooded, green. Petals shorter than sepals, cawed. Stamens 5, alternating with sepals, ca. 0.8 mm; filaments dilated connate at the base. Carpel ca. 1 mm; ovary free, globose, compressed, 2-celled; ovule solitary per cell, ascending with basal funicle; style very short, with 5 branches which are stigmatose within. Fruit separating into 5 orbicular or hemispheric cocci which at length dehisce ventrally. Seeds ca. 1 mm, rounded at the back, black with scattered white glandular prominence.
Flowering period: Almost round the year.
Herbarium specimens examined: NARA DESERT, *Rahmatullah Qureshi*, 12, 14, 31, 47, 48, 95, 106, 1302 (SALUBGH).
Field remarks: A common herb of sandy area.
Distribution: Africa, Arabia, India, Ceylon, Afghanistan and W. Pakistan (Nasir, 1973).

3. Limeum

L., Syst. Nat. ed. 10. 995 (1759); C.B. Clarke, l. c. 664; Jafri, l. c.; Y. Nasir in Nasir and Ali, Fl. Pak., 41: 10 (1973).

Annual or perennial low branched glabrous herbs, sometimes frutescent at the base; branches usually prostrate. Leaves alternate or subopposite, linear-lanceolate, elliptic or obovate, entire or obscurely ciliolate; stipules 0. Inflorescence in dense terminal and subaxillary cymes. Flowers bisexual or unisexual, small, bracteate, greenish. Sepals 5, unequal, ovate, herbaceous or with membranous margin. Petals 3-5, oblong or spathulate, or minute or 0. Stamens 5-10, sometimes imperfect; filaments hypogynous. Ovary free, compressed, 2-celled; ovule solitary in each cell, ascending with basal funicle; style very short, with 2 branches which are stigmatose within. Fruit separating into 2 orbicular or hemispheric cocci which at length dehisce ventrally. Seeds reniform, vertical; embryo annular; radicle inferior.

A genus of 28 species mainly distributed in tropical and S. Africa, Arabia, India and W. Pakistan. This genus represents only 1 species in Pakistan (Nasir, 1973).

Limeum indicum Stocks ex T. Anders in Journ. L. Soc. Bot. 5 (suppl. 1): 30 (1860); Stewart in Ali and Nasir, Ann. Cat. Vasc. Pl. W. Pak. & Kashm., 234 (1972); Y. Nasir in Nasir and Ali, Fl. Pak., 41: 10-11 (1973). **Figure 1B**

An annual or perennial procumbent herb upto 10 cm high. Stem much branched at the base, diffuse, herbaceous, slender, densely grandular hairy, greenish brown in colour; internodes ca. 0.3-1.5 mm. Leaves ca. 3-8 x 2-6 mm in size, broadly elliptic-suborbicular, base cuneate, apex mucronate, margin entire, sparsely glandular and glabrescent above, densely grandular and hairy below, 1-nerved, lateral nerves obscure; petiole ca. 1-4 mm long, slightly dilated at the base. Flowers ca. 2 mm long. Sepals ca. 2 x 1-2 mm in size, imbricate, broadly ovate, margin hyaline, greenish, densely grandular hairy above, glabrous within. Petals much shorter than the sepals, clawed, truncate and 2-dentate at apex. Stamens 7; filaments ca. 1 mm, hyaline, base dilated, tapering above; anthers ca. 0.3-0.5 mm, globose, yellow. Ovary ca. 1 x 0.8 mm in size, pinkish or pinkish brown, glabrous; style ca. 1 mm, filiform, slightly pinkish. Fruit ca. ca. 2 mm, globose, 2-seeded. Seeds ca. 2 x 1 mm in size, concavo-convex, black brown, subglabrous .
Flowering period: March-September.
Herbarium specimens examined: NARA DESERT, *Rahmatullah Qureshi*, 6, 7, 50, 82, 263, 806, 994, 1047, 1130 (SALUBGH).
Field remarks: A very common herb found in all microhabitats.
Distribution: Not common in Punjab Plain but commonly found in san up to 1000 m altitude (Nasir, 1973).

11. Zaleya

Burm. F., Fl. Ind. 110, t. 31 (1768); Jeffrey in Fl. Trop. E. Africa, 28 (1961); Y. Nasir in Nasir and Ali, Fl. Pak., 41: 2 (1973).

Prostrate annual or perennial herbs. Leaves opposite, exstipulate. Inflorescence in axillary globose clusters. Flowers bisexual, sub-sessile, regular. Calyx-tube 5-lobed; lobes with membranous margin, mucronate. Stamens 5-15, free. Ovary 2-locular; styles 2, free. Capsule 4-seeded, dehiscence operculate, the valves separating longitudinally into 2 valves.

A genus of about 6 species; tropical Africa, Asia and Australia. In Pakistan, it is represented by 1 species (Nasir, 1975).

Zelya pentandra (L.) Jeffrey in Kew Bull. 14(2): 238 (1960); Stewart in Nasir & Ali., Ann. Cat. Vasc. Pl. W.Pak. & Kashm., 235 (1972); Y. Nasir in Nasir and Ali, Fl. Pak., 41: 2-3 (1973).

Syn: *Zelya govindia* (G. Don) Nair, Bull. Bot. Surv. Ind. 8: 86 (1966*); Trianthema govindia* Buch-Ham. Ex G. Don in Gen. Syst. 3: 72 (1834); *T. pentandra* L., Mant. 1: 70 (1767).

A prostrate, perennial herb up to 60 cm. Stem somewhat woody at the base, herbaceous above, much branched at the base, fleshy, glaucous, green to pinkish in colour; internode ca. 1-4 cm long. Leaves simple, opposite, unequal, petiolate, ca. 1-2.5 cm long, ca. 0.5-1.5 cm broad, obovate-oblong and elliptic-oblong, base acute-cuneate, apex retuse-apiculate, margin entire, fleshy glabrous; petiole ca. 0.5-1 cm long, dilated and sheathing at the base. Inflorescence axillary umbellate cyme, consisted on 4-8 flowers, peduncle ca. 1-2 mm. Flower ca. 5 x 2 mm in size, bisexual, bracteatc, 5-merous, hypogynous, pedicellate, pedicel ca. 2-3 mm long. Bracts 2, unequal, lanceolate, base rounded, apex acute, ca. 1-4.5 x 1-2 mm in size. Perianth 5, fused at the base, tube obconical, ca. 4-6 x 2-3 mm in size,

persistent, greenish, pubescent. Stamens 5, alternate to perianth, adnate to the corolla sheath; anthers 2-lobed, dorsifixed, ± equal in size, pink in colour. Ovary 2-celled, ca. 2-2.5 mm long, superior, conical-oblong; styles 2, curved, ca. 1.5 x 1 mm in size; stigma bifid. Capsule ca. 5 mm, subterete, cirumscissile with a truncate operculum, pinkish in colour; beak 2-lobed (consisting 2-lanceolar portion), margin mitriform, upper parts 1-2 seeded, basal parts membranous. Seeds 3-4 in each capsule, ca. 2 mm across, sub-reniform, concentrically ribbed, blackish coloured, dotted.
Flowering period: July-November.
Herbarium specimens examined: NARA DESERT, *Rahmatullah Qureshi*, 17, 197, 980 (SALUBGH).
Field remarks: A common herb of sandy soils.
Distribution: Africa, India, Iran and Pakistan (Nasir, 1973).

5. Sesuvium

L., Syst. Ed. Nat. 10 (1058); C.B. Clarke in Hook. F. Fl. Brit. Ind. 2: 659 (1876); Y. Nasir in Nasir and Ali, Fl. Pak., 41: 2 (1973).

Annual, branched herbs with opposite leaves. Petals absent. Flowers solitary. Stamens 5 to many. Ovary 3-5 locular, superior; styles 2-5; stigmas 5. seeds seeds more than 4 (many), reniform. Capsule 3-5 locular; lid of the capsule remaining in one piece, dehiscence circumscile.

A genus of 8 species; usually halophytes found along tropical and subtropical coasts; India, W. Pakistan and Malaya. One species recorded from Pakistan (Nasir, 1973).

Sesuvium sesuvioides (Fenzl) Verdc. In Kew Bull. 12: 349 (1957); Y. Nasir in Nasir and Ali, Fl. Pak., 41: 2 (1973).

Syn: *Diplochonnium sesuvioides* Fenzl, Nov. Stirp. Dec: 7.58 (1839); *Trainthema hydaspica* Edge. In Journ. L. Soc. Bot. 6: 203 (1862); Jafri, Fl. Kar. 117 (1966); *T. polyspera* Hochst. Ex Oliv., Fl. Trop. Afr. 2: 588 (1871).

A prostrate, succulent annual or rarely short lived herb up to 20 cm long. Stem branched, terete and densely papilose with crystalline papillae, pale in colour; internode ca. 0.3-1 cm long. Leaves alternate, subopposite or opposite, ca. 0.5-2 cm long and ca. 4-8 mm broad, elliptic-ovate or sub-oblong, apex obtuse, base truncate, margin entire and downwardly curved, densely papilose, pale green, nerves obscure, fleshy; petiolar sheath dilated, clasping, ca. 2-4 mm long, margin hyaline, papilose. Flowers axillary, solitary or sometimes clustered, sessile. Perianth tube ca. 2 mm, obconic, prominently longitudinally nerved; lobes 5, ca. 2.5-3 x 0.8-1 mm in size, imbricate, triangular-elongate, apex apiculate, margin hyaline, membranous, glabrous within, papilose above, green outside, pink inside, dilated at the base, producing 5 dorsal horns at sinus. Stamens 5-7, free; filaments ca. 0.9-1 mm long, filiform, glabrous, dark pink in colour; anthers ca. 0.3 mm across, oblong, dark pink or red. Carpels 2; ovary ca. 1.2 x 0.8 mm in size, 2-lobed, oval, the beak elongate-conical, hollow; styles 2 ca. 2 mm long, fliliform, white. Capsule ca. 6 mm long, dehiscing by conical cap. Seeds ca. 6-9 in each locule, reniform, dull black, rugose.
Flowering period: December.

Herbarium specimens examined: NARA DESERT, *Rahmatullah Qureshi*, 926, 1129 (SALUBGH).
Field remarks: A common herb of saline soils.
Distribution: Africa, India, and W. Pakistan (Nasir, 1973).

12. *Chenopodiaceae*

Annual or perennial, erect to ascending herbs or shrubs, rarely dwarf trees, mostly xerophytic or halophytic; branchlets terete or angular, ribbed or striate. Leaves well-developed or much reduced, simple, either alternate and well developed or decussate and obscure, estipulate, succulent or not. Flowers small, often usually greenish, bisexual or 1-sexual, small, solitary or in clusters, often arranged in spicate or paniculate inflorescence, bracteate and mostly bracteolate; bracteoles 2, rarely 0. Perianth-lobes (tepals) 3-5, (un)equal, calycine, distinct, imbricate, free or shortly connate, subsucculent or membranous, concealing the flowers, persistent or not. Stamens (1-) 2-5 (rudimentary in female flowers), opposite the tepals, free or basally united into a hypogynous disc, straight or inflexed in bud; anthers 2-celled, parallel or divericate, dorsifixed, dehiscence longitudinal. Ovary superior, more rarely semi-inferior, ovoid, globose or depressed, 1-locular; ovule 1, basal, campylotropous; styles 1-5 or 0; stigmas 1-5, capitate; pistil rudimentary in female flowers. Fruit an indehiscent utricle or achene, rarely a dehiscent capsule, free or adherent to perianth, sometimes additionally enclosed in the bracteoles (valves); segments of fruiting perianth sometimes winged or appendiculate. Seed 1, erect, inverse or horizontal; testa crustaceous, shining; albuminous fleshy or 0; embryo curved, annular or spiral.

A family of ca. 100 genera and more than 1200 species, almost cosmopolitan, mostly of saline and arid areas. In Pakistan it is represented by 35 genera and 106 species (Kassel et al., 2001).

Key to genera:
1a. Branchlets jointed. Leaves obscure. Flowers irregular.
2a. Leaves decussate. Stamen 1.. **1. Arthrocnemum**
2b. Leaves opposite. Stamens 5.. **2. Haloxylon**
1b. Branchlets not jointed. Leaves alternate, well-developed. Flowers regular:
3a. Leaves flat... **3. Chenopodium**
3b. Leaves fleshy, half-terete or absent:
4a. Sepals transversely winged:
5a. Stem jointed... **2. Haloxylon**
5a. Stem not jointed.. **4. Salsola**
4b. Sepals not winged... **5. Suaeda**

1. Arthrocnemum

Moq., Chenopod. Monogr. 111 (1840); Hook. f., Fl. Brit. 5: 11 (1886); Boiss., Fl. Or. 4: 932 (1879); Kassel et al. in Ali & Qaiser, Fl. Pak. 204: 97 (2001).

Halophytic, succulent, much branched shrubs with jointed leafless stems, sometimes rooting at nodes; internodal segments usually club-shaped, each with a pair of opposite, terminal cup embracing the base of the next segment and subtending a group of 3 flowers (bisexual or 1-sexual) ± immersed in the cup. Perianth 3-4-fid, becoming somewhat swollen and spongy in fruit. Stamens 1 (-2). Styles rather along with 2 (-3) stigmas. Fruit a utricle with membranous or indurated pericarp. Seed vertical usually endospermous with arcuate embro.

About 5 species, chiefly of saline marshy places and sea coasts of the Mediterranean and Irano-Turanian regions. Two species are recorded from Pakistan (Kassel et al., 2001).

Arthrocnemum indicum (Willd.) Moq., Chenopod. Enum. 113 (1840); Jafri in Fl. Kar., 96 (1966); Stewart in Ali and Nasir, Ann. Cat. Vasc. Pl. W. Pak. & Kashm., 217 (1972); Kassel et al. in Ali & Qaiser, Fl. Pak., 204: 100 (2001).

Syn: *Salicornia indica* Willd in Ges. Naturf. Fr. Neue Schr. 2: 111, t. 4, f. 2 (1791); *Holosarcia indica* (Willd.) Wilson in Nuytsia 3: 63 (1980).

A suffruticose perennial, diffuse, glaucuous-green shrub up to 150 cm tall. Stem usually prostrate, fairly branched with a woody core; branchlets succulent, terete, jointed, the joints short, constricted at the nodes, purplish with age. Leaves decussate, obscure, those of each pair connate into a cup, ca. 0.5-0.8 x 0.4-0.6 cm, thin towards the margin, tightly enclosing the internodes, 2-lobed, embracing the shortly attenuate base of the higher leaf pair. Stamen included; filament ca. 0.3 mm; anther cell equal, diverícate, ca. 0.2 mm long. Ovary globose, ca. 0.7 mm across; ovule horizontal; stigma ca.1 mm minutely barbed, exserted. Utricle 1 mm.

Flowering period: October-November.

Herbarium specimens examined: NARA DESERT: *Rahmatullah Qureshi*, 196, 639, 855, 892, 957, 1008 (SALUBGH).

Field remarks: A rare shrub of calcarious and gravelly soils.

Distribution: Australia, Malaysia, Sri Lanka, India, Pakistan and tropical East Africa (Kassel et al., 2001).

2. Haloxylon

Bunge, Reliq. Lehm. 292 (1851), Mem. Sav. Etr. Petersb. 7: 468 (1851); Jafri & Rateeb in Jafri & El-Gadi, Fl. Libya 95 (1978); Kassel et al. in Ali & Qaiser, Fl. Pak. 204: 188 (2001).

Arcthrocnemum acutt. P.p. non Schrenk in Bull. Cl. Phys.-Math. Acad. Imp. Sc. Petersb. 3: 211 (1845); *Hammada* Iljin in Bot. Zhurn. 33 (6): 582 (1948).

Small trees or shrubs with spartioid branches, jointed, brittle stems and reduced or obsolete opposite leaves. Flowers borne on short twigs on lower part of older branches, perfect, 5-merous, with a pair of ± membranous bracteoles, solitary in the axils of scale-like bracts. Perianth segments membranous, free with basal tufts of flexuous hairs, developing wings in fruit. Stamens 5; filaments exserted, united at base into a hypogynous disc; anthers oblong-oval, unappendaged; staminodal lobes 5, thin, glabrous, not thickened. Ovary with 2-5 very short, ± sessile stigmas. Seed horizontal with spiral embryo.

About 6 species mostly from Central Asia. Represented in Pakistan by 5 species (Kassel et al., 2001).

Key to species:

1a. Leaves distinct.. **1. H. stocksii**

1b. Leaves reduced to dilated tips of the joints.................................. **2. H. salicornicum**

1. Haloxylon stocksii (Boiss.) Benth. & Hook., Gen. Pl. 3: 70 (1883); Kassel *et al.* in Ali & Qaiser, Fl. Pak., 204: 193 (2001). **Figure 9C**

Syn: *Caroxylon recurvum* Moq. in Dc., Prodr., xiii (2): 175(1849); *Salsola stocksii* Boiss., Diagn. Ser. 2 (4): 75 (1859); *Haloxylon recurvum* (Moq.) Bunge ex Boiss., Fl. Or., iv: 949 (1897).

A perennial, straggling bushy undershrub, turning black after drying, glabrous, pruinose up to 1 m high. Stem fairly branched, woody at the base; branches divaricately spreading, straight or recurved, glaucous-green. Leaves ca. 3-8 x 2-3 mm in size, fleshy, trigonous or half terete, ovate-subulate or ellipsoid, apex obtuse or rounded, margin entire. Flowers ca. 5 mm; floral leaves ca. 2.5 mm long, rotund-ovate, fleshy, with membranous margins; bracteoles 2.5-3 mm long, elliptic-oblong, concave, fleshy, with membranous margins. Perianth-segments fleshy, ca. 4-5 mm long, ovate, apex subobtuse, closing over the utricle; wing membranous, finely veined, ca. 2 mm long, orbicular, attached at the middle of the back of the segment. Seeds ca. 2 mm across, orbicular, flattened.

Flowering period: November-December.

Herbarium specimens examined: NARA DESERT: *Rahmatullah Qureshi*, 194, 197, 631, 890, 895, 908, 969, 1014, 1194 (SALUBGH).

Field remarks: A common shrub of calcarious and gravelly soils.

Distribution: Pakistan, Afghanistan, W. India (Punjab, Rajasthan) (Kassel *et al.*, 2001).

2. Haloxylon salicornicum (Moq.) Bunge ex Boiss., Fl. Or. 4: 849 (1879); Stewart in Ali and Nasir, Ann. Cat. Vasc. Pl. W. Pak. & Kashm., 224 (1972). **Figure 9B**

Syn: *Caryoxylon salicornicum* Moq. in DC., Prodr. 13 (2): 174 (1849); *Hammada salicornica* (Moq.) Iljin in Bot. Zhurn. 33: 583 (1948).

Undershrub or shrub, pale, much branched, almost leafless, 25-60 cm tall, with woody stem; shoots ± ascending rigid, light coloured to white or waxy yellowish when dried. Leaves reduced to minute, short-triangular scales, connate into a short cup, membranous at margins, woolly within. Inflorescence consisting of short, 3-6 cm long, scattered spikes, mostly at the ends of pale greenish , main or lateral shoots of the current season. Bracteoles 2, ovate-concave, woolly at base. Stamens alternating with linear ovate staminodes, papillose at tips; styles short, stigmas 2, strongly papillose on inner side. Fruiting perianth including wings usually 7-8 mm across, somewhat brownish.

Flowering Period: October-November.

Herbarium specimens examined: NARA DESERT: Chundko, *Rahmatullah Qureshi*, 1334 (SALUBGH).

Field remarks: A common shrub of calcarious soils.

Distribution: Egypt, Palestine, Jordan, Iraq, Kuwait, C. & E. Arabian Peninsula, Iran, Afghanistan and Pakistan (Kassel *et al.*, 2001).

3. Chenopodium

L., Sp. Pl.: 218 (1753); Gen. Pl. ed. 5: 103 (1754); Boiss., Fl. Or. 4: 900 (1879); Hook f., Fl Brit. India 5: 5 (1886); Stewart in Ali and Nasir, Ann. Cat. Vasc. Pl. W. Pak. & Kashm., 220 (1972); Bhopal & Chaudhri in Pak. Syst. 1: 42 (1977); Chaudhary, Fl. Kingd. Saudi Arabia 1: 159 (1999); Kassel et al. in Ali & Qaiser, Fl. Pak. 204: 13 (2001).

Blitum L., Sp. Pl.: 4 (1753); *Roubieva* Moq. in Ann. Sci. Nat. Bot. 2, ser. 2, 1: 292 (1834).

An annual or biennial aromatic or non-aromatic, erect or sepals ascending herbs or subshrub; branchlets green or purple, usually ribbed or striate, often with minute, white or yellow vesicles. Leaves alternate, oblong or ovate-lanceolate, herbaceous, base gradually decurrent, margin entire or coarsely toothed, less so, or almost entire, in tender leaves, apex acute, petiolate, gradually passing into bracts on flowering shoots. Flowers actinomorphic, 5-merous, bisexual, sessile, aggregated in small groups, making for a terminal or axillary, slender panicle; bracteoles 0. Flower buds depressed. Perianth-lobes 5, equal, concave, herbaceous, acute, abruptly folded in middle. Stamens 5; anthers parallel. Ovary depressed; ovules erect or horizontal; style short; stigmas 2-5. Utricle enclosed in connivent, herbaceous tepals; epicarp membranous. Seeds somewhat compressed, circular, keeled or not, brownish black, shining or minutely tuberculate, albuminous; embryo annular.

A genus of about 150 species globally distributed in all continents, mainly in subtropical and temperate regions. Eighteen species are reported from Pakistan (Kassel et al., 2001).

Key to species:

1a. Leaves subentire or lobed; seeds smooth,... **1. C. album**

1b. Leaves toothed; seeds finely papilate, dull black.. **2. C. murale**

1. Chenopodium album L., Sp. Pl. 219 (1753); Stewart in Ali and Nasir, Ann. Cat. Vasc. Pl. W. Pak. & Kashm., 220 (1972); Kassel *et al.* in Ali & Qaiser, Fl. Pak., 204: 40 (2001). **Figure 9D**

An annual erect or ascending herb up to 10 ft high. Stem striped; branchlets mealy, green or purplish, glabrous or thinly clothed with white vesicles. Leaves ca. 0.5-6 x 1-3 cm in size, ovate-lanceolate, apex acute or obtuse, subentire, toothed or irregularly lobulate, upper ones smaller and more entire, lateral nerves ca. 4 pairs, faint; petiole ca. 1-3 cm. Panicles ca. 5-10 cm, axillary and terminal. Flowers ca. 2-3 mm, pale green. Tepals 5, ca. 1.5 mm, ovate-rhomboid, base keeled, broadened in the middle, apex obtuse, closing over the thinly membranous utricle. Stamens 5; filaments ca. 0.3-0.4 mm; anthers ca. 0.3 mm. Ovary ca. 0.4-0.5 mm, globose; stigmas 2, ca. 0.6 mm, spreading. Utricle ca. 1.5-2 mm across. Seeds ca. 1.2-1.5 mm across, orbicular, compressed, obscurely keeled with a blunt margin, smooth, shiny.

Flowering period: January-April.

Herbarium specimens examined: NARA DESERT: *Rahmatullah Qureshi*, 1127, 1295 (SALUBGH).

Field remarks: A common weed of sandy soils.

Distribution: Almost cosmopolitan, common in subtropical to temperate zones, more infrequent in the tropics and cooler regions (Kassel *et al.,* 2001).

2. Chenopodium murale L., Sp. Pl. 219 (1753); Stewart in Ali and Nasir, Ann. Cat. Vasc. Pl. W. Pak. & Kashm., 220 (1972); Kassel *et al.* in Ali & Qaiser, Fl. Pak., 204: 40 (2001). **Figure 10A**

Syn: *Chenopodium ilicifolium* Griff., Not. Pl. asiat. 4: 337 (1854).

Erect or ascending annual. Stem erect, much branched especially in basal parts, branches ± spreading, lowermost subopposite, angular, deeply sulcate when dry, clothed with white, powdery vesicles, especially the younger parts. Leaves ca. 1.5-8 x 1-5 cm; blade thin to somewhat fleshy, broadly ovate-rhomboid or elliptic, coarsely dentate, teeth often slightly incurved and acute to obtuse, apex acute to obtuse, base variously attenuate to truncate; uppermost leaves lanceolate, sometimes entire, usually olive green, sometimes yellowish or tinged with red, ± farinose; petiole shorter than or equal to blade. Inflorescence terminal and axillary, mostly leafy, divaricately branched, branches shorter than subtending leaves, glomerules small, ± densely arranged. Flowers bisexual and sometimes female. Perianth segments 5, connate below the middle, prominently keeled near apex, green, sometimes red when old, vesicular hairy. Stamens 5. Stigmas 2. Fruits falling with perianth. Pericarp prominently papillose, firmly adherent. Seeds dull black, 1.1-1.4 mm in diameter, round, very finely papilose.

Flowering & Fruiting Period: January-July.

Herbarium specimens examined: NARA DESERT: *Rahmatullah Qureshi,* 133 (SALUBGH).

Field remarks: A common weed of waste places.

Distribution: From C and S. Europe, N. Africa to Caucasia, Iran, Afghanistan, Turkmenistan, India, Ceylon and to the Arabian Peninsula; introduced in subtropical and tropical C, E and S Africa, America, Australia and Oceanic islands (Kassel *et al.,* 2001).

4. Salsola

L., Sp. Pl. ed. 1: 222 (1753); Gen. Pl. ed. 5: 104 (1754); Moq., Chenopod. Monogr. enum. 134 (1840); Moq. in DC., Prodr. 13,2: 179 (1849); Boiss., Fl. Or. 4: 951 (1878); Kassel et al. in Ali & Qaiser, Fl. Pak. 204: 127 (2001).

Caroxylon Thunb., Nov. gen., 2: 37 (1782); *Salsolaria* Moq., Chenopo. Monogr. enum. 144 (1840), p.p.; *Hypocylyx* Wol. in Stapf, Denkschr. Akad. Wiss. Wien, Math.-Nat. Kl. 51: 275 (1886); *Darniella* Maire & Wieller in Bull. Soc. Hist. Nat. Afrique N. 30: 301 (1939); *Climacoptera* Botsch. in Sborn. Posv. Akad. Sukachevu: 111 (1956); *Nitrosalsola* Tzvelev in Ukrayins'k. Bot. Zhurn. 50: 80 (1993); *Xylosalsola* Tzvelev l.c. Ukrayins'k. Bot. Zhurn. 50: 81 (1993).

Annual, dwarf shrubs, rarely small trees; glabrous or with very diverse types of hairs. Young stems whitish, purplish or fresh green with pale longitudinal lines. Leaves aletrnate, sometimes subopposite to opposite, rarely distichous; usually semiterete or terete and ± succulent, more or less linear, eventually with a spiny apex, isolateral, mostly with 1 large central bundle and numerous smaller peripheral bundles near to the Kranz layer, usually short-lived and functionally replaced in flowering time by bracts and bracteoles; leaf base very variable in shape. Bracts more or less leaf-like and successively shorter. Bracteoles 2, green at least in upper part, usually much shorter and more uniform in length, enclosing at least the lower 2/3 of the flowers. Flowers sessile, solitary or clustered.

A genus of about 100 species centered in the Old World desert belt from the Canary Island through S. Europe and N Africa to S.W. and C. Asia, southward to tropical E. Africa and S.W. Africa; a few species along sea shores, on ruderal and naturally distributed sites penetrating into adjacent semi-humid or even humid areas. Eighteen species are recorded from Pakistan (Kassel *et al.*, 2001).

Salsola imbricata Forssk., Fl. Aegypt.-Arab. 57 (1775) Kassel *et al.* in Ali & Qaiser, Fl. Pak., 204: 157 (2001). **Figure 10B**

Syn: *Chenopodium baryosmon* Schult. in Roem. & Schult. Syst. Veg., 6:269 (1820); *S. foetida* Del. ex Moq., Chenop. monogr. enum. 143 (1840); *Salsola baryosma* (Roem. & Schult.) Dandy in F.W. Andrews, Fl. Pl. Ang. Egypt. Sudan., 1:111, (1950).

A large much branched hoary to subglabrous shrub up to 1 m high. Stem spreading, stout branches slender. Leaves minute, crowded, alternate, fleshy, suborbicular, floral leaves imbricating, turning black when dried. Flowers small, in short cylindrical spikes; bracteoles broadly ovate or orbicular, fleshy, as long as perianth-lobes or shorter. Perianth 4-5-partite, ca. 1 mm long, oblong, apex rounded, silvery-white; wings on the fruiting perianth scarious, ca. 2 mm across, obovate, flat, membranous, margin often overlapping, white, shining. Ovary ovoid or subglobose.

Flowering period: August-October.

Herbarium specimens examined: NARA DESERT: *Rahmatullah Qureshi*, 105, 195, 622, 624, 1245 (SALUBGH).

Field remarks: A common shrub of saline soils.

Distribution: From W and W Sahararian countries throughout the hot desert belt to tropical E Africa, the Arabian Peninsula, S. Iran, Pak. Baluchistan, southern and eastern Afghanistan, NW India i.e. Punjab, Hariana, Delhi, Rajasthan & Gujrat (Kassel *et al.*, 2001).

5. Suaeda

Forssk., Fl. Aeg.-Arab. 70 (1775); Moq. in Ann. Sci. Nat. 23: 303 (1831); Chenopod. Monogr. enum. 120 (1840); Boiss., Fl. Or. 4: 937 (1879); Hook. f., Fl. Brit. Ind. 5: 13 (1886); Chaudhary, Fl. Kingd. Saudi Arabia 1: 174 (1999); Kassel et al. in Ali & Qaiser, Fl. Pak. 204: 104 (2001).

Schoberia C.A. Mey. in Ledeb., Icon. Pl. Fl. Ross. 1: 11 (1829); *Schanginia* C.A. Mey. in Ledeb., Fl. Atalica. 1: 394 (1829); *Brezia* Moq., in DC., Podr. 13,2: 167 (1849); *Chenopodina* Moq. in Dc. l.c. 159.

Annual or perennial subshrubs. Stem woody below, much branched; branchlets leafy, terete, sometimes with purple streaks. Leaves alternate, succulent, (semi)terete, subsessile. Flowers actinomorphic, 5-merous, bisexual or unisexual, sessile, aggregate successively in the axils of leaves/bracts to an axillary spikes or panicle; bracteoles 2-3, scarious, margin entire or toothed, persistent. Perianth-lobes 5, equal, rarely unequal, succulent, concave, shortly united, margin and apex incurved, thinly winged. Stamens 5, opposite to tepals, inflexed in bud, all fertile, or 2 or 3 sterile in male flowers, all sterile (staminodes) in female flowers; anthers parallel. Ovary free or basally adnate to perianth; styles 2-5, spreading, stigmatiferous. Utricle thin-walled, erect or horizontal, concealed by tepals. Seeds erect, horizontal or oblique; testa coriaceous or crustaceous; albuminous 0 or scanty; embryo slender, spiral.

A genus of about 80-90 halophytic species mostly distributed throughout the world, but primarily extra-tropical, growing on wet, moist or dry saline, alkaline and gypsiferous soils in semi-desert, deserts and along sea-shores; in the Old World centered around the Mediterranean and in the semi-deserts of C. and S.W. Asia. Seven species are recorded from Pakista (Kassel et al., 2001).

Saueda fruticosa (L.) Forsk., Fl. Aegypt-Arab., 70(1775); Stewart in Nasir & Ali., Ann.Cat. Vasc. Pl.W.Pak. & Kashm., 227 (1972); Kassel *et al.* in Ali & Qaiser, Fl. Pak., 204: 107 (2001). **Figure 10C**

Syn: *Chenopodium fruticosum* L., Sp. Pl. 221 (1753); *Salsola lana* Edgew. in Hook. J. Bot., 2: 286 (1840); *Suaeda vermiculata* sensu Boiss., Fl. Or. 4: 940 (1879); *S. nudiflora* sensu Burkill, Work. List Fl. Pl. Baluchistan 65 (1909); *S. mesopotamica* Eig. in Palestine J. Bot., Jerusalem Ser. 3: 127 (1945); *S. baluchistanica* Akhani & Podlech in Rech. f., Fl. Lowl. Iraq 194 (1964).

Perennial undershrub up to 1 m high. Stem much branched, pale, glabrous. Leaves o.5-1.5 cm long, ca. 0.5-1.5 mm broad, subsessile, fleshy, linear-oblong or obovate, spathulate or elliptic, apex obtuse, narrowed at the base. Flowers bisexual, axillary, solitary or 2-5-nate; bracteoles membranous, ca. 1.5 mm long, ovate, apex acute, margin entire or slightly denticulate. Perianth in fruit subglobose, ca. 2.5 mm long; segments thick, oblong, concave, apex obtuse, incurved. Styles 3, short; stigmas 3, short, spreading. Utricle obovoid, thickened at apex. Seeds 1 mm across, obliquely ovoid, somewhat beaked, slightly compressed, smooth, shining, and black when ripe.
Flowering period: November-July.
Herbarium specimens examined: NARA DESERT: *Rahmatullah Qureshi*, 79, 1286 (SALUBGH).
Field remarks: A common shrub of saline habitat.
Distribution: From the Cape Verde and Canary Island through N. Africa, the Arabian Peninsula, Jordan, Iraq, S. and C. Iran, Baluchistan, southern and eastermost Afghanistan to Indus and upper Ganges plains, along coastal India at least to Madras, southwards down to Kenya and Ceylon (Kassel *et al.*, 2001).

12. Amaranthaceae

Annual or perennial herb or subshrubs, rarely scandent. Leaves simple, alternate or opposite, exstipulate, entire or almost so. Inflorescence a dense head, loose or spike like thyrse, spike, raceme or panicle, basically cymose, bracteate; bracts hyaline to white or coloured, subtending one or more flowers. Flowers bi- or unisexual (plant dioecious or monoecious), regular, 2-bracteolate, frequently inflorescence ultimate 3-flowered cymules;

lateral flowers of such cymules sometimes modified into scales, spines or hooks. Bract 1, persistent, deflexed with age; bracteoles 1 or 2, usually with an excurrent midrib, falling off with fruit, rarely persistent. Perianth uniseriate, membranous to firm and finally indurate, usually falling with the ripe fruit included, tepals free or somewhat fused below, frequently pilose or lanate, green to white or variously coloured. Stamens as many as and opposite to the petals, rarely fewer; filaments free or commonly fused into a cup at the base, alternating with variously shaped pseudostaminodes or not, sometimes almost completely fused and 5-toothed at the apex with entire or deeply lobed teeth, some occasionally anantherous. Ovary superior, 1-locular; ovules 1-many, erect to pendulous placentation basal; style very short to long and slender; stigmas capitate to long and filiform. Fruit an irregularly rupturing or circumscissile capsule (utricle), usually with thin, membranous walls, rarely crustaceous or a berry. Seeds round to lenticular or ovoid, embryo curved or circular, surrounding the copious endosperm.

A large and almost exclusively tropical family of some 65 genera and over 1000 species, including many cosmopolitan "weeds" and a large number of xerophytic plants (Townsend, 1974).

Key to genera:

1a. Flowers deflexed, spinescent... **1. Achyranthes**
1b. Flowers not deflexed, never spinescent:
2a. Pseudostaminodes present...**2. Aerva**
2b. Pseudostaminodes absent:
3a. Flowers unisexual...**3. Amaranthus**
3b. Flowers bisexual:
4a. Stigma subulate. Fruit a circumscissile dehiscent capsule...........................**4. Celosia**
4b. Stigma recurved. Fruit a crustaceous nut..**5. Digeria**

1. Achyranthus

L., Sp. Pl. ed. 1: 204 (1753); Gen. Pl. ed. 5: 96 (1754); Boiss., Fl. Or. 4: 1879 (1879); Hook. f., Fl. Brit. Ind. 4: 729 (1885); Townsend in Fl. Pak. 71: 33 (1974).

Herbs with opposite, petiolate, entire leaves. Inflorescence a slender spike, terminal on the stem and branches, the flower at first congested and patent, finally laxer and deflexed; bracts deltoid or ovate, the midrib excurrent in a spine. Flowers solitary in the bracts, 2-bracteolate, bisexual. Perianth segments 4-5, narrowly lanceolate, acuminate, 1-3 (-5)-nerved, aristate with the excurrent midrib, indurate in fruit especially at the base. Bracteoles spinous, the spine furnished at the base with short and free to longer and adnate membranous wings. Stamens 2-5; filaments filiform, monadelphous, alternating with quadrate to broadly quadrate-spathulate pseudo-staminodes, these simple and dentate or fimbriate, or furnished with a variably developed dorsal scales. Ovary with a solitary pendulous ovule, the ovary wall very thin in fruit; style slender; stigma small, truncate-capitate. Entire flower with bracteoles falling with the ripening of the cylindrical seed, the deflexed bracts persistent.

About 6 species in the warm temperate and tropical regions of the world (Townsend, 1974).

Achyranthes aspera var. **aspera** L., Sp. Pl. I:204, (1753); Stewart in Nasir & Ali., Ann. Cat. Vasc. Pl.W.Pak. & Kashm., 228 (1972); Townsend in Nasir & Ali., Fl. Pak., 71:33-37 (1974).

Figure 1D

Annual or sometimes perinnating, erect-suberect herb to undershrub, up to 90 cm tall. Stem stout, much branched, 4-angulate, tomentose, often woody below. Leaves ca. 2-8 x 1-6 cm in size, opposite, obovate-suborbicular, base cuneate, apex obtuse-apiculate, margin entire or slightly flexuous, sparingly hairy above, densely tomentose beneath; petiole 1-2 cm long, young often tomentose, older glaberate. Spikes rigid, rapidly increasing in length, up to 60 cm in fruit. Bracts ca. 3 mm, broadly ovate, base truncate, apex acuminate, 1-nerved, persistent; bracteoles 2, ca. 2.5 mm, ovate, concave, midrib excurrent, spinescent, basal wings 1 mm, membranous, thinly ciliate, entirely adnate to midrib, persistent. Flowers sessile, sharply deflexed, appressed to the axis; 3.5-5 mm long, greenish to white. Perianth whitish or pale green to red or purple, segments 5, ca. 3.5 mm, outer one longest, narrowly lanceolate to lanceolate, base truncate to rounded, very acute at margin with a distinct margin and two obscure to distinct lateral nerves, containing fruits disc articulates from the axis above the bracts carrying away with it the spinous bracteoles by the aid of which the perianth clings to body of other objects and is thus dispersed. Stamens 5, alternating with staminodes; filaments filiform, ca. 2 mm long; staminodes fimbriate, prominently appendaged within; anthers 2-celled, ca. 0.5 mm. Ovary obconic, ca. 0.7-0.8 mm; style slender, ca. 1.7 mm; stigma capitate. Utricle ca. 1-3 mm long. Seeds cylindrical, smooth, filling the utricles.
Flowering period: September-December.

Herbarium specimens examined: NARA DESERT: *Rahmatullah Qureshi*, 657, 690, 902, 905, 1166 (SALUBGH).
Field remarks: A common subshrub of sandy soil.
Distribution: Throughout the world in tropical and warmer regions (Townsend, 1974).

2. Aerva

Forssk., Fl. Aeg.-Arab.: 170 (1775); Boiss., Fl. Or. 4: 993 (1879); Hook. f., Fl. Brit. Ind. 4: 727 (1885); Townsend in Fl. Pak. 71: 25 (1974).

Perennial herbs (sometimes flowering in the first year), prostrate to erect or scandent. Leaves and branches alternate or occasionally opposite, margin entire. Flower bisexual or dioecious, sometimes polygamous, 2-bracteate, in axillary and terminal sessile or pedunculate bracteate spikes or panicles, one flower in the axil of each bract. Perianth segments 5, oval or lanceolate-oblong, membranous-margined with a thin to wider green centre, the perianth deciduous with the fruit but bracts and bracteoles persistent. Stamens 5, shortly monadelphous at the base, alternating with subulate or rarely narrowly oblong and truncate or emarginate pseudostaminodes. Ovary with a single pendulous ovule; style very short to slender and distinct; stigmas 2, short to long and filiform (sometimes solitary and capitate, flowers then probably functionally male). Utricle thin-walled, bursting irregularly. Seeds compressed-reniform, firm, black.

About 10 species in the tropics, mostly continental Africa, Madagascar and smaller island i.e. Rodriguez, Mauritius, Socotra (Townsend, 1974)].

Key to varieties:
1a. Leaves up to 25 mm board, narrowly oblong or elliptic oval or obovate.

Outer tepals up to 3 mm long. ....**1. A. javanica** var. **javanica**
1b. Leaves up to 6 mm broad, linear to narrowly linear-oblong, or narrowly oblong-obovate. Outer tepals ca. 2 mm long...........................**2. A. javanica** var. **bovei**

1. Aerva javanica var. **javanica** (Burm.f.) Juss. ex. Journ. A. Schutes. Syst. Veg. ed., 15:565, (1819); Stewart in Nasir & Ali., Ann. Cat. Vasc. Pl. W. Pak. & Kashm., 228 (1972); Townsend in Nasir & Ali., Fl. Pak., 71:26 (1974). **Figure 2B**

Syn: *Celosia lanata* L., Sp. pl. ed. 1: 205 (1753); *Iresine javanica* Burm.f., Fl. Ind., 212 (1768); *I. persica* Burm. f., Fl. Ind. : 212 (1768); *Illecebrum javanica* (Burm. f.) Murr., Syst. Veg. ed. 13: 206 (1774); *Aerva tomentosa* Forsskal, Fl. Aegypt.-Arab. cxxii: 170 & Florula 122 (1775); *Achyranthes javanica* (Burm. f.) Pers., Syn. 1: 259 (1805); *Aerva wallichii* Moq. in DC., Prodr. 13(2): 300 (1849); *A. persica* (Burm. f.) Merrill in Phil. Journ. Sc. 19: 348 (1921).

A perennial hoary-tomentose, erect to scandent dioecious undershrub up to 1 m high. Stem terete, much branched, woody at the base, herbaceous above, tomentose. Leaves simple, alternate, ca. 3-9 x 1-2.5 cm in size, linear-oblong or oblanceolate-spathulate, base cuneate to attenuate, apex obtusely apiculate, margin entire, densely tomentose; petiole ca. 0.5-1 cm long, hairy. Spikes cylindric, terminal, paniculate up to 5 cm. Bracts and bracteoles ovate, ca. 3 mm, apex acute, hyaline. Flowers unisexual, dense. Male flowers: Tepals ca. 3.5 mm, elliptic-oblong, sometimes ovate, apex subobtuse, wooly on above. Female flowers: Tepals 5, ca. 2.5 mm, obovate-oblanceolate or spathulate, base subacute, apex apiculate, tomentose without; outer 2 larger with faini midrib; inner 3 smaller with thick green midrib. Ovary ca. 0.5 mm; style ca. 1.8 mm; stigma deeply forked, as long as style. Utricle orbicular-ovoid, very thin. Seeds lenticular, shining, brown-black.

Flowering period: July-September.

Herbarium specimens examined: NARA DESERT: *Rahmatullah Qureshi*, 12, 13, 613, 656, 685, 824, 836, 847, 887, 894, 923, 932, 939, 1050, 1056, 1075, 1184, 1217, 1228 (SALUBGH).

Field remarks: A very common shrub of sandy soil.

Distribution: It is widely distributed in drier parts of the tropics and subtropics of the Old World, from Burma, India and Ceylon westwards through SW Asia, across N. Africa to Morocco and south to Cape Verde Islands and Cameron through Uganda and Tanzania to Madagascar. It is introduced in Australia and elsewhere (Townsend, 1974).

2. Aerva javanica var. **bovei** Webb., in Hook., Niger Flora (Spic. Gorg): 173 (1894); Townsend in Nasir & Ali., Fl. Pak., 71 : 27-29 (1974). **Figure 2A**

Syn: *A. bovei* (Webb) Edgew. in Journ. L. Soc. 6:206(1862); *A. persica* var. *bovei* (Webb) Chiov., Fl. Somalia : 286 (1929); *A. pseudotomentosa* Blatter & Hallberg in Journ. Bombay Nat. Hist. Soc. 26: 817 (1919).

A weak, tomentose semi-erect, perennial herb up to 70 cm tall. Stem branched from the base, terete, striate, leafy, densely whitish-tomentose. Leaves alternate, linear - narrowly linear to oblong, or narrowly oblong-obviate, up to 6 mm wide, margin plane or involute. Spike sessile, slender, dense and stout up to 10 x 1 cm, often interrupted with oblong or globose lateral clusters, all except terminal spikes frequently shortly cylindrical or oblong;

upper part of stem and branches leafless, the upper spikes thus forming terminal panicles. Bracts 0.75-2 mm, broadly deltoid-ovate, hyaline, acute or obtuse with the obscure midrib ceasing below the apex, densely lanate throughout or only above the base or apex, persistent; bracteoles similar, also persistent. Flowers dioecious. Male flowers smaller the outer tepals 1.5-2 mm, ovate; filaments delicate, the anthers about equaling the perianth; ovary small; style very short; stigma rudimentary. Female flowers with outer 2 tepals 2-3 mm, oblong obovate to obovate-spathulate, lanate, acute to obtuse or apiculate at the tip, the yellowish midrib ceasing well below the apex; inner 3 slightly shorter, elliptic-oblong, more or less densely lanate, acute, with a narrow green vitta along the midrib; style slender, distinct, with the two filiform, flexuous stigma at least equaling it in length, filaments reduced; anthers absent. Capsule 1-1.5 mm, rotund, compressed. Seed 0.9-1.25 mm, round, slightly compressed brown or black, shining and smooth.
Flowering period: September-December.
Herbarium specimens examined: NARA DESERT: *Rahmatullah Qureshi*, 1 (a), 1 (b), 16, 850, 856, 931, 981, 991, 1126, 1247 (SALUBGH).
Field remarks: A rare shrub found in sandy soil.
Distribution: It is distributed from Indian desert westward through Iran and Afghanistan to Palestine, Egypt and the Sehara and south to Niger, the Sudan, Somalia and Arabia (Townsend, 1974).

3. Amaranthus

L., Sp. Pl. ed. 1: 989 (1753); Gen. Pl. ed. 5: 427 (1754); Boiss., Fl. Or. 4: 988 (1879); Hook. f., Fl. Brit. Ind. 4: 718 (1885); Townsend in Fl. Pak. 71: 7 (1974).

Annual or more rarely perennial herbs, glabrous or furnished with short and gland-like or multi-cellular hairs. Leaves alternate, long petiolate, simple and entire or sinuate. Inflorescence basically cymose, bracteate, consisting entirely of dense to lax axillary cymose clusters or the upper clusters leafless and approximate to form a lax or dense "spike" or panicle. Flowers unisexual, dioecious or monoecious, 2-bracteolate. Perianth segments (2-) 3-5, free or connate at the base, membranous, those of female flowers slightly accrescent in fruit. Stamens free, usually similar inflorescence number to the perianth segments, with no alternating pseudostaminodes. Ovule solitary, erect; stigmas 2-3. Fruit a dry capsule, indehiscent, irregularly rupturing or commonly dehiscing by a circumscissile lid. Seeds usually black and shining, testa thin; embryo annular, endosperm present.

Approximately 50 species, chiefly introduced in the warmer regions of the world, about a dozen species being cosmopolitan weeds in such areas (Townsend, 1974).

Key to species:
2a. Prostrate herb. Flowers in axillary clusters.......................................**1. A. graecizans**
2b. Erect or ascending herbs. Flowers in terminal and axillary clusters................**2. A. viridis**

1. Amaranthus graecizans subsp. **thellungianus** (Nevski) Gusev in Bot. Zhurn. 57: 462 (1972); Townsend in Nasir & Ali., Fl. Pak., 71:18 (1974).

Syn: *A. polygonoides* sensu Roxb., Fl. Ind. ed. 2,3: 602 (1832); Lam., Encycl. Meth. 1: 115 (1783), *nom. illegit.*; *A. blitum* L. var. *polygonoides* Moq. in DC., Prodr.

13(2): 263 (1849); *A. polygamus* sensu Hook. f., Fl. Brit. Ind. 4: 721 (1885); *A. thellungianus* Nevski in Act. Acad. Sc. U.R.S.S. 1(4): 311 (1937): *A. lividus* sensu Stewart, Ann. Cat. W. Pl. W. Pakistan. 230 (1972); *A. tenuifolius* sensu Stewart, ibid.: 231 (1972); *A. graecizans* L. var. *polygonoides (*Moq.) Aellen in Rech. F., Fl. Iranica Lf. 91: 11 (1972).

An annual semi-erect herb. Stem up to 15 cm high, simple or branched, striate, sulcate when dry, glabrous or covered with multi-cellular hairs. Leaves ca. 1.5-3.5 x 0.5-2.5 cm in size, linear to rhomboid-spathulate, obtuse, rounded or truncate and often notched at apex, shortly apiculate; petiole ca. 4-16 mm long. Flowers in axillary clusters, throughout the stem, sessile, male and female intermixed. Bracts and bracteoles narrowly lanceolate-oblong, awned, membranous with a strong midnerve. Perianth ca. 1.2 mm long; lobes 3, ovate-oblong, mucronate, membranous, awned, with a strong midnerve. Stigmas 3. Utricle subglobose to ovoid, exceeding the perianth, tapering towards the apex, circumcissile; styles 3, recurved. Seeds compressed, black, shining, faintly reticulate.
Flowering & Fruiting Period: July-December.
Herbarium specimens examined: NARA DESERT: *Rahmatullah Qureshi*, 33, 866, 978, 984 (SALUBGH).
Field remarks: A common herb of sandy loam soil.
Distribution: It is exclusively present in the Indian sub-continent and Middle Asia (Towsend, 1974).

2. Amaranthus viridis L. Sp. Pl. ed. 2, 1405 (1762); Stewart in Ali and Nasir, Ann. Cat. Vasc. Pl. W. Pak. & Kashm., 231 (1972).; Townsed in Nasir and Ali, Fl. Pak., 71: 14-16 (1974). **Figure 2C**

Syn: *A. gracilis* Desf., Table Ecole Bot.: 43 (1804); *Chenopodium caudatum* Jacq., Collect. Bot. 2: 325 (1788); *Albersia caudata* (Jacq.) Boiss., Fl. Or. 4: 992 (1879).

An annual erect or ascending herb up to 35 cm high. Stem much branched, terete or obtuse-angular, thinly pubescent when young, glabrous at length. Leaves ca. 2-7 x 1.5-3.5 cm in size, ovate or rhomboid or oblong from a decurrent, apex acutely emarginate, base obtuse or cuneate to acute, margin entire to undulate; glabrous but sparingly pubescent on stronger nerves; lateral nerves 5-6 pairs, prominent below; petiole ca. 0.5-7 cm. Flowers axillary in lower clusters, upper ones in terminal, rather dense, continuous or interrupted male or female spikes which are 6-7 mm across, slender paniculate spikes, green. Male flowers: ca. 2 mm across; bracts and bracteoles ovate-lanceolate, ca. 1.5 mm, acute, midvein prominent, minutely excurrent; bracteoles shorter; tepals 3, oblong-lanceolate, apex acute, shortly excurrent, midrib faint, concave; those of female flowers narrowly oblong-spathulate; stamens 3, free; filaments ca. 1 mm; anthers ca. 0.8 mm; ovary rudimentary. Female flowers: ca. 1.5 mm across; tepals as in male flower; ovary ca. 0.7 mm; styles 2, ca. 0.7 mm, stigmatose; stigmas 3, on conical top of the ovary, suberect. Utricle ca. 1.2 mm, indehiscent, subcompressed, rugose, brownish, falling off together with the perianth, beak slightly exserted. Seeds ca. 0.8-1 mm across, round, slightly compressed with a blunt margin, shining, dark brown in colour.
Flowering period: August-September.

Herbarium specimens examined: NARA DESERT: *Rahmatullah Qureshi*, 838, 988 (SALUBGH).
Field remarks: A common herb of loamy soil.
Distribution: The most widely distributed species of the genus, found throughout the tropical regions of the world and further into the temperate regions than most of the tropical species (Townsend, 1974).

4. Digera

Forssk., Fl. Aeg.-Arab.: 65 (1775); Boiss., Fl. Or. 4: 994 (1879); Hook. f., Fl. Brit. Ind. 4: 717 (1885); Townsend in Fl. Pak. 71: 21 (1974).

Annual herb with alternate branches and leaves; branches angular. Leaves alternate, entire, petiolate. Inflorescence axillary pedunculate spikes. Flowers small, ternate, the 2 outer reduced to crested scales, the central perfect; bracts and bracteoles persistent. Perianth segments 5, the outer pair opposite and sheathing the remaining flower parts, the inner segments much more delicate and hyaline. Stamens 5, free, without intermediate staminodes-like teeth; filaments filiform; anthers 2-celled; staminodes 0. Ovary 1-celled, truncate, ovule solitary on a curved funicle, radicle, dehiscing; style filiform; stigmas 2, divergent. Fruit a subglobose, hard, compressed crustaceous nutlets enclosed by the persistent perianth and falling together the sterile flower and bracteoles. Seeds erect, globose; testa thin, adhering to the albumen; embryo annular; cotyledons linear; radicle ascending.

A monotypic genus (Townsend, 1974).

Digera muricata (L.) Mart. in Nov. Act. Acad. Caes. Leop.- Carol. 13(1): 285 (1826); Stewart in Ali and Nasir, Ann. Cat. Vasc. Pl. W. Pak. & Kashm., 232 (1972); Townsed in Nasir and Ali, Fl. Pak., 71: 21-23 (1974). **Figure 3A**

Syn: *Achyranthes muricata* L. Sp. Pl. ed. 2: 295 (1762); *A. alternifolia* L. Mant.: 50 (1767); *Digeria arvensis* Forssk., Fl. Aeg.- arab.: 65 (1775); *D. alternifolia* (L.) Aschers. in Schweinf., Beitr. Fl. Aethiop.: 180 (1867).

An annual herb up to 65 cm tall. Stem woody below, much branched, erect or ascending; branches spreading, glabrous, pinkish. Leaves ca. 1.5-7 x 0.5-4 cm in size, elliptic-ovate or lanceolate, apex acute or rounded, base truncate to acute, margin subentire and sometimes reddish, glabrous; petiole ca. 1-4 cm. Inflorescence ca. 5-15 x 0.7 cm in size, slender, lax, thyrsiform spike. Flowers bisexual, sessile, ca. 3 mm across, rosy; bract ca. 1.7 mm, obovate, subtending a central (fertile) and 2 lateral (sterile) flowers; bracteoles 2, ca. 2 mm, lanceolate, each subtending a sterile flower, deciduous. Tepals 5, unequal; outer 2 concave, ca. 2.5 mm, lanceolate-oblong, midrib prominent with 1 or 2 pairs of faint lateral nerves, base rounded to acute, apex obtuse, folded, herbaceous; inner 3, ca. 3 mm, lanceolate-obovate, base obtuse, apex acute, membranous, faintly 1-nerved. Stamens 5, basally connate into a sheath; filaments ca. 2 mm; anthers ca. 0.4-0.5 mm, 2-celled. Ovary ca. 1 mm, subglobose; ovule solitary on a long, curved funicle; style ca. 1 mm, minutely 2-fid, persistent. Utricle ca. 2-2.5 mm across, globose, muriculate, surmounted by a thick rim or corona of firm processes. Seeds yellowish-brown.

Flowering Period: September-October.

Herbarium specimens examined: NARA DESERT: *Rahmatullah Qureshi*, 1165 (SALUBGH).
Field remarks: A common herb of loamy soil.
Distribution: Widespread in southern Asia from tropical Arabia and the Yemen to Afghanistan, India, Ceylon, Malaysia and Indonesia, S. C. and E. tropical Africa and Madagascar (Townsend, 1974).

5. Celosia

L., Sp. Pl. ed. 1: 205 (1753); Townsend in Fl. Pak. 71: 5-6 (1974).

Herbs, usually annual, sometimes rather woody at the base, occasionally scandent. Leaves alternate, simple, entire or somewhat lobed. Flowers small, hermaphrodite in dense terminal and axillary spikes, sessile or shortly pedicellate, white or colored; bracts scarious; bracteoles 2, scarious. Perianth of 5 segments slightly connate below, scarious, oblong or lanceolate, acute or obtuse, striate, erect in fruit. Stamens 5; filaments fused below into a short sheath, the free portions deltoid below and filiform above or ± swollen or with teeth projecting on either side of the anther. Ovary 1-celled, ovoid or subglobose; ovules 2 or more, on long funicles; style 0 or short or long, sometimes elongating in fruit; stigmas simple or 2 or 3, subulate. Fruit a circumscissilely dehiscent capsule membranous or sometimes corky or coriaceous. Seeds 2 or more, black, strongly compressed and shining.

Celosia argentea L., Sp. Pl. ed. 1: 205 (1753); Stewart, Ann. Cat. Vasc. Pl. W. Pak : 231 (1972); Townsend in Nasir and Ali, Fl. Pak., 71: 5-6 (1974). **Figure 2D**

Annual, erect, much branched herb upto ca. 3 ft. tall. Stem erect, ridged, glabrous. Leaves lanceolate-oblong to narrowly linear, acute to obtuse, shortly mucronate with the excurrent midrib, glabrous; lamina of the leaves from the centre of the main stem 2-15 x 0.1-3.2 cm, tapering below into an indistinctly demarcated, slender. petiole; upper and branch leaves smaller, markedly reducing; leaf axils often with small-leaved sterile shoots. Inflorescence a dense (rarely laxer below) many-flowered spike, 2.5-20 x 1.5-2.2 cm, silvery to pink, conical at first but becoming cylindrical in full flower, terminal on the stem and branches, on a long, sulcate peduncle up to c. 20 cm long, which often lengthens during flowering. Bracts and bracteoles lanceolate or towards the base of the spike deltoid, 3-5 mm, hyaline, ± aristate with the excurrent midrib, persistent after the fall of the flower. Perianth segments 6-10 mm, narrowly elliptic-oblong, acute to rather blunt, shortly mucronate with the excurrent midrib, with 2-4 lateral nerves ascending more than halfway up the centre of each segment, margins widely hyaline. Filaments very delicate, free part subequall¬ing or exceeding the staminal sheath, sinuses rounded with no or very minute intermediate teeth; anthers and filaments creamy to magenta. Stigmas 2-3, very short, the filiform style 5-7 mm long; ovary 4-8-ovulate. Capsule 3-4 mm, ovoid to almost globular. Seeds c. 1.25-1.5 mm, sub-reniform, black, shining.
Flowering Period: August-October.
Herbarium specimens examined: NARA DESERT: weed of cotton crop, *Rahmatullah Qureshi*, 1336 (SALUBGH).
Field remarks: A common weed of sandy soil.

Distribution: Practically a pantropical weed, in Pakistan as elsewhere on waste ground, in cultivated fields, along roadsides etc. on sandy or stony soils (Townsend, 1974).

13. Portulacaceae

Annual or perennial, glabrous to subglabrous, usually succulent herbs, rarely undershrubs. Leaves opposite or alternate, simple, entire, fleshy; stipules scarious or modified into hairs, rarely absent. Inflorescence usually a terminal cyme or raceme sometimes flowers solitary. Flowers bisexual, actinomorphic, usually hypogynous rarely perigynous, bracteate or ebracteate. Sepals 2, rarely more (often regarded as bracteoles), free or connate at the base, anterior overlapping the posterior, imbricate, caducous. Petals 4-6, (rarely 2 or 3), free or united at the base, imbricate, entire, fugaceous or marcescent. Stamens from 3-many, antipetalous, free, rarely basally connate or epipetalous; filaments filiform, anthers dithecous, introrse, dehiscing longitudinally. Ovary 2-8-carpelled, syncarpous, superior or rarely semi-inferior, free or partially adnate to the calyx, unilocular; placentation basal or free central; ovules 2-many, campylotropous or amphitropous, often with long and persistent erect funicles; styles 2-8, united, rarely free, stigmas 2-8, elongated. Fruit a membranous or crustaceous, loculicidally or circumscissilly dehiscent, (sometimes opening explosively) capsule, rarely an indehiscent nut. Seeds 2-many, compressed, globose-reniform; endosperm copious and enclosed by more or less coiled embryo.

A family of about 19 genera and nearly 350 species; widely distributed in warmer and temperate regions throughout the world, but is especially centred in South and Pacific North America. One genus and 4 species opccur in Pakistan.

Portulaca

L., Sp. Pl. 445 (1753); Gen. P1. ed. 5. 204 (1754); Bentham & Mueller, Fl. Austral. 1: 168 (1863); Thiselton Dyer in Hook. f., Fl. Brit. Ind. 1: 246 (1872).
Lamia Vand.
Meridiana L. f.

Diffuse, annual or perennial, often trailing or ascending, usually glabrous and fleshy herbs. Leaves simple, alternate or opposite, sometimes subopposite; the uppermost crowded to form a sort of involucre below the flowers, flat or terete-semiterete, fleshy; stipules scarious or modified into a bunch of hairs, rarely absent. Inflorescence a terminal fasciculate cyme or flowers solitary ter¬minal, surrounded by a whorl of leaves. Flowers bisexual, actinomorphic, sessile, perigynous, usually 5-merous, showy, variously coloured; bracts entire or modified into hairs. Sepals 2, fused at the base to form a very short tube partly adnate to ovary and the lobes caducous. Petals 4-6, free or united at the base, inserted on rim of calyx tube, marcescent. Stamens 4-many, free or united at the base, inserted on rim of calyx tube and adnate to petals, rarely epipetalous. Carpels usually more than 3, syncarpous; ovary ovoid, half adnate to calyx, semi-inferior; ovules many, campylotropous; style with 3-8 elongated and radiating stigmas. Fruit brittle with thin membranous wall, subglobose to globose or oblong-conical, unilocular capsule, with transverse (circumscissile) dehiscence. Seeds numerous, minute, reniform or lenticular, testa shining, aculeate, tuberculate or somewhat papillate.

About 200 species, chiefly American but also distributed in other tropical and subtropical regions of the world.

Portulaca oleracea L., Sp. Pl. 1: 445 (1753); Jafri in Fl. Kar. 121 (1966); Stewart in Nasir & Ali., Ann. Cat. Vasc. Pl. W. Pak. & Kashm., 237 (1972).

Syn: *Portulaca aurea* Hort. ex DC.; *P. consanguinea* Schlecht; *P. laevis* Buch-Ham.; *P. officinarum* Crantz; *P. olitoria* Pall.; *P. parviflora* Haw.; *P. suffruticosa* Thw.; *P. viridis* Hort. ex DC.

Annual or perennial, prostrate or erect, c. 25 cm tall, succulent, glabrous, green or purplish green herb. Branches and stem with 3-20 mm rarely up to 50 mm long internodes. Leaves alternate or subopposite, closely crowded below the flowers, spathulate or obovate-oblong to linear-oblong, attenuate at the base, sub-sessile, obtuse or truncate, 3-25 mm long, 1.5-8 mm broad, thick, fleshy, glabrous, glistening white below, green or purplish-green above; stipular appendages usually absent, or rarely every minute and setaceous. Inflorescence usually in the forks of branches, cymose, with clusters of 3-6 flowers subtended by 4-leaved involucre, rarely flowers solitary and terminal. Flowers sessile, yellow, 5-8 mm across, bracteate; bracts membranous, ovate, c. 3 mm long, acuminate, white or somewhat purplish. Sepals subequal, basally united into a short, 2-3 mm long tube, keeled; lobes 2-3 mm long, slightly hooded, margin broad membranous, acute, deciduous. Petals 5, deliquescent, slightly united at the base, obovate, 5-6 mm long, 2.5-3 mm wide, yellow, emarginate with mucronulate notch. Stamens 7-12, basally somewhat united and adnate to petals, filaments c. 2.5 mm long, sensitive to touch, anthers ovoid. Ovary c. 2 mm long, half embedded in calyx tube, ovoid; style 1.5-2 mm long, stigmas 4-5, sticky, c. 1 mm long. Capsule many-seeded, 6-8 mm long, 3-4 mm in diam. Seeds shining black, c. 0.5-0.8 mm, reniform, testa tuberculate.

Flowering Period: Flowering period: August-December.

Herbarium specimens examined: NARA DESERT: a roadside weed, *Rahmatullah Qureshi*, 1337 (SALUBGH).

Field remarks: A common weed of sandy soil and waste places.

Distribution: A cosmopolitan weed in cultivated fields and waste moist places. Probably native of South-West parts of United States and now widely distributed in warm temperate, tropical and subtropical regions throughout the world (Jafri, 1966).

14. Molluginaceae

Annual or perennial diffuse herbs. Leaves decussate to spuriously whorled, or opposite to sub-opposite, simple, those of a pair usually very unequal, membranous or subsucculent; petioles sometimes dilated, sheathing and amplexicaul; stipulate scarious, fugacious, sometimes 0. Inflorescence leaf-opposed, axillary and/or enclosed in petiolar sheath; bracteoles scarious. Flowers small, 5-merous, bisexual, actinomorphic. Calyx-tube obconical, sometimes adnate to the petiolar sheath, sepals 5, imbricate in bud, persistent, rarely subsucculent, petaloid, margins thin. Petals 0. Stamens 3-20, hypogynous or inserted on the top of calyx-tube; filaments free or connate at the base; anthers 2-celled, oblong, dehiscing longitudinally; petaloid staminodes occasional. Ovary superior, 1-5-celled, usually united, rarely free; ovules usually on axile or basal plantation; styles as many as cells, longitudinally stigmatose. Fruit obconical or terete, usually enclosed by calyx-tube, dehiscing loculicidally by 3-5 valves or circumscissile, sometimes indehiscent. Seeds subreniform, compressed, concentrically ribbed or granular, endospermic.

A family of 2 genera and 32 species, tropical to subtropical. Often included in the Azoaceae, from which it differs in being exclusively herbaceous and having sepals free or nearly so, and superior ovary. Represented in Pakistan by 2 genera and 4 species (Nasir, 1973).

Key to genera

1a. Flowers in axillary fascicles. Seeds appendaged..**1. Glinus**

1b. Flowers in terminal paniculate cymes. Seeds not appendaged.......................**2. Mollugo**

1. Glinus

L., Sp. Pl. 463 (1753); Gen. Pl. ed. 5, 537 (1753); Nasir in Fl. Pak. 40: 1 (1973).

Prostrate herbs; branchlets thicken at the nodes. Leaves 3-5 in a cluster at a node, unequal; petiole not dilated at base; stipules scarious. Flowers sub-sessile, 5-20 in an axillary clusters. Sepals 5, unequal, margin whitish, tomentose or glabrous. Stamens 5-12(15); filaments ligulate; staminodes occasional, 2-fid at apex. Ovary globose or oblong, 3-5-celled; ovules (per cell on axile placentae; styles as many as cells, erect or recurved. Capsule globose or oblong, loculicidal, 3-5-valved. Seeds subreniform, granular; strophiole with a long appendage.

A genus of 12 species, tropical and sub-tropical. Represented in Pakistan by 1 species only (Nasir, 1973).

Glinus lotoides L. Sp. Pl. 463 (1753); Stewart in Nasir & Ali., Ann. Cat. Vasc. Pl. W. Pak. & Kaslm., 236 (1972). Nasir in Nasir & Ali., Fl. W. Pak., 40:1-3 (1973). **Figure 16B**

Syn: *Mollugo lotoides* (L.) Kuntze, Rev.Gen. (1264); *M. hirta* Thunb. Prodr. Pl. Cap. 24 (1794); *Glinus dictamnoides* L. Mant. Pl. 243 (1771).

Radically spreading annual procumbent herb up to 60 cm high. Stem much branched, herbaceous, slender; branchlets densely covered with stellate and dendromophic hair. Leaves ca. 0.5-2 x 0.4-1.5 cm in size, obovate, suborbicular or spathulate, unequal, base subacute-cuneate, apex obtuse to slightly apiculate, margin entire, chartaceous, densely stellate-tomentose; petiole, ca. 0.3-0.8 cm, slender, densely stellate-hairy; trichomes ca. 0.5 mm, 7-9 limed, each limb unicellular. Stipules ca. 1 mm, linear, apex acute. Flowers ca. 1 cm across, 5-10 in a cluster, greenish white; pedicel ca. 0.3-0.5 mm. Sepals ca. 6 x 3 mm in size, elliptic-oblanceolate, apex mucronate, basally connivent, erecto-patent, stellate-hairy on above. Stamens 5, subequal; filaments 4-5 mm long, filiform; anthers ca. 0.8 mm. Ovary ca. 4 mm, ovoid; styles ca. 1.5 mm. Capsule ca. 0.8-1 cm, globose. Seeds ovoid-subreniform, ca. 0.5 mm, faintly tuberculate, dark brown to black.

Flowering period: February-April.

Herbarium specimens examined: NARA DESERT: in depression between dunes, *Rahmatullah Qureshi*, 121, 835, 935, 965, 1015, 1024, 1039, 1188 (SALUBGH).

Field remarks: A common herb of sandy loamy soil.

Distribution: Most tropical parts of the world, Ceylon, Africa, India and W. Pakistan (Nasir, 1973).

2. Mollugo

L., Gen. Pl. ed. 5(1754); C.B. Clarke in Hook. f., Fl. Brit. Ind. 2: 662 (1879); Jafri, Fl. Kar. 118 (1966); Nasir in Fl. Pak. 40: 3 (1973).

Erect or diffuse glabrous pubescent or stellately tomentose herbs, often dichotomously branched. Leaves radical and cauline, simple, falsely whorled or alternate, linear, obovate or spathulate, chartaceous; petiole not sheathing at base; stipules scarious. Flowers leaf-opposed and/or terminal, in polychasial, often subracemose to corymbose or umbellate cymes, usually greenish; bracteoles scarious; pedicel stiff or recurved. Sepals 5, subequal, persistent, with membranous margin. Petals 0. Stamens 3-5, sub-hypogynous, alternating with sepals, sometimes intermixed with subulate staminodes; filaments filiform. Ovary free, ovoid, 3-5-celled; ovules many per cell on axile placentae; styles filiform; stigmas 3. Capsule ellipsoid, membranous, 3-5-celled, loculicidally 3-5-valved, included in the calyx. Seeds several (rarely 1) per cell, reniform, granular, estrophiolate; embryo more or less curved.

A genus of 20 species, tropical and sub-tropical. Represented in Pakistan by 3 species (Nasir, 1973).

Key to species:

1a. Leaves radical only..**1. M. nudicaulis**
1b. Leaves both radical and cauline:
2a. Stem almost filiform; branches umbellate. Seeds smooth or nearly so........**2. M. cerviana**
2b. Stem rigid; branches dichotomously arranged.
Seeds prominently tuberculate..**3. M. pentaphylla**

1. Mollugo nudicaulis Lamk. Encycl. 4:234 (1797); Stewart in Nasir & Ali., Ann. Cat. Vasc. Pl. W. Pak. & Kashm., 236 (1972). Nasir in Nasir & Ali., Fl. W. Pak., 40:5 (1973). **Figure 16D**

Syn: *Pharnaceum spathulatum* Spreng., Syst. Veg. I: 948 (1824).

An erect, annual herb up to 6 cm long. Leaves radical, spathulate-oblanceolate, ca.1-2 cm long and ca. 0.5-20 mm broad, membranous, glabrous, base attenuate, apex obtuse-retuse, margin entire, petiole ca. 0.5-1 cm long, green. Cyme terminal, scapes slender, glabrous, peduncle erecto-patent 5-10, ca. 4 cm in size. Bracts 2, ovate-oblong, ca.1-2 x 1 mm across, scarious. Flowers bisexual, white, glabrous, ca.5 x 2.5 mm across; pedicel up to 1.4 cm long. Sepals white, elliptic-oblong to lanceolate, base acute-cuneate to mucronate, margin entire, imbricate, persistent, glabrous, ca. 5 x 2-3 mm in size. Stamens 3-4; filament ca.1 mm long, persistent, anther sub-globose, ca.0.5 x 0.3 mm in size, versatile, free, dehiscing longitudinally. Ovary 3-locular, oblong, 3-lobed, 3-celled, ca. 3 x 2 mm in size; style 3; stigma 3, short, curved. Fruit ca. 4-5 x 2-3 mm, ellipsoid-subglobose, dehiscing by 3-valves longitudinally, glandular and sticky. Seeds ca. 25-30, ca. 1 mm across, muricate, black to brown in colour, shiny, glandular.

Flowering & Fruiting period:: October-November.

Herbarium specimens examined: NARA DESERT: dunes and ridges, *Rahmatullah Qureshi*, 19, 963 (SALUBGH).

Field remarks: A rare herb of sandy soil.
Distribution: S. America, tropical Africa, Ceylon, India and W. Pakistan (Nasir, 1973).

2. Mollugo cerviana (L.) Ser in DC. Prodr. 1:392.1824; Stewart in Nasir & Ali., Ann. Cat. Vasc. Pl. W. Pak. & Kashm., 236 (1972). Nasir in Nasir & Ali., Fl. W. Pak., 40:5-6 (1973).
Figure 16C

Syn: *Pharnaceum cerviana* L. Sp. Pl. 272 (1753); *M. umbellata* Ser. in DC. Prodr. 1. 393 (1824).

Annual erect herb up to 15 cm long. Stem branched in whorls of ca. 8 from the rootstock, nodes thickened, herbaceous; branchlets filiform, glaucous, brownish. Radical leaves rosulate, spathulate or linear-spathulate, apex apiculate, ca. 0.5-1 x 0.2-0.3 cm in size; cauline leaves 5-8 in a cluster, narrowly linear or acicular, base attenuate, apex obtuse-apiculate, margin entire, ca. 1-1.5 cm long, chartaceous, glaucous below; petioles obscure. Cymes terminal, trichotomously branched; peduncle ca. 1.5 cm, slender. Bracts ca. 0.5 mm long, subulate, glabrous. Flowers ca. 2.5 mm across; pedicels ca. 5 mm long. Sepals ca. 1.5-2 mm, subequal, elliptic-oblong, apex obtuse, margin membranous and white. Stamens 5-8; filaments ca. 1 mm. Ovary 3-lobed, 3-celled, ca. 1 mm; styles 3, ca. 0.3 mm. Capsule broadly ellipsoid, ca. 2 mm, glabrous. Seeds ca. 30, minutely granular, blackish brown.
Flowering period: February-April.
Herbarium specimens examined: NARA DESERT: dunes and ridges, *Rahmatullah Qureshi*, 8, 107, 174, 954, 1009, 1053, 1057, 1069, 1201 (SALUBGH).
Field remarks: A common herb of sandy soil.
Distribution: S. Europe, Greece, Rumania, S. Russia, Siberia, Arabia, Galapagos Islands, Australia, tropical Africa, Ceylon, India and W. Pakistan (Nasir, 1973).

3. Mollugo pentaphylla L. Sp. Pl. 89 (1753); Stewart in Nasir & Ali., Ann. Cat. Vasc. Pl. W. Pak. & Kashm., 236 (1972). Nasir in Nasir & Ali., Fl. W. Pak., 40: 6 (1973).

Annual, erect, herb up to 24 cm tall. Stem numerous, with many more or less 4-angular leafy dichotomously arranged branches, glabrous. Radical leaves obovate-spathulat; cauline ones 3-5 in a whorl, linear or elliptic, base cuneate-attenuate, apex obtuse-subacute, margin entire, ca. 1-2.5 x 0.4-0.7 cm in size, chartaceous; petiole ca. 2 mm. Cymes up to 5 cm; peduncle up to 8 cm, filiform. Bracts ca. 1 mm, linear-lanceolate, scarious. Flowers ca. 2 mm across, white; pedicel ca. 5-8 mm, filiform, recurved. Sepals ca. 2 mm, elliptic-oblong, apex obtuse, glabrous, parallel veined. Stamens 3; filaments ca. 1.5 mm. Ovary ca. 1.8 mm, ellipsoid, 3-celled; styles 3, short, linear. Capsule ca. 2 mm, oblong, glabrous. Seeds ca. 15-20, roundish-reniform, covered with raised tubercular points, dark brown in colour.
Flowering period: September-November.
Herbarium specimens examined: NARA DESERT: *Rahmatullah Qureshi*, 107, 872 (SALUBGH).
Field remarks: A rare herb of sandy soil.
Distribution: Malayan Peninsula, Fiji, China, Japan, Africa, Ceylon, India and W. Pakistan (Nasir, 1973).

15. Caryophyllaceae

Annual or perennial herbs, rarely shrubs. Stems with swollen nodes. Leaves simple, opposite, exstipulate. Inflorescence a typically cyme. Flower bisexual, actinomorphic. Perianth biseriate; calyx of 4 or 5 overlapping sepals, free or united at base. Corolla usually of 4 or 5 free petals, often differrentiated into limb and claw. Stamens usually 5 or 10, ripening before the stigma. Ovary superior of 2-5 carpels, syncarpous, unilocular with free-central or basal placentation. Fruit usually a capsule containing numerous seeds, sometimes an indehiscent, 1-seeded nutlet.

Key to genera:

1a. Leaves stipulate, capsule papery..**1. Vaccaria**

1b. Leaves ex-stipulate, capsule not so...**2. Spergula**

1. Vaccaria

Medikus, Philos. Bot. 1: 96 (1789); Gorshkova in Kom., Fl. URSS. 6: 802 (1936); Chater in Tutin et al, Fl. Eur. 1: 186 (1964); Cullen in Davis, Fl. Turk. 2: 177 (1967).

Annual herbs, glabrous. Stem erect, simple or branched. Leaves sessile. Epicalx absent. Inflorescence a lax panicle with flowers in dichasia. Calyx 5-toothed, without commissural veins, but with 5 green wings. Petals 5. Stamens 10. Styles 2. Coronal scales absent. Capsule papery, dehiscing by 4 teeth. Carpophore short. Seeds subglobose.

Vaccaria hispanica (Miller) Rauschert in Feddes Rep. 73: 52 (1966); Ghazanfar & Nasir in Nasir & Ali., Fl. Pak., 175: 109-110 (1986). **Figure 9A**

Syn: *Saponaria hispanica* Miller, Gard. Dict. Ed. 8, in errodis (1768); *S. ampicimus* Miller, l.c. no. 4-nomen invalid. (in erratis in hispanica corrig.) (1798); *Saponaria vaccaria* L., Sp. Pl. 409 (1753); *Vaccaria pyramidata* Medik., Phil. Bot, 1: 96 (1789); *V. oxydonta* Boiss., Diagn ser. 2 (1): 68 (1953); *Saponaria oxydonta* (Boiss.) Boiss., Fl. Or. 1: 525 (1867).

Annual herb, up to 70 cm tall. Stem erect, branched, glabrous. Leaves ca. 5-7.5 x 0.5-2.5 cm in size, lanceolate-ovate, sessile, base clasping, apex acute, margin entire. Inflorescence spreading panicle with the flower born in 3-7 flowered dichasia; pedicel slender. Calyx ca. 1.5-1.8 mm long, triangular teeth, margin scarious, apex of teeth with 5 prominent veins and green wings between them. Petals ca. 1.5-2 mm long, pink, obovate, claw shortly exserted. Stamens 10, ca. 4 mm long. Styles 2, ca. 5 mm long; ovary ca. 3 x 1.2 mm. capsule ca. 12 x 6 mm, ovoid or oblong. Seed ca. 1.5 mm in diam., subglobose, black.

Flowering Period: February-April.

Herbarium specimens examined: NARA DESERT: *Rahmatullah Qureshi*, 1344 (SALUBGH).

Field remarks: A common weed of sandy soil.

Distribution: Common as a weed in cultivated fields, in Chitral, Swat, Hazara, Punjab and Baluchistan (Ghazanfar & Nasir, 1986).

2. Spergula

L., Sp. Pl. 440 (1753); Gen. Pl. ed. 5: 189 (1754); Boiss., Fl. Or. 1: 731 (1867): Edgew. & Hook. f. in Hook. f., Fl. Brit. Ind. 1: 243 (1874).

Annual or perennial herbs. Stems decumbent to ascending, often profusely branching at the base. Leaves opposite and decussate; stipules scarious, not connate at the base, short, axillary branches present. Sepals 5, green with hyaline margins. Petals 5, white, entire. Stamens 5-10. Styles 3-5. Capsule dehiscing by 3-5 valves. Seeds winged or not.

Spergula arvensis L., Sp. Pl. 440 (1753); Stewart in Ali and Nasir, Ann. Cat. Vasc. Pl. W. Pak. & Kashm., 253 (1972); Ghazanfar & Nasir in Nasir & Ali., Fl. Pak., 175: 47 (1986).

Syn: *S. vulgaris* Boem.,Prodr. Fl. Monest. 135. 1824.

An erect herb up to 23 cm tall, Stem green, branched from the base, branches cylindrical, glabrous. Leaves ca. 1.5 x 0.7 cm, linear to subulate, rounded or subacute at the apex, sparsely glandular-pubescent. Inflorescence few-many flowered; bracts ca. 1 mm long, ovate, acute, scarious; pedicels up to 1.75 cm long. Sepals ca. 4 x 2 mm, quincuncial, ovate-elliptic, obtuse, margin scarious, glandular pubescent, green, black moles at sinus of sepals on its abaxial side. Petals broadly obovate, slightly concave, ca. 3 mm long. Stamen 3-10, diathecus, obdiplostemenous. Ovary ellipsoid. Fruit ovoid or subglobose capsule. Seeds ca. 0.8 mm, biconvex, margin winged, brownish-black with paler club-shaped papillae.

Flowering Period: March-April.

Herbarium specimens examined: NARA DESERT: *Rahmatullah Qureshi*, 1343 (SALUBGH).

Field remarks: A common weed of saline soil.

Distribution: Cosmopolitan (Ghazanfar & Nasir, 1986).

16. Polygonaceae

Herbs, shrubs, erect, ascending, straggling/spreading or climbing by tendrils. Leaves alternate, simple, penninerved, petiolate or sessile; stipules conspicuous, generally sheathing. Ochreae membranous, scarious, hyaline, fimbriate, lacerate, ciliate or sometimes entire at the mouth. Inflorescence cymes or disposed in spikes, heads or panicles. Flowers actinomorphic, bisexual, rarely dioecious, sessile or stalked. Bracts entire or forked, glabrous or ciliate. Perianth calycine or corolline, 2-seriate, campanulate; lobes 3-6, equal or unequal, the inner sometimes enlarged or modified with hooks, spines, wings or tubercles, imbricate, persistent. Stamens 4-9 (-16), 1-2-seriate, free, antitepalous, usually alternating with lobed glands; anthers 2-celled, parallel, dehiscence longitudinal; pollen grains 3-4, colpate, 3-colporate, pentaporate or triporate. Carpels 2-3, syncarpous; ovary superior, sessile, 1-locular; ovule 1, basal, orthotropous, sessile or on a funicle; styles 2-4, free or basally connate; stigma capitillate, rarely simple or fimbriate. Disc lobed, annular or 0, usually glandular. Fruit of 1-seeded, hard nutlet, 3-gonous, biconvex, biconcave or compressed, often enclosed by the persistent perianth, shining, brown to black. Seeds erect, albuminous; testa thin; embryo curved.

A family of 55 genera and ca. 1150 species commonly distributed in temperate regions, particularly in Northern Hemisphere. The family represents from Pakistan only 19 genera and 103 species (Qaiser, 2001).

Key to genera:

1a. Shrub; leaves subulate, minute, soon deciduous; stamens 10-16...............**1. Calligonum**
1b. Herbs, leaves of vasious shapes, persistent; stamens 4-8:
2a. Inner fruiting perianth segments accrescent, winged or toothed........................**2. Rumex**
2b. Inner fruiting perianth segments never accrescent, winged or toothed:
3a. Flowers in spicate racemes, ocreae cylindric, truncate or
with a terminal fringe of bristles..**3. Persicaria**
3b. Flowers solitary or in axillary clusters, ocreae bilobed or fimbriate............**4. Polygonum**

1. Calligonum

L., Sp. Pl. 530 (1753); Gen. Pl. ed. 5: 235 (1754); Boiss., Fl. Or. 4: 997 (1879); Hook. f., Fl. Brit. Ind. 5: 22 (1886); Chaudhary, Fl. Kingd. Sadi Arab. 1: 312 (1999); Qaiser in Fl. Pak. 205: 174 (2001).

Shrubs. Stem diffusely and irregularly branched; branches flexuous. Leaves usually absent or very small, linear or filiform, distinct or united with short membranous ochreae. Flowers solitary or in loose clusters, bisexual. Perianth segments 5, subequal, reflexed or rarely spreading, more or less petaloid, nearly free. Stamens 10-18, connate. Carpels 4; ovary 4-angled; styles 4; stigmas 4, capitate. Fruit 4-angled nut, coiled or straight, angles setose beset with membranous, stiffish, filiform bristles, which are often branched or variously winged. Seeds 4-angled, oblong; embryo straight.

A genus of ca. 80 species distributed in Asia, Africa and S.E. Europe. Only 3 species occur in Pakistan (Qaiser, 2001).

Calligonum polygonoides L., Sp. Pl. 530 (1753); Stewart in Nasir & Ali., Ann. Cat. Vasc. Pl. W. Pak. & Kashm., 201 (1972); Qaiser in Fl. Pak., 205: 177 (2001).

A rigid much branched tall shrub up to 2 m high. Stem hard, woody, profusely branched; branches terete, pale, glabrous; flowering branchlets slender; internodes ca. 3-4 cm long. Stipules short, membranous, cup-shaped, obliquely truncate and produced upwards at one side. Flowers ca. 4 mm long, fascicled in the axils of ochreae, pinkish; pedicel ca. 0.5-1.5 cm long. Perianth tube ca. 1 mm; segments ca. 2-3 x 1.5-2 mm in size, subobovate, base cuneate, reflexed in fruit. Ovary tuberculate; style 4, slightly connate at the base; stigma capitate. Fruit ca. 1-1.5 x 1-1.2 cm in size, oblong, densely clothed with reddish-brown bristles; bristles (setae) 2-3 times or more dichotomously branched, dilated at base, arising on 4 pairs of longitudinal wings. Nut oblong, slightly coiled, winged; wings ca. 1-2 mm broad.

Flowering period: March-April.

Herbarium specimens examined: NARA DESERT: *Rahmatullah Qureshi*, 29, 36, 615, 638, 641, 649, 865, 1048 (SALUBGH).

Field remarks: A common shrub mostly found on sand dune.
Distribution: Iran, Iraq, Pakistan, Russia, Turkey, Palestine and Syria (Qaiser, 2001).

2. Rumex

L., Sp. Pl. 333 (1753); Gen. Pl. ed. 5: 156 (1754); Boiss., Fl. Or. 4: 1003 (1879); Hook.f., Fl. Brit. Ind. 5: 58 (1886); Boulos, Fl. Egypt 1: 30 (1999); Chaudhary, Fl. Kingd. Saudi Arab. 1: 308 (1999).

Annual or perennial herbs, rarely shrubs, usually with long, stout roots, sometimes rhizomatous. Leaves alternate, basal and cauline; ochreae tubular, membranous, often entire or pinnately lobed. Flowers bisexual or unisexual, arranged in racemose or paniculate inflorescences; pedicels jointed. Perianth-segments 6, rarely 4, in 2 whorls, the outer remaining small and thin, the inner becoming enlarged and often accrescent in fruit (valves). Stamens 6 in 2 whorls; anthers basifixed. Nutlets 3-gonous, rarely 2-gonous.

About 200 species distributed mainly northern hemisphere (Boulos, 1999).

Rumex dentatus subsp. **klotzschianus** (Meisn.) Rech. f. in Beih., Bot. Centrbl., 49 (1932); Qaiser in Ali & Qaiser., Fl. Pak., 205: 158-160 (2001). **Figure 17C**

Syn: *R. klotzschianus* Meisn. in DC., Prodr. 14 (1856); *R. dentatus* sensu Hook.f., Fl. Brit. Ind. 5: 59 (1890); *R. nipponicus* Franch. & Savat, Pl. Japan. 2: 471 (1879); *R. dentatus* subsp. *halacsyi* (limosus x pulcher) Rech., Pat.,Verh. Zool. Bot. Ges. Wien 49: 105 (1899).

Annual herb, erect, woody, circular, upto 48 cm tall. Stipules ochreate, obtuse, crenate-crenulate, auriculate. Leaves ca. 6-8 x 2.5-6.5 cm in size, oblong-linear, apex obtuse, base rounded or cordate, glabrous; petiole ca. 6.5 cm. Flowers in distinct racemes with leafy or non leafy whorls; pedicellate; pedicel ca. 5 mm long. Perianth segment 6 in two whorls of 3 each; ca.3-4 x 2 mm in size, the inner segments broadly ovate, much enlarged in fruit, reticulately veined with an ovoid-oblong, smooth tuberculate at back and irregularly toothed or pectiolate on the margins; teeths numerous, straight. Stamens 6, ca. 2.5 mm long; anthers basifixed, ca. 1 mm in length Ovary acutely trigonous, ca. 0.8 x 0.6mm; stigmas tufted. Fruit a trigonous nut. Nuts ca. 2.5-1.5 mm, actely trigonous, almost winged.
Flowering Period: January-March.
Herbarium specimens examined: NARA DESERT: *Rahmatullah Qureshi*, 1338 (SALUBGH).
Field remarks: A common weed of sandy soil.
Distribution: Afghanista, Pakistan, India, East Asia (Qaiser, 2001).

3. Persicaria

Mill., Gard. Dict. Abr. ed. (1754); Munshi & Javeid, Syst. Stud. Polyg. Kashm. Himal. 61 (1986); Chaudhary, Fl. Kingd. Saudi Arab. 1: 306 (1999); Boulos, Fl. Egypt 1: 25 (1999); Qaiser in Ali & Qaiser., Fl. Pak., 205: 21 (2001).
Polygonum L. sect. *Persicaria* (Mill.) Meisn., Monogr. Polyg. (1826).

Erect-prostrate, ascending, annual (-perennial) herb. Stem herbaceous, sometimes woody. Leaves narrow, linear-lanceolate or elliptic-ovate, mostly without glands, sometimes gland-dotted. Ochreae tubular, ciliate-partite at the mouth, sometime eciliate, membranous. Inflorescence a spike, raceme or sometimes a head. Ochreolae tubular, membranous, ciliate or partite. Perianth segments 4-5, divided usually upto middle, or above or below, glandular. Glands usually alternating with stamens. Stamens 4-8. Ovary biconvex-trigonous with 2-3 long, filiform, fused or free styles with capitate stigmas. Nuts biconvex-trigonous, glabrous, dark brown - black, shiny.

A genus of c. 150 species, distributed in North temperate regions of both the hemisphere. In Pakistan, it is represented by 21 species.

Key to species:

1a. Annual; inflorescence spikes compact, dense; ochrea ciliate; flowers white.....**1. P. babata**

1b. Perennial; inflorescence branched raceme; ochrea eciliate; flowers pinkish.....**2. P. glabra**

1. Persicaria barbata (L.) Hara in Fl. E. Himal. 70 (1966); Qaiser in Ali & Qaiser., Fl. Pak., 205: 45 (2001).

Syn: *Polygonum barbatum* L., Sp. Pl. 362 (1753).

Erect, 30-60 (-75) cm tall, sometimes decumbent or prostrate simple-branched, annual, herb. Stem glabrous, lineolate-canaliculate. Leaves 1.5-15 x 0.25-3.0 cm, linear to linear-lanceolate, acuminate, glabrous except margin, midrib and veins ciliate, sessile to subsessile, petiole 2-4 (-6) mm long. Ochrea 1.0-3.5 cm long, tubular, brownish, strigose, cilia equalling or larger than the ochreae. Inflorescence 5-10 cm long, many flowered, branched raceme with 1-6 spikes in each raceme, peduncle 0.5-3.0 (-5.0) cm long, glabrous. Flowers white, 1.0-2.0 mm across, pedicel, 0.75-1.5 mm long. Ochreolae 1.5-3.5 mm long, tubular, ovate, compact, not intrerrupted, ciliate, ciliae almost the same size as that of ochreolae. Tepals 5, 1.5-2.5 x 0.75-1.25 cm, oblanceolate-obovate, obtuse, biseriate, unequal, eglandular. Stamens 5-8, filaments long, equal. Ovary 1.0-2.0 x 0.75-1.5 mm, broadly ovate-orbicular, trigonous, styles three, long, filiform and free till half of the length with capitate stigmas. Nuts trigonous, 1.8-2 (-2.25) x 1.2-1.5 mm, glabrous, black, shining.

Flowering period: March-April.

Herbarium specimens examined: NARA DESERT: *Rahmatullah Qureshi*, 1436 (SALUBGH).

Field remarks: A common subshrub of wetland habitat.

Distribution: Africa, W. Asia, Himalayas, India, Pakistan, China extending to Japan and Australia (Qaiser, 2001).

2. Persicaria glabra (Willd.) M. Gómes in Ann. Inst. Segunda Ensef. Habana. 2: 278 (1896); Qaiser in Ali & Qaiser., Fl. Pak., 205: 33 (2001).

Syn: *Polygonum glabrum* Willd. Sp. Pl. 2: 447 (1799).

Erect, 0.5-1.5 m tall, glabrous, branched perennial herb. Stem reddish at base, nodes usually swollen. Leaves 18-22 x 1-3.5 cm, completely glabrous, sometimes red gland dotted at the base, elliptic-lanceolate to lanceolate, acuminate, petiole 0.5-1.35 cm. Ochrea 1-3 cm long, tube-like, completely eciliate. Inflorescence glabrous, pedunculate, branched raceme, 5-10 cm long, very slender. Ochreolae 2-4 mm long, imbricate, glabrous. Flowers heterostylous, pedicel 2.5-4.5 mm long, exerted from ochreolae. Tepals 5, eglandular, white or pink, up to

2.5 mm long. Stamens exserted or included, (6-) 7 (-8), 1.75-4 mm long. Pistil exserted or included, 2.0-4.5 mm long; ovary biconvex, styles 2, rarely 3, connate at the base. Nuts bigonous ± rounded, 1.5-3.0 x 1.5-2.0 mm, glabrous, black, shining.

Flowering Period: April-November.

Herbarium specimens examined: NARA DESERT: *Rahmatullah Qureshi,* 1435 (SALUBGH).

Field remarks: A common herb of wetland habitat.

Distribution: Tropical Asia, Pakistan, India extending upto Malayesia and Philippine; Africa (Qaiser, 2001).

4. Polygonum

L., Sp. Pl. 539 (1753); Gen. Pl. ed. 2: 116 (1754); Hook. f., Fl. Brit. Ind. 5: 23 (1886); Chaudhary, Fl. Kingd. Sadi Arab. 1: 305 (1999); Qaiser in Fl. Pak. 205: 76 (2001).

Annual or perennial, prostrate, subprostrate to erect herbs. Stem simple or much branched at the base or above. Leaves small, simple, alternate, linear-lanceolate, ovate or elliptic, sessile to petiolate. Ochreae small, silvery or whitish hyaline, membranous, partite or ciliate-lacerate. Inflorescence axillary, few-flowered or solitary, or in small clusters. Flowers actinomorphic, bisexual, complete, bracteate. Perianth mostly petaloid, tepals 4-6, elliptic-ovate. Stamens 4-9, filaments short, usually broadened at the base without any interstaminal nectaries. Ovary biconvex or 3-gonous, small; styles 2-3; stigmas capitate. Nuts biconvex or 3-gonous, dark brown to black, shining.

A genus of ca. 60 species commonly distributed in both the hemispheres. In Pakistan, it is represented of ca. 20 species (Qaiser, 2001). Only 1 species is reported from the Nara Desert.

Polygonum plebejum R.Br. Prodr 420 (1810); Stewart in Nasir & Ali., Ann. Cat. Vasc. Pl. W. Pak. & Kashm., 208 (1972); Qaiser in Fl. Pak., 205: 103 (2001).

Syn: *Polygonum herniarioides* Spreng. Syst. Veg. 2: 256 (1825).

A decumbent herb up to 25 cm long. Stem herbaceous, fascicled at the base of root, spreading on all sides of the crown, more or less angled or semi-terete, glabrous, nodulated. Leaves ca. 0.5-2.5 x 0.2-0.5 cm in size, subsessile, linear to oblong or oblanceolate, base angustatus to shortly decurrent, apex acute or obtuse, margin quite entire, somewhat curved towards midrib, fleshy, glabrous, mid nerve impressed above, prominent below, lateral nerves obscure; internode ca. 1-1.5 cm long. Stipules hyaline, white, scarious, obliquely tubular, ca. 4 mm, lacerate and fimbriate to the middle, multicellular hairs are present on mid rib. Inflorescence axillary fascicled-raceme, 3-6-flowered. Flowers ca. 2.5-3 mm across; pedicel ca. 2.5 mm long jointed. Bracts hyaline ca. 1.5 mm, bracteoles ca. 1 mm. Perianth 5-cleft; ca. 3 x 1.5-2.5 mm in size; lobes unequal, outer 3 vertically folded, somewhat spotted without, muricate along the rib, imbricate, inner 2 oblong, slightly connate at the base, smooth, all obtuse or the 2 outer subacute, greenish outside near the base in bud, dark red in the midrib region above when open, persistent. Stamens 8-9, included, alternating with basal, glandular, disc-lobes; filament dilated at the base, ca. 0.5 mm long, slightly pink coloured; anthers 2-lobed, dorsifixed, ca. 0.2 x 0.3 mm in size, darker than the filaments. Ovary 3-gonous, ca. 0.3-0.4 x 1.5 mm in size; style ca. 0.3 mm, broadly 3-fid above, slightly connate at the base; stigma simple.

Nutlets ca. 3 x 1.5 mm in size, 3-gonous, hard, acute at the both end, dark pink in colour with persistent style.

Flowering period: September-December.

Herbarium specimens examined: NARA DESERT: *Rahmatullah Qureshi,* 1241 (SALUBGH).

Field remarks: A very common herb of moist habitat.

Distribution: Pakistan, Iran, Bangladesh, Nepal, Tropical East Africa, Madagascar, Bhutan extending to Australia through Indonesia and Malaysia (Qaiser, 2001).

17. Elatinaceae

Annual or perennial prostrate, ascending or erect herbs or undershrubs, sometimes aquatic or semi-equatic. Leaves opposite or decussate, simple; stipules paired, entire or serrate, persistent. Inflorescence solitary, axillary, or of a few-flowered cymes to dense clusters. Flowers small, regular, bisexual, hypogynous. Sepals 3-5, rarely 2 or 6, free or basally connate, imbricate. Petals as many as sepals, hypogynous, persistent, imbricate. Stamens as many as or twice the number of petals in 1-2 whorls, alternating with the petals, hypogynous, free; anthers versatile, 2-thecous, dehiscence longitudinally. Carpels 3-5, united; ovary superior, 3-5-celled; ovules in two or more rows on axile placentae, anatropous; styles 3-5, rarely 2, free. Capsule globose, septicidal or septifragal. Seeds minute, oblong, straight or curved, with or without little endosperm.

The family of two genera and 40 species distributed in tropical and temperate region. In Pakistan, 1 genus and 3 species are recorded (Ghafoor & Ali, 1972).

Bergia

L., Mant, 2: 152, 241 (1771); Thisleton Dyer in Hook. f., Fl. Brit. Ind. 1: 251 (1872); Ghafoor & Ali in Fl. Pak., 19: 4 (1972).

Diffusely branched, erect or decumbent, annual or perennial herbs or udershrubs, often glandular-pubescent or glabrous. Leaves decussate, oblong-linear to elliptic, crenate-finely serrulate or sometimes entire, shortly petiolate; stipules paired, denticulate or ciliate, persistent. Flowers sessile or pedicellate, usually 3-5-merous, bracteate. Sepals 3 or 5, lanceolate to ovate-oblong, acute, free, keeled with hairy, denticulate or serrulate, hyaline margins. Petals 3-5, mostly ovate-oblong or lanceolate-oblanceolate, nearly equal to sepals. Stamens 3 or 10(-15), in two whorls, rarely more or less, those opposite the petals shorter, the rest with filaments dilated below; anthers small. Carpels 3-5, syncarpous rarely free; ovary ovoid, or subglobose-globose, sulcate, 3-5-locular; styles 3-5, free, straight or recurved, shorter than ovary; stigma simple. Capsule 3-5-locular, ovoid-globose, furrowed, septicidal, subcrustaceous. Seeds sub-cylindric or oblong-ellipsoid-oblong, faintly or strongly reticulate, dark brown to black n colour.

Twenty five species distributed in temperate and tropical regions; represented by 3 species in Pakistan (Ghafoor & Ali, 1972).

Bergia aestivosa Wight and Arn. Prodr. I: 41 (1838); Stewart in Ali and Nasir, Ann. Cat. Vasc. Pl. W. Pak. & Kashm., 487 (1972); Ghafoor & Ali in Nasir & Ali., Fl. Pak., 19: 4 (1972). **Figure 12C**

Syn: *Elatine aestivosa* Wight Icon. T.222 (1839); *Bergia suffruticosa* ssp. *aestivosa* (W. & A.) Jafri., Biologia II: 42 (1965).

An ascending, perennial, up to 50 cm tall subshrub. Stem woody at the base, branches numerous, terete, opposite, decussate, greyish green, thin, long and divaricate; intrnode ca. 4-20 mm long. Leaves opposite, decussate, sub-sessile, narrowly oblong or lanceolate, base attenuate, apex acute, margin minutely serulate-crenate or entire, recurved, pubescent, lower leaves ca. 10-20 x 2-4 mm in size, linear and very small on branches bearing flowers, upper leaves ca. 4-10 x 1-3 mm in size. Stipules 4, ca. 3 x 0.5 mm, laneolate, apex acute, margin sub-crenulate and ciliate, green, setaceous. Inflorescence solitary or loose axillary fasiculate cyme of 2-4 flowers. Flowers ca. 6 x 3 mm. in size, white to pinkish; pedicel 1-4 mm. long. Sepals 5, free, laneolate, apex acute, keeled, denticulate with glabrous membranous margin, greenish veins in the centre, ca. 4 x 1.2-2 mm. Petals 5, free, obovate, apex mucronate, margin entire, glabrous, hyaline, imbricate, ca. 4-5 x 2-3 mm in size. Stamens 10, free, alternately shorter; filaments ca. 2.5-3.5 mm long, shorter one terete, longer one flat, dilated at the base; anthers versatile, 2-lobed, ovate, apex apiculate, base amlexicaule, yellow, ca. 1 mm across. Carpels 5, syncarpous; ovary ovoid, 5 sulcate, ca. 3 x 2 mm in size; style 5, free, clavate, straight, ca. 1.5 mm long; stigma simple.
Flowering period: November-December.
Herbarium specimens examined: NARA DESERT: *Rahmatullah Qureshi*, 433, 968, 1010 (SALUBGH).
Field remarks: A rare herb of moist habitat.
Distribution: W. Pakistan and India (Ghafoor & Ali, 1972).

18. Tiliaceae

Trees, (sub) shrubs or rarely herbs; branchlets with indumentum of stellate, simple or rarely leptoid hairs. Leaves alternate, simple or lobed, palmately 3- or 5-nerved at base, entire or dentate; stipules 2, free, usually small and deciduous. Inflorescence axillary, leaf-opposed or terminal, of cymes or racemes, rarely flower solitary or umbellate. Flowers 4-5-merous, usually bisexual, rarely unisexual (plant then monoecios), hypogynous, regular, bracteate. Sepals 4-5, free or united into a campanulate calyx, valvate, rarely imbricate, caduous, rarely persistent and accrescent. Petals 4-5, rarely absent or sepaloid, free, inserted at the base of torus, often callose and glandular at base, imbricate or valvate. Torus elongate or much condensed, often with tufted indumentum at the ends; androgynophore rarely present (*Grewia*). Stamens 10 or more, free, inserted on the torus, or rarely basally somewhat united into a tube or 5- or 10- adelphous, staminodial or absent in female flower, inserted on the thalamus or androphore, all antheriferous or some without anthers; anther-cell 2, parallel, contiguous dehiscing longitudinally or by an apical pore, rarely divergent and confluent at top. Carpels usually 2-5 9-10) or syncarpous or rarely free, ovary sessile above the torus, superior, 2-10-celled; ovules 1, 2 or more per cell on axile placentae, pendulous from the top or ascending from the base, sometimes few inserted at the middle of the cell, anatropous or subanatropous; raphe ventral or lateral; style single, apex divided or not; stigma as many as carpels, capitate, sometimes lobed-laciniate. Fruit 2-10 or 1-celled by suppression, or many-celled by false septa, armed or unarmed, rarely winged, dry, drupaceous or baccate and indehiscent, or separating into cocci, sometimes loculicidally (rarely septicidally) dehiscent.

Seeds solitary or numerous in each cell, ascending, pendulous or transverse, exarillate; testa usually coriaceous or crustaceous, sometimes pilose; albumen fleshy, copious or thin, rarely 0; embryo straight, rarely curved; cotyledons ovate or cordate-orbicular, foliaceous or slightly fleshy; radicle next the hilum, short or rarely longer than the cotyledons.

A family of 50 genera and ca. 450 species; distributed in tropical and temperate regions, chiefly Southeast and South America (abundant in Brazil). Four genera and 24 species are reported from Pakistan including 3 cultivated ones (Ghafoor, 1974).

Key to species:

1a. Petals glandular at base. Fruit a drupe, indehiscent.....................................**1. Grewia**
1b. Petals eglandular. Fruit a capsule, dehiscent......................................**2. Corchorus**

1. Grewia

L., Sp. Pl. 964 (1753); Gen. Pl. ed. 5: 412 (1754); Benth. & Hook. f., Gen. Pl. 1: 233 (1862); Boiss., Fl. Or. 1: 843 (1879); Masters in Hook. f., Fl. Brit. Ind. 1: 383 (1874); Hutch., Gen. Fl. Pl. 2: 487 (1967); A. Ghafoor in Fl. Pak. 75: 2-3 (1974).

Micrococus L., Sp. Pl. 514 (1753); *Chadara* Forsk., Fl. Aegypt.-Arab. 105 (1775).

Small trees or shrubs, rarely subspreading or climbing; branchlets with stellate-indumentum. Leaves alternate, petiolate, simple, entire or serrate, rarely lobed, coriaceous, basally 3- 5- nerved, cross nerves often prominent below; stipules often caduous, equal to or shorter than petiole, rarely falcate and produced at base. Inflorescence axillary or leaf-opposed, solitary or clustered, pedunculate, umbellate cymes or panicles. Flowers bisexual, pedicellate, yellow, orange, pink or white, 5-merous, bracteate. Sepals 5, free, leathery, mostly coloured and almost glabrous within, valvate, prominently nerved. Petals 5, very rarely absent, free, generally shorter than the sepals, apex emarginate or obtuse, rarely 2-lobed, clawed, claw thick, pitted or glandular near the point of insertion at the base of often glandular hairy torus.

Torus distinct, often with discoid gland sepal at base. Androphore rarely distinct. Stamens 10 or more, free, inserted on the more or less elevated torus or androphore; anthers somewhat orbicular, rarely pilose. Carpels 2-4, syncarpous; ovary superior on top of androphore, 2-4-celled, densely wooly; ovules 2-(per cell on axile placentae, rarely pilose; style subulate; stigma short, 2-4-lobed or lacinate. Drupe 1-4-pyrened, yellow, red or violet, globose, 2-4-lobed, rarely entire, sometimes with crustaceous rind. Seed(s) 1 or 2 per cell, rarely more seeded and often transversely septate (locellate) between he seed, ascending or horizontal; albumin copious, fleshy or rarely scanty, or almost 0; cotyledons flat, foliaceous or fleshy.

A genus with ca. 150 species, distributed in tropical and subtropical Africa, Asia and Australia (Ghafoor, 1974).

Grewia tenax (Forsskal) Fiori, Agric. Colon. 5 (suppl.): 23 (1912); Stewart in Ali and Nasir, Ann. Cat. Vasc. Pl. W. Pak. & Kashm., 474 (1972); A. Ghafoor in Nasir & Ali., Fl. Pak., 130: 7-8 (1974).

Syn: *Chadara tenax* Forsskal, Fl. Aegypt.-Arab. 105 (1775); *Grewia populifoli* Vahl, Symb. Bot. 1: 33 (1790); *G. betulaefolia* A. L. Juss. Ann. Mus. Natl. Hist. Nat. 4: 92. t. 50. f. 1 (1804).

Straggling or erect shrub up to 3 m high. Stem hard, woody, fairly branched, terete; branchlets apically stellate-hairy. Leaves ca. 1.5-3 x 1.5-2.5 cm in size, ovate or rotund, base and apex obtuse, margin crenate or 2-dentate, chartaceous, scabrid above, softly and shortly stellate hairy below, 3-5-nerved; petiole ca. 0.5-1 cm. Stipule ca. 2 mm long, linear. Cymes axillary or leaf-opposed, 2-4-flowered; peduncle ca. 1.5 mm. Flowers ca. 1.5 cm across, white; pedicel ca. 5 mm. Bracts ca. 6 mm long, subulate. Sepals 5, oblong, ca. 1.2 x 0.2 cm in size, stellate-hairy without. Petals ca. 2 mm long, oblong, apex obtuse, ciliate below, glands ca. 2 mm. Torus ca. 1.5 mm, ribbed, glabrous, glands basally discoid. Androgynophore ca. 0.5 mm. Stamens ca. 6 mm. Ovary 4-lobed, 4-celled; ovule solitary per cell; style ca. 8 mm; stigma laciniate. Drupes ca. 1 cm across, 4-lobed, glossy, glabrescent, orange-coloured when ripe. Stones 1-4, 2-celled.
Flowering period: July-September.
Herbarium specimens examined: NARA DESERT: *Rahmatullah Qureshi*, 315, 515, 1225 (SALUBGH).
Field remarks: A rare shrub of hilly tract.
Distribution: Tropical and North Africa to Iran, Pakistan, India and Sri Lanka i.e. Ceylon (Ghafoor, 1974).

2. Corchorus

L., Sp. Pl. 529 (1753); Gen. Pl. ed. 5: 234 (1754); Benth. & Hook. f., Gen. Pl. 1: 235 (1862); Masters in Hook. f., Fl. Brit. Ind. 1: 396 (1874); Hutch., Gen. Fl. Pl. 2: 482 (1967); A. Ghafoor in Fl. Pak. 75: 20 (1974).
Antichorus L., Syst. Nat. ed. 12.2: 264 (1767).

Subshrubs or herbs; branchlets stellate-pubescent or simple hairy. Leaves simple, petiolate, alternate, serrate; basal serrature often appendaged. Inflorescence leaf-opposed or axillary, 1-5 (-8)-flowered, shortly pedunculate cyme. Flowers small, subsessile or shortly pedicellate, 4- or 5-merous, bracteate. Sepals 5, rarely 4, oblong, with a distinct mucro, free, valvate. Petals 5, rarely 4, mostly imbricate, yellow, eglandular. Stamens 8-10 to α, free, inserted on a short torus, anthers versatile. Carpels 2-5, syncarpous; ovary superior, 2-5-celled; ovules α per cell on axile placentae; style short, the apex stigmatose, hollow, crenulate. Capsule elongate, linear-oblong or subglobose and echinate or muricate, ribbed or angled, rarely winged, with an entire or 3-fid beak, loculicidally 2- 5- valved, locules septate or transversely septate between the seeds. Seeds α, pendulous or horizontal, albuminous; embryo usually incurved; cotyledons foliaceous and cordate.

About 100 species, distributed in the tropics and subtropics (Ghafoor, 1974).

Key to species:
1a. Beak of capsule straight, entire:
2a. Prostrate perennial herb; capsule 7-17 mm long; beak 2 mm long........... **1. C. depressus**
2b. Erect annual herb; capsule 30-70 mm long; beak 5 mm long..................**2. C. trilocularis**
1a. Beak of capsule 3-fid; tips spreading:

3a. Capsules short, angular, winged..**3. C. aestuans**
3b. Capsule long, slender, without wings...**4. C. tridens**

1. Corchorus depressus (L.) Stocks in Proc. L. Soc., 1:367 (1848); Stewart in Nasir & Ali., Ann. Cat. Vasc. Pl. W. Pak. & Kashm.,472 (1972); Ghafoor in Nasir & Ali., Fl.Pak., 75:22 (1972).

Syn. *Antichorous depressus* L., Mant. 1:64 (1767); *Corchorus antichorus* Raeusch, Nom. Bot., ed. 3:158 (1797); *C. prostratus* Royle, Ill Bot. Himal., 103 (1834).

A perennial prostrate herb up to 30 cm long. Stem fairly branched from the base; branches twisted, imbricate, woody. Leaves ca. 6-10 x 3-5 mm in size, elliptic or obovate-elliptic to roundish, base rounded or cuneate, margin irregularly crenate-serrate, serratures not appendaged; plicate in bud, usually wrinkled, glabrous; petiole ca. 1-2.5 cm long, very slender. Stipules subulate. Flowers numerous in leaf-opposed cymes, yellow; buds obovoid, apiculate; peduncle and pedicels very short. Bracts ca. 4 mm long, lanceolate-subulate. Sepals ca. 5 mm long, linear-oblong, apex apiculate, margin entire, greenish. Petals ca. 8 mm long, oblong-obovate. Capsules ca. 7-17 x 1.5-2 mm in size, oblong-cylindric; beak 2 mm long, straight or curved towards upward, 4-valved; valves septate between the seeds, pubescent. Seeds ca. 3-gonous, smooth, dark brown.

Flowering period: February-November.

Herbarium specimens examined: NARA DESERT: *Rahmatullah Qureshi*, 205, 601, 605, 875, 882, 1223, 1316 (SALUBGH).

Field remarks: A common herb of sandy loam soil.

Distribution: Central and North West India and Pakistan to north and tropical Africa and Cape Verde Islands (Ghafoor, 1974).

2. Corchorus trilocularis L., Mant. 77 (1767); Stewart in Ali and Nasir, Ann. Cat. Vasc. Pl. W. Pak. & Kashm., 473 (1972); A. Ghafoor in Nasir & Ali., Fl. Pak., 130: 27-28 (1974).

Syn: *C. aestuans* auct. non. Linn (1758); *C. serraefolius* DC., Prodr. 1: 504 (1842); *C. triflorus* Boj., Hort. Maurit. 43 (1837); *C. asplenifolius* E. Mey. Ex Harv. & Sond., Fl. Cap. 1: 229 (1860).

An erect to suberect annual herb up to 100 cm tall. Stem woody at the base, herbaceous above, more or less angular; branchlets glabrescent. Leaves ca. 3-10 x 1.5-4 cm in size, elliptic-oblong or ovate-lanceolate, base cuneate, apex acute or obtuse, margin serrate or crenate, basal filiform serrature appendages, basally 3-4-nerved, glabrous; petiole ca. 0.5-1.5 cm. Stipules ca. 7 mm, filiform. Cymes leaf-opposed, 1-3-flowered. Flowers 4 mm across, yellow; pedicel ca. 2 mm. Bracts and bracteoles filiform, reflexed, ca. 4 mm. Sepals 5, ca. 6 mm, linear-oblong. Petals 5, ca. 6.5 mm, oblong. Stamens 15-25; filaments 5 mm. Ovary ca. 2 mm; style ca. 2 mm. Capsule ca. 3-7 x 0.5 cm in size, mostly 3 in clusters, 3-4-angled, apex acute or shortly beaked, stellate-tomentose when young, scabrid or aculeate when old, valves with transverse partitions; apical beak 5 mm long. Seeds ca. 1 mm across, truncate, 3-gonous, black.

Flowering period: June-October.

Herbarium specimens examined: NARA DESERT: *Rahmatullah Qureshi*, 696, 810, 982, 977, 1006 (SALUBGH).

Field remarks: A common herb of sandy soil.
Distribution: Pakistan, Afghanistan, India (Punjab, Uttar Pradesh, Bihar, Bengal, Peninsular India), Sri Lanka (Ceylon), Sikkim, Bhutan, Australia, tropical and North Africa (Ghafoor, 1974).

3. Corchorus aestuans L., Syst. Nat. ed. 10. 1: 1079 (1758); Ghafoor in Nasir & Ali, Fl.W. Pak., 75: 23 (1974); Jafri, Fl. Karachi., 215 (1966). **Figure 19A**

Syn: *C. acutangulus* Lamk., Encyl.2: 104 (1786); *C. fuscus* Roxb., Hort. Beng. 42 (1814), *nomen nudum*; *C. oppositiflorus* Hassk. in Hoev. & De Driese, Tijdschr. 12: 126 (1845).

Annual herb, up to 80 cm long, stem erect, woody, terete, hispid, sparsely hairy, (unicellular hairs). Leaves 5x3.8 cm, alternate, ovate, 1-2 filiform appendages at base, serrate-dentate, base rounded-truncate, apex acute- obtuse; petiole up to 3 cm long. Stipules filiform, up to 7 mm long. Cymes leaf-opposed, 1-2 flowered, bracts 3, filiform, 5 mm long. Flower 12x14 mm, complete, bisexual, actinomorphic; pedicel ca. 2 mm long. Sepals 5, linear, 9 mm long, margin Papilose at basal part, hooded, awned at the apex. Petals 4, yellow, spathulate, 6x4 mm. Stamens ca.15, unequal in length, up to 3 mm long; anthers ca. 1mm across. Ovary 4x1 mm in size, hairy, unicellular hairs in upward direction, cylindrical, tripartate, septate forming up to 38 chambers; ovules as many as number of chambers in locules; style 2.5 mm long; stigma 1 mm across, papilose. Capsule ca. 1.5 cm long, 6 angular, 3 winged, beak 3, bifid, radiating. Seeds truncate at base, 1.5x1 mm across, blakish – brown.

Flowering Period: July-November.

Herbarium specimens examined: NARA DESERT: *Rahmatullah Qureshi*, 1345 (SALUBGH).
Field remarks: A common herb of flat habitat.
Distribution: Widespread in tropical and subtropical countries of Africa, Asia and North Australia (Ghafoor, 1974).

4. Corchorus tridens L., Mant. 566 (1767); Stewart in Ali and Nasir, Ann. Cat. Vasc. Pl. W. Pak. & Kashm., 473 (1972); A. Ghafoor in Nasir & Ali., Fl. Pak., 130: 25-27 (1974). **Figure 19B**

Syn: *C. trilocularis* auct. non Linn (1767); *C. burmanii* DC., Prodr. 1: 505 (1824); *C. senegalensis* Juss. ex Steud., Nom. Ed. 2. 1: 417 (1840).

An annual erect or suberect herb up to 60 cm tall. Stem much branched, herbaceous, terete or angular, glabrous. Leaves ca. 3-9 x 0.8-2 cm in size, ovate-lanceolate, or elliptic-oblong, base cuneate, apex acuminate, margin crenate-serrate, the basal serratures on each side produced into filiform appendages, basally 3- or 4-nerved, glabrescent; petiole ca. 0.5-1.5 cm long. Stipules ca. 7 mm long, subulate. Cymes leaf-opposed, 1-5-flowered, subsessile; buds ovoid, apex apiculate. Flowers ca. 4 mm across, yellow. Sepals 5, ca. 4 mm, linear, apex apiculate. Petals 5, yellow, ca. 5 mm long, oblong. Stamens 10 or more, ca. 3 mm. Ovary ca. 2 mm; style ca. 1.5 mm. Capsule ca. 1-2 cm long, slender, beak 3-fid, spreading, 3-valved, valves without partitions. Seeds ca. 1.5 mm, truncate at both ends, dark brown.

Flowering period: July-November.
Herbarium specimens examined: NARA DESERT: *Rahmatullah Qureshi*, 529, 652, 671, 680, 916, 1234, 1296 (SALUBGH).
Field remarks: A common herb of flat habitat.
Distribution: Widespread in tropical and subtropical countries of Africa, Asia and North Australia (Ghafoor, 1974).

19. Sterculiaceae

Trees, shrubs or herbs, occasionally woody climbers, mostly stellate hairy. Leaves alternate, petiolate, simple or palmate compound, stipulate. Inflorescence axillary or rarely terminal, cymose, racemnose or paniculate, rarely flowers solitary. Flowers bisexual or unisexual or both, usually actinomorphic rarely zygomorphic, hypogynous. Sepals (3-)5(-7), free or united, valvate, mostly persistent. Petals 5 or absent, free, adnate to the staminal tube at the base, twisted, imbricate or valuate Stamens 5-many, in 2 whorls, outer antisepalous whorl absent or reduced to staminodes, free or monadelphous at the base, anthers dithecous. Carpels (1-)5(-12), syncarpous or sometimes partly to completely apocarpous, ovary superior, sessile or stipitate, locule equalling the number of carpels, each 2-multiovulate, ovules anatropous, placentation axile; style and stigma simple or as many as carpels and free. Fruit dry, dehiscent capsule, schizocarp or etaerio of follicles or rarey flleshy, indehiscent and berry-like. Seeds usually endospermic.

A family of about 68 genera and nearly 1000 species, mainly of tropical and subtropical regions, is represented in Pakistan by 10 genera and 20 species most of which are introduced and commonly grown as ornamentals or roadside plants.

Melhania

Forssk., Fl.Aegypt.-Arab. 64 (1775); Benth. & Hook.f., Gen.Pl. 1: 222 (1862); Masters in Hook. f.,Fl.Brit.Ind.l: 372 (1874); H.Wild,Fl. Zambes. 1: 528 (1961); Hutch.,Gen.Fl.Pl. 2: 550 (1967).

Herbs or subshrubs, covered with stellate hairs. Leaves simple, variously shaped, serrate-crenate. Inflorescence an axillary, few-flowered, umbellate cymes or flowers solitary. Flowers bisexual, actinomorphic, yellow, rarely unfolding; bracts 3, large, variously shaped, usually longer than the sepals, enlarged and peristent in fruit. Sepals 5, united at the base. Petals 5, large, marcescent around the ovary. Stamens 5, alternating with 5, ligulate staminodes, both monadelphous at the base, anther lobes parallel. Carpels 5, syncarpous; ovary sessile, 5-loculed, each locule 1-multiovulate; style small, 5-lobed, lobes subulate, stigmatose on inner side. Capsule loculicidal, 5-loculed, locules 1-many seeded. Seeds endospermous.

A genus of about 60 species, distributed in Africa, Madagascar, tropical Asia. and Australia; only 3 species are reported from Pakistan.

Melhania denhamii R. Br. in Deham & Clapp. Trav. App. 232 (1826); Stewart in Nasir & Ali., Ann. Cat. Vasc. Pl. W.Pak. & Kashm., 485 (1972).

Suffruticose, stellately tomentose herbs. Branches many, spreading, softly stellately tomentose. Leaves ca. 2-5 x 1-2.5 cm in size, ovate-oblong or elliptic, apex obtuse or acute or apiculate, margin crenate-serrate, base rounded and 5-nerved, stellately tomentose above, paler, velvety pubescent and nerved beneath; petiole ca. 1.2-1.5 cm; stipules 4 mm, subulate-filiform, hairy. Peduncles axillary, shorter than the petioles, dividing above into 2-3 pedicels, each of which 4-6 mm long; flowers orange-yellow. Involucral bracts 3, upto 1.5-1.75 cm broad, membranous, accrescent, veined, apiculate, enclosing the capsules. Sepals ovate-lanceolate, 4-6 mm long, stellate villous outside, glabrous within. Petals ca. 5 mm across, orange-yellow, nearly rhomboid, veined, glabrous. Stamens 4, alternating with 4 longer staminodes, the latter dialated and united to form an epipetalous cup below. Capsules ca. 6 mm, subglobose, densely stellately villous. Seeds 1-2 in each cell, angled.

Flowering & Fruiting Period: September-November.

Herbarium specimens examined: NARA DESERT: Chotiari reservoir, *Rahmatullah Qureshi*, 1346 (SALUBGH).

Field remarks: A rare herb recorded from Achhro Thar.

Distribution: Tropical Africa, Arabia, Pakistan (Baluchistan, Sindh), India (Bhandari, 1978).

20. *Malvaceae*

Annual or perennial herbs to shrubs or soft-wooded trees with tenacious inner bark, mucilaginous. Branchlets often stellate-tomentose. Leaves alternate or spirally arranged, petiolate, simple or palmately lobed or partite, entire, usually palminerved, often with nectariferous glands on nerves below. Stipules linear, lanceolate, ovate, rarely auricled, free, sometimes caduous. Inflorescence axillary, solitary or in terminal racemes or panicles (by reduction of upper leaves). Flowers 5-merous, hermaphrodite, rarely polygamous, regular; pedicel accrescent in fruit, often jointed at or below the middle. Epicalyx usually with free or connate segments, persistent, slightly accrescent, rarely 0. Calyx campanulate; usually angular by prominent costae and marginal nerves, slightly accrescent, persistent, sometimes spathaceously, splitting and caduous; sepals (-3) 5 (-6), valvate. Corolla, free, contorted, asymmetric, adnate to the base of the staminal column and falling off with it; petals 5. Stamens numerous, monadelphous; staminal column truncate and antheriferours at apex or 5-dentate at apex and antheriferous throughout; anthers dorsifixed, reniform, 1-celled, splitting longitudinally; pollen grains large, echinate. Pistil (2-) 5-many usually in a single whorl, sometimes in 2 or more whorls; ovary superior, 3- 5- or more celled; ovule(s) 1 or more per cell on axile placentae; style 1 or branched into arms as many, or twice as many, as locules, linear, peltate; stigma(s) capitate or discoid, sometimes clavate. Fruit a dry schizocarp with 1- or more-seeded mericarps or capsular with 3 or more seeds, rarely indehiscent. Seeds with a little endosperm, reniform or obovoid, rarely arilate, glabrous to densely hairy; embryo curved, in reniform seeds, usually straight in obovoid; cotyledons leafy, usually folded or crumpled.

A family of 88 genera and about 2,300 species distributed in tropical, subtropical and temperate regions. In Pakistan, it is represented by 19 genera with 94 specific and infraspecific taxa (Abedin, 1979).

Key to species

1a. Fruit schizocarpic (indehiscent); carpels usually separating:

2b. Epicalyx present; mericap indehiscent:
3a. Epicalyx segments 3..**1. Malvastrum**
3b. Epicalyx segments 5 or more ....**2. Pavonia**
2b. Epicalyx absent; mericarp usually 2 or more seeded..**3. Abutilon**
1b. Fruit a capsule; loculicidally dehiscent..**4. Hibiscus**

1. Malvastrum

A. Gray, Mem. Am. Acad. Sc. II,4 (Pl.Fendler.): 21 (1894), *nom.cons*; Mast. in Hook.f., Fl. Brit. Ind. 1: 321 (1874); Hutch., Gen. Fl. Pl. 2: 563 (1967); Abedin in Fl. Pak., 130: 89 (1979).

Herbs or undershrubs. Leaves elliptic-ovate, entire or lobed. Inflorescence terminal spike or axillary fascicles or flowers solitary. Flowers with 3, 1 near persistent bracts. Calyx campanulate. Petals 5, obovate-cuneate, yellow. Staminal tube branched at the apex. Carpels 5-many; ovary 5-many loculed, each locule uni-ovulate; style branches as many as Carpels; stigma capitate. Mericarp indehiscent, with or without awns, strigose. Seeds glabrous, reniform, usually exalbuminous.

A genus with 3 species native to tropical and subtropical America, 2 of which naturalized in other hemisphere. In Pakistan only 1 species is known (Abedin, 1979).

Malvastrum coromendelianum (L.) Garcke in Bonplandia 5: 297 (1857); Stewart in Ali and Nasir, Ann. Cat. Vasc. Pl. W. Pak. & Kashm., 481 (1972); S. Abdin in Nasir & Ali., Fl. Pak., 130: 89-90 (1979).

Syn: *Malva coromandeliana* L., Sp. Pl. 687 (1753); *M. carpimfolia* Desr in Lam, Encycl. 3: 754 (1789); *M. tricuspidata* R. Br. in Ait.f., Hort. Kew. ed. 2.2: 210 (1812); *Malvastrum tricuspidata* (R. Br.) A. Gray, Pl. Wright, 1: 16 (1852).

An erect herb up to 60 cm tall. Stem terete, more or less woody at the base, herbaceous above, much branched, clothed with stellately hairs, trichomes ca. 0.5-1 mm long. Stipules 2, ca. 5-6 x 0.5-1 mm in size, ovate-rhombic, margin ciliate. Leaves simple, alternate, ca. 2.5-8 x 2-5 cm in size, broadly ovate-elliptic, base cuneate, apex acute-slightly apiculate, margin dentate, lateral nerves ca. 6-pairs, prominent below, stellate hairs on nerves and surface granulate; petiole ca. 0.5-4.5 cm, filiform, base clasping, pinkish-green, densely stellate-pubescent. Inflorescence axillary, solitary or 2-3 in fascicles; pedicle ca. 8 mm long. Bracts ca. 6-7 x 1-1.5 mm in size, elliptic-rhombic. Flowers ca. 1 cm across, yellow. Calyx-tube ca. 4 mm; lobes ca. 4-5 x 3 mm in size. Petals ca. 5-6 x 2-3 mm in size, obovate, apex rounded, base cuneate, margin entire, glabrous. Mericarps 5-10, reniform, ciliate, with 3 spines on the margins. Fruit glabrous, enclosed in enlarged calyx.
Flowering period: August--January.
Herbarium specimens examined: NARA DESERT: *Rahmatullah Qureshi*, 1113, 1304 (SALUBGH).
Field remarks: A common herb of loamy soil.
Distribution: Tropical regions of both New and Old World (Abedin, 1979).

2. Pavonia

Cav., Diss. 2: App. 2 (1786) nom.cons); Mast. in Hook.f., Fl. Brit. Ind. 1: 330 (1874); Hutch., Gen. Fl. Pl. 2: 556 (1967); Abedin in Fl. Pak., 130: 93 (1979).

Annual or perennial herbs, undershrubs or shrubs; branchlets stellate-tomentose, rarely viscid. Leaves petiolate, stipulate, simple or palmately lobed or partite, palminerved, chartaceous. Flowers axillary, solitary, or clustered in racemes or panicles at the end of branches; pedicel jointed. Epicalyx segments 5-16, free or fused. Calyx-tube campanulate; lobes 5. Corolla variously coloured, commonly red, pink or yellow, rarely white or purple; petals 5, rotate. Staminal column rarely longer than petals, dentate at apex and antheriferous throughout. Carpels 5; ovary 5-celled; ovule 1 per cell; stylar branches 10; stigma capitate, papilose. Fruit discoid or globose schizocarp; mericarps 5, indehiscent, 3-gonous, winged or not, prominently nerved, muricate or not, sometimes with 2 or 3 retrosely hairy awns. Seed 1 per mericarp, ascending, reniform, glabrous or hairy.

A genus with ca. 200 species distributed in the warm regions of both the hemispheres. It is represented by 6 species in Pakistan (Abedin, 1979).

Pavonia glechomaefolia var. **karachiensis** S. Abedin in Pak. J. Bot. 11(1): 66 (1979); Abedin in Nasir & Ali., Fl. Pak., 130: 69-98 (1979).

An erect, subshrub up to 45 cm high. Stem slender, minutely stellate tomentose, terete, reddish. Leaves ca. 3.5 x 2-3 cm in size, usually ovate-pentagonal in outline, base cordate, apex acute-obtuse, margins coarsely crenate-serrate, stellate hairy on both surfaces, more densely beneath; petioles ca. 2-3 cm long, pubescent, caduous. Flowers ca. 2.5 cm across, solitary, axillary; pedicels ca. 2.5 cm long, slender, jointed near the apex, stellately hairy. Epicalyx ca. 5-6, spreading, rotate, oblong or ovate-lanceolate; segments ca. 12 x 2.5- 3 mm in size, usually obtuse. Calyx-tube ca. 3 mm long; lobes broadly ovate, 5-nerved. Corolla ca. 10-15 mm long, finely stellate-pubescent outside near the apex. Staminal column glabrous. Carpels 5, ca. 8-9 mm across, obliquely triquetrous, obovoid, apex obtuse, each with a median, dorsal, serrulate crest and 3-4 lateral stout ridges on either side. Capsule indehiscent, stellate hairy. Seeds ca. 3.5 x 2 mm in size, obliquely obovoid, pyriform, glabrous.

Flowering period: November-December.

Herbarium specimens examined: NARA DESERT: *Rahmatullah Qureshi*, 1263, 1307 (SALUBGH).

Field remarks: A common herb of gravelly soil.

Distribution: Known from Karachi (Abedin, 1979).

3. Abutilon

P. Miller, Gard. Dict. Abridg. ed. 4.1. (1754); Boiss., Fl. Or. 1: 835 (1867); Hook.f., Fl. Brit. Ind. 1: 335 (1874); Hutch., Gen. Fl. Pl. 1: 554 (1967); Abedin in Fl. Pak., 130: 52 (1979).

Perennial herbs or shrubs; branchlets simple or stellate-tomentose, cinereous, rarely viscid. Leaves petiolate, stipulate, cordate, ovate, toothed or sometimes shortly lobed, entire, palmately 7- or 9-nerved. Inflorescence axillary or terminal, solitary or in terminal racemes;

pedicel jointed above the middle. Flowers yellow, complete, regular. Epicalyx 0. Calyx campanulate, 5-lobed. Corolla yellow to orange yellow; petals 5, rotate. Staminal column shorter than petals, glabrous or stellate-hairy, truncate and antheriferous, divided at the apex into many filaments, included. Carpels 5- or more celled; ovules 1-9 per cell; styles as many as cells, filiform-clavate, shortly connate at the base; stigmas capitellate-capitate. Schizocarp campanulate or globose, usually exceeding the calyx; mericarps reniform, acute to acuminate or obtuse, dehiscent, as many as ovaries, 2-9 seeds per cell. Seeds reniform, loosely held, stellate-tomentose or glabrous.

A genus with 150 species distributed in the tropics and subtropics of both the hemispheres. In Pakistan, it is represented by 18 specific and intraspecific species (Abedin, 1979).

Key to species:

1a. Carpels 15-20:
2a. Carpels with mucro; seeds reniform..**1. A. indicum**
2b. Carpel without mucro; seeds ovoid to suborbicular....................................**2. A. muticum**
1a. Carpels 8-10...**3. A. fruticosum**

1. **Abutilon indicum (L.) Sweet**, Hort.Brit. ed., 1:54 (1826); Stewart in Nasir & Ali., Ann. Cat. Vasc. Pl. W. Pak. & Kashm.,476 (1972); Abedin in Nasir & Ali., Fl.Pak., 130:69 (1987).

Syn: *Sida indica* L., Cent. Pl. 2:36 (1756); *S. populifolia* Lamk., Encycl. 1: 7 (1783); *Abutilon indicum* var. *populifolium* (Lamk.) W. & A. ex Mast. in Hook. f., Fl. Brit. Ind. 1: 326 (1874); *A. indicum* var. *micropyllum* Hochr., Ann. Ocns. Jard. Bot. Geneve. 6: 20 (1902); *A. badium* Hus. & Baq., Phyton 15(3-4): 229 (1974), *syn.nov.*

An erect perennial undershrub to shrub up to 1 m high. Stem woody at the base, herbaceous above, fairly branched, terete, rigid. Leaves ca. 5-9 x 3-6 cm in size, ovate, apex acuminate, base cordate, 7-9-nerved from the base, margins serrate-crenate, toothed, rarely 3-lobed, velvety on both surfaces; petiole ca. 2-8 cm long; stipules ca. 2.5 mm, filiform, deflexed. Flowers up to 2.5 cm long, axillary, solitary; pedicels ca. 2.5-5 cm long, jointed very near the top, slightly curved at the joints. Calyx-tube ca. 4 mm; lobes 5, ovate, 4 mm long, apex acuminate. Petals 5, ca. 1 cm across, obovate, yellow. Staminal column ca. 6 mm, stellate-hairy at the base, glabrous above; filaments ca. 3 mm long. Carpels 15-20, ca. 11-13 mm long, hairy. Schizocarp ca. 1.5 x 2 cm in size, globose; mericarps 15-20, reniform, acute or somewhat acuminate at the back, densely stellate-hairy, mucro erect, ca. 2 mm. Seeds 3 per cell, ca. 2 mm, ovoid or suborbicular, warty, hairy at the hilum, brow-grey.

Flowering period: March-April.

Herbarium specimens examined: NARA DESERT: *Rahmatullah Qureshi*, 25, 260, 812, 1232 (SALUBGH).

Field remarks: A common subshrub of sandy soil.

Distribution: Tropics and subtropics of New and Old World (Abedin, 1987).

2. Abutilon muticum (Del.ex DC.) Sweet, Hort.ed. 2.65. 1830. Webb, Fragm. Fl. Aethiop-Aegypt. 51.1854.

Syn: *Abutilon asiaticum* Guill. Perr.; *A. malirianum* Hus. & Baq.; *Sida mutica* Del. (Fl.Aegypt.illustr. 20.n.633.1813 nom.nud.) ex DC.

Perennial herb or shrub, 0.5-2 m tall, stellate pubescent. Leaves 2-16 cm across, ovate to orbicular-ovate or orbicular, irregularly and minutely to coarsely serrate or subentire or crenate, usually cordate at base, obtuse to acute or shortly acuminate at apex, 7-9 nerved, stellate pubescent on both sides, scabrous above, more densely hairy and velvety beneath, sometimes 3-angular near the apex; petiole 2-12 (-18) cm long, stellate pubescent, velvety; stipules 6-8 mm long, 1 mm broad, linear, reflexed. Flowers axillary, solitary or fascicled; pedicel 0.5-1.5 cm long, in fruit up to 3.5 cm, articulate in the middle to near the apex. Calyx 7-8 mm long, in fruit up to 10 mm, fused to the middle, pubescent on both sides; lobes ovate to deltoid, acuminate, 4-5 mm broad. Corolla 2-2.6 cm across, yellow to orange-yellow, 2-2.5 times the length of calyx; petals 10-12 mm across, obovate, claw hairy on the margin. Staminal tube 4-5 mm long, stellate pubescent. Fruit usually globose, sometimes truncate, 9-12 mm in diameter; mericarps 27-39, usually obtuse, 6-7 mm long, 5-6 mm broad, separating after dehiscence; mucro absent. Seeds usually 2 in each mericarp, rarely only 1 by abortion, 2 mm across, pubescent, reniform.

Flowering period: April-July.

Herbarium specimens examined: NARA DESERT: *Rahmatullah Qureshi*, 1347 (SALUBGH).

Field remarks: A common subshrub of sandy soil.

Distribution: Tropical Africa, Arabia, India and Pakistan.

3. Abutilon fruticosum var. fruticosum Guill. & Perr. in Guill. et al., Fl. Senegamb. Tent. 1:70 (1831); Stewart in Ali and Nasir, Ann. Cat. Vasc. Pl. W. Pak. & Kashm.,476 (1972); Abedin in Nasir & Ali., Fl. Pak., 130: 55-58 (1979).

Syn: *A. denticulatum* R. Br. in Salt, Voy. Abyssl. Appen. 4. 65 (1814), *nom.nud.*; *Sida amaena* Wall. Cat.n. 1848 (1829), *nom.nud.*; *Abutilon kotschyi* Hochst. ex Webb, Fragm. Ethiop.-Egypt. 52 (1854).

A much branched rigid, perennial, stellate, velutinous, erect subshrub up to 70 cm high. Stem terete, slender, densely covered with glaucous, velvety indumentum. Leaves ca. 1.5-4 x 1-2 cm in size, ovate, base cordate, apex acute or obtuse, margins subentire or crenate-serrate, densely clothed with soft, whitish hairs on both surfaces, upper grey green, lower paler and canescent; petioles ca. 1.5-3.5 cm long; stipules ca. 4 mm long, linear. Flowers solitary on main branches or on axillary shoots; pedicels ca. 1.5-3 cm long, articulated near the apex, unbranched. Calyx-tube ca. 3 mm long, broadly campanulate; lobes 3-4 mm long, triangular-ovate, apex acute or mucronate, margins minutely ciliolate. Petals 5, ca. 7-8 mm long, yellow. Staminal column stellate hairy. Carpels 10, 6-8 mm long in fruit, densely grey, stellate-tomentose, obliquely truncate at the base. Schizocarp 8 x 10-11 mm, broadly cylindric, base rounded, not beaked, splitting down the back. Seeds 2-3 in each carpel, ca. 1.5 mm across, dotted with minute, white, hooked hairs.

Flowering period: October-November.

Herbarium specimens examined: NARA DESERT: *Rahmatullah Qureshi*, 533, 1157 (SALUBGH).

Field remarks: A rare herb of gravelly soils.

Distribution: Tropical Africa, Arabia, India and Pakistan (Abedin, 1987).

4. Hibiscus

L., Sp. Pl. 693 (1753); Gen. Pl. ed. 5: 310 (1754); Boiss., Fl. Or. 1: 839 (1867); Mast. in Hook.f., Fl. Brit. Ind. 1: 334 (1874); Hutch., Gen. Fl. Pl. 2: 554 (1967); Abedin in Fl. Pak., 130: 7 (1979).

Annual to perennial herbs or undershrubs to trees; branchlets stellate or simple-tomentose. Leaves palmately or pinnately lobed, sometimes entire, often with nectariferous glands. Stipules ovate-lanceolate, rarely auricled. Flowers axillary, solitary, often in racemes or panicles; pedicel usually jointed above or below middle, variously coloured. Epicalyx with 3-10 segments, usually free or connate, sometimes adnate to calyx, persistent or sometimes caduous, rarely 0. Calyx-tube campanulate, 5-lobed or parted, valvate, distinctly nerved, persistent, often with nectaries. Petals 5, showy. Staminal column as long as, shorter, or rarely longer than petals, dentate at apex, antheriferous throughout or only at apex. Carpels 5, syncarpous; ovary usually 5-celled; ovules 3 to many per cell; style 1, distally 5-branched, connate below; stigmas discoid or capitate. Fruit a globose or ovoid capsule, loculicidally dehiscent, as many as ovaries. Seeds 3-many per cell, globose or reniform, glabrous or hairy.

A genus with ca. 250 species, distributed mainly in the tropical and subtropical regions of both hemispheres. It is represented in Pakistan by 18 taxa of which 9 are wild including endemic species (Abedin, 1979).

Hibiscus lobatus (J.A Murray) O. Ketze., Rev. Gen. Pl. (32): 19 (1898); Stewart in Ali and Nasir, Ann. Cat. Vasc. Pl. W. Pak. & Kashm.,479 (1972); Abedin in Nasir & Ali., Fl. Pak., 130: 18-19 (1979).

Syn. *Solandra lobata* J. A. Murrey Comm. Soc. Reg. Sc. Goett. 6: 20. T. 1(1785); *Hibiscus solandra* L', Herit., Strip. Nov. 1: 103. T. 49 (1788); *Lagunea sinulata* Hornem., Hort. Hanf. 2: 645 (1851).

An annual, erect subshrub up to 1 m high. Stem branched; branchlets with stellate-pubescent and simple, elongate hairs. Basal leaves orbicular-ovate, base cordate, subcordate or truncate and 3-5-nerved at the base, not lobed; upper ones 3 palmately lobed or lanceolate, margins bluntly toothed or crenate, stellate-hairy above, woolly below, ca. 3-5 x 0.5-1.5 cm in size; petioles ca. 1.5-3.5 cm long; stipules ca. 3 mm, subulate. Flowers ca. 1.5 cm across, axillary, solitary or in lax terminal racemes; pedicels ca. 1.5 cm, jointed below the flower. Calyx-tube ca. 3 mm long; lobes ca. 4 mm long, lanceolate, 3-nerved, hispid with glandular and non glandular hairs. Petals 5, white, obovate, ca. 1.5 cm long, glabrescent. Staminal column equal to petals, antheriferous throughout. Ovary 5-celled; style-arms with discoid stigmas. Capsule ca. 1-1.2 x 0.8 cm in size, exceeding the calyx, ovoid; beak ca. 1-2 mm long, hispid, wrinkled. Seeds ca. 2 x 1.5 mm in size, subreniform or ovoid, irregularly prismatic, glabrous, black.

Flowering period: **September-October.**

Field remarks: A common herb of flat habitat.

Herbarium specimens examined: NARA DESERT: *Rahmatullah Qureshi*,508, 514 (SALUBGH).

Field remarks: A rare herb of sandy soils.

Distribution: Tropical Africa, Madagascar, S., S.E. and Central Asia (Abedin, 1979).

21. Tamaricaceae

Shrub or undershrub, rarely trees or herbs with slender branches, commonly halophytes, rarely xerophytic in nature. Leaves alternate, mostly small, commonly scale-like, quite entire, mostly sessile, rarely subsessile, usually fleshy with punctate glands; stipules absent. Inflorescence solitary axillary or simple racemes, panicles or spikes. Flowers small, hermaphrodite, rarely unisexual then plants dioecious, hypogynous, 4- or 5-merous. Sepals 5, rarely 4, free or connate at the base, much imbricate, persistent. Petals as many as sepals, free or shortly connate at the base, inserted beneath the disc, imbricate, free or united at the base into a tube, elliptic-obovate, rarely ovate or orbicular. Stamens 5-α, inserted on or below the disc, free or basally connate, monadelphous or in 5 fascicles. Anthers bi-celled, versatile, obtuse or apiculate, dehiscing by longitudinal slit; pollen grains tricolpate, with smooth walls. Disc hypogynous or slightly perigynous, 10-glandular, crenate or angled, rarely obsolete. Pistil 1, ovary 1-celled or imperfectly septate, shortly stipitate; placentas 3-5, arising from the base; ovules 2-many on each placenta, erect, anatropous; raphe ventral; micropyle inferior. Styles 3-5, short or 0; stigma capitate. Capsule 3-5-angled, pyramidal, dehiscing by 3-5-valves from apex to the base. Seeds numerous, erect, plumed or winged, densely hairy or pilose, albuminous or not; embryo straight; cotyledons flat; radicle inferior, short.

A small family with 4 genera and ca. 110 species, distributed in temperate regions, usually in sandy tracts and maritime deserts of Asia, Africa and Europe. In Pakistan, it is represented by 4 genera and 36 species (Qaiser, 1982). In the Nara desert only 1 genus and 4 species recorded.

Tamarix

L., Sp. Pl. 270 (1753); Gen. Pl. ed. 5: 131 (1754); Boiss., Fl. Or. 1763 (1867); Benth. & Hook f., Gen. Pl. 1: 159 (1867); Dyer in Hook f., Fl. Brit. Ind. 1: 248 (1874); Jafri in Fl. Kar., 235 (1966); Qaiser in Fl. Pak., 168: 40 (1982).

Trichauus Arn. in Wight & Arn. Prodr. Fl. Penin. Ind. Or. 1: 40 (1834).

Shrubs or trees, 2-12 m tall with prolonged deep root system, densely branched, glabrous, papilose; barks various coloured, mostly brownish to blackish brown. Leaves minute, ca. 0.5-7 mm long, base amplexicaul or sheathing, vaginate or auriculate, glabrous or pilose. Inflorescence simple or compound racemes or panicles, ca. 1-15 cm long, on younger and older branches. Flowers white or rosy, subtended by 1-2 or more bracts; bracts ovate-triangular or linear, base amplexicaul or subamplexicaul, herbaceous or diaphanous, margin denticulate, apex acute or obtuse, shorter than, equal or longer than pedicel. Sepals 4-5-lobed, subequal or equal, free, imbricate, 2 outer, 3 inner (5-merous) or 2 outer, 2 inner (4-merous), denticulate or entire, apex obtuse or acute, various shaped, mostly ovate or orbicular, glabrous or slightly papilose. Petals 4-5, inserted below the glandular-crenate, angled or lobed disc, free or slightly connate at the base, ovate, oblong, obovate or elliptic, obtuse, slightly notched or somewhat cuneate at the apex, equal or inequilateral, pink, violet, purple or white, deciduous or persistent. Stamens 4-10 (-12), diplostemonous or haplostemonous, antisepalous stamens usually slightly larger than antipetalous stamens, inserted on or below the nectiferous disc. Disc various shaped, anthers ovoid, apiculate or obtuse. Stigmas 3-4, rarely 5 or 2, elongate, clavate or divergent, 2-3 times shorter than the ovary. Ovary pyramidal or bottle shaped; placentas very short at the base of the locules. Capsules 3-5-angled, conical or pyramidal, dehiscing by 3-valves, downward to base. Seeds small, compressed, glabrous, hairy from the apex; coma sessile.

A genus with 90 species distributed chiefly in saline areas of deserts, semideserts, steppes or on the mountains along the river and springs in Asia, Europe and N.E. Africa. In Pakistan, the genus is represented by 26 taxa (Qaiser, 1982).

Key to species:

1a. Flowers unisexual..**1. T. dioica**
1b. Flowers bisexual:
2a. Rachis and branches pubescent..**2. T. pakistanica**
2b. plants glabrous:
3a. Stamens 5, haplostamenous:
2a. Leaves vaginate. Bracts triangular..**3. T. aphylla**
2b. Leaves pseudovaginate or amplexicaul. Bracts ovatelanceolate
to trullate-ovate..**4. T. indica**
3b. Stamens 7-10, diplostamenou..**5. T. passernioides** var. **macrocarpa**

1. Tamarix dioica Roxb. ex Roch, Nov. Pl. Ind. Or. 185 (1821).
Syn: *Tamarix dioica* Roxb.; *Tamarix longe-pedunculata* Blatt. & Hallb.

Shrub or under-sized tree with reddish bark, entirely glabrous, leaves vaginate, abruptly acuminate, 1.5-3 mm long, with a broad whitish margin. Plants dioecious. Recemes aestival, simple or loosely compound male racemes, 3-8 cm long, 7-8 mm broad, compact, female raceme as long as the male raceme or somewhat longer, 3-5 mm broad. Flowers purple or purplish pink, subsessile. Bracts semiamplexicaul, triangular, acuminate, 2.5-3 mm long, 1 mm broad. Sepals 5, ovate to broadly ovate or somewhat orbicular, c. 2 mm long, c. 1.5 mm broad, almost entire, obtuse. Petals 5, free, obovate, 2.5 mm long, 1-1.25 mm broad, entire, obtuse. Stamens 5, filaments filiform, 1.75-2 mm long, inserted in between the notches of the disc (mesodiscine disc), notches almost entire to slightly emarginate, anthers obtuse, ± sagittate; stamens abortive in the female flowers. Styles 3 as long as the ovary or slightly shorter, exserted, ovary triquetrous 1-1.5 mm long, abortive or absent in male flowers.
Flowering Period: April-November.
Herbarium specimens examined: NARA DESERT: Chotiari reservoir, *Rahmatullah Qureshi*, 1348 (SALUBGH).
Field remarks: A less common shrub of riparian land.
Distribution: Pakistan, Iran, Afghanistan, India, Nepal, Bangla Desh and Burma (Qaiser, 1982).

2. Tamarix pakistanica Qaiser in Pak. J. Bot., 13:(2):119 (1981); Qaiser in Nasir & Ali., Fl.Pak.,141:13 (1982).
Syn: *Tamarix salina* sensu Baum

Erect shrub, 2-3 m tall with brownish black to blackish gray bark, branches glabrous to subglabrous. Leaves sessile, amplexicaul or pseudo-vaginate, acute or acuminate, 1-2 mm long, 0.5 mm broad. Inflorescence aestival, simple raceme, 3-7 cm long, 5-6 (-7) mm broad,

rachis pubescent to subpubescent. Bracts simple, ovate, triangulate-ovate, 1-1.5 mm long, 0.5-0.75 mm broad, crenato-dentate or entire, pedicel 0.5-1.25 mm long. Calyx pentamerous. Sepals c. 1.5 mm long, 0.75 mm broad, ovate, obtuse with incisodenticulate margin, outer 2 short. Corolla pentamerous, petals elliptic oblong or obovate, 3-2.5 mm long, 1-1.25 mm broad, deciduous. Stamens 5, 3-3.5 mm long, filaments filiform, insertion hypodiscal with epilophic to confluent epilophic disc, anthers cordate, not apiculate, 0.75 mm long. Ovary stipitate, ovate, 2-2.5 mm long after anthesis, styles 3, ± oblong. Capsule pyramidal shaped, c. 5 mm long, 1.5 mm broad.
Flowering Period: January-October.
Herbarium specimens examined: NARA DESERT: Chotiari reservoir, *Rahmatullah Qureshi*, 1349 (SALUBGH).
Field remarks: A rare shrub of wetland habitat.
Distribution: Probably endemic to Pakistan (Qaiser, 1982).

3. Tamarix aphylla (L.) Karst., Deutsche, Fl.,641 (1882); Stewart in Nasir & Ali., Ann.Cat.Vasc.Pl.W.Pak. & Kashm.,489 (1972); Qaiser in Nasir & Ali., Fl.Pak., 141:31 (1982).

Syn. *Thuja aphylla* L., Cent. Plat. 32 (1755); *Tamarix orientalis* Forssk., Fl. Aegypt., 206 (1775); *T. articulata* Vahl. Symb., 2:48, t.32 (1791); *T. aphylla* (L.) Warb. Kent. Sinai 139 (1929).

A large tree up to 14 m tall. Stem branched; branches slender, hoary, with deep, punctate glands; bark reddish-brown to grey, glabrous. Leaves reduced to very minute traingular, teeth, ca. 1-2 mm long, abruptly mucronate, hoary; leaf-bases with short sheaths. Racemes aestival, simple or compound, ca. 2-7 x 0.3-0.4 cm in size, spirally curved. Flowers bisexual, subsessile, ca. 2 mm long, pinkish white; pedicel ca. 0.8 mm long. Bracts ca. 1.3-1.5 x 0.5 mm in size, vaginate, ovate, apex acuminate. Sepals ca. 5, ca. 1.5 x 1 mm in size, broadly ovate-elliptic, apex obtuse, margin entire or minutely toothed, outer somewhat smaller than the inner. Petals 5, ca. 2 mm long, oblong, conniving at the top. Stamens 5; filaments ca. 2 mm long, filiform; anthers cordate, more or less apiculate. Disc 10-lobed, filaments inserted in alternate notches of disc-lobes (mesodiscine), insertion peridiscal. Ovary ca. 2 mm long, conical; styles ca. 1 mm long; stigmas 3-lobed, discoid. Capsule ca. 4 x 2 mm in size, 3-gonous, rounded at the tip.
Flowering period: June-October.
Herbarium specimens examined: NARA DESERT: *Rahmatullah Qureshi*, 609, 616, 821, 976, 986, 993, 1001, 1083, 1296 (SALUBGH).
Field remarks: A common tree of sandy soil.
Distribution: Africa (Morocco, Algeria, Tunisia, Libya, Egypt, Senegal, Sudan, Abyssinia, Eritrea, Somaliland, Kenya), Middle East (Israel, Jordan, Saudi Arabia, Yemen, Iraq, Kuwait, Iran), Pakistan, India and Afghanistan (Qiaser, 1982).

4. Tamarix indica willd., Ges. Naturf. Freunde Berlin Neue Schr., 4:214 (1830); Stewart in Nasir & Ali., Ann.Cat.Vasc.Pl.W.Pak. & Kashm.,491 (1972); Qaiser in Nasir & Ali., Fl.Pak.,141:5 (1982).

Syn: *T. epacroides* Smith. in Rees, Cyclop. 35: 4 (1817); *T. galllica* L. var. *indica* (Willd.) Ehrenb. in Linnaea 2: 268 (1827); *T. gallica* sensu Wight & Arnnott, Prodr.

Fl. Ind. Or. 40 (1843); *T. troupii* Hole in Indian Forester 45: 248 (1919); *T. bengalensis* Baum, Monog. Gen. *Tamarix* 180 (1966).

A perennial tall shrub or small size tree up to 8 m tall. Stem fairly branched, woody; branches slender, pendulous; bark reddish-brown, glabrous. Leaves ca. 2-3.5 mm long, pseudovaginate, sessile, subulate-ovate, apex acute, base subamplexicaul-auriculate, imbricate when young. Racemes ca. 3-13 x 3-4 mm in size, vernal (usually simple) and aestival (densely compound). Rachis minutely papilose or entirely glabrous. Bract immediately beneath the flower, ca. 1 mm long, traingular, apex acuminate, margin more or less irregularly dentate, base sometimes slightly keeled. Flowers ca. 2 mm long, bisexual; pedicel ca. 0.5 mm long. Sepals 5, fused at the base, ca. 0.8-1 x 0.5 mm in size, ovate or trullate-ovate to more or less orbicular, apex acute, margin denticulate and membranous. Petals 5, free, ca. 2 x 1.3 mm in size, elliptic-oblong, apex rounded and often notched, pink. Stamens 5, inserted opposite the lobes; filaments ca. 1 mm long, filiform, glabrous; anthers more or less unequal, apex apiculate. Disc 5-lobed, filaments inserted in alternate notches of disc-lobes (mesodiscine), insertion perihypodiscal. Ovary ca. 1 mm, bottle-shaped, 3-gonous, apex truncate; styles 3, articulated to the ovary. Capsules ca. 4 mm long, conical, somewhat 3-gonous, apex tapering, pale-pink, glabrous.
Flowering period: January-October.
Herbarium specimens examined: NARA DESERT: *Rahmatullah Qureshi*, 21, 45, 57, 650, 813, 983 (SALUBGH).
Field remarks: A common shrub of saline habitat.
Distribution: Bengla Desh, Ceylon, Pakistan, India and Afghanistan (Qiaser, 1982).

5. Tamarix passernioides Del ex. Desv. in Ann. Sc. Nat. Bot. (1):349 (1825); var. **macrocarpa** Ehrenb in Linnaea, 2:276 (1827); Qaiser in Nasir & Ali., Fl.Pak.,141:40 (1982).

Syn: *T. macrocarpa* (Ehrenb.) Bunge, l.c. 79 (excl. sy. *T. arvensis* Zoh. including var. *patentissima* Zoh. & var. *micrantha* Zoh.); *T. pauciovulata* J. Gay ex Coss. in Ann. Sci. Nat. Bot. 4(4): 283 (1855); *nom. nud.*; *T. balansae* J. Gay ex Coss. l.c. *nom. nud.*; *T. pauciovulata* J. Gay ex Batt. & Trab., Fl. Alg. 1(2): 322 (1889); *T. balansae* J. Gay ex Coss. l.c. *nom. nud.*; *T. pauciovulata* J. Gay ex Batta. et Trab., var. *microstyla* maire Bull. Soc. Hist. Nat. Afr. Nord. 22: 30 (1931); *T. pauciovulata* J. Gay ex Batt. & Trab. var. *micrantha* Corti, Fl. Veg. Fezzan 191 (1942); *T. macrocarpa* (Ehrenb.) Bunge, var. *micrantha* (Cotri) Baum, l.c. 143 (1942); *T. macrocarpa* (Ehrenb.) Bunge, var. *micrantha* (Cotri) Baum, l.c. 143.

A much branched medium sized shrub up to 4-5 m high. Stem fairly branched, woody; branches slender, pendulous; bark greyish-brown to blackish-brown, younger parts pubescent, older glabrous. Leaves ca. 1-3.5 x 1-1.5 mm in size, amplexicaul to sub-amplexicaul, ovate, apex acute with punctate glands. Racemes ca. 2-4 x 0.5-0.8 cm in size, mostly aestival; rachis papilose to glabrous. Bracts ca. 1.3-1.8 x 1 mm in size, amplexicaul to semiamplexicaul, triangular, apex acuminate. Flowers ca. 4 mm long, 5-merous, pinkish to purplish pink; pedicel ca. 1 mm long. Seplas 5, slightly connate at the base, ca. 2 x 1 mm in size; outer 2 slightly ovate, apex acute, finely denticulate; inner 3, trullate-ovate, apex obtuse. Petals 5, ca. 4 x 2 mm in size, elliptic-obovate to more or less ovate, margin entire. Stamens 7-10, generally less than 10, inconspicuously alternately long and short; longer stamens ca. 3

mm; shorter ca. 2.5-2.8 mm long; anther-base cordate, apex apiculate, ca. 0.6 mm long, filaments confluent with discal lobes, disc epdiscine. Ovary ca. 2.5 m long, bottle-shaped; styles 3, short; stigma discoid. Capsules ca. 8-10 x 2.5-3 mm in size, ovate-pyramidal, rupturing by 3 valves. Seeds more or less oblong.

Flowering period: January-October.

Herbarium specimens examined: NARA DESERT: *Rahmatullah Qureshi*, 56, 1236 (SALUBGH).

Field remarks: A common shrub of riverine tract.

Distribution: Morocco, Kuwait, Algeria, Egypt, Israel, Jordan, Iraq, Iran, Russia and Pakistan (Qiaser, 1982).

22. Frankeniaceae

Perennial herbs or undershrubs. Stem jointed, often with deposition of calcium carbonate. Leaves simple, opposite, exstipulate. Flowers axillary, solitary or in dichasial cymes, hermaphrodite, regular. Calyx tubular, 4-6 lobed, persistent. Petals 4-6, clawed, free. Stamens 3-24. Carpels usually 3, syncarpous; ovary unilocular; placentation parietal with indefinite ovules. Capsule loculicidal, enclosed within the persistent calyx; seeds many, small.

A family of 5 genera and about 75 species; widely distributed in deserts and sandy coastal areas throughout the world. Closely related to Tamari¬caceae from which it can be distinguished by its opposite leaves, the connate sepals and the flowers which are axillary solitary or in dichasial cymes.

Frankenia

L., Sp. Pl. 331 (1753); Gen. P1. ed. 5. 154 (1754); Boiss., Fl. Or. 1: 779 (1867); Jafri, Fl. Kar. 234 (1966).

Beatsonia Roxb.; *Franca* Mich. ex Adans.; *Nothria* Berg.; *Streptima* Rafin.

Halophytic herbs or undershrubs. Stem with jointed nodes. Leaves opposite, decussate, small. Flowers solitary axillary or in cymes, sessile, pink. Sepals 4-6, induplicate or valvate, persistent. Petals 4-5, imbricate, claw with a ligular scale on the inner side. Stamens 4-6. Stigma 2-3 forked. Capsule unilocular, splitting by three valves.

A genus of about 70 species, mostly subtropical; Mediterranean, drier parts of Canary Islands, Turkey, S. Iran, N. Africa, W. and SW Australia, W. India, W. Pakistan, W. South America and California. Represented in W. Pakistan by only 1 species (Nasir, 1971).

Frankenia pulveulenta L., Sp. Pl. 332 (1753); Nasir in Nasir & Ali, Fl. Pak., 7: 1-2 (1971); Jafri in Fl. Kar., 234 (1966).

Syn: *F. canescence* Presl, Delic. Prag. 61 (1822); *F. intermedia* Costa, Fl. Catal. 31; *F. laevis* Habl. ex Bieb. Fl. Taur. Cauc. 1. 287 (1808).

A procumbent herb up to 15 cm high. Stem diffuse, puberulous. Stem diffuse, puberulous. Leaves spathulate to obovate, 3-7 mm long, 1-3 mm broad, retuse, pubescent beneath; subsessile, petiole base sheathing, ciliate. Flowers sessile, axillary solitary or in dichasial cymes. Calyx tube c. 4 mm long, costate, 5-fid, puberulous; lobes acute c. 1 mm long; petals 5, pink, c. 4 mm long, obovate, apex slightly fimbriate, base with a membranous decurrent

scale attached to the inside. Stamens 6, free; filament c. 2.5 mm long, dilated posteriorly. Ovary c. 2 mm long, ovoid; style 1.5 mm long, linear; stigma 2-3 forked. Capsule c. 2 mm long, splitting by 3 valves; seeds less than 1 mm long, c. 0.5 mm broad, ovoid.
Flowering period: March-May
Herbarium specimens examined: NARA DESERT: saline areas, *Rahmatullah Qureshi*, 1350 (SALUBGH).
Field remarks: A common herb of saline habitat.
Distribution: S. and SE. Europe, Africa, Siberia, Turkey, Iran, Syria, Arabia, Egypt and W. Pakistan (Nasir, 1971). A desert annual with pretty deep pink sessiel flower often found in in the sandy or saline soil.

23. Cucurbitaceae

Vines, rarely prostrate or trailing, annual or perennial herbs or undershrubs; branchlets often scabrid. Leaves alternate, petiolate, simple, usually palmately lobed, sometimes angular, chartaceous, rarely coriaceous, palmately or pedately nerved; tendrils, simple or branched, spirally coiled; stipules 0. Inflorescence paniculate, racemose, umbellate or subumbellate or solitary flowers. Flowers unisexual (monoecious or dioecious), very rarely bisexual, regular, yellow or white; bracts small, rarely large, spathaceous, fimbriate or laciniate. Calyx-tube campanulate, sometimes turbinate, elongate; lobes 5, imbricate, rarely dentate or laciniate. Petals 5, often yellow, imbricate or involute-valvate in bud, campanulate, rarely rotate, connate into a tube, sometimes fimbriate. Stamens basically 5, of which 4 mostly connate in pairs, thus giving false impression of only 3 stamens or sometimes 2 or 1 due to cohesion, inserted at the mouth, or near the middle, or at the base of calyx-tube; filaments free or united, rarely pubescent; anthers connate, confleunt into a capitulum or free, 1-2-celled, rarely all 2-celled, triplicate and 'S'-shaped or straight to oblique; connectives often apically prolonged, crested, 2-fid, 2-partite, ciliate; dehiscence extrose, longitudinally or along curves in the anther-calls; pistillode often prominent. Carpels (2-) 3 (-5), syncapous, ovary inferior, usually uni-locular, rarely multi-locular, adnate to calyx, globose, rarely rostrate; ovules usually numerous, horizontal; placentation parietal, usually 3, often bifurcating; style simple, apically 3-fid; stigma usually 3, entire or lobed, lobes obtuse, fimbriate to papilose; staminodes often present. Fruit berry-like, with a rind (pepo), globose or oblong, fleshy, indehiscent, or dehiscent by valves, smooth or covered with tubercles or prickles. Seeds usually many, often packed in pulp or fibre, ovoid or ellipsoid, often compressed; testa membranous, smooth, granular or tubercled.

A family of about 120 genera and 825 species, mostly distributed in tropical countries, poorly represented in temperate regions. Represented in Pakistan by 17 genera and 32 species and inraspecific taxa (Nizamuddin & Naqvi, 1984).

Key to genera:
1a. Fruits dry, fibrous and operculate...**1. Luffa**
1b. Fruits fleshy, non-fibrous and non-operaculate:
2a. Petioles with a conspicuous stipuliform, fimbriate bract at the base............**2. Ctenolepis**
2b. Petioles without a stipuliform bract at the base:
3a. Anthers conduplicate or sigmoid:
4a. Connective produced beyond the anther loculi......................................**3. Cucumis**

4b. Connective not produced beyond the anther loculi..**4. Citrullus**
3b. Anthers thecae arcuate straight or apically hooked..**5. Mukia**

1. Luffa

Mill., Gard. Dict. Abridg. ed. 4 (1754); Benth. & Hook. f., Gen. Pl. 1: 823 (1867).

Annual, scandent, climbing or trailing, scabrous or hairy. Monoecious or dioecious. Tendrils proximally 3-6-fid, less often bifid. Leaves simple, palmately 5-7-lobed, rarely subentire, lamina sometimes gland-dotted. Petiole eglandular. Probracts at leaf axil, glandular. Flowers medium to large, yellow; male flowers racemose. Calyx-tube shortly campanulate, calyx lobes relatively large, enclosing petals in bud, entire. Petals 5, free. Stamens 3(5), all monothecous or one 1-locular and others 2-locular, filaments inserted on the tube, free; anthers free; connective broad, thecae marginal, much convoluted; female flowers solitary. Ovary smooth, ribbed, tuberculate or spiny, stigmas 3, bibbed; ovules many, horizontal, calyx-tube short. Fruit medium sized to large, subglobose to elongate-cylindrical, beaked, smooth, ribbed or sparsely to densely spiny with short to rather long spines, brownish, dry and fibrous, dehiscent by an apical operculum. Seeds medium-sized, distinctly compressed, oblong-elliptic in outline.

A genus of 6 species, two cultivated throughout the tropics and of doubtful origin (probably Old World), one tropical American and the rest are distributed in the Old World. Four specific and infraspecific taxa are reported from the country.

Luffa echinata Roxb., Fl. Ind. 3: 716 (1832); Stewart in Ali and Nasir, Ann. Cat. Vasc. Pl. W. Pak. & Kashm., 705 (1972).

A climber with bifid puberulous tendrils. Stem sulcate, slightly hairy to glabrous. Leaves reniform, suborbicular, obscurely or deeply 5-lobed, apex round or rarely acute, scabrous on both surfaces, margin minutely denticulate. Petiole stout, puberulous, up to 12 cm long. Flowers white, pedicellate, c. 2.5 cm across; male flowers in 5-12-flowered, up to 15 cm long raceme; calyx tube c. 5.6 mm long, hairy, lobes lanceolate, acute. Petals ovate, 1-1.2 cm long, obtuse, hairy at the base. Stamens 3, filaments united, c. 3-9 mm long, obtuse, hairy at the base, anthers entire or ± bifid. Ovary ovoid. Fruit ashy, oblong, ovoid, 2-5 cm long, densely covered with c. 4-7 mm long bristles. Seeds ovate, black, slightly verrucose, emarginate, c. 4-5 mm long, 3-5 mm broad and c. 2 mm thick.
Flowering Period: August-September.
Herbarium specimens examined: NARA DESERT: Chotiari reservoir, *Rahmatullah Qureshi*, 1434 (SALUBGH).
Field remarks: A common climber found in Achhro Thar.
Distribution: Pakistan, India, Bangladesh and Northern Tropical Africa (Nazimuddin and Naqvi, 1984).

2. Ctenolepis

Hook.f. in Benth. & Hook.f.Gen. Pl. 1: 832 (Sept. 1867); Clarke in Hook. f., Fl. Brit. Ind. 2. 629 (1879); Hutch., Gen. Fl. Pl. 2: 388 (1967); Nazimuddin & Naqvi in Fl. Pak., 154: 37 (1984).

Ctenopsis Naud. in Ann. Sc. Nat. Ser. 5(5): 12 (1866); *Blastania* Kotschy & Peyr., Pl. Tinn. 15. Oct (1867).

Vines; branchlets scabrid. Leaves palmately 5-7-lobed; suborbicular or obovate, with stipuliform bract at the petiole base, ciliate, fimbriate or dentate; tendrils simple. Flowers monoecious, very small, yellow. Male flowers racemose on short peduncle. Calyx-tube shortly obconic and campanulate. Corolla rotate, 5-partite, ovate-ligulate, spreading. Stamens 3, two 2-celled, one 1-celed; cell erect, short ovoid, didymous, connective not produced; filaments inserted on the calyx-tube, free, remote, very short; pistillode 0. Female flowers solitary in the same axil as the males; peduncles short. Calyx and corolla as in the males. Ovary subglobose, smooth; ovule numerous, horizontal; style columnar, basal disc absent; stigmas 2(-3). Fruit small, fleshy berry, subreniform, globose or obliquely subquadrate, red. Seeds few, much compressed, or boat-shaped.

A genus of two species, one tropical African extending to Pakistan and India, the other confined to peninsular India and Ceylon (Nizamuddin & Naqvi, 1984).

Ctenolapis cerasiformis (Stocks.) Hooker.f. in Oliver, Fl. Trop. Africa, ii: 558(1871); Jafri in Fl. Kar.326 (1966); Stewart in Ali and Nasir, Ann. Cat. Vasc. Pl. W. Pak. & Kashm.,703 (1972).

Syn. *Zahneria cerasiformis* Stocks in Kew. Journ. Bot., iv: 49(1852); *Bryonia fimbristipula* Fenzl. Kotschy, Iter. Nub. 205, 231 (1841); *Blastania fimbristipula* Kotschy and Peyr., Pl. Tinn., 15. T. 7(1867); *Ctenopsis cerasiformis* (Stocks) Naud. in Ann. Sci. Nat. ser. 6: 13 (1867); *Ctenolepis fimbristipula* (Kotschy et Peyr.) Broun et Massey. Fl. Sudan 104 (1929); *Blastania cerasiformis* (Stocks) Meeuse in Bothalia, 8 (1): 12 (1962).

An annual trailing, low shrub or climbing herb. Stem slender, elongate, angular, sulcate or striate, glabrous-sparsely hispid, geniculate at nodes; internodes ca. 3-8 cm long; hispid. Leaves ca. 2-5 x 1-3.5 cm in size, palmately 3-5-partite; segments ovate-oblong or obovate-elliptic to lanceolate, narrowed at both ends, base acut-subobtuse, apex mucronate, margins coarsely serrate-dentate, shortly scabrid-setose on both surfaces; central lobe entire or with 2 short lobules, one on each side; lateral lobes smaller; basal often more or less deeply 2-lobed; petioles ca. 1.5 cm long, somewhat flattened, striate-sulcate, shortly setose, hispid. Stipuliform bracts ca. 0.5-2 x 0.3-1.5 cm in size, ovate-oblong, subreniform-suborbicular, subsessile or narrowed, setose-scabrid; cilia ca. 1-1.5 cm long, filiform. Male flowers ca. 3-10, ebracteate; peduncle ca. 0.5-2.5 cm long, filiform, glabrous or puberulous; pedicels ca. 1.3 mm long, persistent with receptacle knob at apex; receptacle ca. 0.5-1 mm across. Calyx ca. 0.6 mm long. Corolla yellowish; segments ca. 0.8-1 mm long, ovate-oblong, apex obtuse, spreading. Femal flowers solitary, rarely 2 in an axil; peduncle ca. 2-5 mm long. Ovary ca. 2 mm across, glabrous. Fruit ca. 1-1.5 cm across, globose, green with white blotches when young, scarlet when mature, smooth, glabrous. Seeds 2, ca. 8-10 x 4-6 mm in size, ovoid, slightly concave on one side, sonvex on other side, smooth, base attenuate to acute, edges sharp.

Flowering period: October- November.

Herbarium specimens examined: NARA DESERT, *Rahmatullah Qureshi*, 1323 (SALUBGH).

Field remarks: A common climber of calcarious soil.
Distribution: Tropical Africa from Northern Nigeria eastward to Sudan and southward to Mozambique, Rhodesia, Yemen and Indo-Pak subcontinent (Nazimuddin & Naqvi, 1984).

3. Cucumis

L., Sp. Pl. 1011 (1753); Gen. Pl. ed. 5. 442 (1754); Benth. & Hook. f., Gen. Pl. 1: 826 (1867); Clarke in Hook. f., Fl. Brit. Ind. 2. 619 (1879); Nazimuddin & Naqvi in Fl. Pak., 154: 41 (1984).

Annual or perennial, prostrate or climbing vines; branchlets stiff-hispid. Leaves angular, 3-7-lobed, ,margin dentate or serrate, chartaceous; tendrils simple. Flower(s) monoecious, male in clusters, females solitary. Calyx-tube narrowly campanulate or turbinate; lobes 5, subulate-linear. Corolla yellow, rotate or campanulate; petals 5, oblong-ovate, apex acute. Stamens 3; filaments free, inserted below the mouth of calyx-tube; anthers oblong, one 1-celled, others 2-celled; cells linear, flexuous-curved; connectives apically crested; pistillodes 3, subulate, glandular. Ovary globose or subterete; placentae 3; ovules α, horizontal; style short, with a basal annular disc; stigma 3-lobed, obtuse; staminodes linear, 3. Fruit globose or ellipsoid-oblong, smooth, sometimes with tubercular prickles when young, with green and white patches, often indehiscent. Seeds α, ovoid, compressed, smooth, mostly without margin.

A genus of about 25 species, mostly tropical African in Distribution. Some are confined to tropical Asia. Two are cultivated almost throughout the world. Represented in Pakistan by 2 wild and 2 cultivated species (Nizamuddin & Naqvi, 1984).

Key to species:
1a. Fruit smooth, glabrous or pubescent..**1. C. melo** var. **agrestis**
1b. Fruit echinate...**2. C. prophetarum**

1. Cucumis melo var. **agrestis** (Naud.) Grebense. in Kulturpfl. Beih 2: 424 (1959); Stewart in Ali and Nasir, Ann. Cat. Vasc. Pl. W. Pak. & Kashm.,703 (1972); Nazimuddin and Naqvi in Nasir & Ali., Fl. Pak., 154: 44-46 (1984). **Figure 12A**

Syn. *Bryonia callosus* Rottl. in Neue Schrift Ges. Nat. Freude Zu Berlin, IV: 210(1803); *Cucumis trigonous* Roxb., Fl. Ind., ii: 722(1832); *C. callosus* (Rottl.) Cogn. in Planzenr., Heft. 88: 129 (1924).

An annual or biennial, trailing, prostrate or climbing, rigid hairy herb. Stem 5-angular, winged, much branched, main branch hoary; branchlets thickly villous; trichomes up to 3 mm, 5-9-celled; internode ca. 2-11.5 cm long, villous. Leaves ca. 3-10 x 2.5-12 cm in size, suborbicular-ovate, base cordate, apex rounded, margin dentate, deeply 5-lobed; lobes obtuse, both surface scabrous or punctate; petiole ca. 0.5-5 cm long, hispid. Tendrils simple, hispid. Male flowers in axillary few-flowered clusters, peduncle ca. 3-5 cm long; flowers ca. 1 cm across, yellow. Calyx-tube ca. 4-5 mm, densely hairy above, glabrous within; lobes 5, free, subulate, apex obtuse-rounded, densely hirsute above, ca. 3-4 mm long. Petals ca. 0.4-0.5 x 0.3-0.4 cm in size, ovate- obovate, base cuneate, apex mucronate, margin undulate, sparsely pubescent within, densely hairy above, 4-5 nerved, prominent above, yellow.

Stamens 3, ca. 3 mm long; anthers flexuous, connectives apically prolonged; pistillode globose, 3-fid, ca. 2 mm. Female flower: Calyx and corolla as in male. Ovary ca. 10 x 5 mm in size, oblong, densely hairy; style ca. 1 mm; stigma ca. 2 mm long. Fruit ca. 2.5-4 x 1-2.5 cm in size, globose or obovoid, rounded, smooth, longitudinally variegated with 7-11 green stripes; pulp bitter when immature but mature one edible. Seeds α, ca. 3.5-4 x 2-2.5 mm in size, oblong or ellipsoid, white, margined.
Flowering period: August-September.
Herbarium specimens examined: NARA DESERT, *Rahmatullah Qureshi*, 35 (SALUBGH).
Field remarks: A common herb of loamy soil.
Distribution: Throughout the Old World tropics in neotropics (Nazimuddin & Naqvi, 1984).

2. Cucumis prophaterum L. Cent. 1:34. (1755); Jafri in Fl. Kr., 322 (1966); Stewart in Nasir & Ali., Ann. Cat. Vasc. Pl.W. Pak. & Kashm., 704 (1972); Nazimuddin and Naqvi in Nasir & Ali., Fl. Pak., 154: 44 (1984).

A perennial, prostrate or climbing, hispid, grey-green perennial herbs up to 45 cm high. Stem much branched, slender, angular, whitish when dry, hispid with aculeate, white hairs, rough and scabrid at length; tendril short, simple, scabrid. Leaves ca. 2-4 cm across, grey-green, rigid, suborbicular or deltoid, base cordate-truncate, 3-5-lobed, margin crisped or dentate. Tendril filiform. Male flowers yellow, solitary or fasciculate; connective-appendages glabrous, ca. 0.5-0.7 mm long. Flowers ca. 3-4 x 2-3.5 cm in size, green with 7-10 longitudinal white stripes; mature fruits yellow, spiny; spines white-tipped. Seed ca. 4-5 x 2-2.5 mm in size, oblong-ellipsoid, emarginate at apex, pale ashy in colour.
Flowering period: August-December.
Herbarium specimens examined: NARA DESERT, *Rahmatullah Qureshi*, 1257, 1314 (SALUBGH).
Field remarks: A rare herb of gravelly soil.
Distribution: Paleotropic in distribution, throughout N. Africa, Arabia to India and Pakistan. In Pakistan, it is mainly distributed in Sindh and Baluchistan (Nazimuddin & Naqvi, (1984).

4. Citrullus

H. A. Schrader in Ecklon et Zeyher Enum. 279 (1836), *nom.cons*; Clarke in Hook. f., Fl. Brit. Ind. 2: 620 (1879); Nazimuddin & Naqvi in Fl. Pak., 154: 11 (1984).
Anguria Mill., Gard. Dict. Abridg. Ed. 4. without pagination (1754), *nom. rejec.* non Jacq. (1760), *nom. illegit.*; *Colocynthis* Mill., Gard. Dic. Abridg. Ed. 4. without pagination (1754), *nom. rejec.*; Ludwig, Inst. Regn. Veg. ed. 2: 139 (1757).

Annual or perennial vines or prostrate herbs, monoecious; branchlets villos-hirsute. Leaves triangular-ovate, deeply 3-5-lobed; lobes lobulatel tendrils simple or 2-3(-4)-fid. Flowers solitary, monecious, yellow, medium sized; bracts spathulate or subulate. Male flower: Calyx-tube broadly campanulate; lobes 5, moderately large. Corolla rotate or broadly campanulate; petals 5, ovate-oblong. Stamens 3, inserted at the base of calyx-tube; filaments free; anthers slightly coherent of free, one 1-celled, others 2-celled; cells linear, triplicate, flexuous; connectives dilated, not apically produced; pistillode glanduliform. Ovary pubescent, ovoid; placentae 3; ovules α, horizontal; style short, stylar disc 0; stigma 3-lobed, thick, reniform; staminodes 3, setose or ligulate. Fruit globose or oblong, greenish-yellow, pericarp firm, mesocarp fleshy, smooth, indehiscent. Seeds many, usually ovate, compressed, smooth or slightly rough.

A small genus of 3 species in Africa and Asia. Represented in Pakistan by one wild and one cultivated species (Nizamuddin & Naqvi, 1984).

Citrullus colocynthis (L.) Schrad. in Linnaea 12:414 (1838); Jafri in Fl. Kr., 323 (1966); Stewart in Nasir & Ali., Ann. Cat. Vasc. Pl.W. Pak. & Kashm., 702 (1972); Nazimuddin & Naqvi in Nasir & Ali, Fl. Pak., 154: 12 (1984). **Figure 11D**

Syn.:*Cucumis colocynthis* L., Sp. Pl. 1011. (1753); *Colocynthis vulgaris* Schrad., Ind. Sem. 1: Figure 99 (1950).

Trailing, scabrid herb, with a long tap root up to 2 m long. Stem diffuse, angular, hirsute; internode ca. 1-4.5 cm. Trichomes ca. 1 mm long, 3-5-celled, double walled. Leaves ca. 2.5-7 x 2-5 cm in size, ovate or narrowly triangular, deeply 3-5-lobed, chartaceous, rigid, densely villous-hirsute below; trichomes 3-5-celled; lobes pinnatifid-sinous, terminal lobe larger, acute; sinous ones obtuse; petioles ca. 1.5-3.5 cm, hirsute; probracts lanceolate-elliptic, ca. 4-5 x 1.5 mm, caduous. Tendril simple or bifid, slender, sparsely villous. Flowers yellow, axillary, solitary, unisexual. Male flower: ca. 1 cm across, campanulate; pedicel ca. 0.5 cm. Calyx-tube ca. 3 mm long; lobes ca. 3-6 mm, linear, apex acute, margin entire, hirsute without, greenish in colour. Petals ovate, apex acute, hirsute, ca. 8 x 4 mm in size, yellow or greenish yellow in colour. Stamens 3; filaments ca. 6-7 mm, filiform, dilated at the base, glabrous, greenish; anthers ca. 5-6 x 3-4 mm. Female flower: ca. 1.5 cm across; pedicel ca. 2.5 cm long, recurved. Ovary ca. 12 x 8 mm in size, obovate, densely villous; trichome ca. 3-6-celled; style ca. 4 mm; stigma 5. Fruit ca. 5-8 cm in diameter, globose, slightly depressed at the top, variegated and green when young, yellow when ripe; epicarp thin, filled with a dry spongy very bitter pulp. Seeds α, ca. 6 x 3 mm in size, obovate or ovate-oblong, compressed, smooth, pale brown, with 2 oblique grooves one on each surface near the slightly narrowed base.

Flowering period: January-April.

Herbarium specimens examined: NARA DESERT, *Rahmatullah Qureshi*, 28, 966, 1012, 1195 (SALUBGH).

Field remarks: A common subshrub of sandy soil.

Distribution: Northern Tropical Africa, Atlantic Islands, North-West India, Pakistan and Australia (Nazimuddin & Naqvi, 1984).

5. Mukia

Arnott, Madras J. Lit. Sci. 17: 50 (1840); Benth. & Hook. f., Gen. Pl. 1: 829 (1867); Clarke in Hook. f., Fl. Brit. Ind. 2: 625 (1879); Hutch., Gen. Fl. Pl. 2: 393 (1967); Nazimuddin & Naqvi, in Fl. Pak., 154: 44 (1984).

Prostrate or climbing herbs; branchlets scabrid or villous. Leaves ovate or deltoid, chartaceous, angular or 3-5-lobed; tendrils simple. Flower(s) monoecious, male and female in clusters, sometimes female solitary. Male flower: Calyx-tube campanulate, villous; lobes 5, linear-subulate. Corolla campanulate, yellow; petals 5, ovate, villous without. Stamens 3, inserted at the calyx-tube; filaments free; anthers free, oblong, one 1-celled, others 2-celled; cells erect, connectives slightly produced; pistillodes 3-lobed. Female flower: Ovary globose, villous; placentae 3; ovules α, horizontal; stylar disc annular; stigma 3-fid, fimbriate. Berries small, subsessile, 1 in axillary clusters, ellipsoid or globose, hispid when young, red almost glabrescent when mature. Seeds horizontally placed, small, ovoid to elliptic, turgid, smoothed or pitted.

About 4 species of Old World, only one occurs in Pakistan (Nizamuddin & Naqvi, 1984).

Mukia maderaspatana (L.) M. Roemer., Syn. Monogr. 2: 47 (1846); Stewart in Ali and Nasir, Ann. Cat. Vasc. Pl. W. Pak. & Kashm., 706 (1972); Nazimuddin and Naqvi in Nasir & Ali., Fl. Pak., 154: 39-40 (1984). **Figure 12B**

Syn: *Cucumis maderaspatana* L., Sp. Pl. 1012 (1753); *Melothria maderaspatana* (L.) Cogn. in A. & C. DC., Monogr. Phan. 3: 623 (1881); *Coccinia cordifolia* (L.) Cogn. in A. & C. DC., Monogr. Phan. 3: 529 (1881).

An annual or perennial, monoecious, scandent climbing or trailing herb up to 4 cm long. Stem herbaceous, pentagonal, rigid, green, bristly hairy; bristle upward directed, up to 2 mm long, ca. 5-7-celled; internode ca. 3-14 cm long. Leaves ca. 2.5-8 x 2.5-10 cm in size, simple, alternate, subtriangulate-ovate, base sagittate, hastate, apex acute-acuminate, margin denticulate, basal sinus 1-3 cm deep, hispid below. Tendrils simple, opposite to leaves. Flowers small, yellow, unisexual in axillary cymes. Calyx-tube villous; lobes linear. Corolla yellow; petals ovate, villous without. Stamens 3; anthers oblong. Ovary villous. Fruit an ovoid or globose berry, juicy, smooth, ca. 1 x 0.8 cm in size, green when young and red when ripe.

Flowering period: July-December.

Herbarium specimens examined: NARA DESERT, *Rahmatullah Qureshi*, 899, 903, 947, 958, 1045, 1242 (SALUBGH).

Field remarks: A common climber of sandy soil.

Distribution: India, Pakistan, China, Africa and Australia (Nazimuddin & Naqvi, 1984).

24. Salicaceae

Dioecious,mostly deciduous,trees and shrubs,rarely prostate subshrubs. Leaves alternate, simple, usually stipulate, entire or variously toothed or lobed. Flowers in spicate or racemose catkin,erect or pendulous, each flower subtended by a bract or catkin scale; perianth cupulate (females of Populus) or reduced to 1 or more glands (Salix), rarely absent. Stamens 2 to many, filaments free or connate (partially or completely). Ovary superior, uni or bilocular, parietal placentation, usually with many ovules. Fruit a capsule, seeds many, small, with long silky hairs.

A family with 3 genera and with about 620 species mostly distributed in Northern hemisphere. In Pakistan the family is represented by 2 genera and c. 32 species.

Populus

L., Sp. Pl. 1034 (1753); Gen. Pl. ed. 5, 456 (1754); Boiss., Fl. Or. 4: 1192 (1879); Hook. f., Fl. Brit. Ind. 5: 637 (1888); Stewart, Ann. Cat. Vasc. Pl. W. Pak. Kashm. 180 (1972).

Deciduous trees. Buds with several unequal scales. Leaves alternate; petiole usually long, terete or flattened, lamina entire or lobed, linear to rhombic ovate, dentate, serrate; often dimorphic on long and short shoots. Flowers appearing before leaves, anemophilous, in elongate, cylindrical, pendent, pedunculate catkin. Each flower in the axil of a laciniate or toothed bract; pedicellate; disc cup-shaped. Stamens 4-many, filaments free. Ovary sessile, unilocular, surrounded at the base or up to three quarters of its length by the disc; ovules many on 2-4 parietal placentas, style short, stigmas 2-4. Capsule 2-4 valved; seeds minute, numerous, with a long dense tuft of silky hairs.
A genus with about 100 species, mostly distributed in north temperate regions of the world. In Pakistan, 6 species recorded.

Populus euphratica Olivier. Voy. Emp. Ottoman. 3: 449 (1807); Stewart in Ali and Nasir, Ann. Cat. Vasc. Pl. W. Pak. & Kashm., 180 (1972).

Syn; *P. diversifolia* Schrenk (1842) & F. & M. (1842); *Balsamifolia euphratic* (Oliv.) Kimura (1939).

A medium sized usually gregarious tree up to 11 m high. Stem woody, fairly branched, cylender, glaucous. Leaves leathery, glabrous, very variable in form and size; on young plants and on long shoots, linear to elongate-ovate, 3-12 x 0.5-4 cm, usually entire, petiole 7.5-15 mm long; on older trees and on short shoots ovate-rhombic, elliptic-orbicular or reniform, 2-5 x 3-7 cm or larger, shallowly dentate in upper part; petiole 1.2-5 cm long. Male catkin 2.5-5 cm long, lax. Male flowers: bract oblanceolate, incised; disc orbicular on a long slender stipe, flat, 8-cleft; stamens 8-12, anthers longer then filaments. Female catkin 5-8 cm long, lax. Female flower: disc membranous caducous, tubular with 8-12 linear segments; stigmas 2-3 ± crescent shaped, narrowed into a short style. Capsule 7.5-12 mm, smooth, glabrous or slightly pubescent when young, 2-3 valved. Pedicel 3.5-5 mm.
Flowering Period: February.
Field remarks: A common tree of riverine tract.
Distribution: Pakistan (Sindh, Baluchistan, Punjab) Kashmir; Afghanistan; Iran; Iraq; Kazakstan, Turkmenistan, Tajikistan, Uzbekistan, China, N. Africa, and Spain (Ali, 2001).

25. Capparidaceae

Annual, or perennial, erect or scandent herbs or shrubs, rarely small trees often glandular hairy. Leaves alternate or rarely opposite, simple or digitately 3-7 (rarely 1-2 or 9-)-foliate; leaflets usually entire; stipules 2-0, often spiny, persistent or caduous. Inflorescence terminal or axillary, often corymbose, sometime solitary or fascicled. Flowers bisexual (rarely dioecious), actinomorphic, sometime with unequal sepals or petals, complete or rarely petals wanting, pedicellate. Sepals 4 (rarely more), free or sometimes connate below, equal or unequal, valvate or imbricate. Petals 4 (rarely wanting or more), sessile or clawed. Stamens

4(not tetradynamous when 6 or didynamous when 4), alike or unequal, usually borne on a short or long androphore; filaments free, sometimes adherent to the gynophore or coherent at the base, usually filiform; anthers oblong, 2-celled, dehiscence longitudinal, basifixed, rarely some of them abortive. Ovary sessile or supported on a gynophore of varying length, unilocular with usually 2 parietal placentas or rarely divided by spurious dissepiments into 2 or more locules; many rarely few (even 1 or 2), usually campylotropous with 2 integuments; style 0- long, with simple stigma. Fruit generally capsular or baccate, often oblong-cylindrical or globose, sometimes torulose or somewhat lobed, 1-seeded. Seeds reniform or angular; endosperm non or scanty; embryo usually incurved; cotyledons folded or convolute; radical often conical and incumbent.

A medium-sized family with about 45 genera and nearly 600 species, mostly in tropical and subtropical regions. It is represented by 7 genera and 23 species (including 4 cultivate spp.) in Pakistan (Jafri, 1973).

Key to genera:

1a. Plants annual or perennial herbs; fruits siliquiform, linear or oval capsule with 2-valves separating from a persistent replum:

2a. Leaves 3-7 foliate; mature flower pink or white; gynandrophorous........**1. Gynandropsis**

2b. Leaves 3-5 foliate; flower yellow; not gynandrophorous...................................**2. Cleome**

1b. Plants perennial herbs, shrubs or climbers; fruits various, but not siliquiform capsule usually indehiscent:

3a. Plants climbers; sepals connate below into a tube; fruit moniliform, fleshy, berry...**3. Maerua**

3b. Plants woody, rigid herbs or shrubs, rarely trees; sepals not connate to form a tube; fruit not moniliform:

4a. Plants armed with stipular spines; stamens indefinite; fruit a globose berry.....**4. Capparis**

4b. Plants unarmed; stamens 4-6; fruit not a globose:

5a. Stamens inserted on a gynophore; fruit cylindrical; raceptacle prolonged into a strap-shaped appendage...**5. Cadaba**

5b. Stamens not inserted on a gynophore; fruit small, compressed, winged; raceptacle not prolonged into a strap-shaped appendage.......................**6. Dipterygium**

1. Gynandropsis

DC., Prodr. 1: 237 (1824); Benth. & Hook. f., Gen. Pl. 1: 106 (1862); Hook. f., Fl. Brit. Ind. 1: 170 (1872); Jafri in Fl. Kar. 136 (1966); Jafri in Fl. Pak. 34: 17 (1973).

Erect or ascending annual, rarely perennating, usually glandular-pubescent herb. Leaves 3-7-foliate, palmately compound. Racemes many-flowered, terminal with foliaceous bracts. Flowers in terminal racemes. Sepals 4, spreading, deciduous. Petals 4, with long slender claws, imbricate or open in bud. Androphore as long as the petals or the gynophore. Stamens 10, filaments inserted on the androphore. Ovary 1-celled, many ovuled, usually on a short gynophore. Fruit capsular, stalked, linear, 2-valved, compressed or subterete, many seeded. Seeds reniform or orbicular; testa rugose or tuberculated.

Gynandropsis gynandra (L.) Briq. in Ann. Conserv. Jard.Bot. Geneve, 17:382 (1914); Jafri in Nasir & Ali., Fl. Pak., 34: 20 (1973). **Figure 8D**

Syn. *Cleome gynandrae* L. Sp. Pl. 671 (1753); *C. pentaphylla* L. Sp. Pl. ed. 2:938 (1763); *C. heterotricha* Burch., Trav. 1: 537 (1822); *G. pentaphylla* (L.) DC., prodr., 1: 23 (1824); *G. heterotricha* (Burch) DC., Prodr., 1: 238 (1824).

An annual, erect, glandular-pubescent herb up to 40 cm tall. Stem woody bellow, herbaceous above, much branched at the base, terete, greyish green, densely glandular hairy; internode ca. 1-2.5 cm. long. Leaves 3-7 foliate, palmately compound, alternate; leaflets obovate-elliptic and rhombic fusiform, middle one ca. 2-4 x 1-1.8 cm in size, laterals ca. 1-3 x 0.5-1.5 cm in size, base cuneate, apex acute-apiculate, margin finely crenate, chartaceous, sparsely glandular hairy, petiole ca. 1-6 cm long, densely glandular-pubescent. Inflorescence corymbose racemes, ca. 12-16 cm long; peduncle ca. 4 cm long, glandular hairy. Floweers ca. 1.2.5 cm across, pedicel ca. 1-2 cm long, bracteate, bracts much smaller than leaves, white. Sepals 4, free, slightly united at the base, valvate, ca. 4-8 x 1-2 mm. in size, elliptic-ovate or lanceolate, base obtuse-cuneate, apex acute-mucronate, margin entire-sub entire, densely glandular hairy, distinctly parallel veined, greenish in color. Petals 4, free, ca. 4-10 x 3-5 mm in size; claw ca. 2-10 mm long, obovate, apex obtuse, base cuneate, white. Androphore ca. 1.5-2.5 cm long. Stamens 6, versatile, filaments ca. 15-20 mm long, subequal, spreading conspicuously from a long androphore, often pinkish or lilac; anthers 2-lobed, dorsifixed, oblong, yellow, ca. 3-4 x 2 mm in size. Gynophore ca. 10-15 mm (usually shorter in flower but longer in fruit); ovary oblong, densely glandular-pubescent, ca. 12 x 2 mm in size; style 1-2 mm long; stigma discoid. Capsule ca. 1.5-5 x 0.2-0.4 cm, linear, narrowed towards both the ends, glandular hairy, finely and obliquely striated, many seeded, dehiscing longitudinally; beak ca. 2 mm long. Seeds α, ca. 2 mm across, cleft closed, concentrically ridges and transversely crested, dark brown.

Flowering period: June-July.

Herbarium specimens examined: NARA DESERT: *Rahmatullah Qureshi*, 420, 1169, 1280 (SALUBGH).

Field remarks: A common herb mostly found in flat habitat.

Distribution: Widely distributed in the tropical and subtropical regions of the world (Jafri, 1973).

2. Cleome

L., Sp. Pl. 671 (1753); Gen. Pl. ed. 5: 302 (1754); Benth. & Hook. f., Gen. Pl. 1: 105 (1862); Boiss., Fl. Or. 1: 410(1867); Jafri in Kew Bull., 173-175 (1957); Jafri in Fl. Pak. 34: 20 (1973).

Herbs or undershrubs, sometimes bushy and woody bellow, usually glandular pubescent. Leaves alternate, petiolate, simple or palmately 3-5 (-9) foliate; leaflets narrow, membranous; stipules 0. Inflorescence in terminal racems, bracteate or ebracteate. Flowers generally small or mediocre, commonly pedicellate, yellow, white or pinkish. Sepals 4, free or shortly united at the base, valvate, deciduous or persistent, often spreading and equal. Petals 4, equal or longer than sepals, subequal, often clawed, imbricate. Receptacle small. Androphore absent or inconspicuous. Stamens 4-18, subequal, sometimes a few sterile (staminodes). Gynophore present or absent; ovary 1-locular; with many ovules on 2 parietal placentas; style present or absent; stigma capitate or discoid. Capsule oblong to linear, usually elongated with two dehiscing valves and persistent replum (rarely sub-dehiscent). Seeds many, reniform to discoid, with open or closed cleft, glabrous or asperous smooth rugulose or rigid.

A large genus with ca. 200 species in the tropical and subtropical regions of the world. Ten species are reported from W. Pakistan (Jafri, 1973).

Key species:

1a. Leaves simple, setose-papilose.. **1. C. scaposa**

1b. Leaves compound, 3-5 foliate, not setose-papilose:

2a. Stamens 6. Capsule not more than 16 mm long.............................. **2. C. brachycarpa**

2b. Stamens more than 11-21. Capsule more than 16 mm long...................... **3. C. viscosa**

1. Cleome scaposa D.C., Proder., 1:239 (1824); Jafri in Fl. Kar., 137 (1966); Stewart in Nasir & Ali., Ann.Cat. Vasc. Pl.W.Pak. & Kashm., 296 (1972). Jafri in Nasir & Ali., Fl. Pak., 34: 30 (1973). **Figure 8B**

Syn: *C. papillosa* Steud. Nomen. ed 2, I:382 (1840); *C. gracilis* Edgew in J. Asiat. Soc. Bengal 16: 1212 (1847).

Erect, setose-papilose, annual herb upto 20 cm high. Stem woody at the base, much branched, slender, setose with long, spreading rough hairs intermixed with short, scabrid eglandular hairs. Leaves simple, ca. 1-2.5 x 0.5-1.6 cm in size, suborbicular, elliptic or ovate base cordate or rounded, apex obtuse or rarely retuse, margin entire, midrib impressed on upper surface, lateral nerves 2-3 pairs-obscure, densely scabrous or papilose; hairs ca. 3-celled; petiole ca. 2-4 mm, pale, upper most leaves subsessile, densely setose-scabrous. Inflorescence lax, elongated racemes increasing upto 10 cm in fruit, often bracteate below. Bracts minute, glandular hairy. Flowers ca. 4 mm across; pedicel ca. 1 mm, filiform. Sepals ca. 1.7 mm long, lanceolate, glandular-eglandular hairy above. Petals ca. 3.5x 1.5 mm in size, ovate, apex obtuse, yellow, glabrous. Stamens 6; filaments ca. 2 mm long, glabrous; anthers ca. 1 mm, oblong. Ovary ca. 2 mm long, sessile, sparsely glandular hairy. Capsule linear, ca. 1-2 x 0.2 cm in size, striated, many seeded. Seeds ca. 0.5-0.6 mm across, granulate, glabrous, brownish black.

Flowering period: August.

Herbarium specimens examined: NARA DESERT: *Rahmatullah Qureshi*, 10, 55, 98, 863, 964, 1070, 1167, 1170, 1221, 1240 (SALUBGH).

Field remarks: A very common herb mostly found in sandy soil.

Distribution: North and Tropical Africa, Egypt, Arabia, W. Pakistan and India (Jafri, 1973).

2. Cleome brachycarpa Vahl in DC. Prodr., I:240 (1824); Jafri in Fl. Kar., 138 (1966); Stewart in Nasir & Ali., Ann. Cat. Vasc. Pl. W. Pak. & Kashm., 295 (1972); Jafri in Nasir & Ali., Fl. Pak., 34:23 (1973). **Figure 8A**

Syn: *C. vahliana* Fresen. in Mus. Senckenb. 5 (2): 110 (1834); *C. ruta* Camb. in Jacquem., Voy. Bot., 19. T. 19 (1844); *C. moschata* Stocks ex T. Anders. in J. L. Soc., Suppl. 5 (1): 5 (1860); *C. brachycarpa* var. *longipetiolata* Sabnis in J. Ind. Bot.

Soc. 4 (1924); and var. *glauca* Blatter & Hallb. In J. Bomb. Nat. Hist. Soc. 26: 221 (1918).

An annual, erect much branched, glandular-pubescent herb up to 20 cm long. Stem woody at the base, branched mostly from the bellow, terete, viscid-pubescent; internode ca. 0.4-1 cm long. Leaves 3-5 foliate, lower leaves petiolate, upper most ternate to simple and sub-sessile, alternately arranged; leaflets ca. 5-20 x 2-8 mm in size, obovate-oblong or linear-oblanceolate, apex acute, or shortly mucronate, base cuneate, margin entire, glandular-pubescent on both sides, lateral nerves obscure, dark green; petiole ca. 2-22 mm., glandular-pubescent. Inflorescence in terminal lax racemes. Bracts 2, leafy, elliptic-oblong, apex mucronate or minutely apiculate, viscid-pubescent, ca. 4 x 2 mm in size. Flowers ca. 15-18 mm across, yellow in colour, fragrant; pedicel up to 18 mm long, glandular hairy, filiform. Sepal 4, ca. 6-7 x 2-2.5 mm in size, valvate, slightly united at the base, elliptic-lanceolate, base acute, apex mucronate, margin entire, glandular hairy out side, light green in color. Petals 4, ca. 16 x 5 mm in size, free, elliptic-oblong, base cuneate, apex mucronate, margin entire, imbricate, yellow in colour, glabrous, thickly reticulately veined, a minute scale-like appendage present at the base towards inner side. Stamens 6; filaments ca. 9-10 mm, filiform, yellow; anthers ca. 5-6 x 1 mm in size, oblong, apex apiculate-sub aristate, versatile, bilobed, extrose, yellow. Gynophore inconspicuous, ca. 1 mm long. Ovary ca. 5.2 x 1.2 mm in size, elliptic, glandular hairy; style ca. 10 mm long, glabrous, persistent, ± curved; stigma capitate. Capsules ca. 10-16 x 2-3 mm in size, oblong, inflated densely glandular hairy; beak ca. 6-12 mm long. Seeds ca. 1.5-5 mm across, sub reniform, finally rugulose, maroon-brown, shiny, glabrous.

Flowering period: November-May.

Herbarium specimens examined: NARA DESERT: *Rahmatullah Qureshi*, 54, 513, 660, 802, 968, 1022, 1028, 1037 (SALUBGH).

Field remarks: A common herb of calcarious soils.

Distribution: Tropical and N. Africa, Egypt, Arabia, Iran, Afghanistan, W. Pakistan and India (Jafri, 1973).

3. Cleome viscosa L. Sp. Pl.672. (1753); Jafri in Fl. Kar., 138 (1966); Stewart in Nasir & Ali., Ann. Cat. Vasc. Pl.W.Pak. & Kashm., 296 (1972); Jafri in Nasir & Ali., Fl. Pak., 34:23 (1973). **Figure 8C**

Syn: *C. icosandra* L., Species. Pl., 938 (1753); *Polanisia viscosa* (L.) DC., Prodr., 1: 242 (1824); P. icosandra (L.) Wight & Arn., Prodr., Fl.Pen. Ind. Orient., 22 (1834).

An annual, erect herb up to 1 meter high. Stem much branched, woody below, herbaceous above, angular, glandular to eglandular-pubescent, trichome ca. 1-2 mm long, violet below and green above; internode ca. 1-2.5 cm long. Leaves cauline and ramal, 3-5 foliate, petiolate, alternate, exstipulate, obovate, elliptic-oblong, base cuneate, apex round-obtuse, margin crenate-sub entire, green, reticulate veined, ca. 2-4 cm long, ca. 1-2 cm broad; petiole ca. 5 cm long. Inflorescence an elongated raceme, up to 25 cm. long, with corymbose flowers at the top and elongated mature fruits below, bracteate. Flowers bisexual, complete, actinomorphic, petiolate, yellowish, ca. 0.5-1 cm across; pedicel, ca. 1.3 cm long; bracts foliaceous. Sepals 4-lobed, ca. 5-6 mm long, 1-3 mm broad, free, valvate, fused below, oblong-lanceolate or elliptic, apex acute-obtuse, base acute-cuneate, margin entire, glandular hairy out side, hairs ca. 1mm long. Petals 4, ca. 11-12 mm long, ca. 2-3 mm broad, obovate, apex obtuse-rounded, base cuneate, margin entire, unequal, irregular, glabrous, the margin curved, veins visible, yellow. Stamens 9-10; filaments free, ca. 7 mm long; anthers fused to form a bundle encircling around the ovary, dorsifixed, 2-lobed, 2-thecous, splitting longitudinally, ca. 4 mm long, 1 mm broad; gynophore absent. Ovary 1-locular, 2-celled, terete, sessile, glandular-pubescent, ca. 7-8 mm long, ca. 2 mm broad; style short, ca. 1-2 mm long. Capsule terete, 3-6 cm long, ribs oblique, glandular-hairy, beak 6 mm long, dehiscing from above; seeds α, ca. 1-1.5 mm across, cleft closed, glabrous, transverse ridges prominent, concentric ones faint, dark brown.

Flowering period: July-September.

Herbarium specimens examined: NARA DESERT: *Rahmatullah Qureshi*, 7 (SALUBGH).

Field remarks: A common weed mostly found in flat habitat.

Distribution: Almost throughout tropical regions of the world (Jafri, 1973).

3. Maerua

Forssk., Fl. Aegypt.-Arab. 104 (1775); Benth. & Hook. f., Gen. Pl. 1: 108 (1862); Hook. f., Fl. Brit. Ind. 1: 170 (1872); Boiss., Fl. Or. 1: 419 (1867); Hutch., Gen. Fl. Pl. 2: 313 (1967); Jafri in Fl. Pak. 34: 15 (1973).

Shrubs, climbing or spreading, rarely small trees, unarmed, often glaucous. Leaves simple, sometimes 3-foliate, alternate or clustered on short lateral shoots, usually short petioled; stipules subulate, minute or indistinct, never spiny. Inflorescence usually corymbs, racemes, sometimes a panicle or flowers 1-4 in the axils of the upper leaves. Sepals 4, valvate, united at the base into a persistent tube lined by the receptacle or disc. Petals 4 or absent, subequal, much smaller than the calyx, inserted on the edges of the calyx-tube, often caducous. Androphore almost equal to or exceeding the receptacle, without an appendage. Stamens usually many, inserted on the torus. Gynophore long, slender, exceeding the stamens; ovary narrowly cylindrical to spheroid, 1-2-locular; ovules usually many on 2-4 parietal placentas. Fruit cylindical to spheroid, torulose or not, glabrous to pubescent, 1-∞-seeded. Seeds large, subglobose; cotyledons incumbent- convolute.

About 90 species mostly in the drier areas of Africa and old world tropics. In Pakistan, it represented 2 species (Jafri, 1973).

Maerua crassifolia Forssk., Fl. Aegypt-Arab. 104 (1775); Stewart in Nasir & Ali., Ann.Cat.Vasc.Pl.W.Pak.& Kashm., 297 (1972); S. M. H. Jafri in Nasir & Ali., Fl. Pak., 34: 16 (1973).

Syn: *M. uniflora* Vahl, Symb. Bot., 1: 36 (1790); *M. bhirtella* Chiov., Fl. Somala, 1: 83 (1929).

A large scabrous, woody, climbing shrub with pale-brown, smooth bark. Leaves ca. 2-6.5 x 1.5-2.5 cm in size, elliptic-oblong, base obtuse, rarely retuse, apex mucronate, glaucous, coriaceous; petiole ca. 0.5-1 cm long, terete. Corymbs terminal and axillary, ca. 2 cm long. Flowers greenish-white, ca. 1.5-2 cm long; pedicel ca. 1-1.5 cm long, glabrous. Bracts ca. 1-2 mm long, ovate, apex acuminate, rigid. Calyx-tube ca. 5-6 mm long, lined by tubular disc with truncate margins; lobes ca. 1.2-1.5 x 0.4-0.6 cm in size, oblong, base acut, apex mucronate with hooded apices, glabrous above, pubescent within, margins villous. Petals ca. 1 x 0.4 cm in size, ovate-lanceolate, apex acuminate, margins slightly undulate, inserted on cup-shaped disc. Stamens many, inserted on ca. 0.5 cm long torus; filaments white, filiform, glabrous, upto 2 cm long; anthers ca. 3-3.5 mm long, basifixed. Gynophore ca. 2.5 cm long, glabrous. Ovary ca. 5 mm long, cylindrical, apex truncate; style 0; stigma sessile, more or less conical, patent. Fruit ca. 5-10 cm long, constricted between the seeds, forming an elongated, twisted and knotted berry, each knot 1-seeded, yellow after ripening. Seeds ca. 3.5-5 x 3-6 mm in size, more or less globose, brown, minutely echinate.

Flowering period: November-April.

Herbarium specimens examined: NARA DESERT: *Rahmatullah Qureshi*, 607, 673, 691, 693, 698, 817, 858, 1036, 1171 (SALUBGH).

Field remarks: A rare shrub recorded from Rohri Hills.

Distribution: Tropical and N. Africa, Egypt, Sinai, Arabia, Palestine and W. Pakistan (Jafri, 1973).

4. Capparis

L., Sp. Pl. 503 (1753); Gen. Pl. ed. 5: 222 (1754); Benth. & Hook. f., Gen. Pl. 1: 108 (1862); Boiss., Fl. Or. 1: 419 (1867); Hook. f., Fl. Brit. Ind. 1: 173 (1872); Jafri in Pak. J. For., 6: 191-202 (1956); Jafri in Fl. Pak. 34: 4 (1973).

Shrubs (erect or straggling) or small tree; branchlets pubescent to stellate-tomentose, sometimes glabrescent. Leaves simple, thin or thick-coriaceous or chartaceous, usually with a pair of stipular, straight or recurved spines. Inflorescence in corymbose racemes, subumbels or in supra-axillary serial clusters, sometimes solitary. Flowers often showy, bisexual, complete, pedicellate; bracts mostly present, caduous. Sepals 4 (-5), free or nearly so, unequal, imbricate, outer pair concave and enveloping the bud, inner pair flat. Petals 4 (-5), imbricate, rather delicate, mostly obovate, clawed, usually caduous after anthesis, often subequal, receptacle (tour) flattish with a small adaxial disk; androphore absent. Stamens (5-) 20-indefinite, free, exceeding petals, glabrous, inserted on the torus at the base of the gynophore. Gynophore as long as stamens, not or hardly increesing in fruit but often thickened. Ovary globose or cylindrical, 1-celled with 2-8 (-10) placentas and few to many ovules; stigma sessile, mostly obscure. Fruit baccate on a thickened stipe (pedicel and gynophore together), with pericarp leathery or corky, rarely dehiscing, oblong, ellipsoid or globose. Seeds 1-many, embedded in pulp, obliquely reniform.

About 250 species in the tropics and subtropics of both the hemispheres, half of which are American. Six species are reported from Pakistan (Jafri, 1973).

Key to species:

1a. Plants leafless, if present then minute, linear, deciduous. Tree or small shrub.
Flowers red or orange red...**1. C. decidua**

1b. Plants leafy. Prostrate subshrub. Flowers white or creamy.......................**2. C. spinosa**

1. Capparis decidua (Forssk.) Edgew. in Journ. L.Soc., 6:184 (1862); Stewart in Nasir & Ali., Ann. Cat. Vasc. Pl. W. Pak. & Kashm., 293 (1972); Jafri in Nasir & Ali., Fl. Pak., 34: 11 (1973). **Figure 7C**

Syn: *Sodada decidua* Forssk., Fl. Aegypt-Arab., 81 (1775); *Capparis aphylla* Roth. Nov. Pl. Sp. 238 (1821); *C. sodada* R.Br. in Denh. & Clapp., Narr. Afr. Appl. 225 (1826).

A densely branched shrub or small tree up to 4 meters high. Stem erect, branched, dichotomously branched, woody bellow, herbaceous above, solid, terete, glabrous, branches green, leafless, crooked spiny; stipules thorney, nearly straight or slightly curved, yellow, ca. 1-7 mm long. Inflorescence ebracteate corymbs on short lateral shoots. Flowers ca. 1-1.8 cm long, ca. 1.1.5 cm broad, tetramerous, orange red in color, pedicel slender, ca. 1-1.4 cm long. Sepals 4, ca. 11-16 x 4-10 mm in size, arranged into two whorls, polysepalous, fused at the base, unequal, posterior one saccate, valvate, ovate-oblong, base acute, apex acute-obtuse, margin floccose-ciliate, greenish white in colour. Petals 4, ca. 18-22 x 5-11 mm in size, free, irregular, unequal, valvate, upper pair hidden in the saccate sepal, oblanceolate-ovate, apex rounded, base obtuse, margin entire, puberulous. Stamens 18, exceeding the petals; filament filiform, encircling around the ovary, red, ca. 1-1.5 cm long; anthers bilobed, basifixed, dithecous, oblong, ca. 4 mm long, 1.8 mm broad. Gynophore ca. 1.5 cm long; ovary ca. 3 mm across with a beak ca. 1mm long, globose; style ca. 1.5 mm long. Fruit globose, ca. 1-2 cm in diam., slightly beaked, glabrous, smooth, fleshy, deep red when ripe and thin pericarp. Seeds ca. 2-6 mm across, reniform.

Flowering period: April and August-September.

Herbarium specimens examined: NARA DESERT: *Rahmatullah Qureshi*, 22, 670, 678, 1168, 1192 (SALUBGH).

Field remarks: A common shrub or small tree distributed in flat and interdunal valleys.

Distribution: North and Tropical Africa, Arabia, eastward to India (Jafri, 1973).

2. Capparis spinosa L. Sp. Pl. 503 (1753); Stewart in Nasir & Ali., Ann. Cat. Vasc. Pl. W. Pak. & Kashm., 294 (1972); Jafri in Nasir & Ali., Fl. Pak., 34: 6-7 (1973). **Figure 7D**

Syn: *C. nepaulensis* DC., Prodr. 1: 246 (1824); *C. marrayana* Grah., Catl. Pl. Bombay. Vic. 9 (1839).

A diffuse, semi erect, perennial shrub; stem hard, woody, branched, branches zigzag often twisted, glaucous, fleshy, up to 50 cm long, internode 1.5-2.5 cm long. Leaves simple, alternate, oblong-suborbicular or broadly ovate, apex retuse to acute, base rounded and cuneate, margin entire, 1-4 cm long, ca.1-3 cm broad, fleshy and cartilaginous, glaucescent; petiole 6-10 mm long; stipulate, stipular spines in a pair, hooked, yellow, 3-7 mm long. Flowers solitary and axillary, ca.3 x 2 cm in size, bisexual, zygomorphic, tetramerous, whitish, receptacle subconical, pedicels 2-5 cm long, becoming thickened in fruit. Sepals 4, subequal, posterior sepal not galeate, ca.1.2-1.5 cm long, ca.0.8 cm broad, polysepalous, imbricate, ovate-oblong, apex acute-obtuse, margin entire. Petals 4, ca. 1.5-1.7 cm long; ca. 0.7-1 cm broad, obovate-oblong, apex retuse rounded, margin entire, sinus pointed ciliate (pubescent) margin and apex. Stamens numerous (57), ca. 1.5-2 cm long, exerted, arranged on the receptacle of the pedicel, base of the filament white, anthers pink or red in coloured, filaments curved in zigzag, anthers bilobed, dorsifixed, lanceolate, dehiscing longitudinally, ca. 4-6 mm long and ca. 0.5-1 mm broad. Carpel 1, arranged on gynophore, ca. 2 cm long; stigma discoid; style subsessile; ovules numerous; pistil ca. 0.5 cm long, ca. 5 mm broad, obovate. Fruit 1.5-3 cm long and ca .0.5-1 cm broad, ellipsoide-obovate, ribbed, red when

ripe (internal structure), often splitting into four valves from the apex when dried, berries on thickened stipe. Seeds globose, smooth, brown ca.4 mm across in size.
Flowering period: April-May.
Herbarium specimens examined: NARA DESERT: *Rahmatullah Qureshi,* 1252 (SALUBGH).
Field remarks: A rare herb of flat habitat.
Distribution: S. Europe eastward to Australia (Jafri, 1973).

5. Cadaba

Forssk., Fl. Aegypt.-Arab., 67 (1775); Benth. & Hook. f., Gen. Pl. 1: 108 (1862); Boiss., Fl. Or. 1: 417(1867); Jafri in Pak. J. For. 8(2): 204 (1958); Hutch., Gen. Flow. Pl. 2: 315 (1968); Jafri in Fl. Pak. 34: 11 (1973).

Unarmed shrubs or small trees. Leaves simple or 3-foliate, petiolate, chartacoeus; stipules minute and inconspicuous. Flowers in terminal racemes or axillary solitary to fascicled, pedicelled. Sepals 2+2, free, petaloid, subequal, upper and lower valvate, enclosing laterals. Petals 2 or 4, clawed. Disc-appendix about as long as the petal claw, spathulate or infudibular, apex generally petaloid and more or less toothed. Stamens 4-6 (-8), exserted, spreading, unilateral on a short androphore or gynophore. Gynophore elongated. Ovary 1 (-2)-locular with many ovules on 2 (-4) parietal placentas. Fruit oblong or subcylindrical, slightly torulose, coriaceous, tardily dehiscent or indehiscent; internal tissues surrounding the seeds often orange coloured; seeds subglobose. About 30 species mostly distributed in africa and a few in Arabia and eastwards to Malesia and northern Australia. Of these only 2 species are recorded from W. Pakistan (Jafri, 1973).

Cadaba farinosa Forssk., Fl. Aegypt-Arab.68 (1775); Stewart in Ali and Nasir, Ann. Cat. Vasc. Pl. W. Pak. & Kashm., 293 (1972); Jafri in Nasir and Ali, Fl. Pak., 34: 12-14 (1973). **Figure 7B**

An erect straggling perennial shrub up to 1.5 m tall. Stem woody, branched, terete, glaucous to puberulent, young twigs farinose with small white waxy sessile scales, becoming glabrous with age; internode ca. 1-6 cm long. Leaves ca. 1-3 x 0.3-0.9 cm in size, uni-trifoliate, alternate, elliptic-oblong or sub ovate, base more or less oblique and cuneate, apex obtuse-apiculate, margin entire, farinaceous; petiole ca. 3-6 mm long, puberulent. Inflorescence shortly subcorymbose racemes, few-flowered. Flowers ca. 1.5-2 cm long, usually creamy yellow in colour; pedicel ca. 18 mm long. Sepals 4, ca. 0.8-0.9 x 0.3-0.4 cm in size, outer sepals oblong-ovate, concave, apex acute, margin entire, covered with scales, inner sepals flattish, less acute, green. Petals 4, ca. 12 x 3-5 mm in size, greenish, spathulate, claw ca. 7 mm. Disk-appendix ca. 10 x 2 mm in size, fimbriate, trumpet shaped, apex petaloid, more or less toothed. Androphore ca. 9 mm long. Stamens 4-5, ca1-1.8 cm long, filaments fused below, free and spreading above. Gynophore ca. 1.8 cm. long, covered with scales. Ovary ca. 10 x 2 mm in size, elliptic-oblong, pubescent; style 0; stigma conical. Fruits narrowly cylindrical, obscurely torulose, farinaceous, many seeded; seeds ca. 2.5 mm in diam., surrounded by orange flesh.
Flowering period: October-December.

Herbarium specimens examined: NARA DESERT: *Rahmatullah Qureshi*, 619, 647, 804, 1172 (SALUBGH).
Field remarks: A common shrub found in gravelly soils.
Distribution: Arabia, Egypt, Tropical Africa and W. Pakistan (Jafri, 1973).

6. Dipterygium

Decaisne in Ann. Sc. Nat., ser. 2.4: 66 (1835); Boiss., Fl. Or. 1: 417 (1867); Hook. f., Fl. Brit. Ind. 1: 164 (1872), under Cruciferae; Jafri in Fl. Pak. 34: 17 (1973).

Undershrub with slender bushy branches terminating in short, lax, usually bracteate, racemes, glabrous, glaucous or rarely glandular. Leaves small, somewhat fleshy, sharply petioled or subsessile. Flowers minute. Sepals 4, free, narrowed at the base. Petals 4, sharlply clawed. Androphore and gynophore indistinct or absent. Stamens 6, equal and not tetradynamous. Ovar more or less 4-winged (wings almost disappearing in fruit), 1-loculed with 1-2-ovules; style elongated with capitate stigma. Fruit small, ellipsoid, indehiscent, subcoriaceous, 1-2-seeded, hardly winged. Seeds not winged.

A monotypic genus with only 1 species in Egypt, Arabia, Sudan and W. Pakistan (Jafri, 1973).

Dipterygium glaucum Decne. in Ann. Sc. Nat. Ser. 2 (4): 66 (1935); Stewart in Nasir & Ali., Ann.Cat. Vasc. Pl.W.Pak. & Kashm., 307 (1972); Jafri in Nasir & Ali., Fl. Pak., 34: 17 (1973).

Syn: *Pteroloma arabicum* Hochst. ex Steud. in Fl. 24. 1: 42 (1841); *Isatis spartioides* Edgew. ex. Hook. f. & Thoms. in J. L. Soc.. 5: 179 (1861); *Dipterygium scabrum* Decne. ex. Boiss. Fl. Orient. 1: 417 (1867).

A small perennial bushy undershrub up to 45 cm high. Stem densely branched, branches straight, rigid, divericate, slender, terete, grooved ending in lax racemes. Leaves ca. 0.5-3 x 0.2-0.8 cm in size, elliptic-ovate, base oblique, apex apiculate, margin entire, chartaceous; petiole ca. 0.5-1.5 cm long, rough with viscid glands. Racemes terminating the branches. Flowers yellow, ca. 4 mm across; pedicel ca. 0.5-1 mm long, filiform, glabrous. Bracts ca. 2.5 mm long, scaly. Sepals ca. 1 mm long, valvate, ovate, apex acute with few glandular outgrowths, margin subentire. Petals ca. 3-3.5 x 2 mm in size, ovate, apex obtuse-subretuse, base shortly clawed, margin scarious; keel ca. 0.5 mm long, basally parallel veined. Stamens basifixed; filaments ca. 2 mm long, filiform, slightly tuberculate, greenish in colour; anthers ca. 1 x 0.5 mm in size, ovate-oblong, base truncate, margin tuberculate. Ovary ca. 1.5 x 0.9 mm in size, elliptic, base cuneate, margin membranous; stigma capitate. Fruit ca. 4 mm long, ellipsoidal, indehiscent, transversely wrinkled, pendulous, compressed, surrounded by a membranous wing, 1-celled, 1-seeded. Seeds erect, not margined.
Flowering period: August-November.
Herbarium specimens examined: NARA DESERT: *Rahmatullah Qureshi*, 39, 125, 247, 608, 612, 669, 677, 703, 815, 941, 1051, 1059, 1273, 1276 (SALUBGH).
Field remarks: A very common bush in the extreme arid regions of the whole Nara Desert on low sand dunes and gravelly soils.
Distribution: Arabia, Egypt, Sudan and W. Pakistan (Jafri, 1973).

26. Brassicaceae

Annual to perennial herbs, rarely woody or shrubby. Leaves alternate or spirally arranged, genrally simple and exstipulate, basal often rosulate. Hairs usually simple or branched, sometimes appressed or absent. Inflorescence usually racemose, corymbs. Flowers usually bisexual, hypogynous, actinomorphic, pedicellate and ebracteate. Sepals 4, erect-closed to spreading-open, in two decussate pairs, inner sometimes saccate at the base. Petals 4 (very rarely suppressed), alternating with the sepals, free, often with distinct claw, usually yellow, white or lilac, cruciform. Stamens 6 (rarely 4, 2 or even 1 due to suppression, very rarely more than 6), tetradynamous, outer pair short and inner 4 long; filaments sometimes winged, toothed or appendaged; anthers 2-thecous (very rarely 1-thecous) and fertile. Lateral nectar glands at the base of the 2 outer short stamens, almost always present and of various shapes; middle glands present or absent. Ovary 2-carpellate, syncarpous, 1-2-celled or with 2-many superimposed celles, 1-∞-ovuled on two parietal placentas; septum false being formed by the placental outer growth; style simple, sometimes absent; stigma often capitate, entire to 2-lobed. Fruit short (silicula) or long (siliqua), dehiscent or indehiscent, usually opening from below by 2 valves which often leave the seeds attached to replum and the false septum, sometimes breaking transversely into 1-few seeded cells, rarely seeds also developing in an indehiscent beak below the style or stigma; valves membranous to coriaceous and thick, plain or inflated, sometimes keeled, winged or appendaged, glabrous or hairy, with 1-many parallel veins (generally mid-vein prominent). Seeds 1-many in 1-2-rows, smooth, granular or reticulated, rarely longitudinally straighted, sometimes winged, usually mucilaginous when wet, exalbuminous; radicle accumbent (when it is bent round so as to lie along the edges of the cotyledons i.e. embryo, plurorhizous) or incumbent (when it lies on the face of 1 cotyledon i.e. embryo notorhizous); cotyledons narrowly linear-spathulate, sessile or stalked, rarely plicate or conduplicate (longitudinally folded).

A large family of about 350 genera and nearly 3000 species primarily of the temperate regions and cooler climate. Of these only 92 genera and ca. 250 species are reported from Pakistan which includes 5 genera and 14 species known from cultivation only (Jafri, 1973).

Key genera:

1a. Erect, canescent herbs; branches densely hairy; fruit dehiscent......................**1. Farsetia**

1b. Prostrate, foetid herbs; branches glabrous; fruit indehiscent......................**2. Coronopus**

1. Farsetia

Turra, Diss. Farsetia, 1, t. 1. (1765); Benth. & Hook. f., Gen. Pl. 1: 140(1862); Boiss., Fl. Or. 1: 157(1867); Jafri in Notes Roy. Bot. Gard. Edin., 22: 209-216 (1957); Jafri in Nasir & Ali, Fl. Pak. 55: 108 (1973).

Undershrub or perennial, rarely annual herbs, often with woody bases and shrubby in appearance, much branched, erect or spreading, hoary with 2-partite appressed white hairs. Leaves linear-oblong, sessile. Racemes lax, ebracteate, elongated in fruit. Flowers small or large, pale yellow, mauve, orange or whitish, shortly pedicelled. Sepals erect, obtuse, not saccate, hoary. Petals slightly longer to about twice as long as the sepals, oblong-linear, clawed, apex rounded, margins often crimped, and limb coiled inwards. Stamens 6; filaments linear; anthers oblong-linear. Lateral nectar glands horseshoe shaped, open towards the inner side; middle absent. Ovary oblong, 10-40-ovuled, hairy; style distinct. Siliquae oblong, compressed, narrow to broad (rarely suborbicular), canescent with 2-partite appressed hairs, 2-locular, dehiscent; gynophore obscure to very short but distinct; valves with distinct mid-vein; septum membranous, with a distinct mid-vein and obscure reticulate venetation. Seeds 1-2-seriate, orbicular, winged, compressed; radicle accumbent.

About 10 species in S.W. Asia, C. and N. Africa, India and Pakistan. In Pakistan it represents 4 species (Jafri, 1973).

Key to species:

1a. Flowers ca. 4.5-5 mm across, white or pink;
sliqua ca. 11-32 x 0.3-0.4 cm..**1. F. hamiltonii**

1b. Flowers 1 cm across, orange-yellow to pale purple;
siliquae 20-48 x 3-4.5 mm...**2. F. jacquemontii**

1. Farsetia hamiltonii Royale, Ill. Bot. Himal. 1: 71 (1834); Stewart in Ali and Nasir, Ann. Cat. Vasc. Pl. W. Pak. & Kashm., 316 (1972); Jafri in Nasir & Ali, Fl. Pak., 55: 111 (1973).

Syn: *F. linearis* Decne.; *Erysimum remotiflorum* O.E. Schulz in Notizblat. Bot. Gart. Berlin, 11: 226 (1931).

Annual, canescent herb, more or less at the base, densely adpressed with 2-partite, white hairs. Stem upto 30 cm high, terete, slender, virgate, somewhat tough. Leaves linear, ca. 1-5 x 0.1-0.4 cm in size, sessile, entire, strigose; petioles ca. 1-3 mm long. Inflorescence 2-20-flowered, lax, ebracteate racemes; buds ca. 3-5 mm long, ellipsoid; pedicels ca. 2-3 mm long, ascending. Flowers ca. 4.5-5 mm across, white or pink. Sepals ca. 4-5 mm long, oblong, sub-obtuse, equal, strigose, the lateral two not succate. Petals ca. 6-7 mm long, oblong, base cuneate, apex rounded. Stamens ca. 3.8-4.5 mm long; anthers obtuse. Stigma 2-lobed. Sliqua ca. 11-32 x 0.3-0.4 cm in size, oblong, compressed, nearly rounded towards both ends, clothed with 2-partite adpressed hairs; valves 1-veined, subtorulose. Seeds 1-seriate, ca. 3 mm across, orbicular; septum membranous with distinct midveins.

Flowering period: August-November.

Herbarium specimens examined: NARA DESERT: *Rahmatullah Qureshi*, 96 (b), 112, 634, 729, 1030, 1142 (SALUBGH).

Field remarks: A common herb mostly found in flat and interdunal valleys.

Distribution: Endemic to the desert area of India and Pakistan (Jafri, 1973).

2. Farsetia jacquemontii Hook. f. & Thoms. in Journ.L. Soc. Bot. 5:148 (1861); Stewart in Ali and Nasir, Ann. Cat. Vasc. Pl. W. Pak. & Kashm., 316 (1972); Nasir in Nasir and Ali, Fl. Pak., 71: 21-23 (1974).

Syn: *Arabis heliophila* DC.; *Farsetia macrantha* Blatter & Hallb.; *Heliophila incana* Burm.f.

Farsetia jacquemontii subsp. **jacquemontii**

Small shrubs usually with woody base, 15-45 cm tall, erect or subspreading, often much branched, hoary with bipartite appressed white, rough hairs; branches usually herbaceous. Leaves 1-6 cm long, (1-) 3-7 (-12) mm broad, linear, sessile. Racemes ca. 15-flowered, lax and up to 20 cm long in fruit. Flowers large, ca. 1 cm across, orange-yellow to pale purple; pedicel up to 5 mm long in fruit. Sepals 6-9 mm long, c.2 mm broad. Petals 10-15 (-20) mm long, c.2 mm broad, oblong-linear, clawed, often inrolled. Stamens ca. 6-8 mm long; anthers c. 2.5 mm long. Siliquae 20-48 x 3-4.5 mm across, oblong-linear, compressed; valves membranous, often pale looking as compared to whitish branches and leaves; style c. 1 mm or less long, slightly thickened with ± distinct somewhat conical bilobed stigma. Seeds usually sub-biseriate (2 or 1 seriate not lacking), almost orbicular, 2.5-3 mm in diam.

Flowering Period: March-June.

Herbarium specimens examined: NARA DESERT: *Rahmatullah Qureshi,* 1351 (SALUBGH).

Field remarks: A common herb mostly found in flat and interdunal valleys.

Distribution: India, Pakistan and Afghanistan (Jafri, 1973).

2. Coronopus

Boehmer in Ludwig; Defin., Gen. Pl., ed. 3: 226 (1760).-*nom. cons.*; Jafri, Fl. Kar., 140 (196) *Senebiera* DC.

Annual or biennial, procumbent or suberect, glabrous or hairy with simple hairs or minute tubercles. Leaves pinnatisect or deeply dissected. Racemes short, capitate, ebracteate, axillary or terminal, sometimes apparently leaf opposed. Flowers minute, shortly pedicelled. Sepals sub-spreading, ovate, not saccate at the base, often lilac or pinkish. Petals minute, whitish, scarcely exceeding the sepals, or absent. Stamens 6, or only 4 or 2 by abortion; filaments simple; anthers minute, subglobose. Nectar glands minute dot-like, between the stamen bases. Ovary subglobose-bilobed, 2-ovuled; stigma minute, almost sessile. Siliculae short, usually with 2 hemispherical lobes breaking away with 1 seed each; valves thick, reticulately pitted or tuberculated; septum very narrow; radicle incumbent.

A genus of 10 species, mostly Eurasian; only 1 occurs in Pakistan.

Coronopus didymus (L.) Smith, Fl. Brit. 2: 691 (1804); Stewart in Ali and Nasir, Ann. Cat. Vasc. Pl. W. Pak. & Kashm., 306 (1972); Jafri, Fl. Kar., 140 (196)

Syn: *Lepidium didymum* L., Mantiss., 92 (1767); *Senebiera didyma* (L.) Pers., Syn., ii: 185 (1805); *Senebiera pinnatifida* DC., Mem. Soc. Nat. Par., 144 (1799).

Annual or biennial, 15-30 cm long, diffuse or procumbent, often somewhat foetid, branched mostly from below, glabrous or hairy. Basal and lower leaves rosulate, pinnatisect, stalked, 6-10 jugate, 5-10 cm long; upper leaves similar or pinnatifid and only 3-5-jugate, 1.5-3 cm long, sessile or subsessile; lobes sinuate toothed, often only on one side. Racemes 30-60-flowered, dense, up to 8 cm long (usually much shorter) in fruit. Flowers minute, c.l mm across; pedicel short, 3-4 mm long in fruit, filiform. Petals present or O. Stamens only 2 or 4, but only 2 fertile. Siliculae shorter than the pedicel, c. 1.5 mm long, 2.5 mm broad, broader than long, bilobed; valves globose, reticulately rugose, glabrous; septum very narrow, inconspicuous, c. 0.3 mm broad; seed c. 1.5 long, reniform, reddish-brown.

Flowering Period: March June.

Herbarium specimens examined: NARA DESERT: *Rahmatullah Qureshi*, 1352 (SALUBGH).
Field remarks: A common weed of moist places.
Distribution: Perhaps a native of S. America, but widely introduced almost throughout the world (Jafri, 1973).

27. *Moringaceae*

Deciduous, unarmed trees, with gummy bark and soft white wood containing myrosin cells. Leaves large, alternate, 2-3 pinnate, imparipinnate; pinnae and pinnules opposite; stipules and stipulues lacking or reduced to stipitate glands at the base of the petiole or white or white streaked with red, bisexual, zygomorphic. Sepals 5, united forming a tube borne on a very short hypanthium, lobes cup-shaped, unequal, deciduous, imbricate, petaloid, reflexed. Petals 5, free, unequal; posterior 2 smaller, lateral ascending, anterior one the largest. Disk lining the hypanthium, flat or saucer shaped. Stamens 5, functional, declinate, unequal in length, antipetalous, alternating with the same number of staminodes, inserted on the margins of the disk; filaments free, villous and thickened at the base; anthers monothecous, dorsifixed, dehiscence longitudinal, forming a head through which style protrudes at dehiscence. Carepls 3, united; ovary oblong, stipitate, superior, curved, villous, unilocular with 3 parietal placentas; ovules numerous in 2 series, pendulous; style 1, tubular, slender, with a single truncate stigma, perforated in the centre. Fruit an elongated capsule, rostrate, 3-valved, torulouse, unilocular, corky or pitted within. Seeds large, 3-winged or wingless, exalbuminous; embryo straight; cotyledons 2, rarely 3, thick; radicle short.

A family with a single genus and ca. 10 species, native of old world tropics, widely introduced. In Pakistan only 2 species found (Qaiser, 1973).

Moringa

Adans., Fam. Pl. 2: 318. 579 (1763); Hook. f., Fl. Brit. Ind. 2: 45 (1876); Jafri, Fl. Kar. 142 (1966); Qaiser in Fl. Pak. 38: 1 (1973).

Characters same as that of the family. It is represented by only 2 species in W. Pakistan (Qaiser, 1973).

Moringa oleifera Lamk. Encycl. 1: 389 (1785); Stewart in Ali and Nasir, Ann. Cat. Vasc. Pl. W. Pak. & Kashm., 332 (1972); M. Qaiser in Nasir & Ali., Fl. Pak., 38: 1 (1973).

Syn: *Gualandinia moringa* L., Sp. Pl. 381 (1753); *Moringa edulis* Medic., Theod. 48 (1786); *M. parviflora* Noronha in Verh. Batav. Gen. 5 (ed. 1): Art. 4. 20 (1790); *M. pterygosperma* Gaertn., Fruct. 2: 314 (1791); *Hyperanthera moringa* Roxb., Fl. Ind. 2: 368 (1824); *Moringa polygona* DC., Prodr. 2: 478 (1825); *M. zeylanica* Pers., Syn. 1: 461 (1830).

A medium sized tree up to 5 m high with younger parts tomentose. Leaves alternate, 3-pinnate; rachis thickened and articulated at the base; pinnae and pinnules opposite, deciduous, rachis articulased and with solitary gland at each articulation. Leaflets 6-9 pairs, ca. 12-25 x 6-12 mm, the lateral elliptic, the terminal obovate, slightly larger than the laterals;

petiole short. Flowers yellowish in large, puberulous, spreading panicles; bracts linear, ca. 3 mm long. Calyx cup-shaped, 5-cleft; the lobes ca. 13-15 x 4 mm in size, linear-lanceolate, reflexed, puberulous above. Corolla ca. ca. 15-20 x 7-10 mm in size, spathulate, unequal, veined. Stamens 5, fertile; staminodes 5, alternating with stamens; filaments dilated and villous at the base. Ovary oblong, villous; style cylindric. Capsules ca. 15-30 x 1.5-2 cm in size, 3-gonous, 9-ribbed, many-seeded, beaked. Seeds ca. 2.5 cm long, globose, 3-angled, with 3 lateral wings which unite at the apex.

Flowering period: February-March.

Herbarium specimens examined: NARA DESERT: *Rahmatullah Qureshi*, 1256, 1313 (SALUBGH).

Field remarks: A rare tree mostly found near settlement.

Distribution: Perhaps indigenous in the sub-Himalayan tracts but commonly cultivated in the Punjab plains, Sindh, Baluchistan and N.W.F.P. and elsewhere throughout India and many other tropical countries (Qaiser, 1973).

28. Primulaceae

Perennial or annual herbs; branchlets angular or winged. Leaves usually entire, mostly opposite or whorled, sometimes basal, exstipulate. Inflorescence panicle, umbel, heads or solitary. Flower actinomorphic, bisexual, 5-merous. Calyx of 5 sepals, persistent. Corolla 5 lobed, rotate, campanulate, contorted. Stamens 5, opposite the corolla-lobes; filaments hairy or glabrous; anthers 2-celled. Ovary superior, globose, unilocular, ovules numerous, style erect. Fruit usually 5-valved capsule, dehiscing by teeth at the tip, or a pyrix, in which the top lifts off like a lid. Seeds minute, numerous.

Anagallis

L., Sp. Pl. ed. 1.148 (1753); Nasir in Nasir & Ali., Fl. Pak., 157 (1984).

Annual or perennials. Stem erect to procumbent or decumbent; diffuse. Leaves opposite or alternate. Flowers solitary axillary, red (or shades of it), blue, rarely white. Corolla rotate or funnel shaped, the lobes variously incised at the apex and fringed with glands or not. Pyxidium globose, operculate. Seeds many, angled, vesiculose.

Anagallis arvensis var. **coerulea** (L.) Gouan, Fl. Monsp. 2930 (1765); Nasir in Nasir & Ali., Fl. Pak., 157: 94-96 (1984). **Figure 17D**

Syn: *Anagallis coerulea* L., Amoen. Acad. 4: 479 (1759).

Herb, up to 10-24.5 cm in length, winged, glabrous, weak. Leaves simple, opposite, obovate, amplexicaule, entire at margin, bluntly obtuse at apex, ca. 1.8x1 cm in size. Flower ca. 8 mm across, actinomorphic, complete, bisexual, pedicel ca.2- 4.3 cm long. Sepal 5, united below, lanceolate – linear, dotted at margin, apex acuminate, imbricate, persistent ca.3.5x 1mm. Petal 5, red color at base, rest blue, margin uneven, ciliate with glandular hairs, glandular reddish, in color, spherical ca.5x3.2mm. Stamens 5, epipetalous, hairy(monoliform) from base to anther, basal hair whitish upper reddish in color, stamen born on disc which cover basal part of ovary attached with lower part of petalsfilament ca.1.5-2 mm, anther ca.

0.5mm, Ovary globose, with few glandular hairs, ca.0.7 mm in diameter, style ca. 2x0.2 mm in size, stigma papillose, ca. 0.4 mm in length, Seed brown.

Flowering period: February-March.

Herbarium specimens examined: NARA DESERT: a weed, *Rahmatullah Qureshi*, 1353 (SALUBGH).

Field remarks: A common weed of moist places.

Distribution: Europe, W. Asia, India, Pakistan (Bhandari, 1978).

29. Neuradaceae

An annual prostrate, wooly, desert herb. Leaves alternate, petiolate, usually with minute stipules or exstipulate. Flowers often small, sometimes large and showy, solitary or in pairs. Calyx-lobes 5, lobes alternating with 5-spiny bracteoles. Petals 5, small. Stamens 10. Carpels 3-10, slightly united at the base and adnate to the throat of calyx-tube (hypanthium) tapering into slightly exserted, persistent styles; ovary inferior, 5-10-loculed; ovule solitary per cell; stigma minute, capitate. Fruit dry, woody, flattened, orbicular, spiny or winged, few seeded. Seeds horizontal, non endospermic, retained in the hard woody, discoid fruit, perforated by the plumule upwards and by the radicle downwards and persists as a collar around the hypocotyle.

A small family of 3 genera and about 10 species, mostly of African desert (Saharo-Sindian region). In Pakistan it is represented by 1 genus and 1 species (Marwat & Siddiqui, 1997).

Neurada

L., Sp. Pl. 441 (1753); Gen. Pl. ed. 5: 199 (1754); DC., Prodr., 2: 548 (1825); Boiss., Fl. Or. 2: 753 (1872); Hook. f., Fl. Brit. Ind., 2: 368 (1878); Marwat & Siddiqui in Fl. Pak., 198: 1 (1997).

Figaraea Viv., Pl. Aegypt. Dec. 11 (1830).

An annual, wooly, prostrate, unarmed, depressed branched herb, with about 18 cm long root. Leaves alternate, petiolate, simple, ovate, lobed; stipules minute or obsolete. Flowers small, axillary, solitary, pedicelled. Calyx-tube flat, dilated, at length conic, spiny and forming with the ripe carpels as orbicular disk; lobes 5, triangular; bracteoles 5, subulate. Petals 5, small, exceeding the calyx, greenish white in colour. Stamens 10, inserted in the contracted mouth of the calyx; filaments subulate. Carpels 10, whorled, united with one another and with the calyx, tapering into slightly exserted; ovule solitary per carpel; styles subulate, spinescent, persistent. Fruit a dry and compressed cone with spinous margins, indehiscent, discoid or 5-gonous, hard, woody, echinate above, smooth and wooly below. Seeds curved, germinating within the carpels, exalbuminous.

Neurada procumbens L. Sp. Pl. 441 (1753); Jafri in Fl. Kar., 145 (1966); Stewart in Ali and Nasir, Ann. Cat. Vasc. Pl. W. Pak. & Kashm., 353 (1972); Marwat & Siddiqui in Nasir & Ali., Fl. Pak., 198: 1-2 (1997).

Syn. *Figaraea aegyptica* Viv., Fl. Aeg. Dec.tab. 1 (1831).

An annual procumbent, tomentose herb up to 30 cm. Stem much branched from the base, terete, herbaceous, branches procumbent or diffuse, densely wooly. Leaves ca. 0.5-2.5 x 0.4-1.5 cm in size, ovate-oblong, with 1-3 (-4) obtuse lobes on each side, base cuneate or rounded, tomentose; hairs short and long, simple and branched; petiole ca. 0.3-1.2 cm long, densely wooly. Stipules 2, ca. 0.3 mm. Flowers ca. 6 mm across, axillary, solitary; pedicel ca. 1-2 mm, slender, densely hairy. Receptacle ca. 5-6 mm across. Calyx-tube flat, spiny, forming a disc with the mature carpels; outer teeth triangular, the inner ones lanceolate, apex acute, ca. 3 x 1-2 mm in size. Petals ca. 2-4.5 x 1.5-2 mm in size, obovate or oblanceolate, base cuneate, apex rounded, margin entire, prominently veined. Stamens ca. 2 mm long; filaments slender, base broad, glabrous, ca. 1.5 mm long; anthers ca. 0.5 mm across, broadly ovate, apex retuse. Styles ca. 2.5 mm, curved, sparingly tomentose, accrescent and hard in fruit; stigma minute, capitellate. Fruit ca. 1-1.5 cm across, orbicular, dry, convex above, flat below, persisting as a collar around the young roots of the seedling even after the plant is fully grown. Seeds ca. 2.5 x 1.5 mm in size, curved, horn like with tapering distal end; testa dark brown, shiny.
Flowering period: September-December.
Herbarium specimens examined: NARA DESERT, Dunal areas, *Rahmatullah Qureshi*, 109, 612, 622, 655, 659, 1013,1020, 1034, 1041, 1155, 1179, 1184, 1222 (SALUBGH).
Field remarks: A less common herb mostly found in crest habitat.
Distribution: Mediterranean, N. Africa, Arabia, Palestine, Syria, Iraq, N. Iran, Afghanistan, Pakistan and NW Indian desert (Marwat & Siddiui, 1997).

30. Mimosaceae

Trees, shrubs or very rarely herbs. Leaves 2-pinnate, rarely 3-pinnate; stipules present, sometimes spinescent. Inflorescence generally a spike or umbel, rarely racemose or globose umbel; bracts small. Sepals usually 5, imbricate or valvate, generally fused together, toothed or lobed. Petals usually 5, valvate, free or fused, corolla lobed, hypo- or slightly perigynous. Stamens few to many, free or monadelphous or adnate to the base of corolla tube; anthers versatile, often crowned by a deciduous gland. Pistil monocarpellary; ovary 1-locular; ovules mostly numerous; style filiform; stigma terminal. Fruit dehiscent or indehiscent, sometimes schizocarpic. Seeds mostly ovate to orbicular; aril rarely present.

A family of about 56 genera and nearly 2800 species mainly distributed in tropics and subtropics of the world. In Pakistan 11 genera and 49 species are recorded. Of them 4 genera and 18 species are native and the rest are introduced (Ali, 1973).

Key to genera:
1a. Stamens 10; anthers gland-tipped.. **1. Prosopis**
1b. Stamens indefinite, more than 10; anthers not gland-tipped.......................... **2. Acacia**

1. Prosopis

L., Syst. Nat. ed. 12. 2: 282, 293 (1767); Mant. 10, 68 (1767); S.I. Ali, Fl. Pak. 36: 1-41 (1973).

Unarmed or armed trees or shrubs with prickles, axillary spines or spinescent stipules. Leaves 2-pinnate; pinnae 1-2 pairs; leaflets few or many paired; stipules small or absent. Petiolar or jugal glands small or obscure. Inflorescence axillary cylindric spikes or globose heads. Flowers 5-merous. Calyx shortly dentate. Petals connate below the middle. Stamens 10, free; anthers with or without a deciduous gland. Ovary sessile or stpitate; ovules numerous. Fruit linear, thick and compressed, straight or falcate, indehiscent; mesocarp spongy. Seeds ovate, compressed.

A genus of about 45 species distributed in tropical and subtropical regions.

Key to species:

1a. Internodes with conical prickles. Petals and ovary glabrous. Podscylindrical, torulose.. **1. P. cineraria**

1b. Internodes without prickles. Petals and ovary villous within. Pods compressed.. **2. P. juliflora**

1. Prosopis cineraria (L.) Druce in Rep. Bot. Club. Brit. Isles., 1913; Stewart in Nasir & Ali., Ann. Cat. Vasc. Pl. W. Pak. & Kashm., 383 (1972); Ali in Nasir & Ali., Fl.Pak., 36: 29 (1973).

Syn: *Mimosa cineraria* L., Sp. Pl., 2:1500 (1763).

A large tree up to 14 cm tall; bark grey, rough with deep fissures or cracks. Branches slender, glabrous, armed with somewhat compressed, straight, conical, scattered prickles, ca. 3-8 mm long. Leaves 5.5-7 cm long; petiole ca. 3-10 mm long; pinnae usually 4-6; leaflets 11-14, ca. 0.4-1.2 x 0.2-0.3 cm in size, elliptic-oblong, base oblique, apex obtuse-mucronate, margin subentire and sparsely ciliate, 1-nerved below, puberulent on both surfaces; petiolule ca. 0.5-1 mm long, pulvinous, green, glabrous. Stipular spines alternate to subopposite, ca. 3-5 mm long, apex tapering, base truncate and pinkish in colour. Inflorescence ca. 5 x 0.6-0.7 cm in size; peduncle ca. 5 mm long. Flowers yellow, ca. 4 x 2 mm. Calyx-tube campanulate, ca. 1-1.2 mm, yellowish. Corolla-lobes ca. 3.5 x 0.8 mm, elliptic, base cuneate and hyaline, apex acute, margin subentire, 1-nerved, yellow. Stamens 10; filaments ca. 2.5-3.5 mm long, filiform, papilose, yellowish; anthers ca. 1 x 0.8, oblong, dorsifixed, tipped with deciduous glands. Ovary ca. 3.5 mm long, cylindric or terete, recurved, greenish, papilose; style ca. 0.5 mm, curved, glabrous. Pods ca. 6-11 x 0.5 cm, linear-moniliform, straight-curved, apex apiculate, pendulous, pale green when unripe, brown when ripened, 10-15-seeded.

Flowering period: April-August.

Herbarium specimens examined: NARA DESERT: *Rahmatullah Qureshi*, 1264, 1270, 1287, 1293 (SALUBGH).

Field remarks: A common tree of calcarious and sandy soil.

Distribution: West Pakistan (Punjab, Sindh & Baluchistan), India (Punjab, Pajputana, Bombay, Madras), Afghanistan, Persia and Arabia (Ali, 1973).

2. Prosopis juliflora (Swartz) DC., Prodr., 2:447 (1825); Stewart in Nasir & Ali., Ann. Cat. Vasc. Pl. W. Pak. & Kashm., 383 (1972); Ali in Nasir & Ali., Fl.Pak., 36: 31 (1973).

Syn: *Mimosa juliflora* Swartz, Prodr. Veg. Ind. Occ. 85 (1788).

A tall shrub 3-4 m tall. Stem hard, woody, fairly branched; branchlets erect to spreading, solid, angular, puberulous, greyish green; internode ca. 1-5.5 cm. Stipular spine 1 on each node, ca. 0.5-5 cm, straight, falcate, apex sharp and brown, younger one puberulous.

Leaves 2-pinnate, ca. 9-16 cm long; petiole pulvinous, ca. 1-2.5 cm, glabrous; petiole ca.1-2 cm, base swollen and brown. Leaflets 11-14 pairs, ca. 1-2.5 x 0.5-1 cm in size, oblong-lanceolate, base oblique, apex apiculate, margins entire-slightly undulate, ciliate; cilia multicellular, lateral nerves 9-pairs, prominent above with sparse dots, densely puberulous below; stipel 1, ca. 6 x 2 mm, ovate, apex acuminate and brown, base truncate, margin entire, glaucous, greenish, on lower side 1 gland is present; petiolule ca. 2 mm. Inflorescence ca. 5.5-6 cm; peduncle ca. 1 cm. Flowers ca. 11 x 2 mm. Sepals campanulate (gamosepalous), cup shaped; tube ca. 1.5, glabrous; lobes free, ca. 1 mm, ovate, apex acute, pubescent; trichomes multicellular, slightly deflexed. Petals 5, free, ca. 6 x 2 mm, elliptic, base cuneate, apex acute, margin entire with multicellular hairs,, lateral nerves obscure, yellow. Stamens 10, free; filaments filiform, 6-10 mm; anthers versatile, 2-lobed, ca. 1.8-2 x 1 mm, oblong, base auriculate, apex retuse, margin entire, yellow. Ovary ca. 2 x 1 mm, linear, densely hairy; styles ca. 0.5 mm long; stigma rounded. Pods ca. 10-15 x 0.5-0.7 cm, 12-18-seeded, linear, straight, base cuneate, slightly curved; beak empty, ca. 1.5-1.7 cm long, recurved. Seeds ca. 11 x 6 mm, elliptic-oblong, shiny, brown, glabrous.

Flowering period: March-June.

Herbarium specimens examined: NARA DESERT: *Rahmatullah Qureshi,* 1238 (SALUBGH).

Field remarks: A common invasive weed of settled area.

Distribution: Indigenous to West Indies and Mexico, widely naturalized Punjab and Sindh (Ali, 1973).

2. Acacia

Mill, Gard. Dict. Arb. ed. 4: (1754); S.I. Ali, Fl. Pak. 36: 2 (1973).

Erect or scandent, armed rarely unarmed shrubs or trees. Leaves usually alternate, sometimes clustered, 2-pinnate, rarely modified into phyllodes; pinnae a few to several pairs, basal ones smallest, terminal ones largest; leaflets even-pinnate, small, opposite, sometimes overlapping, subsessile; foliar glands often present at the insertion of pinnae and on the upper side of the petiole; stipules various, often spinescent; stipulules 0. Inflorescence of terminal and/or axillary spikes or globose heads; bracts and bracteoles deciduous, sometimes bracteoles clustered and involucral. Flowers usually 5-merous, small, bisexual or polygamous. Calyx-tube campanulate, shortly 4- or 5-lobed. Petals 5, exserted, connate at the base, stamens ∞, shortly connate at the base; anthers small. Ovary stipitate or subsessile; ovules 2-∞; style filiform; stigma capitate. Pod ligulate or oblong, often flat, rarely turgid, striate or curved, thin, sometimes fleshy and thick, continuous or septate within. Seeds often transverse or longitudinal, ovoid or compressed.

A genus of about 900 species distributed mainly in tropics and subtropics, especially in Africa and Australia (Ali, 1973).

Key to species:

1a. Flowers in spikes...**1. A. senegal**

1b. Flowers in globose heads:

2a. Plants cupressiform in appearance.. **2. A. nilotica** ssp. **cupresiformis**

2b. Plant not cupressiform in appearance:

3a. Tree; pods moniliform, compressed.................................**3. A. nilotica** ssp. **indica**
3b. Shrub; pods flat, straight...**4. A. jacquemontii**

1. Acacia senegal (L.) Willd., Sp. Pl., 4:1077 (1806); Stewart in Nasir & Ali., Ann. Cat. Vasc. Pl. W. Pak. & Kashm., 380 (1972); Ali in Nasir & Ali., Fl.Pak., 36:6 (1973).
Syn: *Memosa senegal* L., Sp. Pl., 521 (1753).

A small tree up to 3-6 m high. Stem branched; branches flexuous, glabrous-grey, pubescent when young. Stipular prickles in 3, the 2 lateral ones nearly straight or slightly curved upwards, the middle one curved downwards, ca. 7 mm long, dark brown, polished. Leaves ca. 2-3.5 cm long; main rachis pubescent, with a gland between the lowest pair and another gland between uppermost pair of pinnae. Pinnae 3-5 pairs, ca. 1.2-1.8 cm long, shortly stalked, pubescent. Leaflets 8-15 pairs, ca. 1.5-3 x 1.2 mm in size, linear or elliptic-oblong, apex obtuse or subacute, glabrous. Spikes ca. 5-10 cm, axillary; peduncle ca. 0.5-1.5 cm long. Flowers yellow, sessile, ca. 4 mm. Calyx ca. 2 mm long; tube ca. 1.5 mm long; teeth ca. 0.5 mm long, lanceolate, apex acute. Filaments white; anthers yellow. Pods ca. 7-8 x 2.5 cm in size, linear-oblong, thin, flat, attenuated at both ends, reticulated, pale-brown when ripe. Seeds 5-6, ca. 8-12 mm across, suborbicular-lenticular, impressed on sides.
Flowering period: August-October.
Herbarium specimens examined: NARA DESERT: *Rahmatullah Qureshi*, 23 (SALUBGH).
Field remarks: A less common tree only recorded from Achhro Thar.
Distribution: West Pakistan (Sindh, Punjab), India (Punjab, Rajasthan, Delhi), Arabia; widely distributed in Tropical Africa (Ali, 1973).

2. Acacia nilotica ssp. **cupressiformis** (J.L. Stewart) Ali & Faruqi, l.c. 4.
Plants cupressiform in appearance; pods narrowly and regularly constricted between the seeds.
Flowering Period: March-November.
Herbarium specimens examined: NARA DESERT: Chotiari reservoir, *Rahmatullah Qureshi*, 1354 (SALUBGH).
Field remarks: A less common tree only recorded from Achhro Thar.
Distribution: W. Pakistan (Punjab, Sind); India (Deccan, Gujrat) (Ali, 1973).

3. Acacia nilotica subsp. **indica** (Benth.) Brenan in Kew Bull. 12: 84 (1957); Ali in Pak. J. Bot., 1:1-8 (1969); Stewart in Nasir & Ali., Ann. Cat. Vasc. Pl. W. Pak. & Kashm., 380 (1972); Ali in Nasir & Ali., Fl. Pak., 36: 9-11 (1973).

Syn: *A. arabica* (Lam.) Willd. var. *indica* Benth. in Hook. f., Lond. J. Bot. 1: 500 (1842); *Acacia nilotica* var. *indica* (Benth.) A. F. Hill in Bot. Mus. Leafl. Harvard. Univ. 8: 99 (1940).

A large tree with dark brown or black bark. Stem branched; branches slender, terete, pubescent when young; bark on trunk fissured, grey or brown. Leaves ca. 5-10 cm long; pinnae 4-10 pairs, ca. 1.8-5 cm long; main rachis downy, often furnished with glands. Stipular spines ca. 1.5-5 cm long, smooth, whitish, straight, sharp. Leaflets subsessile, 10-25 pairs, ca. 3-6 x 1.5-2 mm in size, linear-oblong, apex subobtuse, glabrous. Inflorescence globose head, ca. 6-15

mm across; peduncle axillary, in fascicles of 2-6, terete; bracteoles 2, above the middle of peduncle, broadly ovate, apex acute, pubescent. Flowers yellow. Calyx-tube ca. 1 mm long; teeth ca. 2 mm. Corolla ca. 3 mm long; lobes triangular. Pods ca. 5-15 x 1.2-1.5 cm in size, moniliform, compressed, constricted at sutures between seeds, densely and persistently grey-downy. Seeds 8-12, ca. 7-8 mm across, suborbicular, compressed, black brown.
Flowering period: March-May and August-October.
Herbarium specimens examined: NARA DESERT: *Rahmatullah Qureshi,* 1178 (SALUBGH).
Field remarks: A fairly common tree found throughout the area.
Distribution: West Pakistan (Sindh, Punjab), India (Madhya Pradesh, Uttar Pradesh, Bengal, Madras, Bombay, Punjab), Tanganyika (Ali, 1973).

4. Acacia jacquemontii Benth. in Hook., Lond. J. Bot. 1: 499 (1824); Stewart in Ali and Nasir, Ann. Cat. Vasc. Pl. W. Pak. & Kashm.,379 (1972); Ali in Nasir & Ali., Fl. Pak., 36: 13 (1973).

An erect shrub up to 2 m high. Stem fairly branched at the base; branches smooth, stiff, zigzag, brown in colour. Stipular spines ca. 2-5 cm long, white, connate at the base. Leaves ca. 2-5 cm long; pinnae 1-4 pairs, ca. 6-12 mm long, petiole ca. 2.5-5 cm long; with a gland between the upper pair of pinnae. Leaflets 5-10 pairs, ca. 1.5-3 x 0.5-1.5 mm in size, linear-oblong, apex obtuse, glabrous. Inflorescence ca. 12-16 m across, globose head; peduncle 2-3, slender, axillary, fascicled. Flowers yellow, sweet-scented, ca. 3 mm. Bracts 2, inserted in the middle of the peduncle. Calyx-tube ca. 1 mm long; teeth ca. 0.3 mm long, deltoid. Corolla ca. 3 mm long; lobes ovate-oblong, apex acute. Pods ca. 3-6 x 1.2-1.5 cm in size, ovate-oblong, base rounded, flat, straight, transversely or reticulately veined, glabrous. Seeds 5-8, compressed.
Flowering period: February-May.
Herbarium specimens examined: NARA DESERT: *Rahmatullah Qureshi,* 81 (SALUBGH).
Field remarks: A very common shrub recorded from the study area.
Distribution: West Pakistan (Sindh, Punjab & Baluchistan), India i.e. Madhya Pradesh, Punjab, Rajasthan (Ali, 1973).

31. Caesalpiniaceae

Trees, shrubs, lianes or stragglers, sometimes armed, rarely hrerbs. Leaves alternate, 1- or 2-pinnate, rarely simple and 2-foliate; pinnae or leaflets numerous. Stipule paired, mostly caduous, stipels may or may not be present. Inflorescence axillary or terminal, rarely leaf-opposed, raceme or panicle, often corymbose, rarely flower(s) solitary or 2 to 3 in a cluster. Flowers usually 5-merous, sometimes 4-merous from the fusion of the 2 upper lobes, bisexual, zygomorphic (rarely actinomorphic). Calyx with a short tube; lobes 5, free, usually imbricate, rarely valvate, unequal, the lower one sometimes larger. Petals 5 (rarely fewer or 0), free or sometimes connate at the base, imbricate in bud, usually with the dorsal one within and overlapped by laterals. Disc enclosed within calyx. Stamens 10 or fewer and replaced by staminodes; filaments usually free, rarely coherent, more or less unequal; anthers variuos. Pistil unicarpellary, ovary superior, sessile or stipitate, uni-locular; ovule(s) 1 to many; style single, undivided. Fruit a legume or indehiscent drupaceous and samaroid, often winged. Seeds various, sometimes arillate, rarely with endosperm.

A family of about 152 genera and 2800 species distributed in tropical and subtropical regions (Ali, 1973).

Key to genera:

1a. Climbing armed shrub; bracteates; pods flattened or turgid.......................**1. Caesalpinia**
1b. Erect unarmed herb; ebracteate; pod terete or compressed................................**2. Senna**

1. Caesalpinia

L., Sp. Pl. 380 (1753); Gen. Pl. ed. 5: 178 (1754); Hutch., Gen. Fl. Pl. 1: 260 (1967); Ali in Nasir and Ali, Fl. Pak., 54: 30 (1973).

Armed stragglers or unarmed trees and shrubs. Leaves large, 2-pinnate; pinnae abruptly pinnate; leaflets opposite, usually numerous; stipules various; stipels 0 or modified into thorns. Inflorescence axillary or terminal, of racemes or panicles; bracts caduous; bracteoles 0. Calyx-tube short; lobes 5, unequal; lower sepal concave, outermost. Petals 5, unequal, spreading, clawed, the uppermost small. Stamens 10, free, declinate, subequal, basally pubescent or villous; anthers uniform. Ovary (sub)sessile; ovules normally a few; style filiform, sometimes clavate at the apex; stigma simple, terminal. Pod oblong or ovoid, thin, flattened or turgid, sometimes winged, indehiscent or dehiscent, smooth or rarely prickled. Seeds 2-8, ovoid-orbicular, transverse.

A genus of 280 species distributed in tropics and sub-tropics of the world (Ali, 1973).

Caesalpinia bonduc (L.) Roxb., Fl.Ind. 2: 362(1832); Stewart in Ali and Nasir, Ann. Cat. Vasc. Pl. W. Pak. & Kashm.,374 (1972); Ali in Nasir & Ali, Fl. Pak. 54: 30-31(1973).

Syn: *Guilandinia bonduc* L., Sp. Pl. 381(1753); *Caesalpinia crista* L., Sp. Pl. 380(1753); *Guilandinia bonducella* (L.) Flaming in As. Research 11: 159. (1810).

An armed scrambling shrub up to 4 m long. Stem hard, woody, branched, densely covered with erect, recurved prickles, ca. 0.2 cm long; branchlets with reddish glandular hairs; internode ca. 5.5-13 cm long. Stipules large, foliaceous, lobed or pinnate. Leaves 15-30 cm long, rachis with 1-2 recurved stipulate spines at the base of each pinnae and scattered straight and recurved prickles between the pinnae; pinnae 6-10 pairs, opposite, 5-20 cm long; petiole ca. 1-4 cm long. Leaflets 6-11 pairs on each pinnae, ± oblique, opposite, ca. 2-4.5 x 1.5-2.5 cm in size, elliptic-oblong, base obtuse-cuneate, apex obtuse-mucronate, margin entire, glabrous above, puberulous below. Inflorescence axillary and terminal pedunculate raceme, ca. 5-7 cm long; peduncle ca. 2-10 cm long, prickled. Flowers bisexual, complete, ca. 1 cm across; pedicel ca. 0.5-1 cm long, yellow. Bracts ca.1.5 cm long, linear-lanceolate. Sepals 5-lobed; tube campanulate, ca. 4-8 mm, gibose at the base, glandular-pubescent; lobes ca. 0.5-0.6 cm long, ca. 0.2-0.3 cm broad, imbricate, unequal, oblong-obovate, base rounded, apex ± mucronate, margin entire, tomentose above, glandular hairy below, yellowish green in colour. Petals 5, ca. 9-13 x 2-4 mm in size, oblanceolate, apex obtuse, base cuneate, the upper smaller and spotted with red colour. Stamens 10, declinate, subequal; filament attenuate, ca. 4-8 mm, base flattened and glandular-villous. Ovary stipitate, globose, ca. 2-ovuled; style ca.

5 mm long, tomentose; stigma simple, dilated with a ring of hairs around. Pod ca. 7.5 x 4.5 cm in size, obovoid, inflated, beaked, densely prickled, prickle up to 0.5 cm long. *Seeds* 1-2, ca. 1-1.7 x 0.8-1.5 cm in size, sub-globose, greyish in colour, shiny.
Flowering period: April-September.
Herbarium specimens examined: NARA DESERT: *Rahmatullah Qureshi*, 953, 971 (SALUBGH).
Field remarks: A rare straggler found near settlement.
Distribution: W. Pakistan (Punjab, Sindh, escape ?); widely distributed in the Tropics of the old and the New World (Ali, 1973).

2. Senna Mill.

Gard. Dict. Abr. (ed. 4) no. 3 (1754); Boulos, Fl. Egypt 1: 359 (1999).

Unarmed trees, shrubs or herbs; leaves paripinnate, often with globose or cylindrical gland on the petiole and the rachis; flowers in few or many-flowered racemes, ebracteate; sepals 5, imbricate; petals 5, yellow. Stamens 10, filaments straight, shorter or twice as long as the basifixed anthers. Pods terete or compressed, indehiscent or tardily dehiscent, valves not twisting. Seeds slightly compressed.

About 350 species mostly distributed in tropics of the New Wrorld (Boulos, 1999).

Key to species:
1a. Leaflets glabrous, glaucous, 3-6 pairs, obliquely obovate to oblong.............**1. S. italica**
1b. Leaflets velvet-hairy, 5-8 pairs, lanceolate-blong...............................**2. S. holosericea**

1. Senna italica Mill., Gard. Dict., ed. 8, no. 2 (1768); Boulos in Fl. Egypt 1: 359 (1999). Figure 7A

Syn. *Cassia aschrek* Forssk., Fl. Aegypt.-Arab. 86 (1775); *C. obovata* Collad., Hist. Nat. Méd. Casses 92 (1816), *nom. illegit.*; *C. obtusa* Roxb. Hort. Beng. 1814 (*nom. illegit*) ex Wt. & Arn. Prodr. 228 (1834); *C. italica* (Mill.) F. W. Andrews, Fl. Pl. Anglo-Egypt. Sudan 2: 117 (1952).

A diffuse perennial erect herb upto 2.5 ft high. Stem ± woody, fairly branched; branches spreading, obtusely angular, striate, pale-green, adpressedly pubescent. Leaves ca. 6-10 cm long; rachis glabrous, glaucous, without glands between the leaflets; petioles ca. 1-3.5 cm long; stipules ca. 0.6 mm long, obliquely lanceolate, apex acute. Leaflets 3-6-pairs, ca. 2-2.5 x 0.9-1.3 cm in size, broadly oblong or obovate, apex mucronate, base obtuse, glaucous beneath. Flowers axillary; racems peduncled, ca. 8-16 cm long, usually longer than the leaves; pedicels ca. 5-9 mm long. Calyx ca. 5-9 mm long, divided at the base; lobes oblong, apex obtuse, membranous, greenish in colour. Corolla ca. 6.5-8.2 mm long, obovate-oblong, shortly clawed, yellow, reticulately veined with darker veins. Stamens 7, unequal; staminodes minute; filaments short. Ovary densely pubescent. Pods ca. 3-4 cm long, flat, thin, papery, glabrous, rounded at the both ends, much recurved, transversely veined with aline of prominent, longitudinal crests down the middle of pods over the seeds, sutures very thin. Seeds 6-12, ca. 6 x 4.5 mm in size, wedge-shaped, truncate or retuse at the apex, finely reticulo-rugose, with a transverse ridge across the middle of each face, dark brown, shiny.
Flowering period: October- December.

Herbarium specimens examined: NARA DESERT: *Rahmatullah Qureshi*, 512, 943, 1007, 1160, 1164 (SALUBGH).
Field remarks: A common herb of calcarious soil.
Distribution: W. Pakistan (Sindh, Punjab, N.W.F.P. Province); India (Punjab, Bombay, Madras); Ethiopia, Somalia Republic; Uganda; Kenya; Tananyika; Bechuanaland; South Africa (Ali, 1973).

2. Senna holosericea (Fresen.) Greuter, Willdenowia 15: 429 (1986); Boulos in Fl. Egypt 1: 359 (1999).

Syn: *Cassia holosericea* Fresen in Flora. 22(1):54. 1839. Cooke, Fl. Bomb. Pres. 1: 449 (1903); Ali & S. Quraishi in S.U. Sci. Res. J. 3: 5 (1967).

Suffruticose, 20-120 cm tall, branches subterete, pubescent. Leaves 5-12.5 cm long, rachis densely hairy, eglandular, petioles 1.2-2 cm long, stipules c. 5-7 mm long, linear-lanceolate, very acute with a small auricle at the base, densely velvet-hairy; leaflets 5-8 pairs, 1.5-2.5 cm long, 7.5-13 mm wide, obovate, oblong, obtuse or retuse, mucronate, velvety pubescent on both surfaces, petio¬lules very short. Flowers in axillary racemes, pedicel very short. Calyx c. 9-10 mm long, divided to the base, segments oblong, obtuse, membranous. Petals c. 1.2-1.3 cm long, ovate, oblong, cuneate, shortly clawed, yellow, conspicuously reticulate with darker veins. Stamens 10, the upper 3 reduced to the small stami¬nodes, the remaining perfect, 2 or 3 lower are largest. Ovary densely hairy. Pods c. 2.5-3.8 cm long, 1.2-1.8 cm wide, flat, thin and papery, recurved, rounded at both the ends, velvety, pubescent, not longitudinally crested. Seeds 4-10, obovate to cuneate, c. 6-7 mm long, 3 mm wide, reticulate-rugose, flattened, glabrous, yellow.
Flowering Period: All the year round.
Herbarium specimens examined: NARA DESERT: *Rahmatullah Qureshi*, 2023 (SALUBGH).
Field remarks: A relatively common herb recorded from Achhro Thar.
Distribution: W. Pakistan (Sind; Lasbella); Arabia; Abyssinia; Tropical Trans¬vaal (Ali, 1973).

32. Fabaceae (Papilionaceae)

Herbs, (sub)shrubs (stragglers or lianes) or trees. Leaves alternate, generally compound, often odd-pinnate or 1- or 3-foliate, rarely simple, usually stipulate; stipels sometimes present; often pulvinate. Inflorescence terminal and/or axillary, sometimes leaf-opposed, in racemes, panicles, umbels, spikes or in a few-flowered clusters, rarely flowers solitary. Flowers irregular, hermaphrodite. Calyx-tube usually campanulate; lobes 5, unequal, imbricate or valvate. Corolla papilionaceous, polypetalous, sometimes united; petals 5, vexillary imbricate; the upper (adaxial) exterior and forming the standard (vexillum); the inner 2 laterals the wings (alae), parallel to each other; the lower 2 interior and connate by their margins into a keel (carina). Disc rare. Stamens 10, rarely 9, monadelphous or diadelphous, rarely free; anthers uniform or dimorphic. Ovary superior, 1-celled; ovules α, sometimes 1 or 2; style straight, often abruptly or gradually curved, bearded or glabrous. Pod dehiscent by 2 or 1 suture(s) or indehiscent, sometime jointed and breaking into 1-seeded segments. Seeds sometimes strophiolate (arillate); embryo with an inflexed radicle; cotyledons accumbent.

A large family with ca. 480 genera and ca. 12000 species, cosmopolitan in distribution (Ali, 1977).

Key to genera:

1a. Stamens 10, 1-delphous:
2a. Styles bearded. Pods turgid..**1. Crotalaria**
2b. Stles not bearded. Pods various:
1b. Stamens 9+1, 2-delphous:
3a. Plants spiny. Leaves simple..**2. Alhagi**
3b. Plants not spiny. Leaves digitaley divided or compound:
4a. Joints of pods turgid, if compressed then corolla not or hardly exserted......**3. Alysicarpus**
4b. Pods not jointed:
5a. Rachis ending in a tendril:
6a. Staminal sheath truncate at the apex..**4. Lathyrus**
6b. Staminal sheath oblique at the apex...**5. Vicia**
5b. Rachis not ending in a tendril:
7a. Fruit spirally coiled...**6. Medicago**
7b. Fruit not spirally coiled:
8a. Climbing or twining herbs:
9a. Leaves gland-dotted beneath; pods 1-2-seeded......................................**7. Rhynchosia**
9b. Leaves not gland-dotted beneath; pods more than 2-seeded:
8b. Erect or prostrate herbs, undershrubs or shrubs:
10a. Hairs medifixed. Anthers apiculate..**8. Indigofera**
10b. Hairs not medifixed. Anthers muticous or obtuse....................................**9. Tephrosia**
11a. Leaves paripinnate; pods 10-20 cm..**10. Sesbania**
11b. Leaves imparipinnate; pods less than 10 cm:
12a. Calyx deeply 2-lipped; pods linear-oblong, flattened.........................**11. Argyrolobium**
12b. Calyx campanulate; pods ovoid or globose..**12. Melilotus**

1. Crotalaria

L., Sp. Pl. 714 (1753); Gen. Pl. ed. 5: 320 (1754); Baker in Hoo. f., Fl. Brit. Ind. 2: 65 (1876); Ali in Fl. Pak., 100: 40 (1977).

Herbs or (sub)shrubs. Leaves simple, 1- or digitately 3-foliate or rarely 7-foliate; leaflets entire; stipules filiform to foliaceous or 0. Inflorescence a terminal, leaf-opposed or axillary racemes, sometimes in subumbels or heads, rarely flowers solitary. Flower often showy; bracts and bracteoles small or absent. Calyx-tube short; lobes unequal, linear, lanceolate or ovate, subequal or connate in 2 lips. Corolla equal to exceeding to calyx, generally yellow, rarely blue or violet; petals clawed; standard orbicular or ovate, shortly clawed, with 2 callus-like appendages, glabrous or pubescent without; wings oblong or obovate, shorter than the standard, clawed, usually with rows of crescent-shaped folds between nerves; keels as long as wings, connate, much incurved, with a beak spirally twisted or not. Stamens 10, monadelphous, connate in a sheath cleft above; anthers dimorphic, alternately short, versatile with a slender filament and long, basifixed with flattened filaments. Ovary sessile or stipitate; ovules 2 or more; style long, incurved or bearded, constricted or not at the bent, often pubescent on the inner side, rarely on both side; stigma small, oblique. Pod subsessile to long-stipitate, oblong or globose, turgid or inflated, continuous within, often dehiscent. Seeds usually many, rarely 1 or 2, ovoid, reniform or oblong, estrophiolate.

About 350 species distributed throughout the tropics and subtropics, most numerous in Tropical Africa. In Pakistan, it is represented by 11 species (Ali, 1977).

Key to species:

1a. Leaves simple; pods ovoid..**1. C. burhia**

1b. Leaves 3-foliate; pod globose..**2. C. medicaginea**

1. Crotalaria burhia Ham. ex Benth. in Hooker, London Journ., v (2): 474 (1843); Stewart in Ali and Nasir, Ann. Cat. Vasc. Pl. W. Pak. & Kashm., 401 (1972).; S. I. Ali in Nasir & Ali., Fl. Pak., 100: 45 (1977). **Figure 13A**

Syn. *C. burhia* Buch-Ham. Ex Benth.var. *tomentosa* Boiss., Fl.Or.2:26 (1872).

A low, rigid, erect to diffuse undershrub up to 2 ft high. Stem hard, woody at the base, herbaceous above, densely branched, 5–angled, hoary with adpressed pubescence, striate. Leaves ca. 4-8 cm long, subsessile, oblong, usually rounded-mucronate, pale green, silky. Racemes terminal, 6-12-flowered, ca. 7.5-8 cm. Flowers ca. 7.5 x 4.5 mm in size; pedicel ca. 1 mm long, 2-bracteolate. Calyx-tube ca. 2 mm long; lobes ca. 5 mm, lanceolate, apex acute, hairy above. Corolla yellow with usually reddish streaks, slightly exserted; standard ca. 7 x 4.5 mm in size, ovate, folded, apex obtuse, with ca. 1 x 2 mm long, hyaline, wooly claw and 2 pockets at the base, veined; wings ca. 6 x 2 mm in size, oblong; keels ca. 7.5 x 5 mm in size, free near apex otherwise united, much incurved, distinctly beaked. Anther lobes heteromorphous; 5 stamens with short lobe alternating with 5 stamens with long lobes. Styles ca. 5 mm long, curved and slightly swollen near the base; hairy on one side throughout the length, more densely so near apex; stigma oblique. Pods ca. 5 x 4 mm in size, densely pilose, obliquely ovoid, beaked, brownish. Seeds 1, rarely 2, ca. 2.5-3 x 1.8-2.2 mm in size, obliquely semilunar, notched at hilum, brown , shiny, compressed.

Flowering period: September-October.

Herbarium specimens examined: NARA DESERT: *Rahmatullah Qureshi*, 3, 38, 92, 202, 629, 637, 688, 704, 861, 1042, 1047, 1118, 1140, 1224 (SALUBGH).

Field remarks: A very common shrub distributed throughout the area.

Distribution: Pakistan (Punjab, Sindh & Baluchistan), India and Afghanistan (Ali, 1977).

2. Crotalaria medicaginea var. **medicaginea** Lamk., Encycl. Meth. 2: 201 (1786); Stewart in Ali and Nasir, Ann. Cat. Vasc. Pl. W. Pak. & Kashm., 401 (1972); S. I. Ali in Nasir & Ali., Fl. Pak., 100: 41 (1977). **Figure 13B**

Syn: *Crotalaria luxurians* Benth. in Hook. Lond. J. 2: 578 (1843).

Annual much branched herb up to 45 cm tall. Leaflets ca. 0.6-2 x 0.2-1 cm in size, obovate or obovate-oblong, retuse or obtuse, apiculate; stipules ca. 2 mm long, filiform, pubescent. Racemes terminal and leaf-opposed, 2- 12 flowered. Bracts linear-subulate, ca. 1 mm long. Calyx-teeth longer than the tube; sepals ca. 3 x 6 mm, lobes acute, persistent. Petals yellow; standard ovate, ca. 5.5 x 3 mm, pubescent on outer side; wings ca. 4.5 x 1.2 mm, keel single, incurved, beaked, ca. 4 x 2 mm. Stamens up to 4 mm long; staminal tube ca. 4x6 mm; anthers dimorphic. Ovary ca. 1 x 0.5 mm, sessile, ovate, having trichomes; style ca. 3 mm long, incurved, bearded; stigma small. Fruit globose, ca. 4 mm in diameter, pubescent, beaked with persistent stylar base. Seeds 2, brown- dark brown, smooth, shiny, reniform.

Flowering Period: May-August.

Herbarium specimens examined: NARA DESERT: *Rahmatullah Qureshi*, 1361 (SALUBGH).

Field remarks: A less common weed mostly found in cotton crop and flat habitat.

Distribution: Pakistan (Punjab, Sindh & Baluchistan), India and Afghanistan (Ali, 1977).

2. Alhagi

Tourn. ex Adans., Fam. 2: 328, 514 (1763); Stewart in Ali and Nasir, Ann. Cat. Vasc. Pl. W. Pak. & Kashm., 383 (1972); S.I. Ali in Fl. Pak., 100: 319 (1977).

Much-branched, rigid subshrub, armed with axillary spines. Leaves simple, quite entire, usually small; stipules small. Inflorescence an axillary racemes; rachis spinescent at the tip. Flowers red, usually few. Bracts small, bracteoles absent. Calyx campanulate; teeth short, subequal. Corolla exserted; standard obovate, with a short claw; wings falcate-oblong, free; keels incurved, obtuse, equal to the standard and wings. Stamens diadelphous, 9+1; anthers uniform. Ovary stalked; ovules many; style filiform, incured, glabrous; stigma small, terminal. Pod linear, 1-9-seeded, jointed, somewhat thick, subterete or compressed, smooth, indehiscent, usually contracted between the seed and divided inside by septa; joints not separating. Seeds reniform.

A genus of about 3 species, distributed in E. Mediterranean, Orient, N. Africa and central Asia (Ali, 1977).

Alhagi maurorum Medic. in Vorles, Churpf.Phys.Oek.Ges., 2:397 (1787); Stewart in Nasir & Ali., Ann.Cat. Vasc.Pl.W.Pak. & Kashm.,383 (1972); S.I.Ali in Nasir & Ali., Fl.Pak., 100: 319 (1977). **Figure 12D**

Syn: *Hedysarum alhagi* L., Sp. Pl. 745 (1753); *H. pseudalhagi* M.Bieb., Fl. Taur.-Cauc.2: 174 (1808); *A. camelorum* Fisch.ex DC., Prodr. 2: 352 (1825); *A. persarum* Boiss & Buhse in Mem.Soc.Nat.mosc.12: 77 91860).

A perennial, deep rooted thorny under shrub up to 100 cm tall. Stem hard, woody at the base, herbaceous above, terete, glabrous to more or less glaucous or puberulent, green; internode ca. 0.5-1.5 cm long, much longer in fruiting stage. Stipular spines ca. 1.2.5 cm long, much longer in fruiting stage, alternate, acicular, pubescent, apex sharp pointed and brown in colour. Leaves simple, alternate, stipulate, ca 0.5-1.2 x 0.3-0.6 cm in size, obovate, base cuneate, apex mucronate, margin entire, puberulous on both surfaces, lateral nerves obscure, green in colour; petiole ca. 1.5-3 mm long, pubescent. Stipules 2, ca. 1 mm long, ensiform, pubescent. Inflorescence an axillary raceme, rachis ca. 1-2 cm long, ending in spine. Flowers irregular, zygomorphic, complete, alternately arranged, purplish, ca. 1 x 0.5 cm in size, pedicel ca. 2-4 mm long, curved, cylindric. Bracts 2, ca. 1-1.5 mm across, ovate. Sepals united, campanulate, tube ca. 4-5 mm long; teeth 2+3, unequal, ca. 1.5-2 x 1-3 mm in size, valvate, obsolete-triangular, apex acute-apiculate, margin entire, glabrous, green. Petals 5, exserted, papilionaceous, dark pink or reddish-violet, irregular, imbricate-descending (vaxillary); standard ca. 1 x 0.5 cm in size, obovate, base cuneate, apex cleft, margin entire, lateral nerves arise from mid nerve and dark pink in colour, claw ca. 2 mm long; wings 2, ca. 20 x 5 mm in size, elliptic-oblong, apex rounded, base oblique, margin entire, glabrous; keels 2, ca. 20 x 6 mm in size, united, elliptic, base cuneate, apex obtuse, margin entire, lateral nerves prominently visible. Stamens 9+1, diadelphous; staminal sheath ca. 1.1 cm long, glabrous; filament ca. 0.3-1.1 cm long, glabrous; anthers ca. 1.5 mm across, dorsifixed, oblong, yellow in colour. Ovary monocapellary, ca. 13 x 1 mm in size, linear, glabrous, green; style ca. 10 mm. long, filiform, glabrous, apex incurved; stigma capitate. Fruit ca. 1-3 x 0.3 cm in size, linear falcate or straight, moniliform, constricted between the seeds, divided by septa, indehiscent, glabrous, brown in colour, 1-9 seeded; beak ca. 5-6 mm long, empty, pointed. Seeds ca. 5 x 3 mm in size, elliptic-reniform, blackish brown, glabrous.

Flowering period: April-September.

Herbarium specimens examined: NARA DESERT: *Rahmatullah Qureshi*, 979, 1244 (SALUBGH).

Field remarks: A common weed mostly found in peripheral areas.

Distribution: Pakistan, Kashmir, Iran, Afghanistan, Russia, Turkey, Iraq, Syria, Palestine, Cyprus & N. Africa (Ali, 1977).

3. Alysicarpus

Necker ex Desvaux., J. Bot. Agric. ser.1 (2): 120 (1813); Baker in Hoo. f., Fl. Brit. Ind. 2: 157 (1876); Ali in Biologia 12: 31 (1966); Ali in Fl. Pak., 100: 341 (1977).

Prostrate or diffuse herb. Leaves alternate, 1- (rarely 3-) foliate; petiole minute; stipules scarious, acuminate, free or connate; stipels 2, prominent. Inflorescence of terminal and/or axillary racemes or panicles. Flowers often in pairs; pedicels often twin; bracts scarious, generally deciduous. Calyx-tube glumaceous, deeply 4-lobed; 2 upper lobes often connate to the apex. Corolla not exserted; petals clawed; standard obovate or orbicular, base with 2 small longitudinal folds within; wings obliquely oblong, adherent to keel; keels incurved, usually with a lateral appendage. Stamens 9+1; anthers uniform. Ovary sessile or shortly stipitate; ovules α; style filiform, incurved at apex, glabrous; stigma terminal or capitate. Pod subterete, constricted between seeds; articles 4-gonous, indehiscent, globose or ovoid, convex or turgid. Seeds suborbicular, strophiole 0.

A genus of about 30 species distributed in Africa, Asia, Australia, Polynesia and Tropical America. Represented in Pakistan by 8 species (Ali, 1977).

Key to species:

1a. Calyx shorter or very slightly longer than the first joint of the pod............**1. A. monilifer**

1b. Calyx much longer than the first joint of the pod.
Pods transversely ribbed..**2. A. scariosus**

1. Alysicarpus monilifer (L.) DC., Prodr. 2: 353 (1825); Stewart in Ali and Nasir, Ann. Cat. Vasc. Pl. W. Pak. & Kashm., 348 (1972); S. I. Ali in Nasir & Ali., Fl. Pak., 100: 345 (1977).

Syn: *Hedysarum moniliferum* L., Mant. 1: 102 (1767).

A prostrate herb, much branched, up to 15 cm high. Branchlets pilose; hairs unicellular, elongate and spreading. Stipules ca. 3-7 x 2 mm in size, scarious. Leaves 1-foliate; petiole ca. 1-3 mm; leaflets ca. 0.5-3 x 0.4-1.6 cm in size, ovate or oblong-elliptic, base obtuse-subcordate, apex obtuse-mucronate, margin entire, chartaceous, subglabrous above and hairy below especially on nerves. Racemes axillary, 1-3 cm, 4-10-flowered. Flowers ca. 6 mm across; pedicel ca. 1-1.5 mm. Calyx-tube ca. 1.5 mm, pilose; upper lobes 1 mm; lateral 3 mm; lower one 4 mm, lanceolate, apex acute, margin entire. Corolla pink-violet; standard ca. 6 x 2 mm, obovate; wings ca. 5 x 2 mm; keels ca. 6 x 1 mm. Staminal sheath ca. 5 mm, glabrous; filaments ca. 1 mm. Ovary ca. 4 mm; styles ca. 2 mm long. Pods ca. 1-2.5 x 0.2-0.3 cm in size, moniliform, 4-8-jointed; joints turgid, erect; hairs downy and aculeate, nerves obscure; apex apiculate; articles subglobose, 4-7, turgid. Seeds ca. 2 mm, ovoid.

Flowering period: October-November.

Herbarium specimens examined: NARA DESERT: *Rahmatullah Qureshi,* 1117 (SALUBGH).

Field remarks: A rare species of calcarious soil.

Distribution: Pakistan, India (Punjab, Uttar Pardesh, Madhya Pradesh, Bengal, Madras, Bombay) & Ethiopia (Ali, 1977).

2. Alysicarpus scariosus Grah. ex Thwaites, Enum.Pl.Zeyl. 88 (1858); Stewart in Ali and Nasir, Ann. Cat. Vasc. Pl. W. Pak. & Kashm., 348 (1972); S. I. Ali in Nasir & Ali., Fl. Pak., 100: 342 (1977).

Syn: *Alysicarpus rugosus* var. *styracifolius* Baker in Hook.f., Fl.Brit.Ind.2: 159 (1878).

A procumbent herb up to 50 cm long. Stem solid, at the base, herbaceous above, terete, fairly branched at the base; branches greyish-green, setose; trichomes ca. 2 mm long; internode ca. 1-2 cm long. Leaves simple, alternate, stipulate, ca. 0.7-2 x 0.5-1 cm in size, elliptic-oblong or orbicular, base obtuse-subcordate, apex rounded, margin entire, glabrous above, apressed to erect pilose below; veins 4-8 pairs, most prominent on below; petiole ca. 0.4-0.7 cm long, pubescent; trichomes ca. 2-3 mm long. Stipules 2, ca. 12 x 4 mm in size, triangular, apex acute, margin ciliated, hyaline, parallel veined. Inflorescence axillary raceme, ca. 1.5 cm long, up to 12-flowered, sessile or short peduncled; peduncles ca. 4 mm long. Flowers bisexual, complete, irregular, zygomorphic, ca. 14 x 8 mm in size; pedicel ca. 2 mm long. Sepals 5, glumaceous, ca. 14 x 3-4 mm in size, imbricate, lanceolate to broadly ovate, base acute-obtuse, apex acute to acuminate, lower upper one hyaline and others green, exceeding to petals, parallel veined which are arising from base to apex, pilose, trichomes white spreading. Petals 5, introse, papilionaceous; standard ca. 11 mm across, obovate, base cuneate, apex emarginate, margin entire, glabrous, parallel veined, 2 purplish spots in middle, yellowish in colour; wings 2, ca. 9 x 2.5 mm in size, obliquely oblong, base oblique, apex rounded, margin entire, purplish spots on margin, adhering to the keels; keels 2, ca. 8 x 3 mm in size, obovate, slightly incurved, appendaged, glabrous, yellowish in colour. Stamens diadelphous, 9+1; staminal sheath ca. 8 mm long, glabrous, greenish; filaments ca. 1.5 mm, filiform; anthers ca. 0.5 mm, ovate, greenish, 2-celled, splitting longitudinally. Ovary ca. 3 x 1 mm in size, linear, sparsely pubescent on upper side, green; style ca. 5 mm long, curved at the apex, glabrescent; stigma ca. 0.2-0.3 mm, capitate. Pod usually included in calyx,

moniliform, apical mucro stout, 2 mm; articles 3-6, transversely ribbed, hooked pubescent. Seeds ca. 2 mm, rugose.
Flowering period: November-March and August-September.
Herbarium specimens examined: NARA DESERT: *Rahmatullah Qureshi*, 654, 905, 1116, 1143, 1289 (SALUBGH).
Field remarks: A less common herb mostly found in moist places.
Distribution: Pakistan, India & Ceylon (Ali, 1977).

4. Lathyrus

L., Sp. Pl. 729 (1753); Gen. Pl. ed. 5. 326 (1754); Baker in Hook.f., Fl.Brit.Ind. 2: 179 (1876); Ali in Biologia 11(2): 1 (1965); Ali, Fl. Pk., 100: 265 (1977).

Annual or perennial herbs, climbing or straggling, stem usually winged. Leaves paripinnate, rachis ending in a tendril, rarely in a bristle or leaflet. Stipules foliacceous, semisagittate or entire. Inflorescence an axillary raceme or solitay. Bracts caducous, usually minute, bracteoles absent. Calyx regular or irregular, calyx tube mostly oblique at the base, teeth subequal. Vexillum broad with a short claw. Wing free or coherent to the keel. Stamens diadelphous (9+1) or monoadelphous, vexillary stamen free or connate with the tube, anthers uniform. Ovary subsessile or stipitate(stalked); ovules many; style dorsally compressed, linear-spathulate, pubescent on upper side, rarely glabrous; stigma terminal or capitate. Fruit compressed or subterete, 2-valved, continuous within. Seeds many.

A genus of about 130 species, widely distributed in the northern hemisphere, S. America and Africa (Ali, 1977).

Lathyrus aphaca L., Sp. Pl. 729 (1753); Stewart in Ali and Nasir, Ann. Cat. Vasc. Pl. W. Pak. & Kashm., 409 (1972); S.I. Ali in Nasir & Ali., Fl. Pak., 100: 276 (1977). **Figure 14B**

Syn: *L. pseudoaphaca* Boiss., Diagn. Pl. Orient. 2: 105 (1843); *L. affinis* Guss., Fl. Sicul. Syn. 2: 853 (1845); *L. polyanthus*Boiss. & Blanche in Boiss., Fl. Orient. 2: 602 (1873); *L. floribundus* Velen., Fl. Belg. 156 (1891).

Annual climbing herb upto 15-45 cm. Stem glabrous, quadrangular, terete. Stipule 2, foliaceous, ca. 1.5-3.5 x 1-2.5 cm in size, ovate-hastate, apex acute-apiculate, parallel veined. Leaves reduced/to simple filiform tendrils. Flowers axillary, solitary, rarely 2; peduncle ca. 4.5 cm long, hairy. Flowers 1-2, ca. 11 x 7 mm; pedicel ca. 2-3 mm. Bracts minute. Calyx 6-8 x 10-13 mm; lobes subequal, lanceolate, 3-nerved. Corolla ca. 0.8-1.2 cm, yellow. Stamens 9+1, ca. 8.5 in length; anthers basifixed, 0.5 x 0.05mm. Ovary 6 x 1.5mm; style 2 x 0.5 mm; stigma terminal, 1.5 x 1mm. Pod 2-3.5 x 0.4-0.6 mm cm, 4 to 6-seeded, oblong-linear, compressed, reticulate-veined. Seeds ca. 2-3 mm, dark brown, smooth.
Flowering Period: December-February.
Herbarium specimens examined: NARA DESERT: *Rahmatullah Qureshi*, 1362 (SALUBGH).
Field remarks: A common climber found in flat habitat.
Distribution: Pakistan; Kashmir; India; Europe, N. Africa; S.W. & C. Asia (Ali, 1977).

5. Vicia

L., Sp. Pl. 734 (1753); Gen. Pl. ed. 5. 327 (1754); Baker in Hook.f., Fl.Brit. Ind. 2: 176 (1876); Ali, Fl.Pk., 100: 265 (1977).

Herbs, usually climbing by means of tendrils or rarely low or sub-erect; leaves paripinnate, the leaf rachis terminating in a simple or branched tendril or a recurved bristle; leaflets numerous, rarely only 1-3 pairs, entire or dentate at the apex, exstipellate; stipules semisagittate; flowers ogten blue, violet or yellowish, either 1 or few in the leaf axils or racemose; bracts very caduceus, mostly minute; bracteoles absent; calyx tube often oblique at the base, obtuse; teeth or lobes subequal or the upper 2 shorter and the lower longer; vexillum narrowed into a broad claw; wings obliquely oblong, adherent to the middle of the keel; keel shorter than the wings; vexillary stamen free or more or less connate with the remainder, sheath with an oblique mouth; filaments filiform; anthers uniform; ovary subsessilept or stipitate; ovules numerous or rarely only 2; style compressed dorsally with a bunch of hairs at the apex or pubescent or pilose all around, rarely glabrous, stigma terminal; fruit compressed, 2-valved, continuous within; seeds globose or rarely compressed, funicle dilated into a thin aril.

Vicia sativa L., Sp. Pl. 736 (1753); Stewart in Ali and Nasir, Ann. Cat. Vasc. Pl. W. Pak. & Kashm., 425 (1972); S.I. Ali in Nasir & Ali., Fl.Pak., 100: 269 (1977). **Figure 16A**

Syn: *V. notata* Gilib., Fl. Lit. Inch. 2: 105 (1782), *nom. inval.*; *V. globosa* Retz., Observ. Bot. 3: 39 (1783); *V. cosentinii* Guss., Fl. Sicul. Prodr. 2: 426 (1828); *V. torulosa* Boreauv, Fl. Centre France, ed. 3, 2: 173 (1857).

Annual herb, 15-70 cm, glabrescent to hairyl. Stem erect to ascending or procumbent, branched, winged. Stipules ca. 4-8 mm, triangular, deeply dentate; unicellular hairs present on margin. Leaves ca. 3-7 cm long; petiole ca. 0.5-1 cm long. Leaflets 4-8 pairs, 1-3 x 0.25-1.2 in size, oblong-obovate, linear or lanceolate, apex mucronate-obtuse, base rounded or truncate, margin ciliate (unicellular hairs present on abaxial side of leaflets, but few on adaxial when plant fully mature); tendril 2-3 branched, ca. 5.7 cm long. Flowers 1-2, axillary, ca. 15 x 7 mm in size; pedicel ca. 5 mm long. Calyx ca. 7 x 6 mm, pubescent; teeth subequal, with unicellular hairs on margin. Corolla purplish to white; vexillum ca.14 x 6 mm; wings ca. 11 x 3.5 mm; keel ca. 10 x 3.5 mm. Stamens diadelphous, dorsifixed, ca. 1 x 0.2 mm; anthers uniform. Ovary stipitate, 9 x 1.5 mm in size; style ca. 1 x 0.25 mm, terete with tuft of unicellular hairs towards distal end; stigma papillose, ca.0.5mm across. Pods linear, ca. 2-6 x 0.5-0.7 cm, torulose or not, pubescent, glabrous when mature. Seeds ca. 2.5-6 mm, sub-globose, brown to black

Flowering Period: December-February.

Herbarium specimens examined: NARA DESERT: *Rahmatullah Qureshi,* 1363 (SALUBGH).

Field remarks: A common climbing herb found in flat habitat.

Distribution: Pakistan; Kashmir; India; Orient; Europe; Russia; Far East (Ali, 1977).

6. Medicago

L., Sp. Pl. 778 (1753); Gen. Pl. ed. 5. 339 (1754); Baker in Hook.f., Fl. Brit. Ind. 2: 89 (1876); Ali, Fl.Pk., 100: 301 (1977).

Annual and perennial herbs or rarely shrubs, glabrous to pubescent, sometimes glandular hairy. Leaves 3-foliate, stipulate. Flowers in axillary racemes, heads or umbels or solitary. Calyx-teeth subequal; corolla yellow or orange, rarely violate or white; standard with major basal vein, usually with more than 3 branches near the base; horn of wing large, ca. 1/3 as long as the wing limb; keel very strongly adherent to the wing by a wing spur in a keel invagination. Androecium with apex of fused portion of anther column arched, alternating free portions of filament often relatively thick. Stigma fungiliform; style always short. Fruits diverse, short or long, straight, recurved or coiled, cylindrical, compressed or flat; sutures spiny, tuberculate, winged, fimbriate or without appendages. Seeds usually smooth, sometimes rugose or papilose.

About 85 species are reported distributed in Mediterranean region, West and Central Asia, Ethiopia, southern Africa (Boulos, 1999).

Medicago polymorpha L., Sp. Pl. 779 (1753); Stewart in Ali and Nasir, Ann. Cat. Vasc. Pl. W. Pak. & Kashm., 413 (1972); S. I. Ali in Nasir & Ali., Fl. Pak., 100: 302 (1977). **Figure 14C**

Syn: *Medicago denticulata* Willd., Sp. Pl. 3: 1414 (1802).

Annual, somewhat spreading. Petiole 1.2-4 cm long, leaflets 10-20 mm long, 7-15 mm broad, obovate to cuneate, obtuse, truncate to retuse, slightly toothed; stipules laciniate. Inflorescence a 2-8-flowered peduncled raceme, peduncle 5-15 mm long. Pedicel c. 5 mm long. Calyx 2.5 mm long, teeth subequal to the tube. Corolla c. 4 mm long, yellow. Fruit 2-12 mm high, with 2-4 (-6) spirals, glabrous, spines hooked in 2 rows, oriented almost parallel to the surface of the disc.

Flowering Period: March-May.

Herbarium specimens examined: NARA DESERT: *Rahmatullah Qureshi*, 1364 (SALUBGH).

Field remarks: A common weed found in flat habitat.

Distribution: Pakistan; widely distributed throughout the world, except for tropical regions and desert (Ali, 1977).

7. Rhynchosia

Louriero, Fl. Cochinch. 425, 460 (1790), *nom. cons*; Baker in Hook. f., Fl. Brit. Ind. 2: 220 (1876); Cooke, Fl Bomb. Pres. (reprint ed.), 1: 412 (1958); Ali in Fl. Pak., 100: 226 (1977).

Vines, stragglers, lianes or rarely subshrubs. Leaves pinnately 3-foliate; terminal leaflets ovate; lateral oblique, deltoid-rhomboid, inequilateral, gland-dotted below; stipules ovate or lanceolate; stipels small. Inflorescence axillary or terminal, racemose or paniculate, rarely solitary or in pairs. Bracts caduous; bracteoles 0. Calyx-tube campanulate; lobes subequal, upper pair connate. Corolla included or exserted; petals clawed; standard obovate or orbicular, with inflexed auricles at base, glabrous or pubescent; wings narrow; keels incurved, not beaked. Stamens 9+1; vaxillary stamen free; anthers uniform. Ovary sessile or stipitate; ovules 1 or 2; style incurved, pubescent below, apically dilated; stigma terminal or capitate. Pod ovoid or oblong, compressed, continuous, sometimes septate within. Seeds 2, rarely 1, subreniform, the hilum parallel to the sutures and the funicle centrical upon it; strophiolate thick, small or not.

A genus with ca. 150 species, distributed mainly in the warmer parts of the world, represented by 7 species in Pakistan (Ali, 1977).

Key to species:

1a. Pods glabrescent..**1. R. minima**
1b. Pods velvety:
2a. Terminal leaflets obcordate; seeds orbicular..**2 R. schimperi**
2a. Terminal leaflets rhomboid-ovate; seeds ellipsoidal..............................**3. R. pulverulenta**

1. Rhynchosia minima (L.) DC., Prodr.2: 385 (1825); Stewart in Ali and Nasir, Ann. Cat. Vasc. Pl. W. Pak. & Kashm.,418 (1972).; S. I. Ali in Nasir & Ali., Fl. Pak., 100: 231-232 (1977).

Syn. *Dolichos minimus* L., Sp. Pl. 726 (1753); *Rhynchosia laxiflora* Camb. in Jacq. Voy.Bot. 44.t.54 (1844).

A perennial, climbing or prostrate herb up to 120 cm long. Stem fairly branched, slender, angular, sulcate, pubescent; hairs 1-4-celled. Leaves alternate, 3-8 cm long; petiole ca. 2-4.5 cm long, more or less glabrous, sulcate; stipules ca. 4 mm long, linear-lanceolate, persistent. Terminal leaflets ca. 1.5-4.5 x 1.5-4 cm in size, rhomboid-ovate or rhomboid-obovate, apex acute-apiculate, base pulvinous, margin entire, more or less glabrous above, pubescent on veins and golden dotted beneath, strongly 3-nerved and cuneate at the base; lateral leaflets same in size; stipulules ca. 5 x 2 mm in size, ovate, base truncate, apex acute, margins ciliate, lateral nerves 2 pairs arising from the base. Flowers axillary in 5 cm long racemes; pedicel ca. 3 mm long. Bracts 2, ca. 2-2.3 x 0.1-0.2 mm in size, linear, greenish, hairy; trichomes unicellular. Calyx 2-lipped; tube ca. 4 mm long; upper 2 lobes ca. 2 x 1.5-2 mm in size, broadly ovate-lanceolate, apex apiculate, margins entire, glabrous within, hairy on out side; lower 3, ca. 4-6 x 2 mm in size, broadly lanceolate-ovate, hairy above; trichomes unicellular, ca. 0.1-0.5 mm long. Corolla yellow, exserted; standard ca. 1 x 0.8 cm in size, obovate, apex retuse; basally parallel veined, hairy above; wings ca. 1 x 0.4 cm in size, oblong or rhombic, apex rounded, base cuneate, glabrous; keels ca. 0.5 x 0.3 cm in size, oblong, glabrous, base clawed; claw ca. 3 mm long. Staminal tube ca. 6 mm long; filaments ca. 2-3 mm long, filiform; anthers ovate, apex apiculate. Ovary ca. 4.5 x 2 mm in size, oblong, densely hairy; style ca. 5 mm long, hairy above the middle. Pods ca. 1.5-2 x 0.5-0.6 cm in size, 2-seeded, somewhat compressed, turgid, slightly recurved, glabrescent. Seeds ca. 3.5-4 x 2.5-3 mm in size, ovate, turgid, pale-olive-green, mottled with grey and black to varied extent, estrophiolate.

Flowering period: January-March.

Herbarium specimens examined: NARA DESERT: *Rahmatullah Qureshi*, 985, 1300 (SALUBGH).

Field remarks: A common herb found in flat habitat.

Distribution: Pakistan, India, Ceylon, Lower Burma, Tropical Africa, West Indies, U.S.A. & Australia (Ali, 1977).

2. Rhynchosia schimperi Hochst. ex Boiss., Fl. Or. 2: 626 (1872); Stewart in Ali and Nasir, Ann. Cat. Vasc. Pl. W. Pak. & Kashm., 418 (1972); S. I. Ali in Nasir & Ali., Fl. Pak., 100: 230-231 (1977).

Syn: *Glycine schimperi* Hochst. & Steud. ex Steud. Nom. Bot. 2 (1): 691 (1840); *Rhynchosia arenaria* Blatt. & Hall. in Journ. Bomb. Nat. Hist. Soc. 26 (1): 243 (1918).

A slender, suberect-erect, perennial herb, densely and softly grey velvety all over, up to 50 cm long. Stem teret, ash coloured. Leaves 3-foliate, ca. 1.5-4 cm long; petioles ca. 0.5-1.5 cm long, softly pubescent, angular; stipules ca. 1-2 mm long, pubescent, setaceous. Terminal leaflets broadly obcordate, ca. 10 x 10-12 mm in size, apex emarginate-mucronate, mucro ca. 1 5 mm long, base cuneate, basally 3-nerved; nerves prominently raised below, densely tomentose with soft hiars on both surfaces; margins rarely slightly recurved; lateral leaflets ca. 8-10 mm across, obliquely suborbiculate, otherwise like the terminal leaflet; petiolule ca. 1 mm long. Racemes 2-3-flowered, axillary, ca. 3-5 mm long; pedicels ca. 2-3 mm long. Calyx ca. 4 mm long; tube ca. 1.5 mm long; upper lobes ca. 1.8-2 mm long; lower ca. 2.5 mm long, all densely hairy, base triangular, apex acute. Corolla ca. 5-6.2 mm long, clawed; standard ca. 8-9 x 6-6.2 mm in size, suborbicular, somewhat emarginated, densely hispid out side; wings ca. 5 mm long, yellow; keel ca. 7 mm long, green. Pods ca. 18-25 x 8-9 mm in size, base attenuate, apex obliquely cuspidate, outer margins more or less falcate, inner constricted between seed, densely covered with persistent tomentum. Seeds 2(1), ca. 5 mm across, suborbicular, slightly compressed, brownish-black with a yellow hilum, more or less round alog the hilum, glabrous, estrophiolate.
Flowering period: September-March.
Herbarium specimens examined: NARA DESERT: *Rahmatullah Qureshi*, 1327 (SALUBGH).
Field remarks: A common herb found in flat habitat.
Distribution: Pakistan India, Rajasthan, Arabia, Egypt (Ali, 1977).

3. Rhynchosia pulverulenta Stocks in Hook. J. Bot. Kew Misc. 4:147 (1852); Stewart in Ali and Nasir, Ann. Cat. Vasc. Pl. W. Pak. & Kashm., 418 (1972); S. I. Ali in Nasir & Ali., Fl. Pak., 100: 232 (1977).

Syn: *R. rhombifolia* Blatter & Hallb. in J. Bomb. Nat. Hist. Soc. 26: 242 (1918); *R. memnonia* auct. Non (Delile) DC. Baker in Hook.f., Fl. Brit. Ind. 2: 224 (1876); *R. elachistantha* Chiov., Ann. Ist. Bot. Roma 8: 100, t. 11, 1 (1903).

Perennial herb woody at the base. Stem 0.3-1.5 n, erect or scrambling, velvety tomentose; stipules 2.5-4 mm, linear-laneolate; petiolules 0.5-2 mm; leaflets 3, the terminal 0.4-3 x 0.6-4 cm, broader than longer, transversely elliptic or rhomboid, silky and with yellow glands on both surfaces; racemes 1-10-flowered; bracts 1.5-2 mm; pedicels ca. 0.5-3 mm. Calyx ca. 3.5-6.5 mm, pubescent, the tube shorter than the teeth. Corolla 4-6.5 mm, pale yellow. Pods ca. 1.5-2 x 0.5 cm, falcate-oblong, densely pubescent with short white hairs and small pale gland dots, 1-2-seeded. Seeds ca. 3.5-4 x 2.5-3 mm in size, ellipsoid, brown, mottled black.

Herbarium specimens examined: NARA DESERT: *Rahmatullah Qureshi*, 1365 (SALUBGH).
Field remarks: A less common herb found on slopes and interdunal valley.
Distribution: Egypt, tropical East and South west Africa, Ethiopia, Somalia, Arabia, Scotra, Pakistan, India (Boulos, 1999).

8. Indigofera

L., Sp. Pl. 751 (1753); Ali in Bot. Notiser 111: 543 (1958); Ali in Fl. Pak., 100: 65 (1977).

Herbs, shrubs or undershrubs often silvery-canescent, densely clothed with appressed medifixed hairs (sometimes mixed with tomentum and laxer hairs). Leaves pinnate or imparipinnate or palmately 3-foliate, unifoliate or simple; leaflets (sub)opposite, entire, chartaceous, sometimes gland-dotted; stipules small, often setaceous; stipels linear-subulate. Inflorescence axillary racemes, spike, heads or solitary. Flowers pedicelled or sessile in the axil of caduous bract; bracteoles 0. Calyx-tube campanulate; lobes subequal, lower one longest. Corolla usually exceeding calyx; petals usually clawed; standard orbicular to ovate, gradually narrowed to the base, glabrous; wings oblong, slightly cohering with keels; keels erect, obtuse, or acuminate, basally with a downward spur on each side. Stamens 9+1; vaxillary stamen free; anthers uniform, usually apiculate. Ovary sessile; ovule(s) α or 1 or 2; style short, incurved, glabrous; stigma capitate, sometimes panicillate. Pod usually oblong-linear, terete or 4-gonous, sometimes globose, 4-3-seeded, erect or deflexed, straight, sometimes curved, torulose or not, septate within, mucronate, usually dehiscent; endocarp often spotted. Seeds globose-ovoid, 4-gonal or terete, estrophiolate.

A genus of more than 300 species, widely distributed in tropics and subtropics of both New and Old World, with centre of distribution in Tropical Africa. In Pakistan, it is represented by 24 species (Ali, 1977).

Key to species:
1a. Plants annual:
2a. Leaves imparipinnate:
3a. Pods quite flat, falcate, papery, 6-8-seeded.......................................**1. I. hochstetteri**
3b. Pods cylindric, turgid, 3-4-seeded..**2. I. sessiliflora**
2b. Leaves simple:
4a. Leaves linear, not more than 2 mm wide; pods globose, 1-seeded...............**3. I. linifolia**
4b. Leaves broadly ovate-cordate, 4-10 mm wide; pods cylindric, 2-seeded......**4. I. cordifolia**
1b. Plants perennial:
5a. Pods 4-6-seeded, spreading; seeds ca. 1.2 mm across, globose-discoid, slightly compressed...**5. I. argentea**
5b. Pods more than 5-8-seeded; seeds oblong, 4-gonous, truncate ....**6 .I. oblongifolia**

1. Indigofera hochstetteri Baker in Oliver, Fl.. Trop.Afr.2:101 (1871); Stewart in Ali and Nasir, Ann. Cat. Vasc. Pl. W. Pak. & Kashm.,408 (1972).; S. I. Ali in Nasir & Ali., Fl. Pak., 100: 76-78 (1977). **Figure 14A**

Syn. *Indigofera anabaptista* Steud.ex Baker in Hook.f., Fl.Brit.Ind. 2: 102 (1876).

An annual, prostrate herb up to 35 cm long. Stem solid, woody at the base, herbaceous above, triangular to circular, greyish-green, appressed hairy; internode ca. 1.5-3.8 cm long. Leaves imparipinnately compound, alternate, ca. 1-2 cm long; petiole ca. 0.2-0.7 cm long. Leaflets 3-5, opposite, ca. 0.5-2 cm long, ca. 6-12 mm broad, oblanceolate or elliptic-oblong, base cuneate, apex apiculate, margin sub-entire, lateral nerves obscure, sparsely pilose on above, densely hairy on below; petiolule ca. 1 mm long, purplish in colour, pubescent. Stipules 2, ca. 4.5 x 1 mm in size, linear to more or less lanceolate, apex acute, margin entire, appressed-pubescent, hyaline. Inflorescence a short peduncled raceme up to 7 cm long, 3-7 flowered; peduncle ca. 0.3-0.5 cm long. Bract I, ca. 3 x 0.8 mm in size, narrowly lanceolate, apex acute, margin sub-entire, hyaline, densely pilose, trichomes ca. 1-2 mm long. Sepals fused at the base, tube ca. 1-1.5 mm, glabrous within, hispid above; lobes 5, free, valvate, unequal, ca. 2-3 x 0.8-1 mm in size, lanceolate, apex acute, margin entire, green in middle with pink spots below the margin, glabrous within, appressed hairy on above, persistent; trichmes ca. 1-1.5 mm long. Petals 5, exserted, papilionaceous, orange red in colour, clawed; standard ca. 7 x 5 mm in size, elliptic, base cuneate, apex apiculate, margin entire, lateral nerves parallel veined, densely pilose on above, glabrous below; wings 2, ca. 5 x 2 mm in size, obovate, apex oblique emarginate, base cuneate, margin entire, glabrous; keels 2, ca. 7 x 3 mm in size, jointed, oblanceolate, base cuneate and spurred, apex rounded and sparsely hairy, margin entire; spur ca. 1 mm long. Stamens diadelphous, 9+1; staminal sheath ca. 5 mm long, glabrous, greenish; filament ca. 1-2 mm long, filiform; anthers ca. 0.5 mm across, dorsifixed, ovate, base auriculate, apex apiculate, margin entire, greenish, bilocular. Ovary ca. 5 x 1 mm, linear, base cuneate, apex acute, appressed-pubescent, trichome ca. 0.4-1 mm long; style ca. 1 mm long, curved on the top, greenish, glabrous; stigma ca. 0.3 mm, capitate. Fruit ca. 1-1.5 long, ca. 4 mm broad, linear, flat, appressed hairy, ridge longitudinal along the centre of each flattened face and a transverse ridge between the seeds, 5-9-seeded.

Flowering period: August-October.

Herbarium specimens examined: NARA DESERT: Saleh pat, Sainewari, Qadan Wari, *Rahmatullah Qureshi*, 659 671, 842, 1120, 1153 (SALUBGH).

Field remarks: A common herb found in interdunal valley.

Distribution: Pakistan (Sindh, Punjab & Baluchistan), Afghanistan, India, Arabia, Sudan, Ethiopia, Somalia Republic, Eritrea, Nigeria, Kenya, Uganda, Tanganyika & Congo (Ali, 1977).

2. Indigofera sessiliflora DC. Prodr. 2: 228 (1825); Stewart in Ali and Nasir, Ann. Cat. Vasc. Pl. W. Pak. & Kashm., 408 (1972); S. I. Ali in Nasir & Ali., Fl. Pak., 100: 74-75 (1977).

Syn: *I. tribuloides* Boiss., Fl. Or. 2: 189 (1872); *I. trigonelloides* auct non Jaub. & Spach) Baker in Hook. f. FBI. 2: 94 (1876).

A caespitose, annual, canescent herb up to 30 cm long. Stem prostrate, procumbent or rarely sub-erect, diffuselybranched from the base, teret, angular, densely covered with adpressed silvery hairs mixed with a few spreading ones. Leaves ca. 1-4.5 cm long, 5-7-foliate; petiole ca. 1-1.5 cm long, hairy; leaflets alternate, ca. 0.5-15 x 3-5 mm in size, elliptic or oblanceolate, apex obtuse-mucronulate, margin entire, densely clothed with medifixed, adpressed, silvery hairs on both surfaces; petiolule ca. 1-2 mm long; stipules 3 mm long, linear, base broad, apex acute-acuminate, hairy out side. Inflorescence a dense, sessile, axillary, villous heads, ca. 5-6 mm across; bracts 2 mm long, ovate-lanceolate, apex acute, hairy. Calyx ca. 3.5-4 mm long; teeth linear, hairy. Corolla as long as calyx, pale-rose in colour. Pods linear-cylindric, torulose, straight, apex acute, densely clothed with medifixed hairs, up to 1 cm long, 3-6-seeded. Seeds 1 mm across, quadrangular globose, pale-brown, shining, glabrous.

Flowering period: February-April.

Herbarium specimens examined: NARA DESERT: Chundko, *Rahmatullah Qureshi*, 262, 806, 869, 883, 927, 1024, 1038 (SALUBGH).

Distribution: Pakistan (Sindh, Punjab), Arabia, Eritrea, Sudan, Nigeria, Senegal (Ali, 1977).

3. Indigofera linifolia (L.) Retz., Obs.4: 29 (1786); Stewart in Ali and Nasir, Ann. Cat. Vasc. Pl. W. Pak. & Kashm.,408 (1972).; S. I. Ali in Nasir & Ali., Fl. Pak., 100: 69-71 (1977).
Syn: *Hedysarum linifolium* L.f., Suppl.331 (1781).

An annual procumbent herb up to 60 cm long. Stem more or less woody at the base, herbaceous above, winged, basally spreading; branches silvery-canescent; hairs appressed; internode ca. 0.5-1.5 cm long. Leaves simple, alternate, ca. 0.5-2.5 cm long, ca. 4-8 mm broad, linear-elliptic, base cuneate, apex acute-mucronate, margin entire, lateral nerves obscure, dark green, sparsely hairy on above, densely appressed-pubescent on below, chartaceous; petiole ca. 0.5-3 mm long, puberulent. Stipules 2, ca. 3-5 x 1-1.5 mm in size, ovate, base obtuse, apex acute, margin ciliate, setaceous above. Inflorescence an axillary raceme, ca. 1-2 cm long; 3-16 flowered, peduncle ca. 3-6 mm long, densely appressed hairy. Flowers bisexual, complete, irregular, zygomorphic, sessile, ca. 8 mm across. Fruit ca. 4 x 3 mm in size, globose, apex apiculate, appressed silvery-pubescent. Seed solitary, globose, ca. 2 mm.

Flowering period: July-October.

Herbarium specimens examined: NARA DESERT: Qadir bek ja kuba, *Rahmatullah Qureshi*, 539, 694, 868 (SALUBGH).

Field remarks: A less common herb found in flat habitat.

Distribution: Pakistan (Sindh, Punjab & Baluchistan), Azad Kashmir, Bangla Desh, India, Afghanistan, Ceylon, Burma, Indochina, Siam, Indonesia, New Guinea, Australia, China, Sudan, Ethiopia & Eritrea (Ali, 1977).

4. Indigofera cordifolia Heyne ex Roth, Nov.Pl.Sp..357 (1821); Stewart in Ali and Nasir, Ann. Cat. Vasc. Pl. W. Pak. & Kashm.,407 (1972).; S. I. Ali in Nasir & Ali., Fl. Pak., 100: 69 (1977). **Figure 13D**

An annual, prostrate diffuse or suberect herb, usually clothed with long, white, medifixed hairs. Stem up to 30 cm long, pubescent when young, the older nearly glabrous. Leaves simple, ca. 0.5-1.5 x 0.4-1 cm in size, broadly ovate-cordate, base more or less truncate, apex subobtuse-mucronate, hairy on both sides; stipules ca. 1.5 mm, setaceous. Racemes axillary subsessile, 4-8-flowered heads. Calyx-tube ca. 1 mm long; teeth ca. 3.5-4 mm long, linear, apex acute, densely hairy outside. Corolla bright red; standard ca. 4 mm long, spathulate, 7-9-nerved. Pods ca. 4-5 mm long, cylindric, oblong, straight or moniliform, densely hairy; beak ca. 1.2 mm long, curved, 1-2-seeded. Seeds ca. 1 mm across, irregularly ovoid, verrucose, truncate at one end, pale-brown.

Flowering period: August-October.
Herbarium specimens examined: NARA DESERT: Chundo, Shadi Shaheed, *Rahmatullah Qureshi*, 651, 653, 841, 1125, 1151 (SALUBGH).
Field remarks: A very common herb found in flat habitat.
Distribution: Pakistan (Sindh, Punjab & Baluchistan), Afghanistan, India, Indonesia (Timor), Australia (n. Australia), Sudan, Ethiopia & W. Tropical Africa (Ali, 1977).

5. Indigofera argentea Burm.f., Fl. Ind. 171(1768); Stewart in Ali and Nasir, Ann. Cat. Vasc. Pl. W. Pak. & Kashm.,406 (1972).; S. I. Ali in Nasir & Ali., Fl. Pak., 100: 72 (1977). **Figure 13C**

Syn. *I. semitrijuga* auct. Non Forskal: Baker in Hook.f., Fl. Brit.Ind. 2: 98 (1876); *I. burmanii* Boiss., Fl. Or.2: 189 (1872).

A rigid, perennial, densely argenteo-canescent undershrubs, diffusely branched from the base. Branches numerous, slender, terete, silvery canescent with scabrous 2-partite adpressed hairs. Leaves ca. 2-3 cm long; imparipinnate; stipules setaceous. Leaflets subsessile, in 3-4, opposite pairs, ca. 5-6.5 x 3-4 mm in size, broadly obovate, apex rounded or rarely emarginate, base cuneate, terminal leaflets shortly stalked, all firm in texture, densely clothed on both sides with medifixed, silvery adpressed hairs; midrib impressed above. Flowers in lax, axillary racemes, ca. 4-5 cm long, 6-12-flowered; pedicels ca. 1 mm long. Calyx-tube 5 mm, canescent above; teeth 1-1.3 mm long, linear-lanceolate, lower longer than others. Corolla crimson-red; standard ca. 5 mm long orbicular, silvery-pubescent above. Pods ca. 1-1.5 x 0.2 cm in size, turgid, clothed with white, adpressed hairs and mixed with numerous, capitate glands, apex shortly mucronate, 4-6-seeded, spreading. Seeds ca. 1.2 mm across, globose-discoid, slightly compressed, pale with a few, irregular, black markings and minute depressions.
Flowering period: Throughout the year.
Herbarium specimens examined: NARA DESERT: Chundko, Salehpat, Rohri Hills, *Rahmatullah Qureshi*, 96, 217, 600, 604, 606, 664, 700, 804, 811, 871, 874, 941, 1011, 1032, 1115, 1141, 1292 (SALUBGH).
Field remarks: A very common herb found in all microhabitats.
Distribution: Pakistan (Sindh, Punjab & Baluchistan), Persia, Arabia, Egypt, Sudan, Somalia, Ethiopia & Libya (Ali, 1977).

6. Indigofera oblongifolia Forsk., Fl.Aegypt.-Arab.137 (1775); Stewart in Ali and Nasir, Ann. Cat. Vasc. Pl. W. Pak. & Kashm., 408 (1972).; S. I. Ali in Nasir & Ali., Fl. Pak., 100: 71-72 (1977).

An annual erect-suberect shrub up to 130 cm long. Leaves imparipinnate, rarely unifoliate, normally 3-5-foliate; leaflets alternate, ca. 1.5-3 x 0.5-1 cm in size, oblanceolate or elliptic-oblong, apex acute-slightly apiculate, margin entire, hairy on both side; petiole ca. 0.5-1 cm long; pubescent; stipule ca. 3-4 mm long. Inflorescence terminal and axillary raceme, ca. 3-8 cm long; peduncle ca. 2-4 mm long. Flowers bisexual, complete, hypogynous, ca. 11 x 3 mm in size; pedicel ca. 2 mm, pubescent. Sepals 5-lobed, fused at the base; tube ca. 2.5 mm long; lobes free, ca. 1-1.5 mm long, ovate- subtriangular, base rounded, apex acuminate, salver-shaped out side, green, persistent. Petals 5, red in colour; vaxillum 6 mm across, obovate, gradually narrowed towards the base, pubescent externally; wings 2, ca. 4.5 x 2 mm in size, suboblanceolate, slightly coherent with the keels, glabrous; keels 2, ca. 5.5 x 2.5 mm in size, gibbous on each side, spurred, jointed together, oblanceolate, apex obtuse, glabrous. Stamens 9+1, diadelphous, vaxillary stamen free, filaments of 9 stamens forming staminal tube around the ovary; filaments ca. 2-4 mm long; anthers ca. 1 mm across, bilobed, dorsifixed, ovate, apex apiculate. Ovary 1-celled, unilocular, ca. 2.5 x 0.25 mm in size, sublanceolate; style ca. 2 mm long, gradually curved, ± pubescent; stigma capitate, ca. 0.2 mm. Pods deflexed, pubescent, 5-8-seeded. Seeds oblong, 4-gonous, truncate.

Flowering period: September- November.

Herbarium specimens examined: NARA DESERT: Tajjal, *Rahmatullah Qureshi*, 278, 421, 808, 820, 1210 (SALUBGH).

Field remarks: A common shrub found in hilly area.

Distribution: Pakistan (Sindh, Punjab & Baluchistan), India, Uttar Pradesh, Ceylon, Java, Jordan, Yemen, Hejaz, Bahrain, Socotra, Sudan, Eritrea, Nigeria, Somaliland, Egypt Senegal, Angola (Ali, 1977).

9. Tephrosia

Persoon, Syn. Pl. 2: 328 (1807), *nom. cons*; Baker in Hook. f., Fl. Brit. Ind. 2: 110 (1867); Ali in Biologia 10: 23 (1964); Ali in Fl. Pak., 100: 58 (1977).

Herbs or undershrubs or shrubs. Leaves odd-pinnate or 1-foliate; leaflets opposite, obovate-oblanceolate or oblong-elliptic, entire, closely nerved; stipules lanceolate or subulate, rarely spinose; stipels minute or 0. Inflorescence a terminal or leaf-opposed racemes, rarely 1-3 in a clusters. Bracts present; bracteoles 0. Calyx-tube campanulate; lobes lanceolate-setaceous; upper lobes subconnate, acute; lower one longest, apex acuminate. Corolla exserted; petals clawed; standard suborbicular, with 2 basal callosities, sometimes auricled, pilose externally; wings obliquely oblong or obovate, slightly adherent to incurved keels. Stamens diadelphous; vexillary stamen becoming free; anthers uniform. Ovary sessile; ovules ∝; style incurved, often flattened, glabrous or pubescent; stigma terminal, penicillate. Pod linear, flattened, continuous within or obscurely septate, dehiscing by both sutures. Seeds reniform, ellipsoid or ovoid, strophiole sometimes present.

A genus of over than 300 species distributed mainly in the tropical and subtropical regions of the world (Ali, 1977).

Key to species:

1a. Flowers axillary, 1-6 together:

2a. Leaves simple...**1. T. strigosa**

2b. Leaves imparipinnate:

3a. Stigma penicillate; claw of the keel less than half the blade......**2. T. uniflora** ssp. **petrosa**

3b. Stigma not penicillate; claw of the keel more than half the blade:

1b. Flowers in axillary racemes:

4a. Pods highly falcate or circinately coiled ....**3. T. falciformis**

4b. Pods straight or slightly curved at the ends:

5a. Leaflets 11-19, calyx-teeth ± as long as the tube; pods velvety tomentose.......**4. T. villosa**

5b. Leaflets 9-21; calyx-teeth much longer the the tube; pods glabrescent.......**5. T. purpurea**

1. Tephrosia strigosa (Dalz.) Sant. & Mahesh. in Journ. Bomb. Nat. Hist. Soc. 54 (3): 805 (1962); Stewart in Ali and Nasir, Ann. Cat. Vasc. Pl. W. Pak. & Kashm.,421 (1972); S. I. Ali in Nasir & Ali., Fl. Pak., 100: 59 (1977).
Syn: *Maconyx strigosus* Dalz. in Hook. Kew. Journ. 2: 35 (1850); *Tephrosia tenuis* Wall. ex Dalz. & Gibs., Bomb. Fl. 61 (1861).
A copiously branched, ± erect slender, annual herb, up to 30 cm high. Stem filiform, adpressedly hairy, branches virry; stipules subulate. Leaves unifoliate, ca. 2-4 x 0.3-0.4 cm in size, linear-elliptic, narrowed at the base, apex apiculate, sparsely hairy above near margins, densely silky beneath with adpressed hairs; petioles ca. 1.5-2 mm long, filiform, hairy. Flowers 1-2 in the axil of leaves; pedicels ca. 6-12 mm long, filiform. Calyx ca. 2.5 mm long, silky; teeth 1 m long, subequal, shorter than the tube. Corolla 4 mm long, bluish or bluish-purple; standard pubescent above. Pods ca. 1.5-2.5 x 0.5 cm in size, somewhat turgid, thinly adpressedly hairy, ca. 6-10-seeded, linear, flat, adpressed hairy. Seeds ca. 1.8 mm in size, suborbicular, slightly turgid, smooth, mottled with brown, separated by internal partitions.
Flowering period: August-October.
Herbarium specimens examined: NARA DESERT: *Rahmatullah Qureshi*, 1366 (SALUBGH).
Field remarks: A rare herb found in southern part (Achhro Thar) of this desert.
Distribution: Pakistan, India; Burma; Laccadive Islands (Ali, 1977).

2. Tephrosia uniflora subsp. **petrosa** (Blatter & Hallberg) Gillett & Ali in Kew Bull. 13: 114 (1958); Stewart in Ali and Nasir, Ann. Cat. Vasc. Pl. W. Pak. & Kashm., 422 (1972); S. I. Ali in Nasir & Ali., Fl. Pak., 100: 63 (1977). **Figure 15C**
Syn: *Tephrosia petrosa* Blatter & Hallberg in J.Bomb.Nat.Hist.Soc. 26: 239 (1918).

A semi-erect perennial under-shrub up to 90 cm tall. Stem solid, woody, herbaceous, branched, quaderangulate, dark greyish-green, setose; internode ca. 2-4 cm long. Leaves imparipinnately compound, alternate, ca. 2-5.5 cm long; petiole ca. 1-2 cm long, hispid. Leaflets 3-7, opposite, ca. 0.8-3 cm long, ca. 6-18 mm broad, oblanceolate-elliptic or oblong, base cuneate, apex emarginate, margin entire, lateral nerves 6-12 pairs, sparsely hairy on above, densely appressed hairy on below; petiolule ca. 4 mm long, greyish in colour, pubescent. Stipules 2, ca. 5-7 x 1-2 mm in size, lanceolate, apex acute, margin entire, appressed-pubescent, green. Inflorescence a short axillary fascicles consisting on 2-3-flowers; flowers ca. 0.9-1 x 0.5 cm in size, pedicel ca. 4 mm long, very hispid. Bract ca. 2 x 1 mm in size, ovate, apex acute, highly hispid, trichome ca. 1 mm long. Sepals campanulate; tube ca. 3 mm, glabrous within, hispid above; lobes 5, free, valvate, unequal, upper lobes subconnate, ca. 2-4 x 2 mm in size, ovate, base rounded, apex acute, margin entire, glabrous within, appressed hairy on above, green, persistent. Petals 5, exserted, papilionaceous, pink in colour, clawed; standard ca. 0.7 x 0.9 cm in size, orbicular, base cuneate, apex emarginate, margin entire, parallel veined, glabrous below, sparsely pubescent on above, claw ca. 4 mm in size; wings 2, ca. 0.7 x 0.4 cm in size, oblanceolate, apex round, base cuneate, margin entire, glabrous; keels 2, ca. 10 x 7 mm in size, jointed, obovate, base oblique, apex rounded, margin entire, glabrous. Stamens diadelphous, 9+1; staminal sheath ca. 12 mm long, curved above, glabrous, greenish; filament ca. 3-4 mm long, filiform; anthers ca. 1 mm across, dorsifixed, elliptic-ovate, greenish, bilocular. Ovary ca. 8 x 2 mm in size, linear, densely hispid; style ca. 4 mm long, curved on the top, filiform, greenish, glabrescent; stigma ca. 1 mm, capitate,

yellowish in colour. Pods ca. 2-4.5 x 0.3-0.4 cm in size, linear, base cuneate, apex acute, brown, salver-shaped, 3-7-seeded. Seeds ca. 8 x 4 mm in size, enclosed in membranous scales, reniform, greyish with black spots, glabrous.
Flowering period: June-August.
Herbarium specimens examined: NARA DESERT: *Rahmatullah Qureshi*, 40 (b), 279, 1136 (SALUBGH).
Field remarks: A very common subshrub found in flat habitat.
Distribution: Pakistan, India; Arabia (Ali, 1977).

3. Tephrosia falciformis Ramaswami in J. As. Soc. Beng. (N.S.) 12: 125 (1916); Stewart in Ali and Nasir, Ann. Cat. Vasc. Pl. W. Pak. & Kashm., 421 (1972); S. I. Ali in Nasir & Ali., Fl. Pak., 100: 60 (1977). **Figure 15B**

A rigid, perennial much branched shrub up to 70 cm tall. Branches angular, densely pubescent or adpressed-white silky. Leaves imparinnate, 6-13 cm long; stipules ca. 3 mm long, linear, persistent; petioles ca. 2-3 cm long, pulvinous, conspicuous. Leaflets 5-17, narrowly oblanceolate or oblong, apex mucronate, base cuneate; terminal leaflet ca. 3-5 x 0.5-1.5 cm in size, lateral ones ca. 2.5-4.5 x 0.3-0.8 cm in size, all argenteo-canescent with adpressed hairs on both surfaces; secondary nerves 11-13 pairs; petiolules of terminal leaflet ca. 3-4 mm long, of lateral leaflets ca. 2-2.5 mm long. Racemes terminal, lax, 10-30 cm long. Flowers usually germinate, one long and other short. Calyx white-silky; tube ca. 2.5 mm long; teeth as long as the tube, lanceolate, apex acute. Corolla purple-red; standard orbicular, base cordate, ca. 8 x 10 mm, densely silvery-white above. Style glabrous, slightly incurved; stigma penicillate. Pods ca. 5-10 cm long, falcate or rarely circinate, flat, densely pubescent; tip mucronate. Seeds ca. 5 mm long, obscurely reniform, glaucous.
Flowering period: June-August.
Herbarium specimens examined: NARA DESERT: *Rahmatullah Qureshi*, 43, 830, 1161, 1189 (SALUBGH).
Field remarks: A less common subshrub found on slopes.
Distribution: Pakistan & India (Ali, 1977).

4. Tephrosia villosa (L.) Pers., Syn. Pl. 2: 329 (1807); Stewart in Ali and Nasir, Ann. Cat. Vasc. Pl. W. Pak. & Kashm., 422 (1972); S. I. Ali in Nasir & Ali., Fl. Pak., 100: 60 (1977). **Figure 15D**

Syn. *Cracca villosa* L., Sp. Pl. 752 (1753).

An annual, bushy, erect herb up to 45 cm long. Stem branched, hard and woody bellow, terete, grey-canescent; internode ca. 2-3.5 cm long. Leaves imparipinnate; rachis ca. 1.2-3 cm long; leaflets 5-13, ca. 0.8-2.4 cm long, ca. 3-12 mm broad, opposite-subopposite, narrowly oblanceolate, apex mucronate, base cuneate, margin entire, chartaceous, parallel veined, sparsely villous above, densely villous bellow; petiole ca.1-1.8 cm long; petiolule 2-3 mm long; stipules 2, ca. 4-6 x 1 mm in size, lanceolate, appressedly pubescent. Inflorescence an elongated axillary raceme up to 5 cm long, lower flowers in fascicles, the lowest often in the axil of the leaf. Bract narrowly triangular. Flower ca. 13 mm long, bisexual, hypogynous. Pedicel ca. 2-3 mm long. Sepals 5, sericeous, lobes lanceolate-subovate, setaceous, ciliate, upper lobes ca. 2 mm, lower one 3-4 mm long, persistent. Petals 5, standard ca. 13 mm

across, orbicular, apex emarginate, base truncate, sericeous externally, veins arising from the centre; wings ca. 13 x 5 mm in size, obovate, glabrous, slightly adherent to incurve keels; keels ca. 10-11 x 2.5-3 mm in size, ovate, beaked, glabrous. Stamens 10, monadelphous; staminal sheath ca. 9 x 2 mm. long; filaments 2-3 mm long; anthers ca. 1 x 0.5 mm in size, dorsifixed, bilobed, elliptic-oblong, yellow in colour. Ovary linear, ca. 7 x 1 mm in size, appressedly pubescent; style ca. 6 mm long, glabrous; stigma penicillate. Fruit ca. 2.5-3.5 cm long, ca. 7-8 mm broad, compressed, linear, 2-valved, sericeous, continuous within, apex slightly incurved, velvety tomentose, densely hairy, spreading. Seeds 5-8, subquadrate, ca. 6 x 3 mm, strophiolate, brownish, shiny, glabrous.
Flowering period: January-September.
Herbarium specimens examined: NARA DESERT: *Rahmatullah Qureshi*, 40 (a), 679, 987, 1029, 1249 (SALUBGH).
Field remarks: A common subshrub found in flat habitat.
Distribution: Pakistan India, Ceylon, Egypt, Tropical Africa, Angola, Uganda, Mozambique, Madagascar, Rhodesia & S. Africa (Ali, 1977).

5. Tephrosia purpurea (L.) Pers. Syn. Pl. 2: 329 (1807); Stewart in Ali and Nasir, Ann. Cat. Vasc. Pl. W. Pak. & Kashm.,421 (1972); S. I. Ali in Nasir & Ali., Fl. Pak., 100: 63 (1977).

Syn: *Caracca purpurea* L., Sp. Pl. 752 (1753); *Tephrosia leptostachya* DC. Prodr. 2: 251. t. 2 (1825); *T. Wallichii* Grah. ex Fawcett & Randle in Journ. Bot. 55: 35 (1917); *T. Hamiltonii* Drumm. ex Gamble, Fl. Madras 1: 320 (1918).

A much branched, erect, perennial herb up to 3 feet high. Stem ± hairy with adpressed hairs. Leaves ca. 10-13 cm long; stipules ca. 7-9 mm long, lanceolate, linear-subulate, erect or reflexed, hairy; leaflets ca. 7-13, ca. 1.8-2.5 x 0.5-2 cm in size, oblanceolate, apex mucronate, subcoriaceous, both sides grey-green, glabrous above, adpressedly pubescent beneath. Flowers in leaf-opposed racemes, reddish purple or bright pink; bracts linear, ca. 2-4 mm long. Calyx ca. 4 mm long; tube ca. 1.5 mm long, teeth ca. 2-3 mm long, lanceolate, apex acuminate. Corolla much exserted, ca. 8 mm long, deep purple; standard pubescent on the back. Style flattened, glabrous; stigma penicillate. Pods ca. 3-4 x 0.4-0.5 cm in size, slightly recurved, glabrescent to softly pubescent, 5-9 seeded.
Flowering period: August-October.
Herbarium specimens examined: NARA DESERT: *Rahmatullah Qureshi*, 1367 (SALUBGH).
Field remarks: A less common subshrub found in flat habitat.
Distribution: Southern Asia, Australia, Tropical Africa, southwards to Natal; introduced to tropical America (Ali, 1977).

10. Sesbania

Adans., Fam. 2: 326 (1763); Gillett in Kew Bull. 17: 91 (1963); Baquar in Pak.J.Bot. 1: 131 (1969); Ali, Fl.Pak., 100: 87 (1977).

Trees, shrubs or herbs. Leaf imparipinnate, leaflets entire, stipels present or absent; stipules caduceus. Inflorescence an axillary raceme. Pedicel present; bract and bracteoles mostly caduceus. Calyx tube truncate or lobed or toothed, teeth subequal, shorter than the tube. Corolla glabrous. Stamens 9+1, vexillary stamens free, anthers uniform. Ovary often stipitate, ovules numerous, style incurved, stigma capitates. Fruit long, dehiscent, transversely septate, 8-50-seeded.

A genus of about 55 species, widely distributed in the tropical and sub-tropics of the world (Ali, 1977).

Sesbania bispinosa (Jacq.) W. F. Wight in U. S. Dept. Bur. Pl. Ind. Bull.no. 137: 15 (1909); S.I. Ali in Nasir & Ali., Fl.Pak., 100: 90-91 (1977).

Syn: *Aeschynomene aculeata* Schreb. in Nov. Act. Cur. 4:134. 1770; *Sesbania aculeata* (Willd.) Poir. in Lam. Encycl. Meth. 7: 128. 1806.

Annual or biennial, erect herb up to 3 m tall, semi-woody. Stem terete, glabrous or sparsely pubescent. Leaves paripinnate, rachis up to 25 cm long, aculeate; leaflets up to ca. 20 x 4 mm, oblong, base sub-oblique, apex obtuse- truncate, mucronate. Stipules ca. 8 mm long, membranous, caducous. Inflorescence 3-6 flowered raceme. Peduncles slender; pedicel ca. 6 mm long. Calyx ca. 6 x 4 mm, glabrous, teeth deltoid. Corolla yellow; vexillum ca. 10.6 x 12 mm, suborbicular; wings ca. 9 x 2.8 mm; keel ca. 10 x 3 mm. Stamens ca. 9 mm long, anthers 0.8 mm long. Ovary ca. 5.2 x 0.6 mm; style 2 mm long. Fruit up to ca. 20 cm long, with ca. 30-38 seeds, falcate, glabrous.

Flowering period: June-September.

Herbarium specimens examined: NARA DESERT: *Rahmatullah Qureshi*, 1368 (SALUBGH).

Field remarks: A common weed found in flat habitat.

Distribution: Pakistan, India, China, S.E. Asia; Mozambique, Rhodesia; Botswana; South West Africa; Lesotho; S. Africa; Madagascar; Mauritius; Kenya; Tanganyika; Jamaica (Ali, 1977).

11. Argyrolobium

Eckl. & Zeyh., *nom. conserve.*; Boiss., Fl. Or. 2: 31 (1872); Baker in Hook.f., Fl.Brit.Ind. 2: 63 (1876); Ali, Fl.Pak., 100: 35 (1977).

Herbs or dwarf shrubs; leaves digitately trifoliate. Flowers in short racemose or umbellate inflorescence, leaf-opposed, rarely solitary. Calyx deeply 2-lipped, the upper lif 2-fid, the lower 3-toothed. Corolla yellow. Stamens united into a close tube. Pods linear-oblong, flattened, dehiscent.

About 130 species, mainly distributed in North, tropical and South Africa, Madagascar, S. Europe, W. Asia and Indo-Pak subcontinent (Ali, 1977).

Argyrolobium roseum ssp. **ornithopodioides** (Jaub. & Spach) Jafri & Ali in S.U. Sci. Res. J. 2: 4 (1966); Stewart in Ali and Nasir, Ann. Cat. Vasc. Pl. W. Pak. & Kashm., 385 (1972); S. I. Ali in Nasir & Ali., Fl. Pak., 100: 38 (1977).

Syn: *A. ornithopdioides* Jaub. & Spach in Ann. Sci. Nat. Ser. 2. 19: 52 (1843).

Prostrate herb. Stem terete, hairy, much branched. Stipules c. 1.5-2 mm, free lateral. Leaves trifoliolate, petiole c. 3-10 mm long. Petiolule less than 1 mm. Leaflets c. 4-13 mm

long, obovate, obtuse, glabrous above, pilose below. Inflores¬cence a 1-4-flowered peduncled raceme, peduncle 0.4- 4.0 cm. Bract c. 1.5 mm long. Calyx c. 4.5-5.0 mm long, bilabiate, pilose. Corolla c. 5 mm long, red. Pod up to 1.9 cm long, 2.5-3 mm broad, 3-9-seeded, pilose.
Flowering Period: April-October
Herbarium specimens examined: NARA DESERT: *Rahmatullah Qureshi*, 1369 (SALUBGH).
Field remarks: A common herb found in hilly area.
Distribution: Pakistan (Sind, Baluchistan); Persian Baluchistan; Arabia, Aden (Ali, 1977).

12. Melilotus

Mill., Gard. Dict., Abr. ed. 4 (1754); Ali, Fl. Pak., 100: 308 (1977).

Annual or biennial herbs; leaves pinnately 3-foliate, nerves often ending in teeth; stipules adnate to the petiole; flowers small, yellow or white, in slender axillary racemes; bracts minute or absent; bracteoles absent; calyx teeth short, subequal; petals free from the staminal tube, deciduous; vexillum obovate or oblong, contracted at the base, subsessile; wings oblong; keel shorter than the wings, obtuse; vexillary free or connate to the middle with the others; filaments not dilated; anthers uniform; ovary sessile or stipitate; ovules few; style filiform, incurved above, stigma terminal; fruit subglobose or ovoid, as long as the calyx, straight, thick, indehiscent or tardily 2-valved; seeds solitary or few, estrophiolate.

About 25 species distributed in temperate and subtropical regions especially Europe, Asia, Mediterranean region, Ethiopia (Ali, 1977).

Key to species:
1a. Corolla white..**1. M. alba**
1b. Corolla yellow..**2. M. indica**

1. Melilotus alba Desr. in Lam., Encycl. Meth. 4: 63 (1776); S.I. Ali in Nasir & Ali., Fl.Pak., 100: 310 (1977). **Figure 14D**

Syn: *M. albus* Medik., Vorles. Churpfälz. Phy.-Ökon. Ges. 2: 382 (1787); *M. argutus* Rchb., Fl. Germ. Excurs. 499 (1832).

Annual erect herb, up to 40 cm tall. Stem much branched, sparsely hairy; stipules ca. 3.5-5 mm, linear-lanceolate, entire. Leaflets ca. 1-2 x 0.5-1 cm in size, oblong-obovate, base cuneate, apex denticulate or apiculate dentate; petiole ca. 0.5-2 cm. peduncle 1-3 cm, elongated in fruit; racemes many-flowered, elongating and lax in fruit; flower ca. 4 x 3.5 mm; pedicel ca. 1-2 mm. Stalk of flower ca. 2.5-3mm long. Bract ca. 2.5 mm long. Calyx 5, toothed, united below, ca. 2.5 x 3.5 mm. Corolla white; standard ca. 4 x 2.5 mm; wings ca. 3.5 x 0.8mm; keel ca. 3.5 x 2.5 mm. Stamens ca. 3 x 1.5 mm; anther basifixed, ca.0.25mmin diameter. Carpel 1, with 1 locule, ovary superior ca. 3.5 x 1mm; style ca. 2 mm long; stigma ca. 0.5 mm across. Pods ca. 3-4.5 mm, globose-ovoid, 1-2-seeded, reticualtely nerved, brown at maturity. Seed ca. 1.5-2 mm, yellowish brown, smooth.
Flowering Period: February-August.
Herbarium specimens examined: NARA DESERT: *Rahmatullah Qureshi*, 1370 (SALUBGH).

Field remarks: A less common weed found in flat habitat.
Distribution: Pakistan, India, Kashmir, Tibet, Persia, Afghanistan, West and Central Asia, Turkey, Arabia, Europe, introduced in Amrica and Australia (Ali, 1977; Boulos, 1999).

2. Melilotus indica (L.) All, Fl. Pedem. 1: 308 (1785); S.I. Ali in Nasir & Ali., Fl.Pak., 100: 308-309 (1977). **Figure 15A**

Syn: *Trifolium M. indica* L., sp. Pl. 765 (1753); *Melilotus parviflorus* Desf., Fl. Alt. 2: 192 (1799); *M. bonplandii* Ten., Index Sem. Hort. Neapol. 14 (1833); *M. tommasinii* Jord. Mém, Acad. Roy. Sci. Lyn. Sect. Sci., ser 2, 1: 266 (1851).

Annual erect herb, up to 60 cm tall, glabrous, hairy when young. Stem fairly branched, quadrangulate, groovd; stipules ca. 3-6 mm, lanceolate, entire or denticulate, acuminate. Petiole ca. 3 cm long; leaflets ca. 0.8-2.4 x 0.3-1.2 cm in size, obovate-oblong, base cuneate, margin sharply serrate-dentate, apex truncate-apiculate. Racemes ca. 1-2.5 cm, 20 to 30-flowered. Bract ca. 2.5 mm long. Flower ca.3 x 1mm; pedicle 2-3 mm. Calyx ca. 1-1.5 x 3 mm in size. Corolla yellow; standard ca. 2.5 x 1 mm, wings ca. 2.5 x 0.5 mm; keel ca. 2.5 x 1 mm. Stamens 9+1; anthers ca. 1.5 x 0.5 mm. Pods ca. 1.5-3 mm, subglobose with reticulate ridges. Seed globose, minutely tuberculate, ca.1.5-2 mm.
Flowering Period: February-May.
Herbarium specimens examined: NARA DESERT: *Rahmatullah Qureshi*, 1371 (SALUBGH).
Field remarks: A very common weed of barren areas.
Distribution: Pakistan, India, Orient, Europe, introduced in warm temperate regions (Ali, 1977), Eurasia (Boulos, 1999).

33. Salvadoraceae

Shrubs or unarmed trees. Leaves decussate, simple, base attenuate, margin entire, petiolate, estipulate (sometimes with rudimentary stipules). Inflorescence densely fascicled or laxly panicled. Flowers actinomorphic or zygomorphic, bisexual or unisexual (plant monoecious or dioecious when flowers unisexual), stalked or subsessile. Calyx-tube campanulate, ovoid or connate at the base; lobes 2 or 4, valvate or imbricate. Corolla-tube shortly united; lobes sub-equal, 4-5-merous, obtuse, imbricate. Stamens 4-5, alternating with corolla-lobes, exserted; filaments free or epipetalous; anthers 2-celled, basifixed, dehiscence lateral longitudinal, introse. Disk of 4 scales or glands between the bases of the filaments or 0. Ovary superior, globular, 1- or 2-locular; ovules 1-2 per cell, placentation basal; style very short or almost 0; stigma broad, truncate or peltate. Fruit a drupe or a berry, globose, supported by persistent calyx and corolla; endocarp crustaceous. Seeds erect, globose, exalbuminous; embryo with thick cotyledons.

A family of 3 genera and 12 species distributed mainly in tropical and subtropical Asia and Africa, in Pakistan it is represented by 1 genus and 2 species (Qureshi, 1972).

Salvadora

L., Sp. Pl. 122 (1753); Gen. Pl. ed. 5. 58 (1754); Clarke in Hook. f., Fl. Brit. Ind. 3: 619 (1882); Cooke, Fl. Bomb. Pres. 2(reprinted ed.): 182 (1958); Rech. f., Fl. Iran. 37: 3 (1966); Qureshi in Fl Pak. 29: 1 (1972).

Shrubs or trees. Leaves opposite, entire, usually pale. Inflorescence axillary and terminal panicles. Flowers bisexual, small, sessile or pedicellate, green. Calyx-tube campanulate; lobes 4. Corolla-tube campanulate; tube short; lobes 4, obtuse, imbricate in bud. Stamens 4, inserted at the base or in the middle of the corolla-tube; filaments slightly flattened. Disc of 4 scales or glands in between the bases of the filaments or 0. Ovary 1-celled; ovule solitary; style very short or almost 0. Drupe globose.

A genus of 4 species distributed in Africa, Ceylon and India. Represented by 2 species in Pakistan (Qureshi, 1972).

Key to species:

1a. Leaves usually 2 cm or more broad, dark-green; lateral nerves 5-6 pairs;
flowers pedicelle.. **1. S. persica**

1b. Leaves usually 1 cm broad, pale-green; flowers almost sessile..................... **2. S. oleoides**

1. Salvadora persica L., Sp. Pl., 122 (1753); Stewart in Nasir & Ali., Ann. Cat. Vasc. Pl. W. Pak. & Kashm., 551 (1972); Saida in Nasir & Ali., Fl.Pak., 29:1 (1987).

A small tree up to 9 m high. Leaves elliptic-oblong, or ovate, ca. 3.5 x 1.5-2.5 cm in size, thick-chartaceous, base rounded to acute, apex sub-acute, apiculate, lateral nerves slightly raised; petiole ca. 2 cm long. Inflorescence drooping, panicled, ca. 5-31 cm long. Flowers actinomorphic, bisexual, greenish-yellow, ca. 3-4 mm across; pedicel 1-3 mm long. Bracts 2, each 1 mm. Calyx cupular, ca. 1 x 1.5 mm; lobes 4, valvate, coriaceous, round, apex obtuse. Corolla cream, ca. 3 mm across, shortly tubular; lobes 4, recurved, ca. 1.5 mm, membranous, apex acute to emarginate. Stamens 4, epipetalous, inserted at the middle of corolla-tube; filaments filiform, ca. 1 mm; anthers semi-circular, ca. 0.7 mm. Ovary 4-lobed, 1 mm, 1-locular; ovule 1, basal; style absent; stigma broadly truncate. Drupe globular, ca. 4 mm across, smooth, red or white on ripening. Seed solitary, 2 mm across; cotyledons fleshy.

Flowering period: March-June.

Herbarium specimens examined: NARA DESERT: *Rahmatullah Qureshi*, 11, 852 (SALUBGH).

Field remarks: A less common shrub found in saline habitat.

Distribution: Widely distributed in the drier parts of West Pakistan, Arabia, Palestine, Syria and Africa (Qureshi, 1972).

2. Salvadora oleoides Decne. in Jacq. Voy. Inde., 140-144 (1844); Stewart in Nasir & Ali., Ann. Cat. Vasc. Pl. W. Pak. & Kashm., 550 (1972); Saida in Nasir & Ali., Fl. W. Pak., 29:3 (1972).

A shrub or small tree up to 7 m tall with a short twisted or bent trunk; branches numerous, stiff, divergent, whitish. Leaves ca. 1-7.5 x 0.4-1.5 cm in size, linear or ovate-lanceolate, or elliptic-lanceolate, apex acute or mucronate, margin entire, pale or whitish-green, coriaceous and somewhat fleshy when mature, glabrous, main nerves indistinct; petiole ca. 2.5-1.6 cm. Inflorescence axillary panicles or branched spikes, ca. 2-4.5 cm long. Flowers greenish-white, sessile, ca. 2-3 mm across; pedicel ca. 1 mm or 0. Calyx ca. 1.5-2 mm long, cleft about half way down; lobes 4, rounded, apex obtuse, margin wavy, glabrous. Corolla ca. 2.5 mm long, deeply cleft; lobes obovate-oblong, apex subacute and recurved. Stamens 4, inserted at the base of the corolla tube. Style absent' stigma peltate. Drupe ca. 5 mm across, globose, yellow when ripe.

Flowering period: December-March.

Herbarium specimens examined: NARA DESERT: *Rahmatullah Qureshi*, 1, 809, 910, 1023 (SALUBGH).

Field remarks: A common shrub or small tree found sandy.

Distribution: West Pakistan, India (Gujrat, Junagarh, Rajputana) and Aden (Qureshi, 1972).

34. Euphorbiaceae

Monoecious or dioecious often poisonous, prostrate, erect or scandent annual, biennial or perennial herbs, shrubs or trees, succulent or not, spiny or unarmed, sometimes with phylloclades, often lacticiferous or not. Indumentum 0 or of simple, branched or stellate hairs or peltate scales, the hairs sometimes urticating. Leaves alternate, rarely opposite or whorled, occasionally all 3, green or scarious and squamiform, petiolate or sessile, stipulate or exstipulate, simple, lobed or compound, entire or toothed, peltate or not, penni/palminerved or glandular or eglandular. Stipules free or connate, sometimes spathaceous, membranous, capilliform, glandular or spiny, sub-persistent or readily caduous. Inflorescence terminal, axillary, lateral or leaf-opposed, cymose, paniculate, racemose, spicate or cyathial, or with the flowers fasciculate or solitary. Flowers unisexual, regular, bracteate, and small to minute. Calyx usually 3-6, imbricate, valvate or open, equal or unequal lobes or free sepals, often dissimilar between the sexes, rarely the female calyx turbinate or spathaceous, sometimes accrescent, minute or 0. Corolla in one or both sexes of 3-6 -lobes, free (rarely united); petals minute, subvalvate or imbricate, rarely 0. Disc in male flowers of 5-6, occasionally more, free or united glands, or annular, cupular or 0; in female flowers hypogynous, usually annular or cupular, entire or lobed, rarely gland free, sometimes 0. Stamens (1-) 3-100 (-1000), free or connate, simple, rarely branched; anthers usually 2-lobed and splitting longitudinally, erect or inflexed in bud, the cells usually parallel and adnate to the connective, sometimes free, vigorously positioned. Pistillode central, columnar or not. Ovary superior, usually sessile, 3-celled; placentation axile, the ovules solitary or paired per cell. Styles 3, free or united, erect or spreading, entire, bifid or laciniate, inner faces stigmatic. Staminodes sometimes present. Fruit usually schizocarpic, often dehiscing into 3 (occasionally less or more) 2-valved cocci having a persistent columella, or else fruit indehiscent and drupaceous. Seeds 1 or 2 per cell, or by abortion 1 per fruit, caruncülate or not, smooth or variously ornamented and sculptured, coccolorous or variously patterned; endosperm usually copious and fleshy; embryo straight, radicle superior; cotyledons usually broad and flat.

A very large family, the sixth largest among the Anthophyta, with 300 genera and 5000 species, sub-cosmopolitan but with the strongest representation in the humid tropics and subtropics of both hemispheres. Represented in Pakistan by 24 genera of which 11 are not native (Radcliffe-Smith, 1986).

Key to genera:

1a. Flowers naked, arranged in flower-like cyathia, milky latex present........... **1. Euphorbia**
1b. Flowers with Perianth, not arragned in cyathia, milky latex absent:
2a. Petals present, at leat in male flower:
3a. Plants stellatey tomentose; leaf margin sinuate...................................**2. Chrozophora**
3b. Plants not stellately tomentose; leaf margin various:
2b. Petals absent:
4a. Leaves palmately 5 to 9-lobed, peltate; stamen numerous; fruit softly echinate...**3. Ricinus**
4b. Leaves entire, not peltate; stamens 2-6; fruit smooth................................**4. Phyllanthus**

1. Euphorbia

L., Sp. Pl. 450 (1753); Gen. Pl. ed. 5: 208 (1754); Benth. & Hook. f., Gen. Pl. 3(1): 258 (1880); Hook. f., Fl. Brit. Ind. 5: 224 (1887); Stewart in Ali and Nasir, Ann. Cat. Vasc. Pl. W. Pak. & Kashm., 447 (1972); Radcliffe-Smith in Fl. Pak., 172: 88 (1986).

Tithymalus [Tourn.] Trew, Herb. Blackw. 1: t. 123 (1754); *Euphorbium* Hill, Fam. Herb. ed. 2: 136 (1755); *Anisophyllum* Haw., Syn. Pl. Saucc. 159 (1812); *Chamaesyce* [diosc.] S.F. Gray, Nat. Ann. Brit. Pl. 2: 260 (1821); *Poinsettia* Grah. in Edin. New Phil. J. 20: 412 (1836); *Agaloma* Raf., Fl. Tell. 4: 116 (1838).

Monoecious herbs, shrubs or small trees of various habit, with milky juice and a simple indumentum, when present. Stem slender and leafy or thick and fleshy and sometimes leafless or nearly so. Leaves often of 3 types, lower, median and upper; lower or stem-leaves usually alternate; median or pseudoumbel leaves whorled, upper or ray-leaves whorled or opposite, free or connate; all or most leaves usually sessile, rarely shortly petiolate, simple, entire or toothed, penni/palminerved. Stipules minute, subulate, interpetiolar and chaffy, glandular, sessile, or spiny or 0. Inflorescence a cyathium, with 1 female flower and several bracteate male flowers enclosed in a gland-bearing involucre; cyathia axillary or arranged pseudodichasially, often in a 'pseudo-umbel' of radiating 'dichasia'. Involucre usually 5-lobed, with 1-5 glands alternating with them. Male flowers each consisting of a single stamen borne directly on its own pedicel; anther-cells subglobose, dehiscing longitudinally. Female flower consisting of a trilocular ovary on a pedicel which usually elongates in fruit; ovules solitary per cell; styles 3, free or connate at the base; stigmas bifid. Fruit 3-celled, dehiscing into 2-valved cocci; endocarp woody or cartilaginous. Seeds carunculate; testa thin, crustaceous, smooth, ornamented or sculptured; albumen thick; embryo straight; cotyledons broad, flat.

One of the six largest genera of flowering plants (the others being *Astragalus*, *Carex*, *Piper*, *Senecio and Solanum*), having some of 2000 species. It is cosmopolitan, but chiefly restricted to tropical to subtropical and warm temperate regions (Smith, 1986).

Key to species:

1a. Large shrub with dendroid form, armed with stipular spines........................**1. E. caducifolia**
1b. Herbs without dendroidal form, unarmed:

2a. Erect or ascending herb. Leaves rhombic...**2. E. hirta**
2b. Prostrate herbs. Leaves not rhombic:
3a. Leaves crenulate; fruits pubescent; seed furrowed..............................**3. E. thymifolia**
3b. Leaves entire or toothed only at the tip; fruits glabrous;
seeds not furrowed...**4. E. clarkeana**

1. Euphorbia caducifolia Haines in Ind. For. 40: 154. 1914. Stewart in Nasir & Ali., Ann. Cat. Vasc. Pl.W. Pak. & Kashm., 447 (1972); Radcliffe-Smith in Nasir & Ali., Fl. Pak., 172: 118-119 (1986).

Syn: *Euphorbia neriifolia* sensu Hook. f., Fl. Brit. Ind. 5: 255 (1887).

A pale green dense, fleshy dendroid shrub up to 3 m high many branches arising from the base. Branches ± erect,ca. 3.5 cm in diam., somewhat cylindrical with small, distant non-confluent tubercles, arranged in 5 spirals; each tubercle with a black areole, possesing a pair of 5-9 mm long glabrous, pointed, divergent spines. Leaves ca. 1-9 x 1-4 cm, ovate or ovate-oblong, margin entire, crisped or undulate, apex acute or cuspidate, fleshy, rigid, glabrous, pale-green, caducous. Flowers solitary or 2-3 together on short fleshy peduncles; when 3, the central usually male, laterals bisexual; pedicles stout, ca. 1-5 mm long; bracts triangular-ovate. Involucre 4-5 mm in diam., lobes broadly oblong, obcuneate, toothed, scarsely fimbriate; glands 5, transversely oblong, glabrous; filaments glabrous or pilose; anthers yellow, subglobose, erecto-patent. Female flowers pedunculate; perianth lobes 3, minute, rounded at apex; styles 3, connate up to half their length. Fruit a capsule, sharpely 3-lobed; lobes transversely plano-convex, glabrous, ca. 10-12 mm across. Seeds 3, ca. 3 mm across, globose, smooth.
Flowering & Fruiting Period: February-March.
Herbarium specimens examined: NARA DESERT: *Rahmatullah Qureshi,* 1358 (SALUBGH).
Field remarks: A less common shrub found only in Achhro Thar.
Distribution: India: in stony ground on coastal plains and hills (Radcliffe-Smith, 1986).

2. Euphorbia hirta L., Sp. Pl. 454 (1753; Jafri in Fl. Kr., 199 (1966); Stewart in Nasir & Ali., Ann. Cat. Vasc. Pl. W. Pak. & Kashm., 449 (1972); Radcliffe-Smith in Nasir & Ali., Fl. Pak., 172: 95 (1986).

Syn: *E. pilulifera* Hook. f., Fl. Brit. Ind., 5: 250 (1887), *auct non* L.

An annual erect to decumbent herb up to 30 cm high. Stem simple or dichotomously branched, clothed with coarsely pilose of yellow, spreading, minute, curved, subadpressed hairs. Leaves ca. 1-4 x 0.5-1 cm in size, opposite, obliquely lanceolate or ovate to rhomboid-oblong, base cuneate on one side of midrib, the other side rounded, apex acute-subobtuse, margins finely serrate, thinly pubescent on both surfaces; nerves distinct; petioles ca. 2-5 mm long, pubescent. Stipules subulate, base broad, rarely pectinate, caduous. Cymes in alternate axils of the subsequent pairs of leaves, ca. 6-12 mm across, more or less globose, dichotomously branched at the apex of a 5-10 cm long, terete peduncle, thinly puberulent with minute curved hairs. Involucres densely crowded, minute, very shortly pedicellate, obconic or cup-like, with 4 glands and 5 acute fringed lobes; glands erect, slightly exceeding the lobes, orbicular, base truncate. Capsule globose-trigonous, ca. 2 mm across, thinly puberulent with minute, adpressed

hairs; style short, 2-armed, truncate. Seeds ca. 1 mm long, oblong, 4-angled, slightly transversely rugose, reddish.
Flowering period: July-November.
Herbarium specimens examined: NARA DESERT: *Rahmatullah Qureshi*, 201, 501, 940, 1055, 1122, 1200 (SALUBGH).
Distribution: A pantropical weed; in grassy plots & roadsides and in flooded area near cultivation (Radcliffe-Smith, 1986).

3. Euphorbia thymifolia L., Sp. Pl., 454 (1753); Jafri in Fl. Kr., 200 (1966); Stewart in Nasir & Ali., Ann. Cat. Vasc. Pl. W. Pak. & Kashm.,453 (1972); Radcliffe-Smith in Nasir & Ali., Fl. Pak., 172: 107 (1986).

Syn: *Chamaesyce thymifolia* (L.) Millsp. Publ. Field Columbian Mus. Bot. Ser. 2: 412 (1916).

Annual, prostrate, pale green herb up to 15 cm long. Stem divericately branched, slender, cylindric, more or less hairy. Leaves ca. 3-6 x 2.5-4 mm in size, opposite, obliquely oblong or elliptic-oblong, apex rounded, base unequal, margins crenulate, glabrous above, glaucous and usually slightly pubescent below; petiole ca. 0.5 mm; stipules fimbriate. Involucres axillary, solitary or 2-3 together in an axil, campanulate; stalk short; glands minute or 0. Male florets: 1-4, ebracteolate; stalk ca. 0.5 mm; anthers ca. 0.2 mm, dehiscence vertical. Female floret: lateral pendulous. Ovary ca. 0.8 x 0.7 mm in size, tomentose; styles 3, forked from the base, erect, ca. 0.7 mm long. Capsule ca. 1.5 mm across, obtusely keeled, adpressed hairy.
Flowering period: December-February.
Herbarium specimens examined: NARA DESERT: *Rahmatullah Qureshi*, 15, 103, 220, 648, 938, 1078, 1212 (SALUBGH).
Field remarks: A common weed of shady places.
Distribution: It is widespread in Tropical Asia and America and introduced into East Tropical Africa (Radcliffe-Smith, 1986).

4. Euphorbia clarkeana Hook. f., Fl. Brit. Ind. 5: 253. 1887. Stewart in Nasir & Ali., Ann. Cat. Vasc. Pl.W. Pak. & Kashm., 447 (1972); Radcliffe-Smith in Nasir & Ali., Fl. Pak., 172: 102-103 (1986).

Syn: *Euphorbia jodhpurensis* Blatt. & Hallb. in J. Bomb. Nat. Hist. Soc. 26: 971 (1920).

A prostrate or decumbent, glabrous annual herb up to 20 cm high with several filiform stems. Leaves ca. 4-10 x 1-4 mm, opposite, leaf-blades narrowly obliquely linear-oblong, entire or toothed at the apex, base almost auricled on one side; petioles very short; stipules rather large, setaceous from a broad toothed base. Cyathia in the axils of leaves on dense lateral shoots and solitary or ± clustered, campanulate, glabrous; lobes lanceolate, toothed, longer than the glands; glandular appendages minute without limbs or 0. Capsules glabrous, less than 2 mm broad; cocci keeled. Seeds more than 1 mm long, ovoid, acutely 4-angled, shrtly bluntly pointed, obscurely transversely rugose, pale-brown.
Flowering Period: More or less throughout the year.

Herbarium specimens examined: NARA DESERT: *Rahmatullah Qureshi*, 1357 (SALUBGH).
Field remarks: A less common herb of sandy soil.
Distribution: Afghanistan, N.W. India. On Sandy soil, clay soil, pebbly ground, in fields and open waste places with *Cressa cretica* (Radcliffe-Smith, 1986).

2. Chrozophora

Neck ex A. H. L. Juss., Euphorb. Tent. 27 (1824) ('Chrozophora'), *nom. cons*; Benth. & Hook. f., Gen. Pl. 3(1): 305 (1880); Hook. f., Fl. Brit. Ind. 5: 408 (1887); Radcliffe-Smith, Fl. Pak., 172: 45 (1986).

Diffuse densely hispid or stellately tomentose, monoecious annual or perennial herbs or undershrubs. Leaves alternate, petiolate, simple, subentire, repand-dentate or sublobate, penni-, subtripli- or palminerved, often plicate or bullate when young, 2-glandular at the base. Inflorescence pseudoaxillary, supra-axillary, lateral or leaf-opposed, spicate, racemose or subpaniculate, distally male, proximally female. Male flowers solitary to their bracts; pedicel short; calyx closed in bud, splitting into 5 valvate lobes; petals 5, distally imbricate and coherent; disc 5-lobed, the lobes alternating with the petals; stamens 3-15, filaments connate into a column; anthers 1-3-seriate, oblong, erect, longitudinally dehiscent; pistillode 0. Female flowers borne on 1-4-flowered peduncles; pedicels long, often elongating and reflexing in fruit; sepals 5, open in bud; petals 5, smaller than in the male flower, or 0; disc as in the male flower; ovary 3-celled, with solitary ovule per cell; styles 3, connate at the base, erect, bifid. Fruit schizocarpic, having a columella; endocarp thinly woody. Seeds usually smooth, ecarunculate but enveloped in a thin pale shiny aril; albumen thick, fleshy; cotyledons broad, flat.

Chrozophora plicata (Vahl) A. Juss. ex Spreng., Syst. Veg. 3: 850 (1826); Stewart in Ali and Nasir, Ann. Cat. Vasc. Pl. W. Pak. & Kashm., 446 (1972); Radcliffe-Smith in Nasir & Ali., Fl. Pak., 172: 47 (1986).

Syn: *Croton tinctorius* sensu Burm.f., Fl. Ind. 304, t. 62. F. 2 (1768); *C. plicatus* Vahl., Symb. Bot. 1: 78 (1790); *Chrozophora rottleri* (Geisel.) A. Juss. ex Spreng., Syst. Veg. 3: 850 (1826); *C. burmanni* Spr., *op. Cit*: 851 (1826); *C. prostrata* Dalz. & Gibs., Bombay Fl.: 233 (1861); *C. parvifolia* Klotzsch ex Schwf., Pl. quaed. Nilot. 10 (1862).

A prostrate or ascending annual herb up to 80 cm long. Stem diffusely branched, fulvous-tomentose. Leaves ca. 2-4 x 0.5-3.5 cm in size, broadly ovate, apex obtuse-subacute, base cordate or truncate, margins sinuate-undulate, subcoriaceous, 3-nerved from the base, lateral nerves 4 pairs; petiole ca. 4-6 cm long, with 2 glands at the base of lamina; stipules deciduous. Racemes terminal-subterminal, densely stellate-tomentose up to 3 cm long; male flower subsessile, crowded in upper axil, female ones pedicellate, lower. Bracts lanceolate. Male flower: ca. 5 mm across. Tepals 10, 2-seriate, valvate; outer ca. 3 mm, basally connate, ovate, concave; inner ca. 5 mm, petaloid, free, lanceolate. Stamens 12-15, 2-seriate, connate, inner longer; filaments thick, ca. 2 mm; anthers oblong, ca. 1 mm. Disc of 5, free, glands. Female flower: ca. 6 mm across. Tepals 10, 2-seriate, unequal, lanceolate, valvate, up to 2 mm, connate, ciliate, acute. Ovary ca. 5 mm across, globose, 3-locular; ovules 3, axile; styles 3, bifurcate, dilated, stigmatiferous. Disc glands 5. Capsule ca. 8 mm across, depressed, 3-lobed, stellate-tomentose. Seed globose, scurfy, grey.

Flowering period: March-July.

Herbarium specimens examined: NARA DESERT: *Rahmatullah Qureshi*, 1253, 1312 (SALUBGH).

Field remarks: A less common herb found in saline soil.

Distribution: Throughout the drier parts of Tropical Africa and Asia (Radcliffe-Smith, 1986).

3. Ricinus

L., Sp. Pl. 1007 (1753); Gen. Pl. ed. 5: 473 (1754); Benth. & Hook. f., Gen. Pl. 3(1): 321 (1880); Hook. f., Fl. Brit. Ind. 5: 457 (1887); Stewart in Ali and Nasir, Ann. Cat. Vasc. Pl. W. Pak. & Kashm., 456 (1972); Radcliffe-Smith in Fl. Pak., 172: 69 (1986).

A large monoecious glabrous annual or perennial herb. Leaves alternate, petiolate, stipulate, peltate, palmately-lobed, the lobes glandular-serrate, penninerved. Petioles glandular at apex and base. Stipule united to form a caducous sheath. Inflorescences paniculate, subterminal or leaf-opposed, ♂ in the lower half and ♀ in the upper or else all ♀; bracts membranous, soon caducous. Male flowers: pedicels articulate, bibracteolate; buds globose, closed, later splitting into 3-5 valvate lobes; petals 0; disc 0; stamens up to 1000 with the filaments variously united, anthers basifixed, the cells globose, longitudinally dehiscent; pistillode 0. Female flowers; pedicels extending in fruit; buds conical; sepals 5, valvate, soon caducous; petals 0; disc 0; ovary 3-celled, with 1 ovule per cell, echinate or smooth; styles 3, ± free, bipartite, papillose-plumose, dark red. Fruit trilobate, echinate or smooth, dehiscing into 3 bivalved cocci. Seeds dorsiventrally compressed-ovoid, smooth, carunculate, testa crustaceous, albumen fleshy; cotyledons broad, flat.

A monotypic genus native to NE Tropical Africa, but now widely cultivated throughout the tropics, subtropics and warm temperate regions, and often escaping and becoming naturalized.

Ricinus communis L., Sp. Pl. 1007 (1753); Stewart, Ann. Cat. Vasc. Pl. W. Pak. & Kashm. 456 (1972); Radcliffe-Smith in Nasir & Ali., Fl. Pak., 172: 69 (1986).

Large erect, shrubs or small trees up to 4 m high, with fistular stem and glaucous-grey or reddish, smooth bark. Young shoots often pruinose. Leaf-blades commonly 7-9-lobed, the median lobe usually 10-20 cm long and 2-5 cm wide, sometimes larger, the lateral lobes progressively smaller; lobes lanceolate, acutely acuminate, coarsely glandular-serrate or biserrate, lateral nerves ca. 10-20 pairs, running to the margins, dark green above, paler beneath; petioles 5-20 cm long; petiolar glands variously-shaped. Stipular sheath ovate, c. 1-2 cm long, leaving a circular scar when fallen. Inflorescence 10-25 cm long; bracts c. 1 cm long, the bracteoles smaller. Male flowers; pedicels c. 1 cm long; calyx-lobes elliptic-ovate, 6-8 x 3-4 mm, acute, yellowish-green; stamens 7 mm long, anthers 0.5 mm. long, pale yellow. Female flowers: pedicels 3-5 mm. long, extending to 2 cm or more in fruit; sepals lanceolate, 5 mm long, acuminate, purplish; ovary trilobate-subglobose, 2 x 2 mm; styles 3-7 mm long. Fruit trilobate, 1-1.8 x 1-1.5 cm, smooth or sparingly to densely covered with narrowly cylindric bristle-tipped fleshy processes 3-5 mm long. Seeds 7-12 x 5-8 x 4-6 mm, shiny, greyish, silvery or beige generally streaked and flecked with brown; caruncle depressed-conic, 1-2 x 2-3 mm.

Flowering & Fruiting Period: Almost round the year.

Field remarks: A less common shrub found in sandy soil near settlements.

Herbarium specimens examined: NARA DESERT: *Rahmatullah Qureshi*, 1359 (SALUBGH).

Distribution: (As for the genus).

4. Phyllanthus

L., Sp. Pl. 981 (1753); Gen. Pl. ed. 5: 422 (1754); Benth. & Hook. f., Gen. Pl. 3(1): 272 (1880); Hook. f., Fl. Brit. Ind. 5: 285 (1887); Stewart in Ali and Nasir, Ann. Cat. Vasc. Pl. W. Pak. & Kashm., 455 (1972); Radcliffe-Smith in Fl. Pak., 172: 22 (1986).

Cicca L., Mant. 1: 124 (1767); *Emblica* Gaertn., Fruct. 2: 122, t. 108 (1791); *Anisonema* Adr. Juss., Eup. Gen. Tent. 19, t. 4, f. 11 (1824); *Kirganelia* Juss., Gen. Pl. 387 (1789).

Key to species:

1a. Annual herb. Stipules peltate; leaves obovate,
wedge-shaped at the base...**1. P. maderaspatensis**

1b. Shrub. Stipules not peltate; leaves elliptic-oblong, obtuse
or rounded at the base...**2. P. reticulatus**

1. Phyllanthus maderaspatensis L., Sp. Pl. 982 (1753); Stewart, Ann. Cat. Vasc. Pl. W. Pak. & Kashm. 455 (1972); Radcliffe-Smith in Nasir & Ali., Fl. Pak., 172: 28-30 (1986).

An erect to asending annual herb up to 50 cm high. Stem glabrous, flowering branchlets sharply angular, compressed, sulcate when dry. Stipules ca. 2 mm long, lanceolate, acuminate, entire, auriculate at base, peltate, membranous, purplish, glabrous. Leaves ca. 2-3 x 1-1.2 cm long; linear-lanceolate or oblanceolate-obovate, rounded or truncate or ± mucornate at the apex, much tapering and wedge shaped at base, lateral nerves 6-7 pairs; petioles ca. 2-3 mm long, terete, glabrous. Flowers axillary, ♂ & ♀ flowers together in the leaf-axil or ♀ more oftely solitary, much larger than males. Male flowers minute, in small clusters; pedicels 1 mm long, sepals 6, ca. 1-2 x 0.8-1 mm, oblong, lanceolate, green with white membranous margin; disc-glands 6, very thin, flat, smooth; stamens 3, filaments connate to a short column, but free at the top, anthers longitudinally dehiscent, bright yellow. Female flowers: sepals 6, biseriate, the 3 outer 2 x 1.5 mm, ovate or elliptic, slightly apiculate, the inner 3 narrower, obtuse, apiculate, dark olive-green, edged with white, often reddish-tinged; pedicels 1.5-2 mm long. Disc-glands 6, free, square, thin, flat; ovary sessile, rounded-trilobate, c. 1 mm diam., smooth; styles 3, 0.5 mm long, ± free. Fruit adpressed globose capsule, ca. 3 mm in diam, 3-lobed, smooth, shiny, olivaceous. Seeds ca. 1-1.5 x 1

mm, brown with 14 rows of shiny black tubercles on the back, and 11-12 concentric rings of tubercles on the sides.
Flowering & Fruiting Period: May-January.
Herbarium specimens examined: NARA DESERT: *Rahmatullah Qureshi*, 1356 (SALUBGH).
Field remarks: A common herb found in sandy soil.
Distribution: Africa, Arabia, Pakistan, India, Sri Lanka, China, Java, Australia; on sandy clay soil, on limestone, stony ground, in forests; arable fields, maize-fields, on waste-ground near villages and by roadsides (Radcliffe-Smith, 1986).

Phyllanthus reticulatus Pair., Encycl. Meth. Bet. 5: 298 (1804); Radcliffe-Smith in Nasir & Ali., Fl. Pak., 172: 32-33 (1986).
Syn: *Phyllanthus multiforus* Willd., Sp. Pl. 4: 581 (1805).

A glabrous or puberulous much-branched shrub up to 2 m. Bark rough, peeling or flaking, brown or grey. Inter-mediate shoots commonly not exceeding 10 cm, leafy, flowering or both; short shoots co-axillary with them. Cataphylls lanceolate, 1-1.5 mm, later hardening and ± spinescent, reddish-brown. Foliage leaves distichous; petioles 1-1.5 mm long; leaf-blades elliptic to elliptic-oblong, 1-2 cm x 4-8 mm, obtuse or rounded at the apex and base, chartaceous, lateral nerves 6-9 pairs, not raised above, slightly raised beneath, tertiary nerves reticulate, usually glabrous, dark geen above, paler beneath. Stipules of the cataphylls triangular-ovate, 1.5 mm long, acuminate, reddish-brown; those of the leaves lanceolate, c. 1 mm long, brownish. Fascicles usually bisexual, with 1-2 ♀ & several ♂ flt per node. Male flowers: pedicels capillary, 2-4 mm long; sepals 5(-6), elliptic to suborbicular-obovate, c. 1 x 1 mm, ± rounded, white with a yellowish-green midrib, sometimes pinkish-tinged; disc-glands 5(-6), free, ovoid, knobbly, fleshy; stamens 5(-6), the 2(-3), outer free, short, the 3 inner usually united at least at the base, longer, filaments stout, anthers triangular, longitudinally dehiscent. Female flowers: pedicels slender, c. 2 mm long; sepals ± as in the ♂ fls.; disc-glands 5, free, ± triangular, flattened; ovary sessile, subglobose, (3-) 4-multilocular, 1 mm diam, smooth; styles very short, free, thick, suberect or indexed, crowded, bifid. Fruit depressed-subglobose, tri to plurilobate, 2.5-4 x 4-6 mm, smooth, baccate, shiny, black, with a dark purplish pulpy pericarp. Seeds irregularly ovoid-trigonous, c. 2 mm long, faintly reticulate, brownish with a pale raphe.
Herbarium specimens examined: NARA DESERT: *Rahmatullah Qureshi*, 1355 (SALUBGH).
Field remarks: A less common shrub found near agricultural fields.
Distribution: Throughout the Old World Tropics (Radcliffe-Smith, 1986).

35. Rhamnaceae

Trees or shrubs, rarely climbers. Leaves simple, alternate or sub-opposite, coriaceous, 3-8 nerved, generally stipulate, evergreen or deciduous, sometimes modified into spines. Inflorescence axillary cymes, corymbs, racemes, spikes or panicles or fasciculate. Flowers bisexual, rarely unisexual, actinomorphic, small, perigynous. Calyx tubular, 5- rarely 4- or 6-lobed. Petals 5, rarely 4- or 6-lobed, sometimes absent, usually clawed and cucullate. Stamens as many as petals, inserted opposite to the petals, often enclosed within their folds.

Disc fleshy, intra-staminal, filling and covering the hypanthium, glabrous, rarely tomentose. Ovary superior to inferior, (1)2-3 (-4)-celled; ovule solitary per cell; style 2-4-fid. Fruit often drupaceous, pyrenes 1-3 (-4), 1-2-seeded or capsule, rarely samaroid. Seed with a large erect embryo; endosperm copious or scanty.

A family of about 58 genera and nearly 900 species, cosmopolitan in distribution but mainly in tropics and subtropics. In Pakistan 6 genera and about 21 species recorded (Qaiser & Nazimuddin, 1981).

Zizyphus

Mill., Gard. Dict. ed. 8 (1768); Benth. & Hook. f., Gen. Pl. 1: 373 (1862); Lawson in Hook f., Fl. Brit. Ind. 1: 632 (1875); Cooke, Fl. Bomb. Pres. 1: 255 (1902); Qaiser & Nazimuddin in Fl. Pak., 140: 9 (1981).

Trees or shrubs, mostly armed with a pair of stipular spines, one usually long and straight and the other short recurved. Leaves mostly distichous to subdistichous, alternate 3(-5)-nerved, usually leathery and oblique at the base, entire serrate or serrulate. Flowers in axillary cymes, 5-merous, regular, greenish yellow, mostly bisexual. Calyx 5-fid, saucer-shaped, lobes keeled. Petals 5, rarely absent, culcullate, disc pitted or flat, 5-10-lobed, mostly lining the calyx tube. Stamens enclosed within the petals. Ovary immersed in disc, 2-4-celled; style 2(3)-partite; stigma papilose, small. Fruit drupaceous, fleshy, succulent, globose-oblong, pyrenes 1-4, rugose or tuberculate, 1-seed in each stone. Seed plano-convex, mostly with very scanty endosperm.

A genus of about 100 species distributed mostly in tropical America, Africa, Mediterranean region, Indo-Malaya, Australia, Pakistan and India. Six species are present in Pakistan (Qaiser & Nazimuddin, 1981).

Zizyphus nummularia (Burm. F.) Wt. & Arn., Prodr., 162 (1834);); Stewart in Ali and Nasir, Ann. Cat. Vasc. Pl. W. Pak. & Kashm., 469 (1972); Qaiser and Nazimuddin in Nasir & Ali., Fl. Pak.,140: 12 (1981). **Figure 18A**

Syn. *Rhamnus nummularia* Burm. f., Fl. Ind., 61 (1768); *Z. rotundifolia* Lam., l.c. 319 Sabnis, Fl. Sind. 17 (1924).

A bushy, much branched from near the base/root shrub up to 6 m high. Stem hard, woody at the base; branches divericate, slender, zigzag, flexuous, tomentose, young branches puberulous, gray; bark light coloured. Leaves ca. 0.5-1 x 0.8 cm in size, orbicular-ovate or ovate-elliptic, base obliquely rounded to sub-cordate, apex apiculate or mucronate, margin dentate, densely whitish tomentose beneath, less tomentose above, lateral nerves 3-paired, prominent below; petiole ca. 0.5-2 mm long, tomentose. Stipular spines in unequal pairs, bigger one straight, sharp and slender, ca. 1.5 cm long, smaller recurved, whitish tomentose when young, ca. 0.3-0.5 cm long. Inflorescence in axillary, sessile pubescent cymes or short clusters. Flowers ca. 0.5 cm across, complete, greenish yellow. Calyx tube ca. 1 mm long, pubescent out side, cleft above half way down; lobes ca. 2-3 x 1-1.5 mm in size, broadly ovate-triangular, apex acute, margin entire, yellowish, glabrous, keeled on the inner face for half their length, lateral nerves obscure. Petals 5, ca. 1-1.5 x 1 mm in size, obovate, base cuneate, apex rounded or truncate, concave, glabrous, yellowish. Stamens 5, included in

petals; filaments filiform, ca. 1 mm long, pinkish, base dilated; anthers ca. 0.5 mm across, ovate, apex apiculate, bilobed, dorsifixed, rupture longitudinally, yellow. Disk 10-lobed, with a pit opposite to each other. Ovary ca. 1.5 mm across, half sunken in the disk, ovoid, glabrous, pinkish in colour; styles 2, connate about half their length. Drupe ca. 0.5-1 cm across, globose, fleshy, smooth, yellow or reddish brown-orange when ripe; stone 1-2 seeded.
Flowering period: March-June.
Herbarium specimens examined: NARA DESERT: *Rahmatullah Qureshi*, 1246 (SALUBGH).
Field remarks: A common shrub found calcarious soils.
Distribution: Palestine, Iraq, Iran, Afghanistan, Pakistan and India (Qaiser & Nizamuddin, 1981).

36. Polygalaceae

Herbs or erect or scandent shrubs, rarely small trees. Leaves simple or occasionally reduced to scales (rarely whorled) or 0, exstipulate. Inflorescence axillary, lateral or terminal, sometimes extra-axillary, in lax or dense racemes, rarely flowers solitary. Flowers 3-bracteate, 5-merous, bisexual, irregular. Sepals 5, unequal, much imbricate in bud; outer smaller; inner ones (wing sepals) petaloid, symmetric or not, larger. Petals 3, basely united and adnate to staminal sheath; lateral petals 2, variously shaped; lower one (keel) boat-shaped, often subapically crested; crests with forked appendages. Stamens 8, monadelphous; filaments free or connate, jointed to the sheath at the same level or not; anthers dehiscing by apically pores. Torus small, rarely expanded into an annular disk. Ovary free, superior, shortly stipitate, laterally compressed, 2-celled; ovules 1 per cell on axile placentae, pendulous; style terminal, variously curved, dilated; stigma lateral, oblique. Fruit a capsule, often overtopped by wing seplas, compressed, often margined, loculicidal. Seeds 1 or 2, oblong, densely appressed-pubescent; caruncle hood-like, 2- or 3-appendaged; appendages sometimes much enlarged.

A family of 12 genera and ca. 800 species, mostly temperate and tropical. It is represented in Pakistan by single genus Polygala (Dar, 1973).

Polygala

L., Sp. Pl. 701 (1753); Gen. Pl. ed. 5: 851 (1754); Boiss., Fl. Or. 1: 468 (1867); Cooke, Fl. Bomb. 1(1): 59 (1901); Jafri in Fl. Kar. 189 (1966); Dar in Nasir and Ali, Fl. Pak., 52: 1 (1973).

Annual or perennial herbs, glabrous or more or less pubescent with simple unicellular upcurved hairs. Leaves alternate rarely opposite or verticellate. Flowers spicate or lateral or terminal racemose. Sepals unequal, the 2 inner larger (wings) usually petaloid. Petals 3, united at the base with the staminal sheath, lower ones keeled and generally crested. Stamens 8; filaments united for their lower half into a split sheath; anthers opening by pores. Ovary 2-celled; ovule 1 in each cell, pendulous. Capsule 2-celled, loculicidal, 2-seeded, compressed. Seeds pilose or glabrous, strophiolate and albuminous.

A large and difficult genus of ca. 600 species, distributed throughout except New Zealand, Polynesia, and arctic zone. Represented in our country by 8-10 species (Dar, 1973).

Key to species:

1a. Bracts persistent; wings with a green midrib; capsule not margined;
seeds with a strophiolate...**2. P. irregularis**

1b. Bracts caduceus; wings reticulately nerved; capsule with a membranous margin;
seeds without a strophiolate..**1. P. erioptera**

1. Polygala irregularis Boiss., Diagn. Ser. I, 1: 8 (1842); Stewart in Ali and Nasir, Ann. Cat. Vasc. Pl. W. Pak. & Kashm., 443 (1972); Dar in Nasir & Ali., Fl. Pak., 52: 10 (1973).

Syn: *P. obtusata* DC. Prodr. 1: 326 (1824).

A low, perennial herb upto 50 cm. stems branching from a more or less woody base, the branches spreading or ascending, adpressedly pubescent, terete, striate, the older with conspicuous bases of the fallen leaves. Leaves very variable in shape and size, ca. 0.5-4 x 0.2-0.5 cm, ± fleshy, obovate-oblong, oblanceolate or narrowly elliptic, lower usually smaller and obovate, apex obtuse, apiculate or retuse, base attenuate into a short, hairy petiole; latter rarely exceeding 2 mm in length, margin entire, often deflexed, surfaces inconspicuously puberulous, midrib prominent beneath. Flowers second in terminal, divaricately-elongate racemes, ca. 5-15 cm long; pedicels ca. 1-3 mm long, hairy, deflexed; bracts very minute, caduceus, scarious, acute, minutely ciliolate, two lateral ovate and slightly smaller than strongly keeled basal; scars of the fallen raised and conspicuous. Sepals 5; 3 outer broadly oblong to nearly oval, ca. 2 mm long, minutely ciliolate on membranous margins, wing-sepals petaloid; ca. 5-6 x 4 mm in size, obliquely ovate, scarious, entire or sometimes very minutely ciliolate, glabrous throughout or inconspicuously hairy along the vein, the veins inconspicuously purple or green, 4-5, diverging from a gibbous base, reticulately branched above, branchlets not reaching the margins. Petals purple; lateral ones 4 mm long, obovate, cuneate, somewhat truncate at the obscurely undulate apex, densely cilate at the base. Fruit a capsule, ca. 6-7 mm long, pale green, very oblique, deeply notched, glabrous, broadly winged; wings gradually narrowed downwards and with close transverse bifurcating nerves. Seeds ca. 3-4 mm long, nearly black, shiny, conical, velvety, hairs short, dirty brown, apical region hairs rather close and adpressed, distal end crowded with transparent papilate which have clavate as well as pointed tips; strophiolate absent, appendages 2, grayish, aculiform, pendant.

Flowering & Fruiting Period: September-December.

Herbarium specimens examined: NARA DESERT: *Rahmatullah Qureshi*, 1372 (SALUBGH).

Field remarks: A less common herbs of plain areas.

Distribution: Tropical Africa, Arabia, Paistan (Baluchistan, Sindh), India (Bhandari, 1978).

2. Polygala erioptera DC., Prod. 1: 326 (1824); Stewart in Ali and Nasir, Ann. Cat. Vasc. Pl. W. Pak. & Kashm., 443 (1972); Dar in Nasir & Ali., Fl. Pak., 52: 9-10 (1973).

Syn. *P. vahliana* DC., 1.c.325; *P. erioptera* var. *vahliana* (DC) Chodar, 1.c.345.

An annual perennial erect herb up to 30 cm long. Stem soft, herbaceous, terete, green, densely pubescent; internode ca. 0.5-1 cm long. Leaves simple, alternate, ca. 0.8-2.5 cm long, ca. 3-4 mm broad, linear, elliptic or oblanceolate, base narrowly cuneate, apex obtuse-apiculate, margin entire, pubescent on both sides, lateral nerves obscure; petiole ca. 1-2 mm long, pubescent. Inflorescence axillary or lateral, solitary or in condensed lax raceme,

up to 7-flowered; peduncle ca. 2 mm long, pubescent. Bracts 3, unequal, ca. 1-2 x 0.5-1 mm in size, ovate, apex acute or ± apiculate, margin sinuate or ciliate, very pilose, 3-nerved, green. Flower bracteate, complete, zygomorphic, ca. 8 x 6 mm in size; pedicel ca. 3-4 mm long, hispid. Sepals 5, outer sepals 3, unequal, ca. 3-4 x 1.5-2 mm. in size, elliptic-obovate, base rounded, apex rounded, margin ciliate and hyaline, green, wing sepals 2, ca. 8 x 4 mm in size, obovate, base oblique, apex rotund, margin ciliate, pilose without, 3-nerved, veinlets anastomosing, greenish white. Petals 3, lateral petals 2, ovate, ca. 5 x 2.5 mm in size, apex acute, base rounded; keel 1, ca. 5 mm across, crests 2-partite, ca. 2 mm long. Stamens 8, monadelphous; staminal sheath ca. 4 mm long; filaments ca. 1 mm long; anthers ca. 0.3 mm across, basifixed, obovate, rupture by apical pores, yellow. Ovary ca. 2 mm across, 2-locular, oblong, apex long ciliate, hispid; style ca. 7 mm long, dilated, glabrous, filiform; stigma capitate, curved.
Flowering period: Almost throughout the year.
Herbarium specimens examined: NARA DESERT: *Rahmatullah Qureshi*, 245, 536 (a), 876, 880, 1154 (SALUBGH).
Field remarks: A less common herbs of plain areas.
Distribution: Arabia, Tropical Asia, and Africa (Dar, 1973).

37. Burseraceae

Shrubs or trees with resin or aromatic oil. Leaves alternate, rarely opposite or fascicled, usually pinnately compound or decompound, l-3 (or more) foliolate, imparipinnate. Flowers in panicles, solitary or sometimes fascicled, bisexual or unisexual (plants then diclino or polygamodioecious), actinomorphic and hypogynous. Calyx 3-6 fid or partite forming a tube, lobes imbricate or valvate. Petals 3-6, free, rarely connate below, imbricate or valvate. Disc annular or cupular, rarely absent, free or adnate to the calyx tube. Stamens in l-2 whorls, as many as or twice the number of petals, free, or rarely connate at the base; anthers usually versatile, dithecous, with longitudinal dehiscence. Carpels 2-5, syncarpous; ovary sessile or stipitate, 2-5-loculed; each locule 2(-1) ovuled; placentation axile; style 1 or absent; stigma 2-5-lobed or undivided. Fruit drupaceous, indehiscent with 2-5 pyrenes or dehiscent. Seeds usually with membranous testa and without endosperm.

A family of 16 genera and about 500 species distributed in tropical regions of both hemispheres. Only 3 genera and 4 species are known from W. Pakistan.

Commiphora

Jacq., Hort. Schoenb. 2.66.t. 249 (1797), *nom. cons.*; Cooke, Fl. Pres. Bomb. l: 212 (1901); Jafri, Fl. Kar. 186 (1966).
Balsamodendron Kunth; *Heudelotia* A. Rich.

Trees or shrubs, often armed or thorny. Leaves alternate or fascicled, compound, l-3 (or more)-foliolate, imparipinnate; leaflets sessile or subsessile, serrate, crenate, or entire. Flowers solitary or in fascicles of 2-5 or in panicles, small, sessile-subsessile, bisexual or unisexual. Calyx cupular, urceolate or tubular, 4 toothed or lobed, persistent. Petals 4, valvate. Stamens mostly 8, inserted on the margin of the annular or cupular disc, filaments usually unequal, dilated at the base. Ovary ovoid, sessile, 2-4-loculed, each locule 2 ovuled; syle short with 2-4 obtusely lobed stigma. Fruit a drupe, globose or subglobose, compressed or not.

A genus of about 185 species found in India, W. Pakistan, Arabia, Tropical and Southern Africa. Of these only 2 species are reported from W. Pakistan.

Key to species:

1a. Leaflets entire, pubescent; petals not reflexed; stamens equal;
fruit with 4 white longitudinal lines.. **1. C. stocksiana**

1b. Leaflets serrate, glabrous; petals reflexed; stamens alternately short and long;
fruit with 2 white longitudinal lines or grooves.................................... **2. C. wightii**

1. Commiphora stocksiana (Engl.) Engl. in DC., Monogr. Phan. 4:17 (1883); Stewart in Ali and Nasir, Ann. Cat. Vasc. Pl. W. Pak. & Kashm., 441 (1972); Nasir in Nasir and Ali, Fl. Pak., 71: 21-23 (1974).

Syn: *Balsamea stocksiana* Engl.; *Balsamodendron pubescens* Stocks

Shrubs or small trees up to 4 m tall; branches covered with papery bark; young shoots pubescent. Leaves alternate, petiole c.l-1.5 cm, 3-5 foliolate, pubescent; lateral leaflets oblong, orbicular or obovate, entire, 0.5-l.3 cm long, 0.3-1 cm broad; terminal leaflet petiolulate, 0.5-1.5 cm long, 0.5-l cm broad, obovate, oval or broadly elliptic, entire. Hermaphrodite and male flowers present on the same plant, sessile, white to red, 3-5 mm long. Bracts 2, opposite. Sepals 4, connate into an urceolate 4-lobed tube, lobes valvate, l-2 mm long, pubescent outside. Petals 4 free, spreading, acute, 3-5 mm long, 1-1.5 mm broad. Stamens 8, free, equal, usually shorter than the petals, 3-5 mm long, anthers basifixed. Disc toothed. Ovary 2-loculed; stigma 2-lobed. Fruit 0.5-l.5 cm long, red when ripe, marked with 4, alternately short and long white lines, subglobose, mucronate, mesocarp yellow, 4 lined, fused at the base on one side and up to the middle (or more) on the other side, dehiscing along the white lines from base upwards.

Flowering Period: April-July.

Herbarium specimens examined: NARA DESERT: *Rahmatullah Qureshi*, 1373 (SALUBGH).

Field remarks: A very rare species of calcarious and sandy soils.

Distribution: West Pakistan (Nasir, 1974).

2. Commiphora wightii (Arn.) Bhandari in Bull. Bot. Surv. Ind. 6: 327 (1964); Stewart in Ali and Nasir, Ann. Cat. Vasc. Pl. W. Pak. & Kashm., 440 (1972); Nasir in Nasir and Ali, Fl. Pak., 71: 21-23 (1974).

Syn: *Balsamodendron mukul* Hook. ex Stocks; *B. roxburghii* Stocks; *B. wightii* Arn.; *Commiphora mukul* (Hook. ex Stocks) Engl.; *C. roxburghii* (Stocks) Engl.

Shrubs up to 4 m tall; branches aromatic, thorny and knotty with papery bark. Leaves sessile, alternate or fascicled, 1-3-foliolate; leaflets glabrous, the terminal sessile or subsessile, obovate, serrate (sometimes serrate only towards the apex), 1-5 cm long, 0.5-2.5 cm broad, lateral when present sessile, serrate or entire, less than half the size of the terminal leaflet. Plants dimorphic, one having bisexual and male flowers, and the other having female flowers with staminodes. Bisexual and male flowers sessile, 3-5 mm long, usually red, sometimes pinkish white. Bracts 2, opposite, glandular hairy. Calyx fused basally with the disc; tubular or urceolate, 1-2 mm long; lobes usually triangular, valvate, glandular hairy outside. Petals reflexed, acute, 3-5 mm long, c. 1 mm broad. Stamens 8, very rarely 10, 3-5 mm long, free, alternately short and long, included, sometimes equalling the petals. Disc conspicuous, toothed; shorter stamens inserted alternately in deeper sinuses. Ovary 2-loculed with sessile 2-lobed stigma. Fruit up to 1 cm long, red when ripe, marked with 2 white longitudinal lines (or grooves), mucronate; mesocarp yellow, rarely orange, 4 lined and fused at the base; epicarp dehiscing from the base upwards on maturation. In female flowers, sepals 2 mm long, petals 3-4 mm long, 1 mm broad; staminodes 8, alternately short and long, 1-1.5 mm long; ovary and fruit the same as in bisexual flowers.

Flowering Period: December July.

Herbarium specimens examined: NARA DESERT: *Rahmatullah Qureshi*, 1374 (SALUBGH).

Field remarks: A very rare species of calcarious and sandy soils.

Distribution: W. Pakistan, India (Nasir, 1974).

38. Zygophyllaceae

Annual or perennial herbs or subshrubs, rarely trees, often xerophytic; branches usually articulate. Leaves opposite or alternate (by suppression), 2-foilate or pinnate compound or multifid, rarely 3-7-foliate or simple, often fleshy or coriaceous, petiolate to subsessile; stipules persistent, coriaceous or spiny; leaflets 1-many, entire, eglandular, petiolulate to subsessile. Inflorescence axillary or subaxillary cymes, rarely in spicate racemes or solitary. Flowers ebracteate, hermaphrodite, rarely unisexual, regular or irregular, hypogynous, white, yellow or red, rarely blue. Sepals 5, rarely 4, free, rarely connate at the base, imbricate (rarely valvate or open). Petals 5, rarely (3-) 4 or 0, free, rarely basally united, often clawed, inserted on disk, imbricate or contorted, rarely valvate. Disk convex or depressed, rarely annular inconspicuous or 0, 10-lobed, eglandular. Stamens 5-15, in 1-3 whorls of 5 each, rarely (3-) 4, inserted on or at the base of disk, outer whorl ovary of antipetalous stamens often connate with the claw, usually unequal in length, free, subulate or filiform filaments with or without scales inside at the base; anthers introse, 2-celled, versatile, dehiscing longitudinally. Carpels (2-3-) 4-5, syncarpous, ovary superior, usually sulcate or winged, sessile or rarely stipitate, 4-5-celled, rarely less or more celled, cells rarely transversely locellate; placentation axile; ovule(s) 1-per cell, often filiform, anatropous, pendulous or ascending; raphe ventral; micropyle superior; style simple, terminal, angular or furrowed, rarely 4-5 and free; stigma simple, capitate or obscurely lobed, rarely sessile or discoid. Fruit usually a loculicidal or septicidal capsule, rarely stony drupe, baccate or schizocarpic. Seeds 1-(per cell, oblong or linear; raphe adnate or free; testa membranous, crustaceous or thick and mucilaginous; albumen scanty (rarely 0); embryo as long as the seed, green, straight or rarely curved; cotyledons oblong or linear, thick or foliaceous; radicle short, straight, superior.

A family of 25 genera and about 240 species, tropical, subtropical and warm temperate in distribution. Represented in Pakistan by 8 genera and 22 species (Ghafoor, 1974).

Key to genera:

1a. Plants spinous, subwoody; fruits pyramidal opening into 5, 1-seeded cocci, tipped with persitent style... **1. Fagonia**

1b. Plants not spinous, usually not subwoody:

2a. Leaves 1-foliate; fruits sharply 5-angled..**2. Zygophyllum**

2b. Leaves not simple, 3-foliate or pinnate:

3a. Leaves 3-foliate; petals 0; fruits neither spiny nor winged.........................**3. Seetzenia**

3b. Leaves pinnate with 3-7 pairs of leaflets; petals present; fruits spiny or winged.. **4. Tribulus**

1. Fagonia

L., Sp. Pl. 386 (1753); Gen. Pl. ed. 5. 182 (1754); Benth. & Hook. f., Gen. Pl. 1: 267 (1862); Edgeworth & Hook f. in Hook. f., Fl. Brit. Ind. 1: 425 (1874); Hutch., Gen. Fl. Pl. 2: 620 (1967); A. Ghafoor in Fl. Pak. 76: 9 (1974).

Diffusely branched, basally suffrutescent, annual or perennial herbs. Leaves opposite, simple or palmately 1-3(-7)-foliate; leaflets entire, mucronate; stipules often spiny, acicular. Inflorescence of solitary axillary flowers. Flowers pedicellate, regular, hypogynous, 5-merous, rosy, purplish-pink or violet, rarely yellow. Sepals 5, free, mucronate, imbricate, deciduous or persistent. Petals 5, free, clawed, imbricate, caduous. Disk short, inconspicuous. Stamens 10, inserted on the disk; filaments filiform, naked; anthers shortly oblong. Ovary sessile, 5-gonous, 5-celled, narrowed into a 5-gonous subulate style; ovules 2, near the base of the cell, collateral, pendulous from ascending funicles; style angular; stigma simple. Capsule deeply 5-angled, pyramidal or globular, septicidally dehiscent into 5, 1-seeded cocci which dehisce along the ventral suture and separate from a horny endocarp. Seeds erect, compressed, broadly oblong; testa mucilaginous; albumen horny; cotyledons broad, flat, ovate.

A genus of about 45 species, distributed in desert and dry arid regions from south west Asia to Indo-Pakistan subcontinent, Mediterranean region, south west Africa, Chile and south west United States (Ghafoor, 1974).

Key to species:

1a. Lower leaves 3 and upper 1-foliate...........................**1. F. indica** var. **schweinfurthii**

1b. All leaves 1-foliate..**2. F. bruguieri** var. **rechingeri**

1. Fagonia indica Burm.f., Fl. Ind. 102 (1768) var. **schweinfurthii** Hadidi in Rech.f., Fl. Iran. 98:6. t. 6 (1972); Ghafoor in Nasir & Ali., Fl. W. Pak.,76:19 (1974).

Annual or biennial prostrate to erect, glandular to glabrous shrublet up to 45 cm tall. Stem woody at base, much branched, terete, striate, glabrous; nodes swollen, internode ca. 1-3.5 cm long. Lower leaves 3-foliate, upper 1-foliate; petioles ca. 0.5-1 cm long. Leaflets ca. 1-3 x 0.2-0.5 cm in size, linear-lanceolate, lateral ones (in 3-foliate) shorter than the central. Flowers ca. 8-10 mm across, pinkish purple; pedicels up to 8 mm long, glandular-glabrous. Sepals ca. 3-4 x 1.5-2 mm in size, ovate, sparsely glandular above, persistent. Petals ca. 4-6 x 2-3 mm in

size, apex obtuse. Stamens ca. 5 mm long. Ovary ca. 1 mm across. Capsules ca. 4 mm across, softly hairy.
Flowering period: Almost throughout the year.
Herbarium specimens examined: NARA DESERT: *Rahmatullah Qureshi*, 535, 868, 896, 904, 1185, 1268 (SALUBGH).
Field remarks: A common species of calcarious soils.
Distribution: India, Iran, Pakistan, Aden, Eritrea, Ethiopia, Sudan, Somalia and Kenya (Ghafoor, 1974).

2. Fagonia bruguieri DC., Prodr. 1:704 (1824) var. **rechingeri** Hadidi in Candollea 27:91 (1972); Ghafoor in Nasir & Ali., Fl.Pak.,76:13 (1974).

Annual or biennial, profusely branched, glandular shrublet up to 40 cm long. Leaves 1-foliate, petiole ca. 1-5 mm long; leaflets ca. 1-3.8 x 0.2-0.6 cm in size, ovate-lanceolate, base distinctly articulate, apex mucronate, succulent, ± glabrous. Stipular spines ca. 1.5-3.5 cm long, awl-shaped, patent to ascending, green, apex pointed and yellow in colour. Inflorescence in solitary and axillary cyme. Flowers mediocre, ca. 12 mm across, pinkish, pedicel ca. 8 mm long, green glandular. Sepals 5, ca. 8-9 x 2-3 mm in size, quincuncialis, ovate, base rounded, apex aristate, margin entire, sparsely glandular above, green, persistent. Petals 5, ca. 10-11 x 4-5 mm in size, imbricate, spathulate, apex retuse and shortly apiculate, base truncate, margin entire, glabrous. Stamens 10, free, inserted on disc; filament ca. 2-7 mm long, filiform; anthers ca. 1-2 mm long, versatile, 2-lobed, elliptic-oblong, yellow in colour. Ovary 5-celled, 5-gonous, ovate, ca. 3 mm across, greenish in colour, densely appressedly hairy; style angular, ca. 3 mm long; stigma simple. Capsule ca. 0.5 cm across, cordate, rounded at the base, pyramidal towards the apex, deeply 5-partite to the axis, dehiscing along the central suture, pubescent; pedicel reflexed, ca. 0.7 cm long. Seeds 5, ca. 5 x 2.5 mm in size, ovoid, acute, flattened, brown, ± puberulent.
Flowering period: March-May.
Herbarium specimens examined: NARA DESERT: *Rahmatullah Qureshi*, 42, 104, 210, 259, 602, 809, 862, 893, 900, 909, 1019, 1076, 1084, 1119, 1191, 1233 (SALUBGH).
Field remarks: A common species of hilly areas.
Distribution: Western Iraq, Iran, Afghanistan and Pakistan (Ghafoor, 1974).

2. Zygophyllum

L., Sp. Pl. 385 (1753); Gen. Pl. ed. 5. 182 (1754); Benth. & Hook. f., Gen. Pl. 1: 266 (1862); Hutch., Gen. Fl. Pl. 2: 619 (1967); A. Ghafoor in Fl. Pak. 76: 27 (1974).

Perennial, rarely annual, diffusely branched, xerophytic or halophytic subshrubs or prostrate herbs. Leaves opposite, usually 1-2-foliate or simple; leaflets cylindrical or flat, usually fleshy or leathery; stipules 2, rarely spiny. Flowers solitary, axillary, pedicellate, regular, hypogynous, 4-5-merous, white or yellow, sometimes with a purple or red basal spot. Sepals 4-5, imbricate, persistent or deciduous. Petals 4-5, free, with or without a claw, imbricate or contorted. Disk fleshy, angular, cup-shaped or concave. Stamens 8-10, inserted at the base of the disk, longer than the petals; filaments filiform, with a scale at the base within; anthers oblong-ovate to rotundate. Carpels 4-5, syncarpous; ovary sessile, 4-5-gonous,

4-5 (rarely 2-3)-celled; ovules 2 in each cell, superposed, pendulous; style angular; stigma minute. Fruit capsular, 4-5-gonous or 4-5-winged, indehiscent or septicidally dehiscent into 5 cocci, or loculicidally 5-valved; the endocarp sometimes spreading. Seeds 1 or more in each cell, pendulous; testa crustaceous; albumen scanty; embryo straight with oblong cotyledons.
A genus of about 100 species, distributed in deserts and steppes from Mediterranean to Central Asia, South Africa and Australia (Ghafoor, 1975).

Zygophyllum simplex L., Mant. 1:68 (1767); Stewart in Nasir & Ali., Ann.Cat.Vasc.Pl.W.Pak. & Kashm., 435 (1972); Ghafoor in Nasir & Ali., Fl.Pak.,76:31 (1974).
Syn: *Z. portulacoides* Forssk., Fl. Aegypt.-Arab. 88 (1775).

Annual sub-erect to procumbent herb up to 35 cm long. Stem dichotomously branched; branchlets herbaceous, striate and papilose, hoary at first with white powdery tomentum then glabrous at length, purple at the base, pale green above. Leaves ca. 0.4-1.5 x 0.1-0.4 cm in size, unequal, subsessile, oblong-cylindric, apex obtuse to slightly apiculate, base cuneate, margin entire, grooved, fleshy; lateral nerves obscure; stipules 2, ca. 1.5 x 0.5 mm in size, lanceolate-triangular, apex acuminate, base truncate, margin hyaline. Flowers ca. 0.5 x 0.3 mm in size; pedicel ca. 3-5 mm long. Sepals ca. 3 x 1-2 mm in size, ovate-obovate, apex cucullate, obtuse, caducous, succulent, scarcely thickened in fruit, grooved. Petals ca. 3 x 1 mm in size, spathulate, spreading, base clawed, margin flat, lateral nerves arising from the base, yellow, glabrous. Disc 10-ribbed, ca. 1 mm long, angled, fleshy. Stamens 10; filaments 2-3 mm long; with 2-partite hyaline staminal scales, base slightly notched. Stigma persistent. Capsule ca. 3 mm across, more or less broadly obovoid, turbinate, rugulose, 5-angled; cocci 3-5 seeded. Seeds ca. 1 x 0.6 mm in size, fusiform, smooth, brown.
Flowering period: May-August.
Herbarium specimens examined: NARA DESERT: *Rahmatullah Qureshi*, 541, 928, 1026, 1085 (SALUBGH).
Field remarks: A common species of saline soils.
Distribution: Pakistan (Sindh, Baluchistan), India, Iran, tropical and N. Africa, Cape Verde Islands and W. Asia (Ghafoor, 1974).

3. Seetzenia

R. Br. in Denth. & Clapperton, Narr. Trav. Afr. App. 231 (1826); Benth. & Hook., Gen. Pl. ed. 1: 266 (1862); Edgeworth & Hook f. in Hook. f., Fl. Brit. Ind. 1: 424 (1874); Hutch., Gen. Fl. Pl. 2: 620 (1968); A. Ghafoor in Fl. Pak. 76: 8 (1974).
Seezenia Nees in R. Br., Verm. Bot. Schrift. 4: 46 (1830).

Perennial prostrate, glabrous to scabrous herbs. Leaves opposite, stipulate, palmately 3-foliate; leaflets obovate, apiculate. Flowers small, bisexual, pedicellate, actinomorphic, hypogynous; disk small, 5-lobed. Sepals 5, cucullate. Petals absent. Stamens 5, inserted on the disk, filaments filiform, without scales at the base; anthers somewhat globose. Carpels 5; ovary sessile, 5-gonous, sulcate, truncate, with 5 uni-ovulate locules; styles and stigmas as many as carpels, free. Fruits a 5-angled, ovoid, pendulous capsule, dehiscent into 5, 1-seeded cocci from persistent central axis, cocci with fleshy exocarp and bony endocarp. Seeds oval; testa thick, 3-layered, outer membranous, middle slimy and inner coriaceous.

A genus of 2 species, distributed in north and south Africa, Middle East, Iran, Pakistan and south west India (Ghafoor, 1974).

Seetzenia lanata (Willd.) Bullock in Kew Bull. 19: 204 (1965); Ghafoor in Nasir & Ali., Fl.Pak.,76: 8 (1974).

Syn: *Zygophyllum prostratum* Thunb., Prodr. Fl. Cap. 189 (1794); *Z. lanatum* Willd., Sp. Pl. 2(1): 564 (1799); *Seetzenia afrricana* R. Br. in Denh. & Clapp. Trav. App. 231 (1826); *S. orientalis* Decne., Ann. Sc. Nat. Bot. Ser. 2, 3: 281. t. 7 (1835).

A prostrate, papillate, glaucous herb up to 10 cm long. Stem and branches terete, articulate, somewhat swollen and lanate at nodes. Leaves 3-foliate, ca. 0.5-0.8 cm long; petiole ca. 1.2-2.5 mm, filiform, papilose; stipules intra-peiolar, ca. 1 x 0.7 mm, triangular, apex obtuse to subacute, tomentose, membranous. leaflets ca. 2.8-6 x 2.5-5.5 mm; medium leaflets obovate, base cuneate, apex apiculate; lateral leaflets oblong-obovate, apex obliquely mucronate, margin subcrenulate, glabrous below, papilose above; petiolule ca. 0.5 mm. Flowers solitary or few-flowered, axillary, ca. 4 x 1.5 mm in size, the buds ± globose, brown; pedicle ca. 5-10 mm, more or less recurved at the apex and longer than petioles, papillate, hairy. Sepals 5, valvate, free, ca. 3.8-4 x 1.3-1.5 mm in size, oblong-elliptic, base cuneate, apex acute, margin crenate, glabrous within, papilose above. Petals 0. Disk 5-lobed. Stamen antisepalous, inserted on the disk; filaments filiform, ca. 2 mm long, base broad and rounded, greenish, glabrous; anthers dorsifixed, ca. 1 mm across, ovate, apex apiculate, glabrous, yellowish. Ovary ca. 3 x 1.5 mm in size, 5-gonous, elliptic-oblong, ribbed, greenish, papilose; style 5, minute; stigma capitate. Fruit an ovoid capsule, ca. 7-8 x 5-6 mm insize, apex obtuse, 5-gonal, pendulous, dehiscing into 5, 1-seeded cocci, separating from the persistent axis, with a narrow succulent exocarp on the back only. Seeds ca. 4.5-5 x 2 mm in size, compressed, 1 in each locule, somewhat pointed at the ends, smooth, polished, brown.

Flowering period: February-May.

Herbarium specimens examined: NARA DESERT: *Rahmatullah Qureshi*, 1325 (SALUBGH).

Field remarks: A common species of saline and calcarious soils.

Distribution: Pakistan, Afghanistan, India, Arabia, Sinai, Palestine, Egypt, Libya and E. Africa (Ghafoor, 1974).

4. Tribulus

L., Sp. Pl. 386 (1753); Gen. Pl. ed. 5. 183 (1754); DC., Prodr. 1: 703 (1824); Benth. & Hook. f., Gen. Pl. 1: 264 (1862); Edgeworth & Hook f. in Hook. f., Fl. Brit. Ind. 1: 423 (1874); Hutch., Gen. Fl. Pl. 2: 618 (1967); A. Ghafoor in Fl. Pak. 76: 21 (1974).

Annual to perennial, prostrate to decumbent, pubescent to glabrescent herbs. Leaves opposite or alternate due to abortion of one leaf, abruptly paripinnate; leaflets 3-10 pairs, opposite, shortly petiolate to sessile, oblique, oblong to ovate or elliptic; stipules lanceolate to subulate. Flowers solitary, axillary or pseudo-axillary, pedicellate, regular, yellow, rarely white. Sepals 5, free, imbricate, pubescent, caduous or persistent. Petals 5, free, spreading, imbricate, obovate, fugacious. Disk annular, fleshy, 10-lobed. Stamens `10 in 2 whorls, rarely 5, inserted on the base of the disk, the longer opposite to the petals, the 5 shorter with a small gland outside; filaments filiform, naked; anthers oblong. Carpels 5; ovary sessile, ovoid or globose, beset with long silky hairs, 5-12-lobed, 5-12-celled; ovules 1-5 in each cell which are obliquely pendulous superposed in axile placentation; style 5-riged, stout, pyramidal filiform; stigma 5-rayed. Fruit 5-angled, 5-12-winged or spinous, pubescent, triangular, indehiscent indurated, transversely loculate 1-seeded mericarps. Seeds obliquely pendulous; testa membranous; embryo exalbuminous; cotyledons oval; radicle short.

About 20 species, distributed in tropical and subtropical countries throughout the world. It is represented by 4 species in Pakistan (Ghafoor, 1974).

Key to species:

1a. Fruit discoid; mericarps winged or spinous (rarely spines reduced or suppressed):
2a. Cocci broadly winged, not spinose..**1. T. longipetalus**
2b. Cocci not winged but spinose..**2. T. terretris**
1b. Fruit pyramidal; mericarps neither winged nor spinous.......................**3. T. ochroleucus**

1. Tribulus longipetalus subsp. **Longipetalus** Viv., Pl. Egypt. Dec. 10.2.f. 5 (1831); Stewart in Nasir & Ali., Ann.Cat.Vasc.Pl.W.Pak. & Kashm., 435 (1972); Ghafoor in Nasir & Ali., Fl.Pak.,76: 23 (1974).

An annual or biennial prostrate to suberect herb up to 45 cm high. Stem much branched, woody below, herbaceous above, spreading, terete, hirsute to scarious, greyish in colour; internode ca. 1-4 cm long. Leaves opposite, two of a pair unequal, up to 24 cm long; leaflets 5-6 pairs, ca. 6-8 x 3 mm in size, silky on both surfaces; stipules ca. 4 mm across, pale-yellow. Flowers ca. 6-8 mm across; pedicel. Sepals ca. 5-5.5 mm long, ovate, apex acuminate, margins scarious, villous above, persistent or deciduous. Petals ca. 4-5 x 2 mm in size, pale-yellow, oblong, base cuneate, margins slightly dentate. Stamens 10. Ovary bristly with long, white hairs; style short, stout. Fruit broadly pyramidal-ovoid, ca. 6-7 mm across; cocci 5, 2-seeded, spines more or less confluent into prominent, vertical, hard wings, ca. 4 mm broad, one on each coccous, the back of the cocci rugose-hirsute.

Flowering period: August-November.

Herbarium specimens examined: NARA DESERT: *Rahmatullah Qureshi*, 4, 52, 83, 200, 209, 219, 244, 261, 402, 610, 611, 617, 686, 849, 959, 1077, 1156 (SALUBGH).

Field remarks: A very common species of sandy soils.

Distribution: Throughout the range of the species (Ghafoor, 1974).

3. Tribulus ochroleucus (Maire) Ozenda & Quezel in Trav. l'Inst. Recherches Sahariennes. 14:74. 1956. Hadidi in Rech., f., Fl. Iran. 98:12. t. 20. f. 1. 1972; in Tackholm, Stud. Fl. Egypt ed. 2. 311. 1974.

Syn: *Tribulus macropterus* var. *ochroleucus* Maire

Prostrate annual herb with dense soft appressed hairs. Leaves alternate 2-2.5 cm long; stipules ovate, 3-4 mm long; leaflets 4-6 pairs, sessile, unequal sided, oblong-elliptic, 4-5 mm long, c. 3 mm broad, acute. Flowers pale yellow, 10-12 mm across; pedicels 6-8(-10) mm long. Sepals lanceolate, c. 4 mm long, acute. Petals obovate, 5-6 mm long, 3-4 mm broad, obtuse. Stamens 10, filaments as long as sepals. Ovary subglobose, densely hairy, style

short. Fruit 10-15 mm in diameter, pyramidal; mericarps without spines and wings, densely pilose.
Flowering Period: April-August.
Herbarium specimens examined: NARA DESERT: *Rahmatullah Qureshi*, 1375 (SALUBGH).
Field remarks: A less common species of sandy soils only recorded from Achhro Thar.
Distribution: Pakistan, Afghanistan, Iran, North Africa, Chad, Niger and Somalia (Ghafoor, 1974).

2. Tribulus terrestris L., Sp. Pl. 387 (1753); Stewart in Nasir & Ali., Ann.Cat.Vasc.Pl.W.Pak. & Kashm., 435 (1972); Ghafoor in Nasir & Ali., Fl.Pak., 76: 26 (1974). **Figure 19D**

Annual hirsute, decumbent herb up to 60 cm long. Stem woody below, herbaceous above, branched, spreading, terete, hirsute to scarious, greyish in colour; internode ca. 1-4 cm long. Leaves paripinnate, opposite, ca. 2.5-3 cm long; petiole ca. 0.5-0.8 cm long; leaflets 4-6 in pairs, opposite, almost sessile or short petiolule, ca. 1 mm long, oblique, oblong-ovate, base obtuse-cuneate, apex acute to mucronate, margin entire, ca. 0.5-1 cm in size, villous. Stipule ca. 2-3 in clusters, lanceolate, or subulate, apex acute, base rounded, margin entire, green, dandromorphic hairs on upper side, ca. 7-8 x 3 mm in size, trichomes ca. 2-4 mm. long. Flowers ca. 1-1.5 cm across, solitary and axillary, pedicel ca. 1.5 cm long. Sepals 5, free, imbricate, ovate, lanceolate, base obtuse, apex acute, margin entire, ca. 8-10 x 2-3 mm in size, appressedly hairy on above, green in color. Petal 5, free, imbricate, obovate, base cuneate, apex obtuse, ca. 7-8 x 3-4 mm in size, golden yellow in colour. Disc annular, fleshy, 10-lobed. Stamens 5+5, subtended by 5-glnadular scales, antipetalous; filaments 5-6 mm long; anthers 2-lobed, versatile, globose, yellow, ca. 0.9 mm across. Ovary 5-celled, ovoid, sessile, hirsute, ca. 4 mm across; style ca. 1.5 mm long; stigma 5-lobed, decurrent. Schizocarp ca. 1 cm in diam., 5-angled, woody, mericarp densely crested and tuberculate on dorsal side, densely hairy, each with pair of unequal downwardly directed spines, ca. 3-5 mm long, separating into 5-cocci. Seeds solitary per coccus.
Flowering period: July-September.
Herbarium specimens examined: NARA DESERT: *Rahmatullah Qureshi*, 543, 645, 1016, 1207 (SALUBGH).
Field remarks: A common weed species of wasteland areas.
Distribution: Tropical and subtropical countries in Asia, Africa, S. Europe, North Australia and introduced in New World tropics (Ghafoor, 1974).

39. Oxalidaceae

Herbs, rarely shrubs. Leaves digitately 3 foliolate. Flowers solitary or subumbellate, bisexual, pentamerous, hypogynous. Sepals 5, free. Petals 5, contorted. Stamens 10, obdiplostemonous, connate at the base. Ovary quinquilocular, placentation axile. Styles 5, persistent, free. Fruit a loculicidal capsule. Seeds usually many.

A tropical and subtropical family of about 3 genera and 875 species (Stewart, 1972).

Oxalis

L., Sp. Pl. 433 (1753); Gen. Pl. ed. 5, 198 (1754); Edgew. and Hk. f. in Hk. f. Fl. Brit. Ind. 1: 435 (1874); Y.J. Nasir, Fl. Pak., 4: 1 (1971).

Bulbous herbs with acidic juice. Aerial stem reduced, creeping or rhizomatous. Leaves digitately compound, showing sleep movements. Flowers regular. Stamens monadelphous at the base; filaments 5 long and 5 short, alternating with one another. Ovary 5 locular with one or more ovules in each locule. Fruit capsular. Seeds with an elastic testa.

A large genus with about 800 species, mostly distributed in North and South America, South Africa and fairly well represented in Europe and Asia.

Oxalis corniculata L., Sp. Pl. 435 (1753); Stewart in Ali and Nasir, Ann. Cat. Vasc. Pl. W. Pak. & Kashm., 431 (1972); Y.J. Nasir, Fl. Pak., 4: 2 (1971).

Syn: *Oxalis foliosa* Blatter; *Oxalis repens* Thunb.; *Oxalis villosa* M. Bieb.

Creeping and pubescent herbs mostly rooting at the nodes. Leaflets 4-15 x 8-32 mm across, obcordate, pilose-tomentose. Flowers solitary to 2-5 flowered axillary umbels. Pedicel 5-15 mm long, deflexed in fruit. Bracts 3-5, linear. Sepals 5, linear lanceolate, pilose. Petals 5, yellow. Filaments glabrous. Carpels 5, pubescent; styles longer than the shorter stamens. Capsule 1-2.5 cm long, subcylindric, pubescent. Seeds 1.5 mm long, brown, transversely ribbed.

Flowering Period: March December.

Field remarks: A common species of moist habitas.

Distribution: A cosmopolitan weed (Nasir, 1971).

40. Geraniaceae

Annual or perennial herbs. Leaves alternate, rarely opposite, lobate; lobes digitate, variously lobes in turn, stipulate. Inflorescence solitary, axillary or subumbellate. Flowers 5-merous, bisexual, regular or slightly irregular. Sepals 5, persistent, imbricate. Petals 5, imbricate, nectar glands alternating with the petals. Stamens 10-15; filaments free or connate at the base; staminodes frequent. Ovary superior, 5-celled; ovule(s) 1 or 2 per cell on axile placentation. Capsule splitting into five 1-seeded mericarps, often united to long awns.

Five genera and about 750 species, cosmopolitan, but mostly in temperate areas. In Pakistan, it represents by 4 genera and 35 species (Nasir, 1983).

Monsonia

L., Mantissa, 14 (1767); Boiss., Fl. Or. 1: 897 (1867); Jafri, Fl. Kar., 181 (1966); M. Zohray, Fl. Palest. 2: 24.3 (1971); Y.J. Nasir, Fl. Pak., 149: 38-39 (1983).

Annual or perennial. Leaves toothed or divided, stipulate. Flowers umbellate or solitary, 5-merous. Sepals and petals imbricate, the latter alternating with 5 glands. Stamens 15, all fertile. Ovary 5-lobed, 5-celled. Fruit a schizocarp with 5 mericarps; mericarps with a tail as long as beak.

A genus of 25 species distributed in Africa, especially the South, Madagascar, SW Asia, N.E. India and Pakistan. It represented by 3 species in Pakistan (Nasir, 1983).

Monsonia heliotropoides (Cav.) Boiss, Fl. Or. 1: 987 (1867); Stewart in Ali and Nasir, Ann. Cat. Vasc. Pl. W. Pak. & Kashm., 431 (1972); Nasir in Nasir & Ali., Fl. Pak., 149: 39 (1983).

Syn: *Geranium heliotropoides* Cav., Monad. Class, Diss. Dec. 4: 220. tab. 113. Figure 2 (1787); *Monsonia hispida* Boiss., Diagn. Pl. Or. Nov. Ser. 1, 8: 120 (1849); *M. asiatica* Vicary, J. s. Soc. Beng. 16, 2: 1161 (1847).

A small annual hoary herb up to 15 cm long with a woody root-stock, much branched from the root. Leaves radical, ca. 1-2.5 x 0.8-1.6 cm in size, ovate-rhomboid, apex obtuse, base cuneate-cordate, softly villous with long, white hairs on both surfaces; nerves impressed; petiole ca. 2-4 cm long, hirsute and pilose with long, spreading hairs; stipules subulate, hairy. Flowers ca. 6-8 mm across, umbellate at the apex of 8-10 cm long, slender scape, the latter distinctly clothed with long, white, bulbous-based, spreading hairs; pedicels ca. 3-6, each 1-1.5 mm long, sparsely hirsute. Bracts ca. 3-4 mm long, lanceolate. Sepals ca. 4-5 mm long, elliptic-oblong, becoming slightly enlarged and obovate in fruit, apex mucronate, margins membranous, canescent above. Petals ca. 2.5-2.7 mm long, obovate, hairy near the base, clawed. Stamens 15; filaments slightly 5-delphous, base ciliate. Carpels ca. 5 mm long, stipitate, densely hairy, obliquely truncate, with 2-3 wrinkles below the apex; beak ca. 7-8 cm long, slightly downy above, clothed with long silky hairs within.

Flowering period: August-October.

Herbarium specimens examined: NARA DESERT: *Rahmatullah Qureshi*, 867, 881, 925, 1033 (SALUBGH).

Field remarks: A rare species of sandy soils.

Distribution: N. Africa, S. Iran, Jordan, Arabia, Pakistan & India (Nasir, 1983).

41. Asclepiadaceae

Erect, twining or straggling herbs or shrubs; latex usually milky. Leaves decussate or whorled, simple, entire, petiolate or subsessile, sometimes reduce or 0, exstipulate. Flowers bisexual, regular, 5-merous, pedicellate, in axillary or terminal umbels, umbellate, racemes, panicles or cymes, bracteate; bracteoles present or not. Calyx-lobes with(out) glandular scales within, (un)equal, valvate or imbricate. Corolla gamopetalous, rotate, tubular, campanulate or rarely funnel-form; lobes equal, valvate, imbricate or overlapping. Stamens 5, alternating with corolla-lobes; filaments distinct or connate into staminal column; anthers 2-celled, connivent around, or adnate to stigma (gynostegium), the cells dehiscing by slits, tips often produced into membranous flaps; pollen grains granular, united in tetrads and discharged to translators, or when massive, contained in 2 pollinial bags, each derived from anther cells of 2 adjacent anthers, united by caudicle, into a hard, bilobed, glandular receptacle (pollen carrier), which rests on the angles of style-apex. Pollinia erect, horizontal or pendulous; translators spoon-shaped or spathulate, alternating with stamens narrowed above, derived from and attached to style-apex. Corona single or double, corolline, staminal or both, variously shaped. Pistil superior of 2, free, 1-celled carpels, oblong-cylindric, flat along inner face; ovules α, marginal or parietal; style enclosed within staminal column, free at the base, connate and dilated above; stigma usually flat, 5-gonous, enlarged, included or exserted, corners sometimes horny. Follicles of 2 mericarps, inflated or uniformly cylindric, dehiscing by ventral suture. Seeds α, usually overlapping, flat or plano-convex, slightly margined, with a tuft of silky hairs at the narrowed end; albumen scanty.

A family with 175-180 genera and 200 species distributed mainly in the tropical and subtropical regions of the world. In Pakistan it is represented by 23 genera and 41 species (Ali, 1983).

Key to genera:

1a. Erect shrubs:

2a. Leaves ovate; flowers more than 1.5 cm across, white; follicles in pairs........**1. Calotropis**

2b. Leaves linear; flowers less than 1.5 cm across; yellowish; follicles solitary..**2. Leptadenia**

1b. Twining or climbing plants:

3a. Corolla divided half-way down, white with purple vein..........................**3. Oxystelma**

3b. Corolla divided to the base; greenish..**4. Pentatropis**

1. Calotropis

R. Br., Mem. Wern. Nat. Hist. Soc. 1:39 (1811); Hook. f., Fl. Brit. Ind. 4: 17 (1883); Boiss., Fl. Or. 4: 54 (1879); Cooke, Fl. Pers. Bomb. 2: 214 (1904); Rechinger. f., Fl. Iran. 73: 6 (1970); Ali in Ali & Jafri, Fl. Libya 9: 6 (1977); Ali in Nasir & Ali, Fl. Pak. 150: 6 (1983).

Large erect shrub or small tree. Leaves large, opposite, decussate, broad, subsessile, more or less fleshy. Flowers in terminal or axillary umbellate or subracemose lateral pedunculate cymes. Calyx of 5 distinct sepals, glandular inside, broadly ovate. Corolla 5-lobed, broadly campanulate or subrotate; lobes broad, naked, valvate. Corona of 5 fleshy laterally compressed lobes radiating from the gynostegium, each with a recurved vesicle at the base and a pair of auricles at the apex. Anthers short and broad, with short broad membranous appendages inflexed over the depressed pentagonal summit of the style-apex; pollen masses solitary in each anther-cell, pendulous, attached to the pollen-carriers by short slender caudicles. Follicles short, thick, not echinate. Seeds comose.

A genus of 4 species distributed in tropical and subtropical Africa and Asia and Central and South America. Represented in Pakistan by 1-2 species (Ali, 1983).

Calotropis procera (Willd.) R. Br. in Ait., Hort., Kew., 2:78 (1811); Stewart in Nasir & Ali., Ann. Cat. Vasc. Pl. W.Pak. & Kashm., 566 (1972); Ali in Nasir & Ali., Fl. Pak. 150: 7-9 (1983). **Figure 3C**

subsp. **hamiltonii** (Wight) Ali, l.c. 289.

Syn. *Asclepias procera* Willd., Sp. Pl., I:1263 (1798); *Calotropis hamiltonii* Wight, Contrib. Bot. Ind. 53 (1834); *C. procera* auct. non (Ait.) Ait. f.: Hook. f., Brit. Ind. 4: 18 (1883).

A tall shrub or small tree up to 3 m tall with milky latex. Stem woody below, herbaceous above, branched, tomentose, fleshy, terete; internode ca. 3.5-9.5 cm long. Leaves sessile, opposite and decussate, ca. 4.5-7.5 x 2.5-4.5 inches in size, elliptic or obovate-oblong, apex apiculate-cuspidate, base slightly cordate or amplexicaule, margin entire, chartaceous, thick, glaucous-green, lateral nerves alternately arranged. Inflorescence a long peduncled umbellate panicle, subterminal. Flowers pedicellate, complete, regular, bisexual, hypogynous, pentamerous except pistil, ca. 1.5 cm across. Sepals 5, slightly connate at the base; lobes free, quincuncialis, inferior, lobes ovate, apex acute, base rounded, ca. 11-12 x 6-7 mm in size; a ring of scales are present inside the sepals, lanceolate. Petals 5, gamopetalous, lobes erect, inferior, valvate, ovate, ca. 1.5 x 0.8 cm in size, pink or whitish with purple spots within, pale silvery out side, lathery; corona 5, fleshy, laterally compressed, radiating from the staminal column, base terminating in an up-curving spur, apex of the corona bifid, without auricles, ca. 0.7 x 0.4 cm in size. Stamens (Pollinia) 5, epipetalous; pendulous, filaments connate in a fleshy tube around the ovary (staminal tube), the apex of the staminal tube united with the much-dilated stigmatic disc, to which the anthers are also coherent forming the pentagonal gynostegium (gynandrous); anthers short, broad lipped with inflexed membranous flaps, 2-celled, the pollen of each cell agglutinated in small ovoid pendulous waxy pollen mass (Pollinia), the pollinia of the adjacent cells of the neighbouring anthers connected by means of short stalks or caudicles to a distinct dark coloured dot-like glands, the corpusculum, which lies at the angle of gynostegium. Ovary 2-celled, apocarpous, unilocular, ca. 3 mm; styles 2, free, meeting up in stigmas which are united to form angular disc, 0.9 cm long. Follicle ca. 8 x 4 cm, oblong, recurved. Seeds numerous, broadly ovoid, flat, ca.9 x 6 mm in size; coma tuft of silky hairs.

Flowering period: July-September.

Herbarium specimens examined: NARA DESERT: *Rahmatullah Qureshi*, 918, 1208, 1284 (SALUBGH).

Field remarks: A common species of calcarious and sandy soils.

Distribution: Pakistan, India, Afghanistan, Iran (?), Iraq (Ali, 1983).

2. Leptadenia

R. Br., Mem. Wern. Nat. Hist. Soc. 1:34 (1811); Boiss., Fl. Or. 4: 62 (1879); Hook. f., Fl. Brit. Ind. 4: 63 (1838); Cooke, Fl. Pers. Bomb. reprint ed. 2: 236(1958); Ali in Nasir & Ali, Fl. Pak. 150: 51 (1983).

Erect or twining shrubs with much branched rush-like stems. Leaves opposite, linear to lanceolate, small or 0. Flowers small, in umbellate lateral sessile or pedunculate many-flowered cymes. Calyx deeply 5-fid; lobes ovate, subacute. Corolla greenish yellow, deeply divided, rotate or with a very short campanulate tube; lobes 5, pubescent or bearded inside, valvate in bud. Corona double, the corolline of 5 short transverse or rounded fleshy lobes tipped with a tuft of hairs or with a subulate hairy point, inserted at the sinuses of the corolla, the staminal corona of a raised undulate fleshy ring at the base of staminal-column. Staminal-column arising from the base of the corolla; anthers incumbent on the top of the style, without terminal appendages; pollen masses 1 in each anther-cell, suberect, pellucid at the apex, attached in pairs by short caudicles to the pollen carriers. Follicles ovate-oblong to oblong, tapering towards apex, smooth. Seeds comose.

A genus of ca. 10 species, distributed in tropical Asia and Africa. Represented in Pakistan by only 1 species (Ali, 1983).

Leptadenia pyrotechnica (Forsk.) Decne. in Ann. Sci. Nat. ser.2, 9:269 (1838); Stewart in Nasir & Ali., Ann. Cat. Vasc. Pl. W.Pak. & Kashm., 569 (1972); Ali in Nasir & Ali., Fl. Pak. 150: 52 (1983). **Figure 3D**

Syn. *Cynanchum pyrotechnicum* Forssk., Fl. Aegypt-Arab., 53 (1775); *Leptadenia spartium* Wight & Arn in Wight, Contrib. 48 (1834).

An erect, rush-like, often leafless green shrub, attaining 3-4 m height. Stem fairly branched, woody at the base, herbaceous above, cylindric, green, glabrous, with watery sap. Leaves when present ca. 2-25 x 2-6 mm in size, subsessile, narrowly linear to linear-lanceolate, apex acute, glabrous, caduous. Flowers in lateral umbellate cymes, ca. 1 cm long; flower ca. 3 mm across, greenish-yellow; pedicel ca. 1.5 mm. calyx-tube ca. 1 mm; lobes valvate, hairy above; trichomes moniliform. Corolla-tube funnel-shaped, ca. 1 mm, glabrous within green; lobes ovate, apex acute, ca. 2 x 1 mm in size, hairy above; trichomes unicellular. Outer corona of 5 small scales; staminal corona of raised, undulate, fleshy ring. Follicles ca. 7-14 x 0.5-0.8 cm in size, terete, lanceolate, tapering to a slender beak, glabrous. Seeds ca. 5-8 mm long, ovate-lanceolate, glabrous; coma ca. 2.5-3.5 cm long.
Flowering period: August-January.
Herbarium specimens examined: NARA DESERT: *Rahmatullah Qureshi*, 2, 44, 630, 638, 829, 838, 843, 847, 849, 924, 1049, 1072, 1079, 1196, 1214, 1238 (SALUBGH).
Field remarks: A common species of sandy soils.
Distribution: Pakistan, India, Iran, Arabia, Egypt, Sudan, Somalia, Chad, Libya and Algeria (Ali, 1983).

3. Oxystelma

R. Br., Mem. Wern. Nat. Hist. Soc. 1:40 (1811); Benth. & Hook. f., Gen. Pl. 2: 749 (1876); Boiss., Fl. Or. 4: 59 (1879); Hook. f., Fl. Brit. Ind. 4: 17 (1838); Cooke, Fl. Pers. Bomb. reprint ed. 2: 216(1958); Ali in Nasir & Ali, Fl. Pak. 150: 21 (1983).

Twining glabrous herbs or undershrubs. Leaves opposite, ovate-lanceolate to linear-lanceolate. Flowers large and handsome, in lax racemose or subumbellate cymes or solitary; pedicels filiform. Calyx small, 5-lobed; lobes lanceolate, acute, glandular hairy inside. Corolla divided half-way down, white with purple vein, shallowly campanulate to almost rotate, glabrous outside and densely pubescent ring at the mouth (sometimes described as a corolline corona) within, and a broad saucer-shaped; limb 5-lobed; lobes deltoid, acute, valvate at the base, the tip overlapping to the right in bud. Corona of 5 erect, ovate or lanceolate, acute lobes with 2 keels inside arising from staminal column. Stamens inserted at the base of the corolla-tube; filaments short, connate; anthers erect, terminated by a short inflexed membranous appendage; pollen masses pendulous, waxy, elongate-clavate, compressed, attached in pairs by their attenuated ends to the short ovoid pollen-carriers. Style apex truncate or convex, not exserted beyond the anthers. Follicles often solitary, oblong and obtuse or lanceolate and acute. Seeds comose.
A genus with 4 species, distributed in Old World Tropics, represented in Pakistan by only 1 species (Ali, 1983).

Oxystelma esculentum (L.f.) R. Br. in Mem. Wern. Soc. I:40 (1811); Stewart in Nasir & Ali., Ann. Cat. Vasc. Pl. W.Pak. & Kashm., 569-70 (1972); Ali in Nasir & Ali., Fl. Pak. 150: 21 (1983). **Figure 4A**

Syn. *Periploca esculenta* L.F., Suppl. 168 (1781); *Asclepias rosea* Roxb. Fl. ind. 2: 40 (1832); *Oxystelma secamone* auct. non (L.) Karst. (1880-83); *Sarcostemma esculentum* (L. f.) Holm in Ann. Miss. Bot. Gard. 37: 482 (1950).

A perennial twining herb or undershrub with milky sap up to 5 m high. Stem fairly branched, slender, herbaceous, glabrous; internodes ca. 7-11 cm. Leaves ca. 4-9 x 0.4-0.8 cm in size, linear-lanceolate, apex acute, base truncate, margin entire, subcoriaceous, pale green, glabrous; petiole ca. 0.5-1.5 cm long. Flower(s) usually paired, axillary, solitary or in lax racemes. Peduncle ca. 5-6 cm long. Bracts deciduous. Flower ca. 2.5 cm across; pedicel ca. 1 cm. Calyx cupular; lobes ca. 4 mm, equal, imbricate, ovate, apex acute, chartaceous, glabrous. Corolla pink within, with purplish reticulations, white outside, ca. 2.5 cm across, rotate-angulate; lobes triangular, united at the mid-half, valvate, ca. 2 cm long, shortly overlapping to right in bud, chartaceous, ciliate, apex acute, recurved. Corona double; outer corolline, annular, pubescent at the base of corolla within; inner staminal, 2-keeled, basally inflated, ca. 5 mm; staminal column ca. 7 mm. Pollinia pendulous; pollinial bags oblong, ca. 1.3 mm; caudicle ca. 0.2 mm, united into the receptacle with a membranous flap; receptacle ca. 0.3 mm. Ovaries (oblong) globose, ca. 3 mm; placenta bifurcate; styles ca. 4 mm. Follicles ca. 4-8 x 1-1.5 cm in size, inflated, lanceolate-oblong, apex acute. Seeds ca. 2.5-3.5 x 2 mm in size, ovate, base rounded, apex pointed; coma up to 2 cm, silky, dull white.
Flowering period: December.
Herbarium specimens examined: NARA DESERT: *Rahmatullah Qureshi*, 840, 1433 (SALUBGH).
Field remarks: A less common species of wetland habitats.
Distribution: Pakistan, India, Burma, Indonesia, Ceylon, Iraq and Egypt (Ali, 1983).

4. Pentatropis

Wight, Contrib. Bot. Ind. 53 (1834); Boiss., Fl. Or. 4, 59 (1879); Hook. f., Fl. Brit. Ind. 4: 19 (1883).

Twining glabrous herbs or subshrubs. Leaves opposite, fleshy, ovate to oblong. Flowers in lateral sessile to subsessile umbelloid cymes. Calyx lobes 5, ovate-lanceolate acute. Corolla divided to the base; greenish, suffused with purple, 5-lobed, lobes longer than the united portion, twisted in bud. Corona in one series, consisting of 5 laterally compressed segments, attached to the staminal column, produced at the base into a spur and the apex into a slender tip. Pollinium 1 in each locule, pendulous, without a pellucid margin. Follicles lanceolate, narrowed into a beak, glabrous.

A genus of c.6 species, ranging from Africa through Arabia, India, Madagascar to Australia, represented in Pakistan by 1-2 species.

Pentatropis nivalis (J.F.Gmel.) Field & J.R.I.Wood, Kew Bull. 38 (2): 215 (1983).

Syn: *Asclepias spiralis* Forssk.; *Pentatropis cynanchoides* R. Br.; *Pentatropis spiralis* (Forssk.) Decne.

A much branched climber with rough corky bark. Leaves 1.2-3.8 x 6-18 cm, linear, oblong, ovate or elliptic, acute or obtuse, mucronate, fleshy; petiole 2.5-13 mm long.

Peduncle up to 5 mm long. Pedicel 5-12.5 mm long, bracts minute. Calyx lobes c. 1.5 mm long. Corolla lobes 8-13 mm long, glabrous outside, minutely puberulous inside. Corona lobes c. 1.5 mm long. Follicles 5-7.5 cm x 8 mm. Seeds c. 4.5 mm long, ovate, flattened, minutely irregularly crenate at the lower end, coma 1.8-2.5 cm long.
Flowering Period: February.
Herbarium specimens examined: NARA DESERT: *Rahmatullah Qureshi*, 840, 1432 (SALUBGH).
Field remarks: A less common species of sandy soils recorded from Achhro Thar.
Distribution: Pakistan, India, Afghanistan, Iran, Arabia and Tropical Africa (Ali, 1983).

42. Solanaceae

Herb, shrubs, erect, trailing or scandent (rarely small trees). Leaves alternate, rarely clustered, simple, blade entire or lobed, rarely pinnately divided, peninerved, margin entire or irregularly toothed, petiolate, exstipulate. Inflorescence solitary or an axillary, lateral racemes, cymes, corymb, umbels or panicles, sometimes extra-axillary in Solanum. Flowers actinomorphic, hermaphrodite, rarely zygomorphic, 5-merous, sessile or pedicellate; bracts and bracteoles often 0. Calyx cupular, 5-lobed, campanulate or tubular, valvate, persistent and enlarged in fruits. Corolla sub-rotate, campanulate, salver-form or trumpet-shaped; lobes 5 or more, usually equal (rarely unequal), , distinct or connate into a circular limb, plicate, valvate or induplicate-valvate in bud, spreading or reflexed. Stamens 5 (rarely 4), alternating with corolla lobes, equal or didynamous, inserted below or above the middle of the tube, exserted or included, free or connivent into a cone; anthers oblong, rarely cordate, parallel, dehiscence longitudinally (introse) or porous (apical); connectives narrow. Ovary superior, obliquely placed, 2-5-locular; ovules (; placentation axile (peltate); style linear, usually as long as corolla; stigma capitate or obscurely 2-5-lobed. Disc often nectariferous. Fruit a berry or a capsule; seeds α. Seeds compressed, discoid, or (sub)reniform, rarely angular; albumen fleshy; testa smooth or pitted, scaly, often crustaceous; embryo curved or straight.

A family of ca. 84 genera and 3000 species, widely distributed in tropical and temperate regions of both the hemispheres but chiefly W. and Southern S. Ameriac. It is represented in Pakistan by 14 genera and ca. 52 species of which 27 species are native, 6 naturalised and the other exclusively or found occasionally as escapes (Nasir, 1985).

Key to genera:
1a. Fruit a dehiscent or sub-indehscent spiny capsule.......................................**1. Datura**
1b. Fruit a berry:
2a. Shrubs with thorny twigs; corolla tubular...**2. Lycium**
2b. Herbs without thorny twigs; corolla rotate of campanulate:
3a. Fruiting calyx inflated, enclosing the berry:
4a. Flowers solitary, pedicellate...**3. Physalis**
4b. Flowers clustered, sessile...**4. Withania**
3b. Fruiting calyx not inflated; berries not enclosed in the persistent calyx............**3. Solanum**

1. Datura

L., Sp. Pl. 179 (1753); Gen. Pl. ed. 5: 83 (1754); Boiss., Fl. Or. 4: 292 (1879); Clarke in Hook f., Fl. Brit. Ind. 4: 242 (1883); Jafri in Fl. Kar., 300 (1966); Nasir in Fl. Pak., 168: 40 (1985).

Annual herbs or subshrubs. Leaf-blade entire or lobed, chartaceous, lateral nerves ca. 6 pairs, glabrous or pubescent, margin entire, apex acute, petiolate. Flower(s) solitary, axillary, bisexual, erect, purple or white, showy. Calyx tubular, herbaceous, 5-lobed, upper part deciduous, lower part persistent. Corolla trumpet-shaped; tube cylindric, gradually widened from middle; lobes 5, connate into a circular limb, plaited, 5- or 12-toothed. Stamens 5, inserted near the middle of tube, decurrent below, included; filaments filiform; anthers oblong, puberulent, dehiscence longitudinal. Ovary conical, echinate, 2-locular, spuriously 4-locular; ovules α, on bifurcate placentae; style elongate, filiform; stigma 2-lobed. Capsule globose, armed, irregularly bursting; calyx base persistent; epicarp thick. Seeds numerous, circular, compressed, rugose; embryo peripheric.
A genus of 10 species, distributed in Mexico, Australia, Pakistan and India. Represented in Pakistan by 4 species, of which none are native ((Nasir, 1985).

Datura fastuosa L., Syst., Nat. ed. 10, 2: 932 (1759); Nasir in Nasir & Ali, Fl. Pak., 168: 44-45 (1985).

Syn: *D. metel* L., Sp. Pl. ed. 1: 179 (1753), *nom. confusum*; *D. alba* Rumphius ex Nees., Trans. L. Soc. London 17: 53 (1837); *D. fastuosa* var. *alba* (Nees) C. B. Clarke in Hook.f. Fl. Brit. Ind 4: 243 (1883).

An erect subshrub up to 85 cm. Stem woody below, herbaceous above, terete, softly pubescent, greyish-green; internode ca. 2.5-7 cm. long. Leaves ca. 3-14 x 2-10 cm in size, elliptic-ovate to angulate, base obliquely truncate-cordate, apex acute to apiculate, margin frequently lobed, rarely entire, sparsely hairy or puberulous on dorsal side, densely hairy on above; lateral nerves ca. 6-9 pairs; petiole ca. 2-7 cm long, greyish green, softly pubescent; trichome ca.1.5 mm long, greyish green. Inflorescence solitary and axillary cyme. Flower bisexual, actinomorphic, complete, white, ca. 16-17 x 2 cm in size; pedicel ca. 1-1.3 cm long, densely hairy. Sepals 5-lobed, tubular; lobes free, ca. 1.3-2 x 0.3-0.5 cm in size, lanceolate, base rounded, apex acute, margin entire, glaucous within, hairy above, greenish in colour; tube ca. 7.5 cm long. Petals trumpet-shaped, white above, greenish below; tube ca. 15 cm long; limb 4 cm long; teeth 5, ca. 1 cm long, spreading, lanceolate, apex acute-acuminate. Stamens 5, adnate to the base of petals tube; filaments ca. 13.5 cm long, filiform, base decurrent, greenish; anthers ca. 1 x 0.3 cm in size, basifixed, elliptic-oblong, apex retuse, margin more or less hairy, dehiscing longitudinally, white. Ovary ca. 0.5 cm across, conical, echinate, 2-locular; style ca. 12 cm long, elongated; stigma ca. 0.2 cm across, bilobed. Capsule ca. 3-4 cm across, globose, thickly armed with spines, irregularly bursting; spines short, stout, ca. 1 cm, blunt; calyx base persistent, pericarp thick. Seeds ca. 0.4-0.5 x 0.3 cm in size, reniform, compressed, rugose.
Flowering period: September-May.
Herbarium specimens examined: NARA DESERT: *Rahmatullah Qureshi*, 1258, 1311 (SALUBGH).

Field remarks: A common species of calcarious soils.
Distribution: Tropical and subtropical regions of Asia, Mediterranean, Africa and America (Nasir, 1985).

2. Lycium

L., Syst. Nat. ed. 1 (1735); Gen. Pl. ed. 5: 88 (1754); Clarke in Hook f., Fl. Brit. Ind. 2: 240 (1883); Jafri in Fl. Kar., 299 (1966); Nasir in Fl. Pak., 168: 31 (1985).

A perennial subshrub. Stem hard, woody; branchlets often sharply ending. Leaves small, alternate or often fascicle, linear, terete or flat. Pedicels usually solitary (rarely fascicled at the nodes); flowers small. Calyx campanulate, at first 5-merous, then irregularly 3-5-lobed or 2-lipped, in fruit not or scarcely enlarged. Corolla tubular-infundibuliform; lobes 5 (4), imbricate in bud, spreading in flower. Stamens 5 (4) on the corolla tube; anthers exserted or included, dehiscing longitudinally. Ovary 2-celled; ovules few or many; style filiform; stigma subcapitate. Berry small, globose or oblong. Seeds many, few or solitary, compressed, suborbicular; embryo peripheric.

A genus of 10 species, distributed in Mexico, Australia, Pakistan and India. It represented 4 species of which non are native (Nasir, 1985).

Lycium edgeworthii Dunal in DC., Prodr. 13, 1: 525 (1852); Nasir in Nasir & Ali, Fl. Pak., 168: 35 (1985).

Syn: *L. europeum* auctt. non L., Jafri, l.c. 300; *L. indicum* Wight, Ic. Pl. Ind. Or. 4: 2119, t. 1403 (1848).

Woody shrubs, up to 1.5 m high, with sharp spines. Leaves ca. 6-22 x 1.5-9 mm in size, lanceolate-obovate or obovate-elliptic, base attenuate to short petiole. Flowers solitary or paired; pedicels ca. 1-6 mm long, sparsely hairy or subglabrous. Calyx ca. 2.5-3 x 1.5-2.5 mm in size, campanulate or tubular; lobes 5, acute. Corolla ca. 7-17 mm long, infudibuliform, white. Filaments subequal, 3 longer one exserted, 2 smaller ones included, glabrous at the base. Berries ca. 3-4 mm across, globose. Seeds ca. 5-6, reniform, reticulately muricate.
Flowering period: October-January.
Herbarium specimens examined: NARA DESERT: *Rahmatullah Qureshi,* 1262 (SALUBGH).
Field remarks: A common species of rocky habitat.
Distribution: S. Iran, Pakistan (Sindh, Baluchistan & Punjab) and India (Nasir, 1985).

3. Physalis

L., Sp. Pl. 182 (1753); Nasir in Fl. Pak., 168: 25 (1985).

Annual or perennial herbs, glabrous to pubescent. Leaves usually alternate, petioled, entire or lobed, texture soft. Flowers solitary. Calyx 5-cleft, enlarged, bladder-like and 3-angled in fruit. Corolla campanulate, white or yellow. Stamens 5, attached near the base of corolla tube. Ovary bilocular. Fruit a globose berry, surrounded by the inflated calyx. Seeds many, compressed.

Physalis minima L., Sp. Pl. 183 (1753); Stewart in Nasir & Ali., Ann. Cat. Vasc. Pl. W. Pak. & Kashm., 642 (1972). **Figure 18C**

Syn: *Physalis divaricata* D. Don, Prodr. Fl. Nep. 97 (1825).

A diffuse annual from 15-45 cm tall, subglabrous to pubescent. Leaves 3-8 5(-11) x 1.5-4 (-7) cm, ovate. sinuate, repand or sinuate-dentate to subentire, acute or acuminate, base cordate to oblique. Petiole up to 40 mm long, slender. Flowers solitary axillary. Pedicel less than 10 mm long, somewhat elongating in fruit. Calyx c. 2.5 mm long, campanulate. inflated, globular-avoid, membranous and up to 25 mm in fruit, pubescent. Corolla 5 mm long, shortly tubular, yellow; lobes acute, pubescent. Stamens subincluded. Anthers c. 1 mm long; filaments ± 2 mm long. Ovary 1.1 mm long. ovoid. Style linear, stigma subcapitate. Berry globose, 10 mm broad, orange. Seeds c. 2 mm long, subreniform, compressed; minutely reticulate-undulate. brownish-yellow.

Flowering Period: August-October.

Herbarium specimens examined: NARA DESERT: *Rahmatullah Qureshi*, 1262 (SALUBGH).

Field remarks: A rare species of moist soils.

Distribution: Afghanistan and eastward to Nepal (Nasir, 1985).

4. Withania

Pauquy, Diss. Bellad. 14 (1825), nom. cons.; Boiss., Fl. Or. 4: 287 (1879); Clarke in Hook f., Fl. Brit. Ind. 4: 239 (1883); Cooke, Fl. Bomb. Pres. reprinted ed. 2: 340 (1906); Jafri in Fl. Kar., 298 (1966); Nasir in Fl. Pak., 168: 28 (1985).

Syn: *Physaloides* Moench, Meth. 473 (1794).

Unarmed perennial shrubs, often hoary-tomentose. Leaves alternate, entire. Flowers usually axillary clusters, subsessile or shortly pedicellate. Calyx campanulate, 5-toothed, accrescent in fruit. Corolla campanulate; lobes 5, slightly larger than the calyx, valvate in bud. Stamens 5, attached near the base of the corolla; anthers dehiscing longitudinally. Ovary 2-celled; ovules α; style linear; stigma shortly bifid. Berry globose, enclosed in the large calyx. Seeds many, discoid, small, compressed; embryo peripheric.

An Old World genus with 10 species. Represented in Pakistan by 2 species (Nasir, 1985).

Key to species:

1a. Plant mealy tomentose; leaves elliptic-lanceolate or oblanceolate;
flowers unisexual; berry with accresent calyx.................................. **1. W. coagulans**

1b. Plant hoary tomentose; leaves ovate, obovate or oblong; flowers bisexual;
berry loosely enclosed in the persistent calyx.................................**2. W. somnifera**

1. Withania coagulans (Stocks) Dunal in DC., Prodr. 13, 1: 685 (1852); Stewart in Nasir & Ali., Ann. Cat. Vasc. Pl. W.Pak. & Kashm., 645 (1972); Nasir in Nasir & Ali, Fl. Pak., 168: 28 (1985).

Syn: *Puneeria coagulans* Stocks, J. Roy

A densely packed shrub up to 1 m tall. Shoots rigid, greyish-green, stellate-tomentose. Leaves 3-8 x 1.4-3.5 cm, elliptic-ovate to elliptic-lanceolate, obtuse or acute, cuneate, leathery, nerves faint. Petiole 5-10 mm long. Bisexual flowers: calyx cupular-campanulate, stellate-tomentose; lobes 1.5-2.5 mm long, acute. Corolla exceeding calyx, campanulate, stellate-tomentose outside, yellowish Within. Stamens included. Stigma subexserted. Male flowers: stamens ± subexserted. Anthers c. 3.5 mm long; filaments 2.0 mm long. Female flowers: stamens included. Anthers smaller than in male flowers. Style glabrous. Berry globose, 10-12 mm broad, red, included in fruiting calyx. Seeds c. 2.5 mm broad. reniform, brown, minutely rugose-reticulate to ± smooth.

Flowering Period: Jan.-April.

Herbarium specimens examined: NARA DESERT: *Rahmatullah Qureshi,* 1262 (SALUBGH).

Field remarks: A rare species of sandy soils.

Distribution: Iran, Afghanistan, Pakistan and India (Nasir, 1985).

2. Withania somnifera (L.) Dunal in DC. Prodr. 13(1) 453 (1852); Stewart in Nasir & Ali., Ann. Cat. Vasc. Pl. W. Pak. & Kashm., 645 (1972); Nasir in Nasir & Ali, Fl. Pak., 168: 30 (1985). **Figure 18D**

Syn: *Physalis somnifera* L. Sp. Pl. 182 (1753).

A perennial erect undershrub up to 2 m high. Stem hard, woody at the base, herbaceous above, terete, branched, branchlets greyish green, hoary-tomentose; internode ca. 2-4.5 cm long. Leaves cauline and ramal, ca. 3-8 x 1.5-5 cm in size, elliptic-ovate, base truncate-acute, apex acute-subacute, margin entire, chartaceous, softly appressed pubescent, lateral nerves ca. 6-8 pairs; petiole ca. 1-3 cm long, softly hairy. Inflorescence in axillary, solitary or in few flowered fascicles umbellate cyme. Flowers hypogynous, bisexual, complete, actinomorphic, campanulate, ca. 1 cm long; pedicel ca. 0.5 cm. Calyx-tube ca. 4 mm long, puberulous without; lobes 5, free, lanceolate, apex acute, softly tomentose, ca. 4-5 mm, accrescent and inflated in fruit, persistent, green in colour. Petals 5-lobed, ca. 6 mm; lobes triangular-oblong, base truncate, apex acute, margin entire and appressedly pubescent, valvate, connate bellow the middle, puberulous without, greenish yellow in colour. Stamens included, alternate with petals; filament ca. 5 mm across; anthers ca. 2 mm, dorsifixed, bilobed, oblong. Ovary ca. 3 mm, ovoid or globose, glabrous; style ca. 5 mm long, filiform; stigma capitate. Berry ca. 0.8 cm across, orange red when ripe. Seeds (sub) trigonous; testa pitted.

Flowering period: September-December.

Herbarium specimens examined: NARA DESERT: *Rahmatullah Qureshi*, 225, 1186, 1274 (SALUBGH).

Field remarks: A less common species of moist soils.

Distribution: Canary Islands, Mediterranean, Africa, Iraq, S. Iran, Syria, Turkey, Palestine, Arabia, Pakistan and India (Nasir, 1985).

3. Solanum

L., Sp. Pl. 184 (1753); Gen. Pl. ed. 5: 224 (1754); Clarke in Hook f., Fl. Brit. Ind. 4: 229 (1883); Jafri in Fl. Kar., 294 (1966); Metthew, Fl. Tamilnadu Carnatic 1: 1057 (1983); Nasir in Fl. Pak., 168: 3 (1985).

Armed or unarmed, erect/spreading herbs or shrubs, rarely climbers. Leaves alternate or appearing opposite (from adnation), those of the same node unequal, blade entire, sinuate, angulate or pinnatifid, sometimes prickly, margin entire or toothed. Flowers stalked, aggregate in axillary, extra-axillary or terminal umbels, corymbs, cymes or racemose panicles. Calyx-lobes 5, free or connate, rarely enlarged in fruit. Corolla white, blue or purple, sub-rotate or rarely campanulate, tube short; lobes 5, plicate. Stamens 5, attached in the mouth of short tube, connivent around style, brightly coloured (usually yellow), exserted; anthers oblong, with bases adnate to one another, often narrowed upwards, dehiscing by apical pores. Ovary globose or conic, 2-celled; style columnar, exserted or included; stigma acute, capitate or 2-lobed. Berry pulpy, globose. Seeds α, triangular or circular, compressed.

About 1400 species, widespread in tropical and temperate regions of both the hemispheres. Represented in Pakistan by 15 species, of which 8 are native (Nasir, 1985).

Key to species:

1a. Plants prostrate and armed with prickles; sepals prickly; berries yellow, more than 1 cm across...**1. S. Surattense**

1b. Plants erect and unarmed; sepals not spiny; berries black, 5-7 mm across.......**2 S. nigram**

1. Solanum surattense Burm.f., Fl. Ind., 57 (1768); Stewart in Nasir & Ali., Ann.Cat.Vasc.Pl.W.Pak. & Kashm.,644 (1972); Yasin in Nasir & Ali., Fl.Pak., 168:10 (1987).

Syn: *S. xanthocarpum* Schrader & Wendl. Sert. Hannov. 1: 8. T. 2 (1795); *S. jacquini* Willd. Sp. Pl. 1: 1041 (1798); *S. diffusum* Roxb. Fl. Ind. 1: 568 (1832).

A perennial, procumbent, diffuse prickly herb up to 60 cm. Stem woody at the base, herbaceous above, terete, zigzag, stellate tomentose when young, greenish-brown, straight yellow prickles all around the stem; internode ca. 2.5-7 cm long. Leaves ca. 3.5-8 x 3.5 cm in size, sub-opposite, ovate-oblong, pinnately 5-11 sub-equal lobed, base obliquely truncate-attenuate, apex obtuse-subacute, chartaceous, lateral nerves ca. 5-11, stellate hairy, and armed with long, sharp prickles on midrib and nerves, ca. 0.5-2 cm long; petiole ca. 2.5-5.5 cm long, prickly. Inflorescence 4-5-flowered, extra-axillary cymes, ca. 4.5 cm long; peduncle ca. 2.5-3 cm long; pedicel ca. 1-1.2 cm long. Flower complete, bisexual, actinomorphic, bluish purple in colour. Sepals 5-lobed, free above, fused at the base, ca. 8 x 1-2 mm in size, imbricate, linear-lanceolate, base acute-obtuse, apex acuminate, margin entire, herbaceous, prickly and densely stellately hairy, reticulately veined, green, persistent. Petals rotate, shallowly 5-lobed, tube ca. 5 mm long; lobes ca. 1.7 x 1 cm in size, bluish purple, broadly ovate-triangular, base rounded, apex acute, margin dentate, stellately pubescent on dorsal side, puberulent within. Stamen 5; filament ca. 2-5 mm long; anther ca. 0.9 x 0.2 cm in size, linear-oblong, basifixed, yellow in colour, opening by apical pores. Ovary ca. 4.5 x 4 mm, obliquely globose; style ca. 1.4 cm long, glabrous; stigma acute, incurved. Berry ca. 1.5 cm across, spherical-globose, smooth, drooping when unripe, variegated with green and white, yellow when mature. Seeds smooth compressed, reniform.

Flowering period: March-April and July-September.

Herbarium specimens examined: NARA DESERT: *Rahmatullah Qureshi*, 84, 973, 996, 1209, 1254 (SALUBGH).

Field remarks: A common species of calcarious soils.

Distribution: N. Africa, S. & S.E. Asia, Australia and Polynesia (Nasir, 1985).

2. Solanum nigrum var. nigrum L. Sp. Pl. 186 (1753); Yasin J. Nasir in Nasir & Ali., Fl. Pak.,168: 6-7 (1985).

An annual herb up to 1.5 m tall. Stem woody at the base, herbaceous above, fairly branched, trigonous to angular; branchlets zigzag, sparsely scabrous; trichomes ca. 0.5 mm, unicellular, dark green to brown in colour. Leaves ca. 2.5-9.5 x 1.3-5.5 cm in size, broadly ovate or ovate-lanceolate or rapandus, apex retuse-apiculate, base obliquely angustatus, margin crenate-undulate, chartaceous; lateral nerves 3-6 pairs, prominent below, venation campatodromous, hairs on nerves; petiole ca. 0.5-5 cm long, elongated. Umbel ca. 1.5 cm, usually 9-flowered, extra-axillary; peduncle ca. 1-2 cm long; densely hairy; trichomes 2-3-celled. Flowers ca. 1 cm across, salver-shaped, white; pedicel ca. 8 mm. Calyx-tube cupular, ca. 2-3 mm long, herbaceous; lobes unequal, ca. 1-2 x 1-1.5 mm in size, oblong, apex rounded-subacute, margin entire and hairy, green, persistent, glabrous within, hairy above; hairs ca. 2-3-celled. Corolla rotate, white; tube ca. 2 mm, greenish, glabrous; lobes spreading, ca. 6 x 3 mm in size, ovate, basally decurrent, apex acute, margin hairy, ca. 2-3-celled, lateral nerves obscure. Stamens 5; filaments white, ca. 1-2 mm; anthers ca. 2.5 x 1 mm in size, lanceolate, apex cleft, base sagittate, yellow. Ovary ca. 1 mm across, globose, glabrous, green; style ca. 4 mm long, filiform, densely hairy above the middle; stigma capittellate, ca. 0.5 mm. Berries ca. 8-9 mm across, globose, glabrous, shiny, juicy, green when young and purplish black when mature. Seeds ca. 2 x 1.5 mm in size, reniform, minutely pitted.

Flowering period: Almost round the year.

Herbarium specimens examined: NARA DESERT: *Rahmatullah Qureshi*, 1266, 1299 (SALUBGH).

Field remarks: A common weed of plain areas.

Distribution: Cosmopolitan, being absent from the arctic and sub arctic regions (Nasir, 1985).

43. Convolvulaceae

Annual or perennial herbs, shrubs, lianas, twiners and creepers; branchlets terete or ribbed, rarely winged; trichomes stellate, simple or 0. Leaves alternate, simple, entire, lobed, dissected or rarely palmately compound, petiolate, exstipulate. Flowers actinomorphic, 5-merous, bisexual, generally showy and evanescent, solitary or in cymes, capitate clusters, panicles, rarely in axillary clusters or racemes; bracts persistent or deciduous. Calyx-lobes 5, free, or slightly connate at the base, subequal or unequal, imbricate, entire, persistent, sometimes enlarged in fruit. Corolla funnel/salver-form, campanulate, rarely rotate, often plicate in bud; lobes 5, either connate or plaited or distinct and imbricate. Stamens 5, alternating with corolla-lobes, (un)equal, included or exserted, base usually dilated and glandular-hairy; anthers oblong, parallel or sagittate, dehiscence longitudinal; pollen grains spinulose, smooth or colpate. Carpels 2, rarely 3; ovary superior, conical or globose, 1-2-locular, rarely 3-locular; ovules 2 per cell, basal (axile); style filiform or stout, included or exserted, 1 0r 2(3); stigma capitate, globose or oblong, rarely recurved or linear. Disc annular, hypogynous. Fruit 2-4-valved or an indehiscent capsule, rarely a berry. Seeds 4 or fewer, hairy, glabrous or winged; albumen scanty; cotyledons plaited.

A pantropical family with 40 or more genera and about 1200 species and in Pakistan, this family represented 13 native and cultivated genera and 54 species (Austin & Ghazanfar, 1979).

Key to genera:

1a. Styles 2, free or more or less united at the base:
2a. Corolla lobes imbricate; stamens and style exserted.................................... **1. Cressa**
2b. Corolla lobes induplicate-contorted; stamens and style not exserted..............**2. Seddera**
1b. Style 1:
3a. Stigma filiform, terete or subclavate..**3. Convolvulus**
3b. Stigma capitates... **4. Ipomoea**

1. Cressa

L., Sp. Pl. 223 (1753); Gen. Pl. ed. 5: 104 (1754); Choisy in DC., Prodr. 9: 439 (1845); Clarke in Hook. f., Fl. Brit. Ind. 4: 225 (1883); Austin & Ghazanfar in Fl. Pak. 126: 30 (1979).

Much branched , perennial, subshrubby erect or diffuse herbs; often grey-pubescent. Leaves small, sessile, entire. Flowers sessile in the upper axils or in terminal clusters or spikes; bracts 2, beneath the calyx, not enlarging in the fruit. Sepals 5, subequal, small, obovate, not enlarging in the fruit, thin imbricate. Corolla infundibuliform; tube short; lobes 5, ovate, distinct, imbricate, spreading. Stamens 5, exserted; filaments filiform; anthers oblong. Ovary 2-celled; ovules 2 per cell; styles 2, distinct from the base; stigmas large, capitate. Capsule 2-4-valved, 1-4-seeded. Seeds glabrous, brown; cotyledons linear, plicate.

A small genus of 5 species found in the tropical and sub-tropical regions of both the hemispheres. Represented in Pakistan by 1 species (Austin & Ghazanfar, 1979).

Cressa cretica L., Sp. Pl. 223 (1753); Jafri in Fl. Kr.,271 (1966); Stewart in Nasir & Ali., Ann. Cat. Vasc. Pl.W. Pak. & Kashm., 574 (1972); Austin and Ghazanfar in Nasir & Ali, Fl. Pak. 126: 30-31 (1979). **Figure 11B**

A biennial to perennial small herb upto 25 cm high. Stem herbaceous, much branched at the base, cylindric, green, densely appressed hairy; trichomes ca. 0.2-0.6 mm long, unicellular, hyaline. Leaves cauline and ramal, ca. 4-6 x 2-4 mm in size, closely alternate, always with short, axillary branches, ovate, apex acute, base truncate, sometimes oblique, 1-nerved, lateral nerves obscure, both side clothed with unicellular hairs; trichomes upto 0.7 mm long, acicular, hyaline, crystals are also present. Flowers in racemes of 1-3-flowered, alternately arranged, ca. 0.5 x 0.2 cm in size, white; bracts 2, ca. 1.5 x 0.5 mm in size, oblong-lanceolate, apex acute, base clasping, margin entire, densely hairy. Calyx-lobes free, imbricate, ca. 2.5 x 1.6 mm in size, obovate, apex acute, base cuneate, margin entire and incurved, green, glabrous within, hairy outside. Corolla-tube ca. 3 mm long; lobes 5, ca. 1.5-2 x 0.6-1 mm in size, imbricate, elliptic-oblong, apex subacute and hairy, glabrous within, hairy above, reflexed, white. Stamens alternate to petals, adnate to corolla-tube, extrose; filaments ca. 4-5 mm long; anthers ca. 2 x 0.8 mm in size, apex apicualte, granulate (pusticulate). Ovary ca. 2 x 1 mm in size, globose, pubescent above, slightly greenish; styles 2, ca. 3.5 mm long,

sparsely hairy below; stigma ca. 0.5-0.6 mm across, capitellate. Capsule ellipsoid-globose, ca. 5 x 3 mm in size, 2-4-valved. Seeds glabrous.
Flowering period: Almost throughout the year.
Herbarium specimens examined: NARA DESERT: *Rahmatullah Qureshi*, 625, 825, 1081, 1218, 1229, 1317, 1320 (SALUBGH).
Field remarks: A common species of saline soils.
Distribution: Widely distributed in the xeric tropical zones of both the hemispheres (Austin & Ghazanfar, 1979).

2. Seddera

Hochst., Flora 27, Beil. T, t. 5 (1844); Austin & Ghazanfar in Nasir & Ali., Fl. Pak., 126: 61 (1979).

Herbs or subshrubs, with prostrate or erect branches, the old sometimes become spinescent. Leaves mall, variable, usually with midi-fixed trichomes. Flowers 5-merous, axillary or in terminal spikes or dense few-flowered cymes. Sepals acute or btuse, subequal. Corolla small, funnel-shaped, lobes mostly short. Styles 2-fid, usually to the base; stigmas more or less peltate, sometimes bilobed. Fruit capsular, 4-valved.

A genus of about 20 species (Austin & Ghazanfar, 1979).

Seddera latifolia Hochst. et Steud. in Fl. 27. Beil: 8 t. 5 B & C (1844); Stewart in Nasir & Ali., Ann. Cat. Vasc. Pl.W. Pak. & Kashm., 577 (1972); Austin & Ghazanfar in Nasir & Ali., Fl. Pak., 126: 61 (1979).

Syn: *Cressa latifolia* (Hochst.) T. Anders. in J. L. Soc. Bot. 5: Suppl. 1. 25 (1860); *Breweria latifolia* C.B. Clarke in FBI. 4: 224 (1883); *Bonamia latifolia* (Hochst. et. Steud.) Santapau in J. Bomb. Nat. Hist. Soc. 47 (2): 342 (1947).

Erect subshrub with many stems from a woody base, clothed with grey pubescent. Leaves ca. 4-27 x 3-8 mm, elliptic-oblong, obtuse, silky, narrowed at the base. Flowers 1 to 3 together in upper axils, sessile. Bracteoles 2, linear-lanceolate. Calyx silky; sepals ovate, apex acuminate. Corolla slightly exceeding the calyx; lobes 5, with hairy bands between the folds. Styles free nearly to the base. Capsules 3-6 mm in diam., ovoid, acute, splitting into 4-valves. Seeds 2, ovoid, dark brown, glandular.
Flowering & Fruiting Period: October-February.
Herbarium specimens examined: NARA DESERT: *Rahmatullah Qureshi*, 1431 (SALUBGH).
Field remarks: A less common species of sandy soils.
Distribution: Eeast Africa, Socotra, Arabia, Pakistan and India (Austin & Ghazanfar, 1979).

3. Convolvulus

L., Sp. Pl. 153 (1753); Gen. Pl. ed. 5: 76 (1754); Boiss., Fl. Or. 4: 84 (1879); Clarke in Hook. f., Fl. Brit. Ind. 4: 217 (1883); Austin & Ghazanfar in Fl. Pak. 126: 7 (1979).

Herbs or undershrubs, erect, prostrate or twining; sometimes spinescent. Leaves simple, alternate, entire, toothed or lobed, often pubescent. Flowers bracteate, in terminal or axillary cymes, involucrate heads or solitary. Calyx of 5, free, equal to subequal sepals. Corolla campanulate or infudibuliform; limb plicate, subentire. Stamens 5, attached near the base of corolla, included; filaments filiform, often unequal, gradually subulate. Ovary 2-4-locular; ovules 4; style single, filiform; stigmas 2, linear or oblong or cylindrical, distinct. Fruit a capsule, 2-celled, 4-valved or irregularly rupturing, glabrous or pubescent. Seeds 1-4, smooth or tuberculate, glabrous or pubescent.

A cosmopolitan genus of about 250 species, mostly temperate and subtropical in distribution, rarely in the tropics. Represented in Pakistan by 21 wild and cultivated species (Austin & Ghazanfar, 1979).

Key to species:

1a. Corolla 4-5 times long as calyx; plants quite glabrous............................**1. C. arvensis**
1b. Corolla smaller, equal to or twice as long as calyx; plants densely hirsute:
2a. Flowers clustered in leaf axils..**2. C. prostratus**
2b. Flowers solitary dense globose, villous heads.................................**3. C. glomeratus**

1. Convolvulus arvensis L. Sp. Pl. 153(1753); Stewart in Ali and Nasir, Ann. Cat. Vasc. Pl. W. Pak. & Kashm., 572 (1972); Austin & Ghazanfar in Nasir and Ali, Fl. Pak.126: 28-29 (1979). **Figure 10D**

A climbing or prostrate annual or perennial herb upto 60 cm high. Stem woody at the base, much branched, twining or trailing and twisted, angular, glabrous or hairy. Leaves ca. 1-6 x 0.5-4 cm in size, ovate-lanceolate to oblong, apex obtuse to apiculate (mucronate), base hastate or sagittate; petioles ca. 0.5-2.5 cm long. Flowers in axillary, 1-3-flowered cymes; peduncles ca. 1.5-8 cm long, slender; pedicels ca. 0.8-2 cm long. Bracts linear. Calyx-lobes ca. 5-6 mm long, broadly elliptic-oblong, base obtuse to truncate, apex mucronate. Corolla ca. 1.5-2 cm long, funnel-shaped, pink or nearly white, glabrous within, with hairy mid-petaline bands outside. Stamens unequal. Ovary glabrous; stigmas ca. 2.5 mm, filiform. Capsules ca. 5-6 mm across, ovoid-globose, glabrous. Seeds subtrigonous, brown, tuberculate.
Flowering period: All the year.
Herbarium specimens examined: NARA DESERT: *Rahmatullah Qureshi*, 689 (SALUBGH).
Field remarks: A common climber found in plain areas.
Distribution: Throughout the temperate and tropical regions of the world, except Australia. Very commonly found throughout Pakistan (Austin & Ghazanfar, 1979).

2. Convolvulus prostratus Forssk., Fl. Aegypt. Arab. 203(1775); Austin & Ghzanfar in Nasir and Ali, Fl. Pak.126: 23-24 (1979). **Figure 11A**

Syn. *C. microphyllous* Sieb.ex Spreng., Syst. Veg. 1: 611(1825); *C. parviflorus* Spreng., Syst. 1:611 (1825); *C.pluricaulis* Choisy, Conv. Or. In Mem. Soc. Phy.Hist. Nat. Geneve, 6:477 (1833); *C. pluricaulis* var. *macra* C.B. Clarke in Hook. F., Fl. Brit. Ind. 4: 218 (1883).

A perennial prostrate or procumbent herb up to 45 cm high. Stem suffruticose at the base, often floriferous from the base, hairy or ferruginously pilose. Leaves sessile, ca. 1-2.5 x 0.3-0.8 cm in size, linear-oblong or elliptic-lanceolate, apex subulate to obtuse, base tapering,

margin entire, villous on both surfaces, lower leaves usually larger and spathulate. Flowers sessile or subsessile, axillary or on short lateral branches, 1-4-flowered; peduncles up to 2 cm long. Bracts ca. 3-3.5 mm long, linear-oblanceolate or lanceolate, apex acuminate, hairy. Calyx-lobes ca. 6.5 mm long, ovate-lanceolate, apex acute, hairy on both sides, green. Corolla ca. 10-12 m long, pink-white, funnel-shaped, with hairy bands outside; limb shallowly 5-lobed, deltoid, apex acute, with tuft of hairs at the apex of each. Stamens unequal. Ovary glabrous, seated on a cup-shaped disk; stigmas ca. 4 mm long, filiform. Capsules ca. 3.5 mm long, ellipsoid or subglobose, somewhat scarious, glabrous. Seeds 2-4, ca. 2.5 mm long, brown or black, pubescent.
Flowering period: November and January.
Herbarium specimens examined: NARA DESERT: *Rahmatullah Qureshi*, 51, 59, 105, 111, 206, 251, 544, 661, 603, 665, 695, 697, 801, 807, 814, 819, 827, 851, 879, 917, 921, 1005, 1071, 1150, 1183, 1319 (SALUBGH).
Field remarks: A common prostrate herb found in flat habitats.
Distribution: Egypt to Pakistan and India (Austin & Ghazanfar, 1979).

3. Convolvulus glomeratus Choisy in DC., Prodr. 9:401 (1845); Stewart in Ali and Nasir, Ann. Cat. Vasc. Pl. W. Pak. & Kashm., 572 (1972); Austin & Ghazanfar in Nasir and Ali, Fl. Pak.126: 24-26 (1979).

A biennial or perennial, decumbent or creeping herb upto 90 cm high. Stem branched from vertical root, slender, pubescent. Leaves ca. 2-4 x 0.3-1.5 cm in size, elliptic-oblong, base rounded or cordate, apex mucronate, margin entire, hairy on both surfaces; petiole ca. 0.5 cm long, slender, sparsely hairy. Flowers in axillary, solitary dense globose, villous heads; peduncles ca. 3-7 cm, slender, hairy. Bracts ca. 1x 0.7 cm in size, ovate-lanceolate, apex acute, margin densely ciliate and hairy, foliaceous. Calyx-lobes unequal, 2+3, ca. 1.2 cm long, ovate-lanceolate, apex acuminate, very densely clothed with long silky hairs, the 2 outer sepals broader than the 3 inner. Corolla tubular, ca. 1.2 cm long, the upper part sparsely hairy outside; lobes 5, broadly deltoid shallow, apex acute, with small tuft of hairs at the apex of each. Stamens equal. Ovary glabrous; stigmas ca. 1.5-2 mm long, clavate or elliptic-oblong. Capsules ca. 5 mm across, subglobose, glabrous. Seeds smooth, not tuberculate.
Flowering period: December-August.
Herbarium specimens examined: NARA DESERT: *Rahmatullah Qureshi*, 412, 523, 1162, 1203, 1219, 1321 (SALUBGH).
Field remarks: A less common prostrate herb found in flat habitats.
Distribution: N.E. Africa (Sudan, Egypt), Arabia, Iran, India, Pakistan and Afghanistan (Austin & Ghazanfar, 1979).

4. Ipomoea

L., Sp. Pl. 159 (1753); Choisy in DC., Prodr. 9: 348 (1845); Clarke in Hook. f., Fl. Brit. Ind. 4: 196 (1883); Austin & Ghazanfar in Fl. Pak. 126: 34 (1979).
Calonyction Choisy; *Exogonium* Choisy; *Mina* Llave & Lexarza; *Pharbitis* Choisy; *Quamoclit* Miller.

Herbs or shrubs, often twining, sometimes prostrate, erect, or floating. Leaves petiolate, entire, lobed, or divided. Inflorescences mostly axillary, cymose, 1- to many-flowered, rarely paniculate; bracts various. Flowers small to large. Sepals persistent, equal to unequal, ± enlarged in fruit. Corolla variously colored, rarely yellow, funnelform, campanulate, or salverform; limb 5-lobed to entire, midpetaline bands well defined. Stamens included or exserted; filaments filiform, usually unequal, dilated and pubescent basally; anthers ovate or linear, longitudinally dehiscent, not twisted; pollen globular, pantoporate, finely spiny. Disc ring like. Ovary 2-4-loculed, 4- or 6-ovuled. Style 1, filiform, included or exserted; stigma capitate, or 2- or 3-globulose. Capsule globose or ovoid, 4- or 6-valved. Seeds 4(-6) or fewer, glabrous or pubescent.

Approximately 500 species: widely distributed in tropical to warm temperate regions, especially of North and South America

Key to species:

1a. Annual aquatic, repent and rooting or floating plant. Sepals 7-8 mm; corolla 3.5-5 cm ....**1. I. aquatic**

1b. Erect shrub. Sepals 5-6 mm; corolla 7-9 cm.. **2. I. carnea**

1. Ipomoea aquatica Forssk., Fl. Aegypt. Arab. 44 (1775); Stewart in Ali and Nasir, Ann. Cat. Vasc. Pl. W. Pak. & Kashm., 574 (1972); Austin & Ghazanfar in Nasir & Ali., Fl. Pak., 126: 43-44 (1979). **Figure 11C**

Syn: *I. reptans* Poir. in Lam., Encycl. Meth. Bot. Suppl. 3: 40. 1814.

Herbs annual, terrestrial and repent or floating; axial parts glabrous. Stems terete, hollow, rooting at nodes. Petiole 3-14 cm, glabrous; leaf blade variable, ovate, ovate-lanceolate, oblong, or lanceolate, 3.5-17 x 0.9-8.5 cm, glabrous or rarely pilose, base cordate, sagittate or hastate, occasionally truncate, margin entire or undulate, apex acute or acuminate. Inflorescences 1-3(-5)-flowered; peduncle 1.5-9 cm, base pubescent; bracts squamiform, 1.5-2 mm. Pedicel 1.5-5 cm. Sepals subequal, glabrous; outer 2 ovate-oblong, 7-8 mm, margin whitish, thin, apex obtuse, mucronulate; inner 3 ovate-elliptic, ca. 8 mm. Corolla white, pink, or lilac, with a darker center, funnelform, 3.5-5 cm, glabrous. Stamens unequal, included. Pistil included; ovary conical, glabrous. Stigma 2-lobed. Capsule ovoid to globose, ca. 1 cm in diam., woody, tardily dehiscent or indehiscent. Seeds densely grayish pubescent, sometimes glabrous.

Flowering Period: May-September.

Herbarium specimens examined: NARA DESERT: *Rahmatullah Qureshi*, 1430 (SALUBGH).

Field remarks: A less common herb found in wetland habitats.

Distribution: **A native of the Old World tropics (Austin and Ghazanfar, 1979).**

2. Ipomoea carnea subsp. **fistulosa** (Martius ex Choisy) D. F. Austin, Taxon. 26: 237 (1977); Austin & Ghazanfar in Nasir & Ali., Fl. Pak., 126: 40-41 (1979).

Syn: *Ipomoea fistulosa* Martius ex Choisy in A. de Candolle, Prodr. 9: 349. 1845; *Batatas crassicaulis* Bentham; *I. crassicaulis* (Bentham) B. L. Robinson.

Erect Shrubs up to 3 m high; axial parts puberulent, later glabrescent; sap milky. Branches terete or angular, stout, becoming sparsely lenticellate with age. Leaf blade ovate or ovate-oblong, 6-25 X 4-17 cm, densely puberulent or adaxially subglabrous, base cordate or truncate, margin entire, apex acuminate, mucronulate; petiole 2.5-15 cm; midvein 2-glandular abaxially at base, lateral veins 7-9 pairs. Inflorescences few to several flowered; peduncle stout, 5-10 cm; bracts early deciduous, ovate. Pedicel 1-1.5 cm. Sepals ovate or nearly circular, broadly rounded, abaxially puberulent, equal or inner ones longer, 5-6 mm. Corolla lilac or pink, darker inside, funnelform, 7-9 cm; tube and midpetaline bands mealy outside. Stamens included; filaments unequal; anthers linear, base sagittate. Pistil included; ovary puberulent. Style base puberulent; stigma 2-lobed. Capsule pale brown, ovoid, 1.5-2 cm, 4-valved, apiculate. Seeds black, ca. 1 cm, brown sericeous-pubescent.

Flowering Period: July-September.

Herbarium specimens examined: NARA DESERT: *Rahmatullah Qureshi*, 1429 (SALUBGH).

Field remarks: A less common shrub found in wetland habitats.

Distribution: **A native of America tropics (Austin and Ghazanfar, 1979).**

44. Cuscutaceae

Twining herbaceous parasites. Stem slender, yellowish-brown or reddish-purple or with violet spots, attached to host by haustoria; spirally twining round the host, sometimes interlaced. Leaves reduced to small membranous scales. Flowers small, white to yellowish-white or with violet spots usually arranged in cymose clusters or umbel-like cymes. Sepals mostly fleshy, gamosepalous, 4-5 (-6)-lobed. Corolla tubular, 4-5 (-6)-lobed, often with minute scales. Scales membranous or petaloid, equalling the number of stamens, attached to the corolla tube below each filament, simple or bifid, fimbriate, dentate or fimbriate-ciliate, forming a fringed or fimbriate ring-like structure inside the corolla tube. Stamens sessile or with filament, 5(-4), inserted in the throat, alternating with the corolla lobes. Ovary 2-celled; 2-ovules per cell; style 1-2, free or united or obsolete; stigmas capitate or linear-elongated or oblong. Fruit a capsule, with or without intrastylar opening, 4-seeded or less by abortion, remaining closed or opening with a definite or irregular line of circumcission near the base. Embryo without cotyledons.

The family includes a single cosmopolitan genus **Cuscuta**, which has about 170 species, several of which are important weeds of agricultural crops. This family represented 14 species in Pakistan (Rajput & Tahir, 1988).

Cuscuta

L., Sp. Pl. 124 (1753); Gen. Pl. ed. 5. 60 (1754); DC., Prodr. 9: 452 (1854); Benth. & Hook. f., Gen. Pl. 2: 881 (1873); Clarke in Hook. f., Fl. Brit. Ind. 4: 225 (1883); Stewart in Ali and Nasir, Ann. Cat. Vasc. Pl. W. Pak. & Kashm., 578 (1972); Rajput & Tahir in Fl. Pak., 189: 1-2 (1988).

Leafless, twining, parasitic vines. Stem slender, sometimes filiform; branchlets much branched, glabrous, yellowish (achlorophyllous). Flowers sessile or shortly stalked, aggregated in lateral fascicles or in short racemes; bracts minute. Calyx-lobes 5, equal, ovate, concave, shortly decurrent. Corolla campanulate; lobes 5, equal, imbricate, reflexed, truncate, acute or acuminate. Stamens 5, opposite scales, exserted; anthers oblong; pollen grains smooth. Ovary globose, 2-locular, each 2-ovulate; styles 2; stigmas acute or capitellate. Capsule subglobose or ovoid, dry or succulent, circumscissile or irregularly beaking up, 2-4-seeded. Seeds glabrous; albumen fleshy; embryo slender, spiral; cotyledons 0 or obscure.

Cuscuta chinensis Lam., Encycl. Meth. Bot. 2: 229 (1786); Stewart in Ali and Nasir, Ann. Cat. Vasc. Pl. W. Pak. & Kashm.,418 (1972).; Rajput and Tahir in Nasir & Ali., Fl. Pak., 189: 20-21 (1988).

Stem slender, ca. 0.5-0.8 mm boad, much branched, fleshy, greenish-yellow or brown, glabrous, smooth. Flowers solitary or in fascicles arranged in racemes, ca. 2 x 1.5 mm; pedicels ca. 1.5 mm, greenish-yellow, fleshy, glabrous. Bracts ca. 1.5 x 1 mm in size, ovate-suborbicular, margin entire, hyaline, fleshy. Calyx-tube ca. 0.5 mm, glabrous, greenish; lobes 5, imbricate, ca. 1 x 0.8-1.5 mm in size, broadly ovate-oblong, apex obtuse, margin entire, tubercular keel on the back, nerves obscure, hyaline, persistent. Corolla cream; tube campanulate, ca. 1 mm; lobes 5, imbricate, united from near the middle, ca. 1 mm across, triangular-ovate, apex acute-shortly acuminate, membranous, reflexed, persistent. Scales 5, ca. 1 x 0.8 in size, attached near the middle of corolla tube, ovate-oblong, apex subobtuse, white, fimbriate. Stamens 5; filaments ca. 1.5 mm long, broad at the base, alternating with corolla lobes, glabrous, hyaline; anthers ca. 0.5 mm across, basifixed, ovate, apex retuse. Ovary ca. 1 mm across, globose, tuberculate or puberulent, greenish; styles 2, subequal, filiform, thin, divergent, ca. 0.4-0.6 mm, distinct; stigmas 2, capitate, yellowish. Capsules dry, globose, 3-3.5 mm across, depressed, hyaline, ca., irregularly dehiscent, surmounted by calyx, corolla and styles. Seeds 2-4 per capsule, ca. 1.5-1 mm in size, subreniform, with narrow oblong hillum.

Flowering & fruiting: February-April.

Herbarium specimens examined: NARA DESERT: *Rahmatullah Qureshi*,974, 1271 (SALUBGH).

Field remarks: A less common stem parasite found associated with *Zizyphus nummularia* and leguminous species.

Distribution: Ethiopea, Sokotra and Afghanistan, easward to Ceylon, Australia and China (Rajput & Tahir, 1988).

45. Boraginaceae

Annual, biennial or perennial herbs, shrubs or trees, very rarely trees; branchltes generally scabrous or strigose to sericeous; hairs appressed, subappressed or setose. Leaves alternate or decussate, simple, penninerved, margin entire or distantly toothed, petiolate or subsessile, exstipulate. Inflorescence normally a dichotomous cyme with helicoid branches, sometimes a simple unilateral raceme or spike or an irregularly 3-chotomous panicle (rarely of 1-2 flowers in the axils of the leaves). Flowers actinomorphic, hermaphrodite (rarely polygamous), 4- or 5-merous, stalked or sessile; bracts leafy, minute or 0. Calyx inferior, persistent, lobes 4 or 5, equal or unequal, valvate or (sub)imbricate, connate fully or partly, herbaceous. Corolla slaver-form or tubular-campanulate, rarely sub-rotate; lobes 4 or 5, equal, imbricate or twisted in bud, spreading or reflexed later, with or without glandular spots on the throat. Throat glabrous or with scales (appendages) or a zone of hairs. Stamens as many as, and alternating with corolla-lobes, inserted below or above the middle of tube, included or exserted; filaments filiform or dilated at the base; anthers 2-celled, ovate, oblong or linear, obtuse, or with produced connectives, parallel, basifixed or dorsifixed, introsely or laterally dehiscent. Ovary superior, 2- or 4-celled, each with 2 or 1 ovule(s) respectively; ovules erect or pendulous from inner angle; style simple or divided, terminal or gynobasic, sometimes disciform above; stigma capitellate, conical or pointed or 2-lobed. Fruit drupaceous or dividing into 2-4 nutlets, depressed or oblong, variously pubescent. Seeds erect or oblique, straight or curved; testa membranous; albumen copious, scanty or 0.

A family of about 100 genera and 2000 species distributed in temperate, especially Mediterranean and tropical regions. Thirty two genera and 135 species exist in Pakistan (Nasir, 1989).

Key to genera:

1a. Ovary not lobed; style terminal.

2a. Style twice forked; a small tree or large shrub; fruit drup............................**1. Cordia**

2b. Style simle, entire; herbs or shubshrubs; fruit of 4 nutlets...........................**2. Heliotopium**

1b.Ovary deeply 4- lobed; style gynobasic or subterminal:

3a. Calyx large, accrescent in fruit; anthers connivent into a cone; the connectives much produced and twisted at the apex...**3. Trichodesma**

3b. Calyx small, not accrescent in fruit; anthers not connivent into a cone; the connectives not much produced; style bifid.......................................**4. Arnebia**

1. Cordia

L., Sp. Pl. 190 (1753); Gen. Pl. ed. 5.84 (1754); C.B. Clarke in Hook. f., Fl. Brit. Ind. 4: 136 (1885); Nasir in Nasir & Ali, Fl. Pak. 191: 5 (1989).

Trees or shrubs, often polygamous, rarely subscandent. Leaves alternate (subopposite), petiolate, exstipulate, entire or toothed. Inflorescence cymes, spikes or heads. Flowers actinomorphic, polygamous. Calyx tubular or campanulate, smooth or ribbed; lobes 3-5, short, often irregular, persistent. Corolla infundibuliform, hypocrateriform or campanulate, usually white or orange coloured; lobes 4-5(-8), imbricate, reflexed or spreading. Stamens as many as and alternating with the corolla-lobes, exserted or included; filaments often hairy at the base. Ovary undivided, 4-celled, ovule solitary in each cell, erect; style terminal, elongate, 2-partite, the branches again 2-partite; stigma capitate or clavate. Fruit usually with viscid pulp, drupaceous, of 4 or fewer 1-seeded cells; putamen often boney. Seeds exalbuminous; cotyledons plicate.

A genera of ca. 250 species, mainly of tropical America. In Pakistan, Nasir (1989) reported 6 species.

Key to species:

1a. Leaves oblanceolate-oblong; calyx ribbed, minutely pubescent to tomentose...**1. C. gharaf**

1b. Leaves orbicular, broadly ovate, elliptic or obovate; calyx not ribbed, Glabrous...**2. C. myxa**

1. Cordia gharaf (Forssk.) Ehren. ex Asch., Sitz-ber. Ges. Naturf. Freunde . 46 (1879); Stewart in Ali and Nasir, Ann. Cat. Vasc. Pl. W. Pak. & Kashm., 584 (1972); Yasin J. Nasir in Nasir and Ali, Fl. Pak., 191: 6-8 (1989).

Syn: *Cornus gharaf* Forssk., Fl. Aegypt.-Arab. 95 (1775); *Cordia rothii* Roem. & Schult., Syst. Veg. 4: 498 (1819);

Tree or shrub up to 6 m tall. Leaves 6-10.5 x 2-4 cm, oblanceolate-oblong, entire to ± undulate, coriaceous to scabrid, cuneate, obtuse, nerves impressed above, strigose to pubescent, especially in the axils of the nerves on the under surface. Calyx 3.5 mm long, tubular-campanulate, pubescent, accrescent, shallowly lobed; lobes obtuse. Corolla c. 5 mm long, white; lobes ± ligulate-obuse, reflexed. Filaments c. 1.8 mm long, glabrous. Anthers c. 1 mm long. Drupe 11 mm long, ovoid, mucronate, reddish-brown, longitudinally striate.

Flowering Period: April-June.

Herbarium specimens examined: NARA DESERT: *Rahmatullah Qureshi*, 725, 1134, 1278 (SALUBGH).

Field remarks: A less common small tree found near settlements.

Distribution: N. Africa, Arabia, Pakistan, India, Sri-Lanka (Nasir, 1989).

2. Cordia myxa L., Sp. Pl. 190 (1753); Stewart in Ali and Nasir, Ann. Cat. Vasc. Pl. W. Pak. & Kashm., 584 (1972).; Yasin J. Nasir in Nasir and Ali, Fl. Pak., 191: 8 (1989).

Syn: *C. obliqua* willd., Phytogr. 1: t. 4. Figure 1 (1794); *C. obliqua* var. *tomentosa* (Wall.) Kazmi in Arn. Arb. 51 (2): 143 (1970).

A deciduous tree up to 6 m tall with rusty pubescent branches. Stem hard, woody trunk; bark dark-coloured, rough, fissured; branchlets usually glabrous. Leaves alternate, ca. 7-12.5 x 5-11 cm in size, orbicular, broadly ovate, elliptic or obovate, base rounded or cordate or sometimes shortly cuneate, margin entire or sinuate-dentate, glabrous and more or less scabrous above without white discs on the upper surface; basal nerves 3 (rarely 5); petiole ca. 0.7-3.5 cm long. Flowers white, polygamous (male and bisexual) in large lax terminal and axillary pedunculate cymose panicles; peduncles ca. 2-5 cm long; pedicel short; buds pyriform. Calyx ca. 0.4-0.7 cm long, glabrous or nearly so outside, pubescent inside, entirely closed over the corolla in bud, afterwards irregularly splitting on the expansion of the flower into 5 shallow lobes. Corolla-tube as long as the calyx; lobes 5, oblong, apex obtuse, sometimes emarginate, narrow, recurved, equalling the tube. Filaments hairy at the base. Drupe ovoid, ca. 1.2-2., cm long, pinkish-cream coloured or nearly black, shining, minutely rugose, apiculate with the remains of the style, supported by the saucer-shaped irregularly 5-10-lobed more or less longitudinally striate glabrous calyx.

Flowering period: March-April.

Herbarium specimens examined: NARA DESERT: *Rahmatullah Qureshi*, 725, 1134, 1278 (SALUBGH).

Field remarks: A less common small tree found near settlements.

Distribution: Pakistan, India and Sri-Lanka (Nasir, 1989).

2. Heliotropium

L., Sp. Pl. 130 (1753); Gen. Pl. ed. 5.63 (1754); Jafri in Fl. Kar., 276 (1966); Nasir in Nasir & Ali, Fl. Pak. 191: 18 (1989).

Annual or perennial, prostrate or erect herbs or undershrubs, villous or scabrid, rarely glabrous. Leaves alternate, penninerved, sericeous or softly wooly, base cuneate or decurrent, margin entire, sometimes revolute, apex obtuse to acute, petiolate or subsessile. Inflorescence often unilateral along the branches of forked terminal scorpioid cymes, sometimes 1-sided raceme or spike, (sub)terminal, axillary or leaf-opposed. Flowers small, 5-merous, sessile or shortly stalked; bracts minute or 0. Calyx more or less deeply 5-cleft; lobes subequal, almost free, linear-lanceolate, imbricate, strigose or wooly, shortly decurrent, acute, spreading or enclosing drupe. Corolla white, with or without yellow throat, campaulate or salver-form; tube cylindric; lobes 5, subequal, broadly ovate, plicate in bud, imbricate or induplicate, spreading in flower, acute, incurved. Stamens 5, subsessile, inserted in corolla-tube, included; anthers elongated, ovate-lanceolate, free or apically connivent, acuminate or apiculate. Ovary globose, imperfectly 4-celled; ovules 4, attached to inner angle; style short, erect; stigma surmounted by a pointed or conical, stigmatic disc usually present. Fruit somewhat fleshy, comprising of 4 dry nutlets drupe, free or combined in pairs; nutlets usually smooth, glabrous or appressed-hairy. Seeds straight or curved.

A large and complicated genus of ca. 270-276 species, distributed in temperate and tropical regions of both hemispheres. In Pakistan, Nasir (1989) reported 23 species in about 11 sections.

Key to species:

1a. Plants glabrous..**1. H. curassavicum**
1b. Plants villous-scabrid:
2a. Inflorescence bracteate, sometimes bracts much reduced......................**2. H. strigosum**
2b. Inflorescence ebracteate:
4a. Leaves elliptic-oblong or obovate...**3. H. europaeum**
4b. Leaves lanceolate, linear-lanceolate or elliptic-lanceolate:
6a. Corolla-lobes broad, oblong, ovate or roundish, subvalvate or imbricat.......**4. H. crispum**
6a. Corolla-lobes narrow, caudate-acuminate, inflexed in corolla-tube when young,
later patent..**5. H. subulatum**

1. Heliotropium curassavicum L. Sp. Pl. 130 (1753); Jafri in Fl. Kar., 278 (1966); Stewart in Nasir & Ali., Ann.Cat.Vasc.Pl.W.Pak. & Kashm., 589 (1972); Nasir in Nasir & Ali, Fl. Pak. 191: 23 (1989).

Syn: *H. glaucum* Salisb., Prodr. 113 (1796); *H. glaucophyllum* Moenich, Meth. Suppl. 147 (1802); *H. curassavicum* var. *obovatum* and *virens* (E. Mey. ex. DC.) DC., Prodr. 9: 538 (1845).

A decumbent, glabrous, perennial halophytic herb up to 55 cm high. Stem fleshy, glabrous, less woody at the base, much branched, spreading all around, more or less terete; branchlets softly scarious; internode ca. 20 mm. Leaves simple, cauline and ramal, exstipulate, subsessile, ca. 1.5-7 x 0.4-1.4 cm in size, linear-lanceolate or narrowly spathulate,

apex subobtuse, base cuneate-decurrent, margin entire, succulent, appressed-sericeous on either side, 1-nerved, lateral nerves obscure; petiole ca. 0.1-0.3 cm long. Inflorescence terminal, and axillary scorpoid cymes, up to 6.5 cm long; peduncle ca. 2 cm long. Flowers ebracteate, sessile, complete, bisexual, actinomorphic, arranged alternately, grouped into two, in 2-ranked spike, white in colour. Calyx-lobes 5, divided nearly to the base, valvate, oblong-ovate, apex acute, base cuneate, margin entire, ca. 1.5-2 x 0.5-1 mm in size, fleshy, glabrous, green, persistent. Petals-tube ca. 1 mm; lobes 5, alternate to sepals, twisted, glabrous, white, ca. 1-1.5 x 0.5 mm in size, somewhat greenish to off white. Stamens 5; anthers oblong, ca. 1 x 0.3 mm in size. Ovary sub-globose, ca. 1 mm long; style very short; stigma conic, shortly bifid. Fruit ca. 2 mm across, globose, margin thickened, lobed, attached to the fruiting axis by the base. Seeds 4, ca. 1.3 x 1.5 mm in size, glabrous, brown or pale-brown, rugulose.
Flowering period: January.
Herbarium specimens examined: NARA DESERT: *Rahmatullah Qureshi*, 906, 912 (SALUBGH).
Field remarks: A less common herb found in saline soils.
Distribution: America, Europe, Africa, Pakistan, India and Australia (Nasir, 1989).

2. **Heliotropium strigosum** subsp. **strigosum** Willd. Sp. Pl., I:743 (1812); Jafri in Fl. Kar., 282 (1966); Stewart in Nasir & Ali., Ann.Cat. Vasc. Pl.W.Pak. & Kashm., 589 (1972); Nasir in Nasir & Ali, Fl. Pak. 191: 25 (1989).

A decumbent, perennial, scrabidly appressed hairy herb up to 20 cm long. Stem herbaceous, much branched, branches arranged alternately, terete, pubescent. Leaves subsessile, 6-20 mm long and ca. 1-3 mm broad, linear-lanceolate, obtuse at base, apex acute, margin entire, covered by erect trichomes, ca.0.5-1 mm long; greyish-white in colour. Spikes terminal, ca. 2.5-6 cm long. Inflorescence arranged alternately on spike, lower ones pedicelled, ca.0.5-1 mm long, upper ones sessile; bracteate, ca. 2-5 mm long, lanceolate in shape, base truncate, apex acute, margin entire, strigose on abaxial side. Flowers white, complete, actinomorphic, bisexual, hypogynous ca.3 x 2 mm in size. Sepals 5-lobed, ca.1.5 x 0.5-1 mm in size, fused bellow, free above, ovate-lanceolate, base acute-rounded, apex acute, margin entire, strigose on outer side, persistent, imbricate, green. Petals-tube ca.1.5 mm long; lobes valvate, lanceolate, 0.5 x 0.2 mm broad, apex acute, upper surface hirsute, white in colour. Stamens 5, epipetalous, alternate with petals, ca.0.5 x 0.3 mm in size; anthers 2-lobed, dorsifixed, lanceolate, syngenesus. Ovary globose, imperfectly 4-celled, ovules 4; style short, ending in a circular disc surmounted by a pointed stigma. Fruit dried drupe, ca. 1.5 x 2 mm in size, ovoid, 4-lobed, depressed at the top, dark brown to blackish. Nutlets 4, conical, rounded at the base, sparsely grey hairy, ca.1.5 mm across.
Flowering period: July-September.
Herbarium specimens examined: NARA DESERT: *Rahmatullah Qureshi*, 699, 803, 1124, 1133, 1147, 1187 (SALUBGH).
Field remarks: A common herb found in calcarious soils.
Distribution: Afghanistan, Pakistan eastward to Nepal, Burma, Malya (Nasir, 1989).

3. **Heliotropium europeum** L. Sp. Pl., 130 (1753); Jafri in Fl. Kar., 277 (1966); Stewart in Nasir & Ali., Ann.Cat.Vasc.Pl.W.Pak. & Kashm., 590 (1972); Nasir in Nasir & Ali, Fl. Pak. 191: 39 (1989).var. **lasiocarpum** (F. & M.) Kazmi in J. Arn. Arb. 51: 176 (1970).

Syn: *H. lasiocarpum* Fish. & Mey. Index Sem. Hort. Petrop. 4: 38 (1837); *H. eichwaldii* Steud. var. *lasiocarpum* (Fish. & Mey.) Clarke in Hook. f. Fl. Brit. India 4: 150 (1883); H. *eichwaldii* auct. non Steud. ex DC. (1845); *H. ellipticum* auct. plur. non Ledeb. 1831. non R.Br. (1810); *H. ellipticum* var. *lasiocarpum* (Fisch. & Mey.) M. Popov in Acta Hort. Petrop. 42: 220 (1931).

Annual erect herb up to 35 cm long. Stem woody at the base, herbaceous above, branched, terete, green to greyish-green; clothed with white subpatent tubercle-based hairs, trichomes ca. 0.5-1.2 mm long, densely upward directed, unicellular; internode ca. 1-5 cm. Leaves cauline and ramal, alternate, ca. 1-4.5 x 0.5-1.5 cm in size, elliptic-oblong, base truncate- attenuate, apex acute or rounded, margin entire, green, pubescent; petiole ca. 0.5-2.5 cm long. Inflorescence in terminal and axillary ca. 1-4 cm long spike; peduncle 1-5 cm long. Flowers white, sessile, alternately arranged, terminal ones in hook form, complete, actinomorphic, bisexual, ca. 9 x 5 mm in size. Sepals 5, free, valvate, ca. 0.5-5.5 x 1.5-2 mm in size, lanceolate, apex acute, base obtuse, margin entire, light green in colour, pubescent; trichomes glandular and villous. Petals 5-lobed; tube narrowed at the throat, pubescent, ca. 7 x 3 mm in size; lobes ca. 2 mm long, obovate-orbicular, apex rounded, sinus sharp pointed, nerves visible. Stamens 5; anthers ca. 2 x 1 mm in size, lanceolate, apex acute, base obtuse, margin entire, basifixed, 2-lobed, dehiscing longitudinally. Ovary ca. 1 x 1.5 mm, conical or subglobose; style gynobasic, ca. 1 mm; stigmatic appendages ± bifid, ca. 2 x 1 mm in size. Drupe ca. 4-5 mm across, 4 one-seed cocci. Nutlets densely pilose, minutely verrucose, fleshy, ca. 4 x 3 mm in size.
Flowering period: November-January.
Herbarium specimens examined: NARA DESERT: *Rahmatullah Qureshi*, 236, 816, 1183 (SALUBGH).
Field remarks: A common herb found in calcarious soils.
Distribution: Trans Jordan, Syria, Iraq, Iran, Russia, Afghanistan, Pakistan, India (Nasir, 1989).

4. Heliotropium crispum Desf., Fl. Atlant., 1:151-4 (1798); Stewart in Nasir & Ali., Ann.Cat.Vasc.Pl.W.Pak. & Kashm., 589 (1972); Nasir in Nasir & Ali, Fl. Pak. 191: 42 (1989). **Figure 6B**

Syn: *Lithospermum hispidum* Forsk. Fl. Aegypt.-Arab. 39 (1775); *Heliotropium undulatum* auct. non Vahl (1790); *H. undulatum* Vahl, var. *ramosissimum* Lehm. Asperif. 24. t. 40 (1831); *H. ramosissimum* (Lehm.) DC. Prodr. 9: 536 (1845); Clarke in Hook. f. Fl. Brit. India 4: 150 (1883

A perennial, erect herb or subshrub up to 50 cm tall. Stem woody below and herbaceous above, branched, terete, greyish-green, covered with short stiff trichomes; internode ca. 1 cm long. Leaves simple, subsessile, ca. 0.7-4.2 x 0.1-0.4 cm in size, linear-lanceolate, base cuneate, apex acute, margin crisped-crenate or undulate, strigose. Inflorescence terminal or axillary scorpoid cymes, ca. 0.5-2 cm long; peduncle ca. 1-1.5 cm long. Calyx-tube ca. 1 mm; ca. 1-2 x 0.5-1 mm in size, ovate, margin entire, imbricate, hairy on dorsal side, glabrous on ventral side, green. Corolla-tube rotate, ca. 1-2.5 mm; lobes imbricate, linear-ovate, pubescent out side, glabrous within, white-slightly yellowish, lobes ca. 0.5-1 mm long. Stamens 5; anthers lanceolate, ca. 1.2-1.8 x 0.5-0.8 mm in size, yellow in

colour. Ovary globose, glabrous, ca. 1 mm across; style terminal, ca. 0.3 mm; stigma gynobasic, subconical. Fruits covered with hairs, surface rugulose. Nutlets 4, not margined, when young in pairs, free in maturity.
Flowering period: Januar-April.
Herbarium specimens examined: NARA DESERT: *Rahmatullah Qureshi*, 26, 197, 216, 844, 860, 920, 926, 972, 1132, 1211 (SALUBGH).
Field remarks: A very common subshrub found in calcarious soils.
Distribution: Senegal, N. Africa, Arabia, Iraq, Trans Jordon, Iran, Ifghaniatan, Pakistan & India (Nasir, 1989).

5. **Heliotropium subulatum** (Hochst. ex DC.) Vatke., Oesterr. Bot. Zeistchr. 25: 166 (1875); Jafri in Fl. Kar., 278(1966); Stewart in Ali and Nasir, Ann. Cat. Vasc. Pl. W. Pak. & Kashm.,591 (1972); Nasir in Nasir & Ali, Fl. Pak. 191: 46 (1989). **Figure 6C**

Syn: *H. zeylanicum* auctt., iii: 94 (1789); *Tournefortia subulata* Hochst. in DC., Prodr., ix: 528 (1845); *Messerschmidia hispida* Benth. in Royle, Illustr. Bot. Himal. Mount. 1: 306 (1839); *Tournefortia edgeworthii* DC., Prodr. 9: 527 (1845); *T. royleana* DC., l.c.

An annual or perennating, erect herb up to 40 cm tall. Stem erect, branched, woody at the base, herbaceous above, terete, highly strigose; trichomes erect, 1-2 mm long; internode ca. 1 cm long. Leaves ca. 2-4 x 0.2-1.2 cm in size, cauline and ramal, subpetiolate, alternate, obovate-oblanceolate to oblong, base cuneate, apex acute, margin entire, highly strigose, dark green in colour; petiole 0.5-1 cm long. Inflorescence subterminal, 1-ranked, forked into 1-2, in 5-10 cm long spike; peduncle 1-4 cm long. Flower complete, actinomorphic, bisexual, hypogynous, sessile, yellowish, ca. 8 x 2 mm across; pedicel 0.3-2 mm long. Sepals 5, quincuncialis, subequal, polysepalous, ovate, base rounded, apex acute, margin ciliate, villous on outer side, green in colour. Petals 5-lobed, gamosepalous; lobes free, spreading, caudate-acute, margin entire, yellowish; sinus pointed. Stamens 5, inserted in the swollen part of the petals, encircling around the ovary, alternate to petals, lanceolate, bilocular. Ovary globose, 4-celled, 4-ovuled; style 1, disciform above; stigma conical, pointed and pubescent.
Flowering period: October-December.
Herbarium specimens examined: NARA DESERT: *Rahmatullah Qureshi*, 546, 667, 924, 1152 (SALUBGH).
Field remarks: A less common herb found in sandy soils.
Distribution: N.E. and C. Africa, Pakistan, India (Nasir, 1989).

3. Trichodesma

R. Br., Prodr. Fl. Nov. holl. 496 (1810); Clarke in Hook. f., Fl. Brit. Ind. 4: 153 (1883); Jafri in Fl. Kar. 282 (1966); Nasir in Nasir & Ali, Fl. Pak. 191: 87 (1989).

Perennial, erect, scabrid herbs. Leaves decussate or alternate, oblong-lanceolate, chartaceous, distantly penninerved, white-glandular above, each gland with a bristle, variously scabrous below, base acute to auriculate, margin entire, gradually tapering above, apex acute, subsessile or petiolate. Inflorescence solitary or in racemose terminal unilateral cymes, simple or bifid. Flowers 5-merous, long-pedicellate; bracts 0. Calyx cupular; lobes 5, triangular-lanceolate, shortly united basally, base rounded or hastate, apex acuminate, angled, enlarged in fruit. Corolla salver-form or sub-rotate to funnel-form; tube cylindric, as long as, or scarcely longer than calyx, with(out) glandular spots along the throat; lobes 5, broadly ovate to rounded, twisted to left, abruptly acuminate, apiculus recurved or not. Stamens 5, inserted in the middle of tube, connivent into a cone, exserted or included; anthers lanceolate, pilose or sericeous, (sub)sessile; connectives produced, sometimes twisted. Ovary conical 4-lobed, each 1-ovuled; style erect, slender; stigma minute, simple or obscure. Fruit of 4, almost free nutlets, around a 4-angled, pyramidal carpophore; nutlets smooth on the outer, pitted and/or rugose along the inner face. Seeds obovoid, exalbuminous.

A genus of 40-45 species distributed in tropical and subtropical Africa and Australia. In Pakistan, only 4 species are reported (Nasir, 1989).

Trichodesma indicum (L.) R.Br., Prodr. Fl. N. Hol. 496 (1810); Stewart in Ali and Nasir, Ann. Cat. Vasc. Pl. W. Pak. & Kashm.,604 (1972); Nasir in Nasir & Ali, Fl. Pak. 191: 88 (1989). **Figure 6D**

Syn. *Borago indica* L., Sp. Pl. 137 (1753); *Trichodesma hirsutum* Edgew. in J. Asiat. Soc. Bengal. 21: 175 (1853); *T. amlexicaule* Roth, Nov. Sp. 104 (1821); *T. indicum* var. *amlexicaule* (Roth.) T. Cooke, Fl. Bomb. Presid. Reprint. ed. 2,2: 281 (1967).

An annual, erect bushy and hispid herb up to 20 cm tall. Stem hard below, herbaceous above, cylindrical, bristly with tubercle-base hairs; internode ca. 2-5 cm long, bristle ca. 0.7-1 mm long, transparent. Leaves opposite, sessile, dark green in colour, lanceolate-oblong, base auriculate, apex acute-obtuse, margin entire, scabrous along nerves bellow, chartaceous, ca. 2-6.5 cm long, ca. 0.5-1.5 cm broad. Inflorescence solitary and axillary racemes. Flowers pedicellate, complete, actinomorphic, bisexual, hypogynous, ca. 1.4 x 1 cm; pedicel 1-2 cm long, strigose. Sepals 5-lobed, valvate, 3-ribbed, hastate or sub cordate, base auriculate, apex acuminate-cuspidate, margin ciliate, sparsely hirsute within, ca. 1.2-1.5 cm long, ca. 0.5 cm broad, green in colour. Petals 5-lobed, gamopetalous, lobes free above, united below to form tube, to ca. 0.7 cm long, 5-gonous, inflated over stamens, 4 unequal spots toward one side, pinkish white, lobes broadly ovate, base obtuse, apex shortly acuminate, margin entire, apiculus not recurved, ca. 1 x 0.4 cm in size, salver shaped. Stamens 5, attached to the middle of corolla tube, connate into a cone around the style, included, sericeous on the back side, lanceolate, apex caudate, base rounded, connective 1 mm, twisted, ca. 6-9 x 0.5-1 mm in size. Ovary conical, 4-lobed, each 1-ovuled, ca. 1 mm across; style sub-terminal, ca. 1 cm long; stigma simple, minute, ca. 0.2 mm across.

Flowering period: November-December.

Herbarium specimens examined: NARA DESERT: *Rahmatullah Qureshi*, 507, 519, 520, 844, 903, 906, 1159, 1226 (SALUBGH).

Field remarks: A less common herb found in calcarious soils.

Distribution: Afghanistan, Pakistan, India, Phillipines, Mauritius (Nasir, 1989).

4. Arnebia

Clarke in Hook. f., Fl. Brit. Ind. 4: 176 (1883); Cooke, Fl. Bomb. Pres. Repr. ed. 2,22: 288 (1967); Jafri in Fl. Kar. 284 (1966); Nasir in Nasir & Ali, Fl. Pak. 191: 50 (1989).

Annual or perennial erect or diffuse hispid herbs; root often staining red. Leaves simple, alternate. Inflorescence a terminal scorpoid cyme; bracts foliaceous. Flowers dimorphic, yellow or violet, sessile or very shortly pedicellate, heterostylic. Calyx 5-fid or 5-partite; lobes lanceolate or linear, not or slightly enlarged in fruit, accrescent. Corolla-tube slender, straight, the throat naked inside, salver-form, white, yellow or shade of blue; lobes 5, imbricate, obtuse, spreading, throat glabrous. Stamens 5, much below the mouth of the tube in the long-styled; anthers small, oblong, obtuse, included or slightly exserted. Ovary deeply 4-lobed; ovules erect, with short funicles; style gynobasic, 2-partite; stigma capitate. Nutlets 4, erect, oblong or acuminate from a broad base, rugose, more or less tuberculate; scar basal, large, triangular, flat, shortly produced up to the inner face.
A genus of 25 species distributed in the Mediterranean, tropical Africa and Himalya. In Pakistan it is represented by 11 species (Nasir, 1989).

Arnebia hispidissima (Sieber) DC., Prodr., x: 94 (1846); Jafri in Fl. Kar., 284 (1966); Stewart in Ali and Nasir, Ann. Cat. Vasc. Pl. W. Pak. & Kashm.,582 (1972); Nasir in Fl. Pak. 191: 65 (1989); Nasir in Nasir & Ali, Fl. Pak. 191: 65 (1989).

Syn: *Anchusa hispidissima* Sieber. in Lehm., Icon., 1: 23(1821); *A. asperima* Delile, Fl. Aegypt. Illust. 7 (1813), *nom. nud.*; *Lithospermum hispidissimum* (Sieber.) Lehm., I.c.Nov. Strip. t. 39 (1821); *L. vestitum* Wall. Cat. no. 941 (1829), *nom. nud.*; *Arnebia asperrima* (Del.) Hutchinson & Dalz., Fl. W. Africa. 2: 201. 608 (1913); *Arnebiola migiurtina* Chiov., Fl. Somala., 227, t. 24, f. 1 (1929).

An annual or perennial erect or decumbent herb up to 20 cm high. Branchlets diffuse from the base, clothed with long, whitish, spreading, bulbous based stiff hairs. Leaves sessile, alternate, ca. 1.5-5 x 0.2-0.8 cm in size, linear-lanceolate, apex acute to subobtuse, margin entire. Flowers subsessile, in simple, dense, secund spikes or racemes, yellow. Bracts foliaceous, ca. 2.5-3 mm long, linear-lanceolate, densely hispid. Calyx ca. 6 mm long, very hispid, divided to the base or nearly so; lobes lanceolate, apex acute, unequal. Corolla yellow, pubescent outside; tube ca. 8 mm long, slender; lobes ca. 2.5 mm long, oblong, apex obtuse, without purple spots at the base of the lobes. Stamens inserted much below the mouth of the corolla in the long-styled, at or in the throat and half-exserted in the short-styled forms; anthers ca. 1 mm long, oblong; filaments very short. Style slender, 2-lobed; stigma small, flattened on the inside, rounded on the outside. Nutlets 4, ca. 1.5-2 mm long, ovoid or pyramidal-ovoid, broadest at the base, tuberculate, pink to brown.
Flowering period: November-March.
Herbarium specimens examined: NARA DESERT: *Rahmatullah Qureshi*, 548, 692 (SALUBGH).
Field remarks: A rare herb found in calcarious soils.
Distribution: N. & E. Africa, Arabia, Pakistan (Nasir, 1989).

46. Verbenaceae

Erect, prostrate/straggling herbs or subshrubs or trees, rarely woody climbers; branchlets and twigs 4-gonous or terete, not prominently nodose or articulated. Leaves opposite, decussate or ternate, rarely whorled or alternate, exstipulate, simple, rarely compound, digitate or pinnate, peninerved, margin entire or variously toothed, petiolate or sessile. Inflorescence definite and centrifugal and centripetal, spicate, racemiform or panicled to umbellate or a head, very rarely solitary and axillary. Flowers zygomorphic or actinomorphic, generally 5 rarely 4 or 6-8-merous, bisexual, subsessile or pedicellate, hypogynous; bracts deciduous or persistent; bracteoles minute, often 0. Calyx gamosepalous, mostly tubular or campanulate, (2) 4-(7)-toothed or lobed to subentire, rarely truncate, valvate, persistent, hardly enlarged in fruits. Corolla tubular, often with as many lobes as the calyx and sometimes bilabiate, cylindric, often dilated/inflated above, or sub-rotate; lobes 4-8, (un)equal, imbricate or valvate. Stamens 4 (2 in *Stachytarpheta*) and didynamous, or as many as corolla lobes, included or exserted; inserted on the corolla tube and alternating with the lobes; staminodes often present; anthers 2-celled, parallel or divaricate, dehiscence longitudinal, introse. Gynoecium 2 (4-5)-carpellary, syncarpous, somewhat 4-lobed and 2-4-ovuled, each with 1 ovule in each apparent cell on axile placentation; style simple, terminal, erect or curved; stigma terminal or lateral, entire or 2-fid. Fruit schizocapic or drupaceous with ± hard endocarp, dry or fleshy when capsular, indehiscent or dehiscent into 1-2(-10)-celled pyrenes; seeds few, ovoid or oblong, exalbuminous.

A large family of ca. 75 genera and nearly 2500 species, maily of tropical and subtropical regions, is presented form Pakistan by 17 genera and ca. 35 species (Jafri & Ghafoor, 1974).

Phyla

Loureiro, Fl. Cochinch. 63, 66 (1790); Greene in Pittonia 4: 48 (1899); Patzak & Rech. f. in Rech. f., Fl. Iran. 43: 3 (1967); Jafri & Ghafoor, Fl. Pak., 77: 10 (1974).

Procumbent or prostrate, perennial herbs with mostly medifixed pubescence opposite, decussate, serrate, coriaceous leaves. Inflorescence of short, ovoid to oblong-cylindrical, congestedly flowered, spicate heads. Flowers very small, sessile, zygomorphic, bisexual; bracts short, obovate. Calyx-tube 2-4-dentate, ovoid-campanulate, membranous, persistent. Corolla-tube straight to curved, slightly exceeding the calyx; limb obliquely sub-bilabiate, 4-fid, upper lip entire to 2-lobed, lower lip 3-fid. Stamens 4, didynamous, included or slightly exserted. Ovary 2-locular, with 1 ovule in each locule; style short with oblique stigma. Fruit a small, dry schizocarp, enclosed by or sometimes adnate to calyx, separating into two, 1-seeded pyrenes at maturity.

A small genus of ca. 10 species, maily distributed in North and Central America, with 1-2 species in warmer parts of Asia and Africa; only one species observed in Pakistan (Jafri & Ghafoor, 1974).

Phyla nodiflora (L.) Greene., Pittonia 4:46 (1899); Stewart in Nasir & Ali., Ann.Cat.Vasc.Pl.W.Pak. & Kashm., 607 (1972). **Figure 19C**

Syn. *Verbena nodiflora* L., Sp. Pl. 20 (1753); *Zapania nodiflora* (L.) Lam., Tab. Encyc., i, 59 (1791); *Lippia nodiflora* (L.) L.C. Rich. ex Michaux, Fl. Bor. Amer., ii: 15 (1803).

A widely creeping perennial herb up to 60 cm high. Stem prostrate, rooting at the nodes; branches slender, ± clothed with adpressed, medifixed white hairs. Leaves opposite, thick textured, spathulate, oblanceolate or obovate, sometimes elliptic or cuneiform, ca. 5-25 x 3-15 mm in size, apex rounded or obtuse, base cuneate, decurrent into a petiole, margin rather regularly sharply serrate above the middle with sharply acute or acuminate teeth, entire at the base, 2-4-nervedminutely strigose-puberulent on both surfaces. Flowers sessile, densely packed in long, pedunculate, axillary heads, the latter at first globose, afterwards elongate and becoming spicate and oblong in fruits; peduncles solitary in each axil, ca. 1-2 cm long; stalk 5 cm long, sparsely adpressed-peberulent; bracts closely imbricate, obovate or rhomboid, base cuneate, apex mucronate, acuminate, concave, glabrous or finely ciliate. Calyx ca. 2 mm long, hyaline-membranous, copular, deeply 2-lobed, slightly 2-carinate; keels puberulent; lobes lanceolate. Corolla purplish white, ca. 2-2.5 mm across; salver-form, slightly strigose outside, 2-lipped; tube ca. 1.5 mm; lobes 5, spreading, sub-orbicular, apex obtuse, 2+3, ca. 1-1.2 mm; upper lip 2-fid, erect; lower lip 3-fid; mid lobe the largest. Stamens 4, attached in 2 pairs at the throat of tube; filaments ca. 0.4 mm, curved; anthers ca. 0.2 mm, ovate. Ovary globose, ca. 0.5 mm, 2-celled; ovule solitary per cell, basal; style ca. 0.5 mm; stigma su-capitate, oblique. Fruit ca. 1.5 mm long, globose-oblong, dry splitting into 2, 1-seeded, plano-convex, glabrous pyrenes.

Flowering period: **Almost round the year.**

Herbarium specimens examined: NARA DESERT: *Rahmatullah Qureshi*, 1326 (SALUBGH).

Field remarks: A common herb found in wetland habitats.

Distribution: Throughout tropical and subtropical regions (Jafri & Ghafoor, 1974).

47. Labiate (Lamiaceae)

Annual to perennial, sometimes suffruticose herbs, subshrubs or shrubs, often aromatic and usually with some sessile oil glandular hairs. Plants glabrous or with an indumentum of simple, branched to stellate-dendroid hairs. Stems often 4-gonous; branchlets decussate, sometimes terete. Leaves decussate or whorled, undivided to pinnate, petiolate or sessile; without stipules. Inflorescence of variously arranged cymes, often in condensed false whorls of flowers (verticillasters). Verticils axillary or terminal, simple or superposed into spurious racemes, spikes or the clusters rarely 1-sided in elongate, secund spikes. Flowers subsessile or pedicellate, zygomorphic hermaphrodite but male sterile in gynodioicious plants. Bracts decussate or whorled, subtending the verticils, persistent or caduous. Calyx turbinate, campanulate or tubular; lobes 5, valvate or imbricate, usually 2-lipped to actinomorphic: with a 3-lobed upper lip and a 2-lobed lower lip, or a 1-lobed upper lip and a 4-lobed lower lip, or a 1-lobed upper and lower lip, enlarging in fruit or not, veins 5-20; tube glabrous within to densely bearded. Corolla gamopetalous, zygomorphic and bilabiate to actinomorphic, variously arranged; upper (adaxial) lip entire or emarginate with a 3-lobed lower (abaxial) lip of 1 lobe and lower lip 4-lobed; upper lobe wanting and lower lip apparently 5-lobed; or lobes subequal, 4 or 5. Stamens variously inserted in the corolla, 4, often with 2 (usually anterior pair) longer and 2 (usually posterior pair) shorter (didynamous); or 2 (the anterior or lower pair in almost all cases), with usually 2 (posterior or upper pair) staminodes; anthers thecae 2-locular or 1-locular (by fusion or not), parallel, divergent or separated by an elongated connective, basifixed, introse. Pollen tricolpate or 6-colpate, rarely tetracolpate. Ovary deeply 4-grooved, superior of 2 carpels, 4-locular, 1-ovulate loculi; style simple, filiform, generally arising from base of ovary divisions, gynobasic, rarely inserted above the base of ovary lobes (terminal); stigma equally or unequally bifid; ovules erect, anatropus. Disc subentire or 4-lobed, sometimes the interior lobe enlarged. Fruit a schizocarp of 4, sometimes less, dry, variously shaped nutlets enclosed at the base of enlarged, persistent calyx. Nutlets basilar or laterally ventral, rarely 1 or 2 aborted, scarred; epicarp chartaceous, erect or transverse with or without little endosperm; embryo straight or curved; fruiting pedicel usually elongate and recurved.

A wide spread family (cosmopolitan) of about 180 genera and possibly over 3500 species (Hedge, 1990).

Key to genera:

1a. Scented undershrub with 2 perfect stamens..**1. Salvia**

1b. Non aromatic herb with 4 perfect stamens..**2. Leucas**

1. Salvia

L., Sp. Pl. 23 (1753); Gen. Pl. ed. 5: 15 (1754); Hook. f., Fl. Brit. Ind. 4: 658 (1885).

Arischrada Pobed.

Schraderia Heister ex Medikus

Stenarrhena D. Don

Shrubs, subshrubs, perennial, biennial or rarely annual herbs, usually aromatic. Stems erect, usually with simple eglandular hairs below and glandular hairs above. Leaves undivided, crenate, deeply lobed to pinnatisect, petiolate or subsessile. Inflorescence of few- to many-flowered verticillasters, distant or approximating. Bracts usually distinct from leaves, sometimes showy. Calyx tubular to infundibuliform, bilabiate, slightly or much expanding in fruit, 10-15-veined; upper lip shortly or scarcely 3-toothed; lower lip equally 2-toothed. Corolla bilabiate; upper lip falcate or ± straight, entire or bifid; lower lip 3-lobed, median lobe usually larger and concave; tube straight or curved, ventricose or not, invaginated and with a small internal scale (squamula) or not, annulate or not. Stamens 2, with a large fertile theca at upper end and a usually much smaller fertile theca or a sterile dolabriform plate at the lower end; connectives clearly separating the thecae; stamens articulating at the point of attachment of filament and connective (in ours); staminodes usually present, very small. Style bifid with unequal flattened or terete lobes. Nutlets 4, glabrous ellipsoid to ± spherical, smooth, veined or not, usually mucilaginous on wetting.

About 800 species throughout the Old and New World. SW Asia and C. Asia are important centres for this very distinct genus. The structure of the stamens is quite different from any other genus in the Labiatae.

Salvia santolinifolia Boiss., Diagn. Pl. Or. Nov. 1, 5: 13 (1844); Stewart in Ali and Nasir, Ann. Cat. Vasc. Pl. W. Pak. & Kashm., 632 (1972); Hedge in Nasir & Ali., Fl. Pak., 192: ???? (1992).

Dwarf, suffruticose much branched undershrub up 30 cm. Stems with an often gnarled woody base, eglandular below with often retrorse long hairs, glandular above. Leaves borne ± erect to erect-spreading linear in outline with a sinuate-convolute margin and apparently or indeed pinnatifid, 5-13 x 1.5-3.5 mm, white canescent below with glandular and eglandular hairs and sessile oil globules, petiolate. Inflorescence with usually several, 1-2-

flowered distant or approximating verticillasters. Bracts and bracteoles present. Pedicels in flower 1.5 mm elongating to 2.5 mm and recurved in fruit. Calyx tubular-campanulate, 3-4 mm in flower and c. 5 mm in fruit, with a dense indumentum of very long white spreading eglandular hairs, shorter capitate glandular hairs and sessile oil globules; upper lip of ± connivent acuminate-spinulose teeth. Corolla mauve, pink to lilac often with darker markings, 5-6 mm; upper lip straight much shorter than lower. Lower theca fertile; staminodes prominent. Nutlets smooth, black, ovoid-trigonous, c. 1.5 x 1 mm.
Flowering Period: February-May.
Herbarium specimens examined: NARA DESERT: *Rahmatullah Qureshi*, 1428 (SALUBGH).
Field remarks: A less common bush found in calcarious soils.
Distribution: Afghanistan, Persia, Pakistan, India (Bhandari, 1978).

2. Leucas

Burm. ex R. Br., Prodr. Fl. Nov. Holland 504 (1810); Benth. in DC., Prodr. 12: 523 (1848); Hook. f., Fl. Brit. Ind. 4: 680 (1885); Rech. f. Fl Iran. 150: 318-320 (1982); I.C. Hedge in Fl. Pak., 192: 155 (1990).

Annual or perennial, erect or diffuse herbs or subshrubs with a pilose to villous eglandular indumentum of simple hairs. Leaves undivided, ovate, oblong, linear to lanceolate, entire to crenate-serrate, petiolate at least on lower leaves, usually gland-dotted. Inflorescence many-flowered cymes, aggregated into terminal and/or axillary multi-seriate pseudoverticils (verticillasters), distant from each other. Bracts numerous, linear-subulate, ovate-lanceolate or filiform, ciliate, shorter than to as calyces. Flowers white, sometimes with cream, rose or green markings. Calyx tubular, turbinate sometimes funnel-shaped; mouth straight or oblique, 8-10-ribbed, pilose to villous within, 8-10-toothed, equal or unequal, valvate. Corolla rather small; tube included in calyx, cylindric, glabrous or annulate within; lobes 5, 2-lipped, 1+4; upper lip usually shorter than lower, rarely subequal, hooded or ± straight, enclosing the stamens, densely white-bearded, lower lip deflexed, broad and flat, lateral lobes shorter than the connate mid (2)lobes, membranous. Stamens 4, didynamous, ascending into the upper lip, included or partly exserted; anthers divaricate, at length confluent; thecae 2-locular. Ovary 4-celled, cells subtrigonous; style gynobasic, flat and broadened towards apex; stigma obliquely 2-fid. Disc 4-lobed, alternating with 4 locules, or anteriorly enlarged. Nutlets 4, oblong, subtrigonous, obtuse to truncate-rounded at apex; fruiting calyx enlarged.

A palaeotropic genus of about 80 species with two main centres: one in Africa, especially east tropical Africa; the other in the Indian sub continent (Hedge, 1990).

Leucas aspera (Willd.) Link. Enum. Hort. Berol. 2: 113 (1822); Stewart in Ali and Nasir, Ann. Cat. Vasc. Pl. W. Pak. & Kashm., 615 (1972); Hedge in Nasir & Ali., Fl. Pak., 192: 159- 161 (1992).

Syn. *Phlomis aspera* Willd., Hort. Berol. 2: 621 (1809); *P. plukenetii* Roth,Nov. Sp. 261 (1821); *Leucas plukenetii* (Roth) Spreng., Syst. Veg. 2: 743 (1825).

An annual scabrid, erect or diffuse herb up to 40 cm high; hairs spreading. Leaves ca. 2-6 x 0.3-1 cm in size, linear-lanceolate, base cuneate, apex acute, margin entire-serrate,

chartaceous, puberulous; petiole ca. 0.3-0.5 cm long. Verticils terminal and axillary; bracts linear-lanceolate, apex acute, bristle-tipped, equalling calyx. Calyx tubular, ca. 6 mm, upper half strigose without, glabrous below; mouth oblique, 10-teethed; teeth unequal, posterior the longer, linear, ca. 1 mm. Corolla-tube ca. 6 mm, obliquely annulate within; upper lip ca. 3.5 mm, villous without; lower lip ca. 8 mm. Stamens 4; filaments pairs ca. 4.5 and 6 mm; anthers ca. 0.6 mm. Ovary ca. 0.5 mm; style ca. 8 mm. Nutlets oblong, smooth, brown.
Flowering Period: June and again November.
Herbarium specimens examined: NARA DESERT: *Rahmatullah Qureshi*, 530, 1121, 1158 (SALUBGH).
Field remarks: A rare herb found in calcarious soils.
Distribution: Kashmir to Sikkim, Indo-China, Malaysia (Hedge, 1990).

48. Scrophulariaceae

Erect or prostrate, marsh or aquatic herbs, or small shrubs, rarely root parasites. Leaves all decussate or alternate, often both, rarely ternate or whorled, simple or pinnatifid, petiolate or subsessile, exstipulate. Flowers irregular, rarely sub-irregular, 4-5-merous, axillary, solitary or in terminal, lax racemes, cymes or pedunculate capitate clusters, subsessile or pedicellate; pedicels elongating in fruit; bracts and bracteoles present or 0. Calyx-lobes (4)5, equal or unequal, connate or free, valvate or imbricate, herbaceous. Corolla tubular, 2-lipped; lobes (sub)equal or unequal, imbricate, rarely rotate, membranous; upper lip outer, rarely inner in bud. Stamens 2 or 4, with or without staminodes, included or exserted; filaments equal, or if didynamous, upper pair shorter; anther 2- or 1-celled, sessile or stipitate, distinct, contiguous, parallel or diverticate, rarely transverse, dehiscence longitudinal, introse; connectives broad; staminodes minute, linear, simple or forked. Ovary superior, globose or conical, 2 or imperfectly 4-locular; ovules α, on swollen, axile placentae, anatropous or amphitropous; style terminal, simple or dilated; stigma 2-lamellate, 2-lobed, capitate or simple. Disc annular, cupular or unilateral, glandular, hypogynous. Capsule oblong-lanceolate, cylindric or compressed, loculicidal or septicidal, or both, 2-4-valved, or irregularly rupturing; calyx persistent, usually not enlarged; epicarp thin. Seeds α, minute, angular, truncate or globose; testa smooth, reticulate or rugose; septa and valves persistent.

A family of over 200 genera and nearly 2500 species distributed throughout the world (Jafri, 1966).

Key to genera:
1a. Stamen..**1. Anticharis**
1b. Stamens 4:
2a. Corolla-tube spurred at base..**2. Kickxia**
2b. Corolla-tube not spurred at base:
3a. Upper 2 stamens with perfect anthers; lower 2 reduced to staminodes.................**3. Bacopa**
3b. Stamens all perfect, didynamous..**4. Schweinfurthia**

1. Anticharis

Endl., Nov. Strip. Dec. 22, Gen. 682 (1839); Cooke in Fl. Bomb. Pres., 2: 351 (1906); Jafri in Fl. Kar., 302 (1966).

Small erect viscid-hairy herbs. Leaves alternate, entire. Flowers axillary, usually solitary, with 2 bracts on the pedicel. Calyx 5-partite; sepals narrow, subvalvate. Corolla-tube enlarged at the throat lobes 5, subequal. Stamens 2 perfect; filaments filiform; anthers versatile, curved; staminodes 0. Style filiform; stigma obtuse or notched. Capsule ovoid or oblong, acuminate; loculicidally and septicidally dehiscent; valves with inflexed margins, separating from the axis, lying bare the placentaniferous column. Seeds numerous, small, oblong or obovoid, straightly ribbed; embryo straight; cotyledons ovate.

A genus of about 9 species distributed in tropical and South Africa, Arabia and Indo-Pak subcontinent (Cooke, 1906).

Key

1a. Leaves ovate, not exceeding 1.8 cm in length, pedicels less than twice as long as the calyx...**1. A. glandulosa**

1b. Leaves narrowly linear, exceeding ca. 2.5 cm in length, pedicels more than twice as long as the calyx..**2. A. linearis**

1. Anticharis glandulosa Aschers. in Monatsb. Akad. Wiss. Berl, 800 (1866); Stewart in Ali and Nasir, Ann. Cat. Vasc. Pl. W. Pak. & Kashm., 645 (1972).

An erect, branched leafy herb up to 20 cm high, densely clothed with viscid glandular hairs. Leaves 1-3 x 0.5-1.5 cm in size, ovate-oblong, or oblong-lanceolate, apex subobtuse, margin entire, densely glandular-pubescent, narrowed into a short flattened often obscure petiole, which is up to 5 mm long. Flowers solitary (very rarely 2 together), axillary; pedicel slender, glandular hairy; peduncle ca. 4-5 mm long. Bracts 2, about the middle of pedicel, linear-spathulate, glandular hairy, reaching ca. 5 mm long. Calyx ca. 5 mm long, divided at the base; lobes oblanceolate or more or less spathulate, apex acute-obtuse, margin ciliate, glandular-hairy. Corolla ca. 1-1.5 cm long, pink, rosy or blue; lobes broad, rounded. Filaments short, filiform, glabrous; anthers versatile, curved, the segments unequal. Ovary ovoid, glabrous, seated on a somewhat cup-shaped disk; style glabrous. Capsule ca. 8-10 mm long, narrowly ovoid, acuminate, pubescent. Seeds oblong, truncate, longitudinally striately ribbed, ca. 0.5 mm.

Flowering period: October-May.

Herbarium specimens examined: NARA DESERT: *Rahmatullah Qureshi*, 946, 955, 967, 1018 (SALUBGH).

Field remarks: A less common herb found in hilly areas.

Distribution: Found in Upper Egypt (Cooke, 1906).

2. A. linearis (Benth.) Hochst. ex Aschers. in Monatsb. Akad. Wiss. Berl. 882 (1866); Stewart in Ali and Nasir, Ann. Cat. Vasc. Pl. W. Pak. & Kashm., 615 (1972).

Syn: *Doratanthera linearis* Benth. in DC., Prodr., x: 347 (1846).

An erect, much branched, glandular pubescent annual upto 35 cm long. Branches ascending. Leaves 2-5 x 0.2-0.4 cm in size, linear to narrowly lanceolate, hairy on both surfaces. Peduncles solitary, axillary, filiform, deflexed above in fruit, glabrous near the base and glandular the apex; bracts 1-3 mm long, inserted little above the middle of peduncle, opposite subopposite, linear, glandular, hairy. Calyx 5 mm long, divided near the base,

segments linear, acute, glandular hairy. Corolla pale-purple, exceeding the calyx, divided scarcely ¼ down way. Style glabrous. Capsules ca. 7 mm long, ovoid, acuminate, pubescent, many seeded. Seeds oblong-lanceolate, longitudinally striately ribbed.
Flowering period: October-May.
Herbarium specimens: NARA DESERT: *Rahmatullah Qureshi*, 1057, 1175 (SALUBGH).
Field remarks: A rare species confined to hilly area and observed in depression of Rohri hills.
Distribution: Found in Upper Egypt (Cooke, 1906).

2. Kickxia

Dumort. Fl. Belg. 35 (1827); Jafri in Fl. Kar., 303 91966).

Annual climbers. Leaves usually opposite or whorled below and alternate above, triangular-hastate. Flowers small, axillary, solitary on long slender filiform pedicels. Calyx 5-partite, segments imbricate. Corolla personate, 2-lipped; tube spurred in front (shorter than corolla, 2-5 mm long); upper lip erect, 2-lobed; lower lip spreading, 3-lobed. Stamens 4, didynamous, included, ascending; anther with distinct parallel cells. Ovary 2-celled; ovules many; style filiform; stigma minute. Capsule ovoid or globose, 2-celled, both cells are the anterior only opening by an apical pore. Seeds angled or flattened, wingless or discoid and winged.

A genus of few species (Jafri, 1966).

Kickxia ramosissima (Wall.) Janchen in Oesterr. Bot. Zeitech., lxxxii: 152 (1933); Jafri in Fl. Kar. 303 (1966); Stewart in Ali and Nasir, Ann. Cat. Vasc. Pl. W. Pak. & Kashm., 649 (1972). (Local Name: *Kanodi*)

Syn: *Linaria ramosissima* Wall. Pl. As. Rar., ii: 43, t. 153 (1831); *Elatinoides ramosissima* (Wall.) Wetstt. in Engl. Prantl. Pflanzen. 4 (36): 58 (1891).

A perennial decumbent herb. Stem much branched, delicate; branches filiform, thin, up to 45 cm high. Leaves alternate, triangular-hastate, apex acute, margin entire, ca. 6-40 x 3-25 mm in size, the lower ones 5-7-lobate, the upper lanceolate-sagittate, glabrous or the lower sometime pubescent. Flowers small, yellow, about 1 cm long, solitary on axillary filiform pedicels; pedicel ca. 1-3.5 cm long, terete, ebracteate. Calyx-tube ca. 2-3 mm; lobes linear-lanceolate, membranous towards the base, apex acute. Corolla 5-10 mm long, pubescent outside; spurr ca. 2.5 mm long; upper lip ca. 0.3 mm, oblong, obtuse; the lower lip ca. 0.5 mm long, oblong, obtuse, yellow in colour. Stamens didynamous; filaments with few long hairs at the base, more or less papilose above. Ovary globose; style stout, ca. 0.5 mm long, glandular-pubescent. Capsule globose-ovoid, opening on one side by a valve and other side by a subapical pore, ca. 3 mm across. Seeds small, brown, echinate.
Flowering period: November-January.
Herbarium specimens examined: NARA DESERT: *Rahmatullah Qureshi*, 560, 1176 (SALUBGH).
Field remarks: A rare herb found in hilly areas.
Distribution: India, W. Pakistan, Afghanistan and Iran (Jafri, 1966).

3. Bacopa

Aublet, Hist. Pl. Gui. France 1: 128. t. 49 (1775)-*nom. Cons.*

A genus of 56 species (Mabberley, 1997). One species is recorded from the study area.

Bacopa monnieri (L.) Wettestein in Engl. & Prantl. Pflanzenfan. 5 (36): 77 (1891).

Syn: *Lysimachia monnieri* L., Cent. Pl., ii: 9 (1756); *Gratiola monnieri* L lc, 120; *Monniera cuneifolia* Michaux, FL. Bor. Amer. I: 22 (1803); *Herpestris monnieri* (L.) HB. & K.,Nov. Gen. et Sp., ii: 366 (1816); *Bacopa monnieri* (L.) Wettest. In New Man. Bot., 724 (1908); *Brami monnieri* (L.) Pennell, Scrop. South U.S. in Proc. Aca. Nat. Sc. Philad., *71*: 243 (1920).

A glabrous, somewhat succulent, creeping herbs, rooting at the nodes with numerous prostrate branches, each 10-30 cm long, leaves obovate-oblong, obtuse, sessile, decussate, 1-2.5 x 0.3-0.8 cm in size, ± fleshy, entire, punctuate, obtuse. Flower axillary, solitary, pale violate; peduncles 1-1.5 cm long, often much longer and deflexed in fruiting stage; bracteoles 5 mm long, linear, obtuse. Calyx divided to base; posticous segment broadly ovate, 6 x 3-4 mm, obtuse, the other 4 segments slightly sorter than the anticous segments acute, slightly narrower. Corolla gamopetalous, funnel like, white or pinkish with purple blotches, 10-12 mm long, lobes subequal, rounded and glabrous. Stamens 4, didynamous, included. Ovary 2-chambered.

Flowering & fruiting period: January-June.

Herbarium specimens examined: NARA DESERT: *Rahmatullah Qureshi*, 1195, 1206 (SALUBGH).

Field remarks: A common herb found in wetland habitat.

Distribution: All warm countries including W. Pakistan and Afghanistan (Jafri, 1966), Ceylon, India, Rajputana, (Bhandari, 1978).

4. Schweinfurthia

A. Braun, Sitzb. Ges. Naturf. Fr. xx: 24 (1866); Cooke in Fl. Bomb. Pres., 2: 354 (1906); Jafri in Fl. Kar., 304 (1966).

Annual or perennial, glabrous herbs. Leaves usually alternate, quite entire. Flowers small, axillary. Calyx 5-partite; segments imbricate, the upper the largest. Corolla-tube large, subsaccate at the bas, not spurred; limb 2-lipped, the upper lip erect, 2-lobed, the lower 3-lobed, the middle-lobe narrower than the lateral; palate closing the throat. Stamens 4, fertile, didynamous (the 5th rudimentary); anthers 2-celled, the cells divericate after flowering. Style filiform; stigma minute. Capsule subglobose, fragile, 2-celled, the upper cell small, 3-4-seeded or empty, the lower longer and broader, larger, many seeded, bursting irregularly. Seeds obconic, ribbed or winged, truncate at the ends.

A genus of about 3 species distributed in N.E. Africa, Arabia, Afghanistan, Baluchistan and Sind (Cooke, 1906).

Schweinfurthia papilionacea (Burm. f.) Boiss., Fl. Or. Iv: 387 (1879); Jafri in Fl. Kar. 304 (1966); Stewart in Ali and Nasir, Ann. Cat. Vasc. Pl. W. Pak. & Kashm., 659 (1972). **Figure 18B**

Syn: *Antirrhinum papilionaceum* Burm. f., Fl. Ind., 121, t. 39, f. 2 (1789); *A. glaucuma* Stocks in Wight, Icon. T. 1459 (1850). *Linaria sphaerocarpa* Benth. in DC., Prodr. X: 287 (1846); *Schweinfurthia sphaerocarpa* (Benth) A. Braun. in Montsb. Akad. Wiss. Berl., 875 (1866).

A glaucous glabrous sub-succulent, erect herb up to 30 cm long, from a perennial root. Stem ascending, succulent, glaucous, much branched at the base. Leaves ca. 1-3 x 0.5-2.5 cm in size, ovate-spathulate, or orbicular-obovate, apex subacute-apiculate, margin entire, attenuated at the base into a short petiole, sparsely clothed on both sides with very minute hairs, rarely glabrous. Flowers axillary, ca. 12 mm, deflexed in fruit. Calyx 5-partite; upper segments broadly ovate-subcordate, apex acute, veined; rest lanceolate, apex acute, margin entire. Corolla-tube ca. 5 mm long, white, with purple veins; upper lip 2-lobed, oblong, apex obtuse, ca. 4-5 mm long; lower lip 3-lobed, spreading, the lobes oblong, apex obtuse, ca. 7 mm long. Capsule obliquely globose, 2-celled, lower cell many seeded, upper cell few-seeded. Seeds obconic, truncate at both ends, acutely winged, pale.

Flowering period: November-February.

Herbarium specimens examined: NARA DESERT: *Rahmatullah Qureshi*, 668, 800, 805, 886, 1177, 1204, 1215, 1239 (SALUBGH).

Field remarks: A less common shrub found in calcarious soils.

Distribution: Rajputana, W. Pakistan and Afghanistan (Jafri, 1966).

49. Orobanchaceae

Perennial root-parasites, devoid of chlorophyll, herbaceous, scapigerous, usually fleshy with somehow tuberous rootstock; stem simple, stout or slender; leaves alternate, scaly or small fleshy. Inflorescence in racemes or spikes or 1-flowered. Flowers bisexual, hypogynous, irregular, blue, violet, yellowish to brown. Bracts large, scaly; bracteoles much narrower than bracts, 1-2, scale-like or 0, usually free, rarely slightly to conspicuously connate with calyx-tube. Calyx spathaceous or 2-lipped or with 2-4 (-5) free or connate lobes, persistent, scaly. Corolla hypogynous, curved, usually distinctly 2-lipped (occasionally subequally 5-lobed); upper lip often vaulted; throat usually with 2 villous folds. Stamens 4, didynamous, inserted on the corolla-tube; filaments filiform, often hairy at the base, included, rarely subexserted; anthers 1-2-celled, the cells frequently spurred at the base, opening by slits or apical pores, 1 cell often imperfect. Disk glandular or unilateral or 0. Ovary superior, 2 (rarely 3) connate carpels, 1(-2)-celled, with numerous ovules on (2-)4(-6) parietal placentae; style simple, often long with curved tip and somewhat 2-lobed, capitate stigma. Fruit an ovoid to oblong capsule, 1-celled, dehiscing into 2 incompletely separating valves, glabrous, usually many seeded. Seeds minute; testa usually pitted or tuberculate; albumen fleshy; embryo ovoid, often undivided.

A family of about 14 genera and nearly 200 species mainly distributed in the warm temperate zones of the Old World. In Pakistan, it represents by 4 genera and about 22 species (Jafri, 1976).

Cistanche

Hoffmg. & Link, Fl. Port. 1: 318, t. 63 (1809); Hook. f., Fl. Brit. Ind. 4: 323 (1885); Cooke, Fl. Bomb. Pres. (rep. ed.) 2: 386 (1958); Jafri in Fl. Pak., 98: 4 (1976).
Phelipaea, sect. *Cistanche* Walp., Rep. Bot. Syst. 3: 461 (1845).

Erect, fleshy, scaly, root-parasites, often conspicuously swollen at the base. Spikes often large, congested. Flowers large, outcurved, bracteate and usually 2-bracteolate. Calyx broadly tubular-campanulate; lobes 4-5, rarely fissured posteriorly, subequal, obtuse. Corolla tubular-infudibuliform, curved, dilated above; limb spreading, oblique, with 5 broad subequal lobes. Stamens subexserted, inserted low down on the corolla-tube; anthers equally 2-celled, much shorter than the filaments, often villous, bases obtuse. Ovary 1-celled; ovules many on 4 parietal placetae; style long, decurved at the tip with broad stigma. Capsule ovoid, laterally compressed, 4-valved to the base. Seeds numerous, subglobose, reticulately punctuate.

A genera of about 15 species of Afro-Asian region. Its representation in Pakistan is only a single species (Jafri, 1976).

Cistanche tubulosa (Schrenk) Hook. f., Fl. Brit. Ind. 4: 324 (1884); Jafri in Fl. Pak., 98: 4 (1976).

Syn: *Phelipaea tubulosa* Schrenk, Pl. Spec. Aegypt-Arab. 23 (1840); *C. lutea* auct. non Hoffmg. & Link: Wight, Illust., 2: 180, t. 158b., f. 4 (1831); *C. tubulosa* var. *tomentosa* Hook., l.c.

A fleshy, glabrous-puberulous, yellowish to yellow-brown, erect herb with a purple tinge. Underground part much swollen, tuberous and attached to the root of the host plant. Stem up to 1 m high, mostly broader (up to 6 cm) at the base, unbranched, furrowed. Scales ca. 2-3.5 x 1-1.7 cm in size, triangular-broadly linear, apex acute. Bracts ca. 1-1.5 cm long, oblong-lanceolate, apex acuminate, margin membranous, often purplish; bracteoles ca. 5-8 mm long, linear-sublanceolate, margin membranous. Calyx ca. 1.5-1.8 cm long, including lobes ca. 1/3 as long as the tube; lobes laterally overlapping, subequal, apex rounded, margins membranous. Corolla ca. 3-5 x 1.5-2 cm in size, the lower half tubular, erect, the upper half campanulate, bent outwards from the base middle, yellowish with purplish or rarely whitish lobes; limb shortly 5-lobed, the lobes rounded, reflexed. Filaments wooly at the base; anthers subexserted, wooly, rounded or blunt at the apex. Placentae 4, subequidistant; style exserted, curved below the stigma. Capsules ca. 2-2.5 cm long, ovoid-oblong, laterally compressed, beaked, many seeded. Seeds ca. 1-1.2 mm long, subglobose, deeply pitted, dark coloured.
Flowering period: December-January.
Herbarium specimens examined: NARA DESERT: *Rahmatullah Qureshi*, 1427 (SALUBGH).
Field remarks: A rare herb found in hilly areas.
Distribution: N. Africa, Arabia, W. Asia to Pakistan, India and C. Asia (Jafri, 1976).

50. Acanthaceae

The members of this family are perennial or somewhat annual herb, shrub or rarely trees. Stem erect, prostrate or climbing; branchlets 4-gonous or terete, often nodose, suffrutescent, armed or unarmed. Leaves simple, opposite, decussate, sometimes appearing whorled, entire, penninerved, petiolate or sessile, exstipulate. Inflorescence cymes, often in dense to lax spike, racemes, panicles or rarely solitary. Flowers bisexual, zygomorphic, sessile to pedicellate. Bract(s) 1, 2 or more, often coloured, large, imbricate, persistent in fruit, rarely absent. Sepals gamosepalous, 4-5-partite, unequal or unequal, valvate or imbricate, free or basally connate, persistent in fruit, sometimes spinescent or reduce to truncate or 10-15-dentate minute tube. Petals gamosepalous, 2-lipped or subequal 5-lobed, tube cylindrical or ventricose; upper lip erect or suberect, entire or 2-fid rarely absent or fimbriate; the lower lip more spreading, 3-fid, lobes imbricate or contorted. Stamens 2 or 4, didynamous, epipetalous, with (1-3) or without staminodes, exserted or included; anthers dorsifixed, sometimes stamens pairs unequal, cells parallel, divergent or superposed, appendage or bearded, introse. Ovary superior, 2-celled, conic or oblong-cylindric, sessile or minutely stipitate, 2-locular, placentation axile, rarely parietal, ovules 2-10 (-many) in each locule, in 1 or 2 vertical rows, rarely collateral, anatropous; style simple, filiform, mostly as long as corolla; stigmas minutely 2-fid, capitate. Fruit a loculicidal capsule, 2-valved, clavate to elliptic-ovoid or oblong-cylindric, terete to compressed at right angles to the septum, beaked or not, splitting through the septum. Seeds compressed, lenticular or discoid rarely globose or angular, supported by upcurved retinacula, reduced papillate, testa smooth, rugose, echinate, tuberculate, glochidiate, papilose or appressed-hairy, albumen scanty, cotyledons flat.

It is a large family of about 250 genera and 2500 species mostly distributed in the tropics and sub-tropics extending to Mediterranean, Australia and Southern U.S.A. in Pakistan, it is represented by 18 genera and nearly 60 species (Malik & Ghafoor, 1988).

Key to species:

1a. Corolla 1-lippid, the upper lip absent, 3-lobed; 2-seeded........................ **1. Blepharis**
1a. Corolla not 1-lippid, the upper lip absent, 5-lobed; 4-seeded......................**2. Barleria**

1. Blepharis

Juss., Gen. Pl. 103 (1789); Nees in DC., Prodr. 11: 265 1847); Benth. & Hook. f., Gen. Pl. 2: 1089 (1876); Clarke in Hook. f., Fl. Brit. Ind. 4: 478 (1885); Malik & Ghafoor in Fl. Pak. 188: 4 (1988).

Perennial, prostrate-erect herb or undershrub; intenode elongated. Leaves sessile or petiolate, opposite or pseudo-whorled, simple, entire or spinescent toothed. Inflorescence in axillary clusters among the leaves, or strobilate spikes or solitary; bracts closely imbricate, 4 in opposite pairs, margin dentate with stiff, elongate hairs, outer pair smaller than successive inner pairs, concave, puberulous; bracteoles lanceolate or spathulate, longer than bracts, hairy. Sepals unequally 4-lobed, free, 2+2, imbricate, outer pairs longer, lanceolate, persistent. Corolla-tube cartilaginous, shorter than lobes; upper lip obsolete; lower lip 3-lobed apically, imbricate, deflexed, puberulous, sericeous at throat. Stamens 4, didynamous, attached at the throat of the tube, all fertile, exserted; filaments inserted the base of the corolla tube, subequal, rigid, flattened, anterior pair produced near connectives; anthers oblong, 1-celled with white hairs, sterile cell bearded. Ovary oblong-cylindric, 2-ovuled, 4 or 2-ovuled; style stout with a basal ring; stigma 2-fid. Capsule compressed, oblong or ellipsoid, 2 or 4-seeded, shining. Seeds flattened, (sub)orbicular, beset echinate-papilose, retinacular rigid.

A genus of about 100 species, distributed mainly in the Old World tropics, Mediterranean, S. Africa and Madagascar. Three species are present in Pakistan (Malik & Ghafoor, 1988).

Blepharis sindica Stocks ex. Anders in Journ., L. Soc. Bot., 9: 500 (1867); Jafri in Fl. Kar., 312 (1966); Stewart in Ali and Nasir, Ann. Cat. Vasc. Pl. W. Pak. & Kashm., 675 (1972). **Figure 1A**

Syn: *Acanthodium grossum* auct. non. Nees 1830; Wight, Icon. Pl. Ind. Or. Tt. 1535 & 1536 (1850).

A small dichotomously branched under shrub upto 40 cm high, with woody root-stock. Stem usually very short, sometimes 0; branches ash-coloured, slender, subterete, clothed with very short hairs. Leaves sessile in a whorl of 4, unequal, ca. 2.5-10 x 0.3-0.7 cm in size, linear, base acute, sometimes with a few spinous teeth near the base, finely apiculate, margin revolute, clothed with short stiff hairs beneath, midrib prominent below. Flowers subsessile, in hairy spikes, ca. 2.5 x 0.4 cm, in the forks of branches, solitary or few or many together on the top of a very short stout woody peduncle, usually numerous in the lowest fork, solitary in the upper. Bracts 4-fariuos, 2 x 1 cm, the lowest the smallest, ovate, apex acute-acuminate, spine-tipped, margin recurved, dentate, spinous, 5-nerved, densely hairy on both sides. Bracteoles 2, jointed at the base, ca. 9 x 1 mm, linear-narrowly lanceolate, apex acuminate, densely hairy and ciliate; cilia ca. 1-1.5 mm long, unicellular, hyaline reticulately veined and with a strong midnerve. Sepals 4-lobed, divided almost to the base; larger segment ca. 1.2 x 0.5 cm in size, ovate, apex shortly and bluntly acuminate, base truncate, 7-nerved, the 3 midnerves parallel and continued beyond densely ciliate at base into 3 bristly teeth; the opposite calyx segments ca. 0.7 x 0.3 cm in size, of similar shape, the 3 parallel veins less strong, not produced into teeth at base; lateral segments ca. 0.8 cm long, ovate, apex acute, 1-nerved. Corolla-tube ca. 0.4 cm long, constricted below the limb, conical, glabrous within; limb ca. 0.8 x 05 cm; middle lobe of lower lip larger, ca. 1-2 x 2-2.5 mm, oblong-subovate, truncate, twice as long and twice as broad as the rounded lateral ones; densely on adexial side, glabrous within. Stamens 4; filaments ca. 3 mm long, 2 filaments filiform, 2 flattened and ends on a projection of ca. 1 mm long, base unicellular hairs; anther-cell bearded, dorsifixed, ca. 2 x 1 mm, ovate, apex mucronate and ends on 2-fid thick walled hairs; margin densely ciliate. Ovary ca. 4 x 1.5 mm, elliptic, compressed, glabrous, apex papilose; styles ca. 7 mm, cylindric, gradually tapering, the middle sparsely hairy; hairs unicellular. Capsule ca. 9 x 4.5 mm in size, ellipsoid, compressed, narrowed at both ends, glabrous, brown, shiny. Seeds ca. 3.5-4 mm across, ovoid, compressed, covered with long brownish hair.

Flowering period: July-September.

Herbarium specimens examined: NARA DESERT: *Rahmatullah Qureshi*, 505, 1149 (SALUBGH).

Field remarks: A common bush found in calcarious soils.

Distribution: Endemic to Southern Pakistan (Sindh, Baluchistan) and SW India i.e. Rajasthan, Rann of Kutch, Saurastra (Kamal & Ghafoor, 1988).

2. Barleria

L., Sp. Pl. 636 (1753); Gen. Pl. ed. 5. 283 (1754); Benth. & Hook. f., Gen. P1. 2: 1091 (1876); Clarke in Hook. f., Fl. Brit. Ind. 4: 482 (1884).

Armed or unarmed herbs or shrubs. Leaves petiolate to almost sessile, simple, entire, cystoliths present on both sides. Flowers often large and showy yellow, blue-purple or white, in axillary or terminal spikes; bracts entire or serrate, often spinescent or ciliate rarely absent; bracteoles linear or absent. Calyx deeply 4-lobed; lobes imbricate, unequal, narrow, occasionally spit tipped. Corolla infundibuliform or salver-shaped; tube slender, straight incurved; limb subequally 5-lobed or obscurely 2-lipped, upper lip 4-lobe lobes patent, imbricate, rounded or ovate. Stamens 4, all fertile and ± exserted, or 2 posterior ones staminodial and partly sterile or absent, anthers dithecous, oblong, muticous, emucronate. Ovary oblong-conical, 2 or 4-ovuled; style thickened at apex; stigma minutely 2-fid or entire. Capsule ovoid-oblong elliptic-obovoid, flattened, ± stipitate, pointed above, seminiferous below the middle, 2 or 4-seeded. Seeds flat, ovate or orbiculate, covered with hygroscopic hairs, rarely glabrous, on hooked funicles.

A genus of nearly 230 species mostly distributed in tropical and subtropical countries of the Old and New World.

Barleria acanthoides Vahl, Symb., 1: 47 (1790); Stewart in Ali and Nasir, Ann. Cat. Vasc. Pl. W. Pak. & Kashm., 674 (1972).

Syn: *Barleriacanthus acanthoides* Oerst.

Prickly, rigid, perennial subshrub to shrub up to 2 ft high. Stem much brached, branches hoary or velvety-tomentose. Leaves ca. 7 x 3 cm in size, obovate-oblong or obovate, obtuse or emarginated, mucronate, base attenuate, margin entire, pubescent. Spines up to 17 mm long, inter-petiolar, simple or branched. Flowers solitary or in 1-8-flowered, unilateral racemes. Bracts straw-colored, spinous-tipped. Outer calyx-lobes reticulately veined. Corolla infundibuliform; tube ca. 7.5 cm long, pubescent outside. Filaments bearded at the base only. Capsules up to 1.5 cm long, ellipsoid, glabrous, brown. Seedsca. 3.5 mm in size, broadly ovoid, compressed, silky hairy, brownish.

Flowering & Fruiting Period: October-March.

Herbarium specimens examined: NARA DESERT: *Rahmatullah Qureshi,* 1426 (SALUBGH).

Field remarks: A less common herb found in sandy soils.

Distribution: Egypt, Sudan, Ethiopia, Somalia, Djibouti, Pakistan (NWF Punjab, Sind) and N.W. India (Malik & Ghafoor, 1988).

51. Bignoniaceae

Trees or climbing shrubs. Leaves opposite, 1-3-(odd) pinnate; leaflets (sub)opposite, sessile or petiolate, penninerved, margin entire or variously toothed, lower or the lateral leaflets smaller and often oblique, terminal one large and symmetric; rachis stout, usually swollen at the base. Flowers zygomorphic, 5-merous, bisexual, aggregate in terminal or axillay, corymbose, cymose or racemose panicles, bracteate. Calyx campanulate, cupular or spathaceous; lobes obscure or 5, valvate, equal or 2-labiate, sometimes reduced to teeth. Corolla tubular or funnel-form; tube cylindric, ventricose from above the middle; lobes 5, equal or 2-labiate, imbricate, usually crisped. Stamens 4, inserted between the cylindric and ventricose part of tube, didynamous, with(out) a rudimentary; anthers 2-celled, parallel or divericate, dehiscence longitudinal. Ovary superior, 2-locular; ovules $\propto$, 2-α-seriate; style elongate; stigma 2-lamellate, equal. Disc fleshy, annular or cupular. Fruit an elongate, 2-valved, capsule, terete, compressed or 4-gonous; epicarp thick; septum cylindric or narrow, deciduous with placentae. Seeds compressed, discoid or 3-gonous, often prominently winged, exalbuminous.

A family of ca. 120 genera and ca. 800 species, mostly tropical South America, but also in the tropics and subtropics of the old world. In Pakistan, it is represented by 3 native genera and 13 frequently cultivated species (Nasir, 1979).

Tecomella

Seeman, Ann. Mag. Nat. hist. ser. 3,10: 30 (1862); Cooke, Fl. Bomb. Pres. 2: 328 (1905); Talbot, For. Fl. Bomb. Pres. & Sind, 2: 307 (1911); Jafri in Fl. Kar. 309 (1966); Nasir in Fl. Pak. 131: 4 (1979).

Small trees. Leaves opposite, simple. Calyx and corolla 5-lobed; throat of corolla wide. Stamens 4,exserted, glabrous; anthers 2-locular. Disk cupular. Capsule linear, dehiscing longitudinally. Seeds with a broad transverse hyaline wings.

A monotypic genus found in Arabia, S. Iran, E. Afghanistan and Indo-Pak (Nasir, 1979).

Tecomella undulata (Sm.) Seem. in Ann. & Mag. Nat. hist. ser. 3. 10: 30 (1862); Stewart in Ali and Nasir, Ann. Cat. Vasc. Pl. W. Pak. & Kashm., 669 (1972); Yasin J. Nasir in Nasir & Ali, Fl. Pak., 131: 5 (1979).

Syn: *Bignonia undulata* Sm. Exot. Bot. 1: 35 (1805); *Tecoma undulata* G. Don, Gen. Syst. 4: 223 (1838).

A medium sized tree or small shrub up to 3 m high. Stem hard, woody at the base, fairly branched; branches drooping, glabrous, minutely stellate hairy when young. Leaves simple, ca. 5-12 x 1.5-2.5 cm in size, usually sub-opposite, oblong, apex obtuse-emarginate, slightly cuneate at the base, margins entire -more or less undulating; midrib prominent below; petioles ca. 1-2 cm long, terete, glabrous. Inflorescence in few-flowered cymes, arranged in racemose manner, terminating short lateral branches; pedicels ca. 1-2 cm long, terete, minutely stellately hairy, pedicel-bases persistent. Calyx ca. 8-9 mm long, campanulate, greenish to yellow; lobes broadly ovate, apex obtuse-mucronate, often recurved, with few black dots outside. Corolla ca. 5-7 cm long, yellow or orange-yellow, campanulate; lobes 5, subequal. Stamens exserted; filaments glabrous, anterior ca. 3-3.5 cm long; posterior ca. 2.5 mm long. Ovary on a yellow annular disc; style up to 4 cm long, glabrous; stigma 2-lamiate, elliptic, ca. 3.5 mm long.

Flowering period: January-April.

Herbarium specimens examined: NARA DESERT: *Rahmatullah Qureshi*, 719, 1272 (SALUBGH).

Field remarks: A rare small tree herb found calcarious soils.
Distribution: Arabia, S. Iran, E. Afghanistan, Pakistan & NW. India (Nasir, 1979).

52. Asteraceae

Annual or perennial, erect, ascending, prostrate or climbing herbs, shrubs, rarely trees, sometimes lacticiferous. Leaves alternate or opposite, at times both, simple or compound, peninerved, petiolate or subsessile, exstipulate. Inflorescence a centripetal capitulum of several (rarely 1 or a few) florets, aggregated on the flattened peduncle (receptacle), surrounded by bracts. Florets (4-)5-merous, bisexual or unisexual, rarely neuter, tubular and regular, radiate or ligulate and aggregated in any one of the following types: (i) all tubular and bisexual-capitula homogamous, disciform; (ii) all tubular, outer female, inner bisexual-capitula heterogamous, disciform; (iii) outer radiate and female, inner tubular and bisexual-capitula heterogamous, radiate; (iv) all ligulate and bisexual-capitula homogamous, ligulate; and exceptionally, (v) all unisexual-capitula monoecious. Capitula 1 to several, variously aggregated in simple clusters, cymes, racemes, panicles or in glomerules. Bracts (Phyllaries) usually α, multi-seriate to uni-seriate, herbaceous or scarious, sometimes spinescent, spreading or reflexed after fruiting; bracteoles 0 or reduced to paliaceous scales or bristles on the receptacle. Calyx-tube adnate to ovary; limb of scales, hairs or awns, persistent in fruit (pappus), rarely 0. Corolla regular, tubular-campanulate or obliquely extended into a 2-4-toothed strap (radiate) or a 5-toothed strap (ligulate). Stamens (4)5, epipetalous, alternating with corolla-lobes, usually included; filaments filiform, free; anthers 2-celled, oblong-linear, parallel, usually flattened, connate by margins, base entire or tailed; connectives usually produced. Ovary inferior, narrowly oblong-linear, 1-locular; ovules 1, basal; style running through anther-tube, usually 2-fid above; style-arms acute, obtuse or truncate, glabrous or variously pubescent. Disc minute, annular, epigynous. Achenes hardly enlarged, compressed, ribbed or angled, variously pubescent, rarely rostrate. Seed 1, erect, exalbuminous; testa membranous; cotyledons plano-convex; radical short.

A large family of about 1000 genera and 8000 species distributed throughout the world (Cooke, 1903).

Key to genera:

1a. Involucral bracts spinescent or aristate:
2a. All florets bisexual; achenes elongate, subterete or 4-gonous........................**1. Echinops**
2b. Ray florets neuter; disc florets bisexual; achenes striate and
pitted between ribs... **2. Amberboa**
3a. Heads dioecious (male & female separate, but not on the same plant)
or polygamo-dioecious; purple or greenish yellow:
4a. Involucral bracts of female heads united into an ovoid, 2-beaked,
herbaceous utricle clothed with hooked spines. Florets apetalous...................**3. Xanthium**
4b. Involucral bracts of female heads free, spine-tipped.
Florets petaliferous...**4. Cirsium**
3b. Heads bisexual; florets orange yellow..**5. Carthamus**
1b. Involucral bracts neither spinescent, nor aristate:
5a. Heads homogamous, either male or female or bisexual:
6a. Achenes columnar, truncate at both ends; heads cylindric...............................**6. Launaea**

6b. Achenes compressed at both ends; heads ovoid, campanulate or cylindric........**7. Sonchus**
5b. Heads heterogamous, outer florets usually female, inner ones bisexual or male:
7a. Pappus absent... **8. Eclipta**
7b. Pappus present:
8a. All florets yellow. Plants with offensive odour. Pappus 1-2-seriate...................**9. Iphiona**
8b. Florets pinkish or white. Plants without odour. Pappus 1-seriate....................**10. Pluchea**

1. Echinops

L., Sp. Pl. 814 (1753); Hook. f., Fl. Brit. India 3: 385 (1881); Cooke, Fl. Pers. Bomb. reprint ed. 2: 112 (1905); Shetty & Singh, Fl. Ind. in: Fl. Rajasthan, 1: 408 (1987); Jafri in Fl. Kar., 342 (1966).

Thistle like herbs more or less albo-tomentose. Leaves alternate, pinnatifid, spinous. Heads in globose involucrate solitary balls, blue or white, sessile or shortly stipitate on a common receptacle, 1-flowered; flowers bisexual, all fertile. Involucre oblong; bracts 1-seriate, rigid, pungent or some spinescent, outer shorter, inner spathulate, innermost linear or lanceolate, sometimes all connate into a tube with 1 rigid spine on the outer side. Receptacle minute. Corolla regular; tube slender; limb cylindric, 5-partite with slender segments. Filaments glabrous; anther-bases sagittate, the auricles connate; tails short, fimbriate or entire. Style arms thick and with a thick basal ring, at length more or less spreading. Pappus coroniform, of many short subpaliaceous free or more or less connate bristles. Achenes elongate, subterete or 4-gonous, villous (rarely glabrous).
A genus of about 70 species distributed S. Europe, N. Africa and Asia to Japan (Cooke, 1906).

Echinops echinatus DC. in Wight, Contrib., 24 (1834); Jafri in Fl. Kar., 342 (1966); Stewart in Nasir & Ali., Ann.Cat. Vasc. Pl. W.Pak. & Kashm., 742 (1972).

Annual rigid, spiny, white-wooly herb up to 130 cm high. Stem woody below, 7-13-angled, white-tomentose. Leaves simple, alternate, lanceolate, coriaceous, pinnatifid base decurrent, sometimes auriculate, subsessile, ca. 9-46 x 1-8 cm in size; lobes alternate-subopposite, triangular or oblong, spine-tipped, apex acute. Capitula 1-flowered, aggregated in dense, globose, axillary or terminal or leaf-opposed; subsessile or pedunculate glomerules. Receptacle minute; involucre tubular or oblong. Phyllaries 1-seriate; outer filiform, ca. 10 mm, barbellate; inner ones obovate, usually connate into a hard tube, ca. 12 mm, with unequal, dentate spiny tips, sometimes with 1 rigid spine. Florets white or bluish-white, disciform, bisexual, each floret with a single up to 4 cm long spine, with double spines of which one is 4 cm and other is 2 cm or with triple spines each about 4, 3.5 and 2.5 cm long. Pappus 1-seriate, equal, barbellate, ca. 2 mm. Corolla ca. 7 mm across; tube cylindric, ca. 5 mm, pubescent; lobes deeply cleft, oblong, ca. 6.5 x 1 mm, curled back, apex acute. Stamens 5, exserted; anthers linear, ca. 4.5 mm, base auricled, tails imbricate; connectives acute. Ovary linear, broadened above, ca. 6 x 1 mm in size, pubescent; style ca. 1.2 cm, shortly 2-fid; stigma stout, acute, with a thick, pubescent basal ring. Achenes elongated, more or less angled, ca. 5 mm long, densely silky, surrounded by the connate hardened inner involucral bracts.

Flowering period: August-October.
Herbarium specimens examined: NARA DESERT, *Rahmatullah Qureshi*, 1261, 1289 (SALUBGH).
Field remarks: A common herb found in sandy soils.
Distribution: India, Pakistan, Afghanistan (Jafri, 1966).

2. Amberboa

A. Less., Syn. Comp. 8 (1832), *nom. cons*; Jafri in Fl. Kar., 344 (1966).

Herbs, with usually dissected alternate leaves, minutely scabrid or so. Heads heterogamous, purple or violet. Bracts α-seriate, innermost narrow, acute; outer shorter, awned or spinescent. Receptacle densely bristly; outer flowers 1-seriate, neuter; disk flowers bisexual, fertile. Anther-bases sagittate, auricles connate, shortly tailed. Achenes striate and pitted between ribs. Pappus α-seriate, bristly, outermost gradually shorter.
A genus of about 4-5 species distributed in Indo-Pak subcontinent and Afghanistan (Jafri, 1966).
Amberboa ramosa (Roxb.) Jafri in Scientis III: 29 (1959); Jafri in Fl. Kar., 344 (1966). **Figure 4B**

Syn: *Centaurea divaricata* Wall. Cat. 2984 (1831), *nom. nud.*; *Carduus ramosus* Roxb., Fl. Ind. III:107 (1832); *Tricholepis procumbens* Wt. Icon. Pl. ind. Or. 3. pt. IV: 12. t. 1139 (1846); *Volutarella divaricata* Benth. in Benth. et Hook. Gen. Pl. 2: 476 (1873), *pro parte*; *Amberboa divaricata* (DC.) O. Kuntze, Rev. Gen. Pl. 1: 305 (1891); *Volutarella ramosa* (Roxb.) Santapau, P. Saur. 22 (1953); *Oligochaeta ramosa* (Roxb.) Wargenitz in Verroffent. Geobot. Inst. 37: 323 (1962).

Annual, forkedly branched, straggling or prostrate or sub-erect herb up to 75 cm high. Stem dichotomously branched, longitudinally striate, angled, glabrous or with a few, scabrous points, produced by the dried glands. Leaves ca. 1.5-5 x 0.5-2 cm in size, sessile, lobed to subentire, often with basal rosette; lower leaves oblong or obovate or lyrate; upper sinuately pinnatifid, entire or toothed with mucronate lobes, crisped, pubescent. Heads ca. 1-2.5 x 0.5-1.5 cm in size, ovoid-oblong; peduncles grooved, pubescent, scabrous with a few, small, foliaceous bracts. Involucral bracts upto 1.7 cm long, elliptic-oblong, spinescent; outer bracts smaller, recurved; inner longer, erect. Corolla pale-purple, ca. 1.2-1.5 cm long; lobes ca. 3 mm long, linear, apex acute. Achenes ca. 5 x 1.7 mm in size, acutely angled, pitted between the angles, base narrow. Pappus hairs numerous, unequal, silvery-brown, upto 10 mm long; inner 3 or 4 flattened, upto 15 mm long.
Flowering period: January-February.
Herbarium specimens examined: NARA DESERT: *Rahmatullah Qureshi*, 265, 632, 636, 1035, 1255 (SALUBGH).
Field remarks: A rare herb found in calcarious soils.
Distribution: India, West Pakistan and Afghanistan (Jafri, 1966).

3. Xanthium

L., Sp. Pl. 987 (1753); Hook. f., Fl. Brit. India 3: 303 (1881); Cooke, Fl. Pers. Bomb. reprint ed 2: 93 (1905); Jafri in Fl. Kar., 339 (1966).

Annual coarse rough herbs, unarmed or with 3-fid spines. Leaves alternate, toothed or lobed. Heads monoecious (female & bisexual), axillary, the bisexual in the upper axils, globose, many-flowered, sterile, tubular, 5-toothed; the femal 2-flowered, apetalous, fertile. Heads of bisexual flower: Involucre short; bracts few, 1-2-seriate, narrow. Receptalcle cylindric, with hyaline, cuneate or linear-spathulate paleae enclosing the flower. Corollas tubular, enlarged into a 5-toothed limb. Anthers distinct, bases obtuse, apex mucronate, inflexed; filaments monadelphous. Style undivided, slender, slightly dilated at the apex, achenes rudimentary. Heads of female flower: Involucral bracts united into an ovoid 2-beaked herbaceous 2-celled utricle (with 1 flower in each cell), enlarging in fruit, with hooked bristles and with sometimes a few small free outer bracts. Corolla 0. Style-arms exserted from the beaks of the involucre. Pappus 0. Achenes completely enclosed in the indurated cells of the enlarged involucre (like seeds in a capsule), compressed, obovoid, thick.
About 4 species of probably American origin (Cooke, 1905).

Xanthium strumarium L., Sp. Pl. 987 (1753); Jafri in Fl. Kar., 339 (1966); Stewart in Ali and Nasir, Ann. Cat. Vasc. Pl. W. Pak. & Kashm., 801 (1972). **Figure 6A**
Syn: *X. indicum* Koenig ex Roxb. Fl. Ind. 3: 601 (1832).

An annual erect unarmed subshurb up to 3 m high. Stem hard, woody at the base, herbaceous above, stout, sulcate; tender parts spinulous, hispid-pubescent. Leaves ca. 4-12 cm across, shallowly 3-5-lobed, deltoid-ovate or suborbicular, apex acute, base subcordate or acute, unequally coarsely dentate, 3-nerved from the base, chartaceous, hispidulous-scabrid above and below ; petiole ca. 4-10 cm long. Capitula in terminal and axillary racemes; barren heads at the apex and fertile heads axillary; receptacle ca. 7 mm long. Phyllaries (sub)biseriate, ovate, acute, ca. 12 x 1 mm in size, with 2 erect or somewhat curved beaks, pubescent, thickly clothed with usually hooked prickles, hard, tough, 2-celled. Florets of male flowers numerous. Paleae spathulate-rhomboid, acute, ca. 2 x 0.7 mm in size, pubescent. Corolla-tube ca. 1 mm across, campanulate; lobes 6, ca. 1.5 mm long, free near apex, acute, glandular-pubescent without. Stamens exserted; filaments connate; anthers ca. 1.3 mm, free, erect, spreading after anthesis, oblong, apex acute, incurved. Ovary sterile; style rudimentary, ca. 1.5 mm. Female capitula solitary, ovoid, 2-flowered, ca. 8 mm long. Achenes ca. 12 mm long, oblong-ovoid, compressed, glabrous, enclosed in prickly glandular-hispidulous envelope, forming an utricle; persistent style-base tapering to an acute point.
Flowering period: September-December.
Herbarium specimens examined: NARA DESERT: *Rahmatullah Qureshi*, 823 (SALUBGH).
Field remarks: A less common herb found in sandy soils.
Field remarks: A common weed of waste places.
Distribution: India, Pakistan, Ceylon and warmer parts of the world (Jafri, 1966).

4. Cirsium
Mill.

Perennial or biennial prickly herbs, simple or few-branched. Leaves alternate, serrate-dentate or pinnatifid, often very spinescent. Heads large, heterogamous, many flowered, discoid, long-peduncled, florets uniform, bisexual. Involucre globose or ovoid, bracts in many series, appressed, erect, spreading or recurved, prickly tipped or unarmed. Receptacle flat or convex, densely bristly. Corolla slender, purple, pink, white or yellow, tube slender, limb equal or obliquely 5-lobed. Filaments hairy or rarely glabrous; anthers sagittate at base, auricles connate, lacerate. Style branches short, obtuse. Achenes glabrous, obovoid, obtusely 4-angled, glabrous, smooth or ribbed, truncate apically, base areolate, nearly straight. Pappus copious, bristles in many series, slender, feathery, deciduous with a basal ring.

There are about 250 species distributed in Eurasia, northern and eastern Africa and North America.

Cirsium arvense (L.) Scop. Fl. Carn. ed. 2. 2: 126 (1772); Stewart in Nasir & Ali., Ann. Cat. Vasc. Pl.W. Pak. & Kashm., 733 (1972).

Syn: *Serratula arvense* L. Sp. Pl. 820 (1753); *Cnicus arvensis* (L.) Hoffm. Deutschl. Fl. ed. 2. 2: 130 (1804).

Erect, thistle like herbs with subterranean, stolon and cottony stem. Leaves sessile, ca. 8-15 x 2-4 cm in size, oblong-lanceolate, pinnatifid to toothed, amplexicaul at base, spinescent on margins, white-tomentose beneath. Heads peduncled, solitary or in corymbs, dioecious, purple; male 1 cm in diam., globose with closely arranged; females ca. 2-2.5 x 1.3-5 cm, ovoid with loosely arranged involucral bracts. Receptacles flat, with filiform, bristly pales. Achenes smooth. Pappus hairs plumose, dull-brown.

Flowering & Fruiting Period: January-May.

Herbarium specimens examined: NARA DESERT: *Rahmatullah Qureshi*, 1425 (SALUBGH).

Field remarks: A common weed of sandy soils.

Distribution: A very common on wet or poorly drained soil (Stewart, 1972).

5. Carthamus

L., Sp. Pl. 830 (1753); Hook. f., Fl. Brit. India 3: 386 (1881); Shetty & Singh, Fl. Ind. in: Fl. Rajasthan, 1: 400 (1987).

Annual herbs. Stem woody, branched. Leaves simple, alternate or opposite, sessile, margins spiny. Involucral bracts multi-seriate; outer bract foliaceous. Heads campanulate, orange-yellow. Florets bisexual, tubular, homogamous. Corolla-tube campanulate; lobes 5. Stamens 5, syngenecious. Ovary obovoid; styles elongate. Achenes 4-angled. Pappus absent. Three species are recorded from Pakistan (Stewart, 1972).

Carthamus oxycantha Bieb. Fl. Taur. Cauc. 2: 283 (1808); Stewart in Nasir & Ali., Ann.Cat. Vasc. Pl. W.Pak. & Kashm., 728 (1972). **Figure 4C**

Thistle-like erect, annual herb up to 2 feet high. Stem corymbosely branched, circular, glaucous. Leaves alternate to somewhat opposite, sessile, oblong or oblanceolate, base semi-amplexicaul, apex acute, margin spinulose-toothed, curved toward midrib; cauline leaves ca. 7-14 x 2.5-3 cm in size; ramal leaves ca. 1-3 x 0.5-1.2 cm in size, turned yellow after drying; each leaf bears 17-20 spines on margin; spines ca. 5-20 mm long. Involucre 5-seriate, 14-22 in number, ca. 1.5-2.5 x 0.5-1.5 cm in size; outer bracts foliaceous, ovate-oblong, exceeding the head, white below the contracted portion, green above it, spine-tipped, spine yellow, ca. 10-18 mm long; inner most lanceolate. Heads ca. 1.5-2 cm long, somewhat campanulate, orange-yellow. Florets bisexual, tubular, homogamous, yellow. Corolla-tube ca. 18 mm, campanulate; lobes 5, ca. 8 mm, narrowly lanceolate, apex acute-obtuse, margin entire, glabrous. Stamens epipetalous, ca. 12 mm long. Ovary ca. 2 mm, obovoid; style elongate. Achenes ca. 7-10 x 4-5 mm in size, obovoid, apex truncate, smooth, shining. Seeds white gray to dark brown with black streaks and spots.

Flowering & fruiting period: May-August.

Herbarium specimens examined: NARA DESERT: *Rahmatullah Qureshi*, 857, 891, 911, 1027 (SALUBGH).

Field remarks: A common weed mostly found in sandy soils.

Distribution: Baluchistan, NWFP; Swat, Lower Hanza, Punjab, Jammu (Stewart, 1972).

6. Launaea

Gassini, Dic. Sci. Nat. 25: 61, 321 (1822); Cooke, Fl. Pers. Bomb. reprint ed. 2: 112 (1905); Jafri in Fl. Kar., 347 (1966).

Perennial glabrous herbs usually with yellow juice. Leaves mostly radical, sinuately-lobed or pinnatifid; margin often with cartilaginous or subspinous teeth. Heads pedunculate or subsessile, solitary, fascicled, racemose or peniculate, homogamous. Flowers all ligulate, yellow. Involucre campanulate or cylindric. Bracts (-seriate, imbricate, often with scarious margins; innermost bracts subequal; the outer gradually shorter, the outermost often very short. Receptacle flat, naked. Corollas ligulate, truncate and 5-toothed at the apex. Anther-bases sagittate with acute or shortly setaceo-acuminate, auricles. Style-arms slender. Pappus copious; hairs (-seriate, simple, very slender, white, a few inner sometimes longer and stronger, all connate at the base into a deciduous ring. Achenes narrow, usually columnar, not contracted at the base or apex, truncate4 at both the ends, sometimes winged, 4-5-ribbed.

A genus of about 20 species distributed in Europe (Mediterranean region), East Indies, Africa and the Canary Island (Cooke, 1906).

Key to species:

1a. Heads terminal, on paniculately branched flowering stem.
Achenes 6-7 mm long..**1. L. resedifolia**

1b. Heads in clusters on the nodes of branches. Achenes 2-4 mm long...... **2. L. procumbens**

1. Launaea resedifolia (L.) Kuntze, Rev. Gen. Pl. 1: 351 (1891); Stewart in Nasir & Ali., Ann. Cat. Vasc. Pl.W.Pak. & Kashm., 763 (1972); Jafri in Fl. Kar., 348 (1966)

Syn: *Scorzonera resedifolia* L., Sp. Pl. 1198 (1753); *Zollikaferia chondrilloides* DC. Prodr. 7: 183 (1838); *Launaea chondrilloides* (DC.) Hook. f., Fl. Brit. Ind. 3: 415 (1881).

Perennial herbs with mily sap up to 2 ft high. Stems dichotomously branched, terete, sulcate, woody at the base. Radical leaves sub-rosulate, ca. 4-15 x 2-5 cm in size, lanceolate, runcinate-pinnatifid, spinulose-toothed; cauline leaves sessile, semi-amplexicaul, narrow,

pinnatifid; pedunculate leaves ovate or deltoid, entire, bract-like. Heads on sub-divaricate, bracteates peduncles, 1-2 cm long, yellow. Outermost involucral bracts 3 to 4-seriate, much smaller and deltoid-ovate, inner ones oblong-lanceolate with ciliolate tips. Achenes 6-7 mm long, cylindric, 3 to 4-toothed at the base, ribbed, scabrulous on ribs. Pappus white.
Flowering & Fruiting Period: Almost throughout the year.
Herbarium specimens examined: NARA DESERT: Rohri hills near Shadi Shaheed, *Rahmatullah Qureshi*, 110, 970, 1000 (SALUBGH).
Field remarks: A rare species of sandy soils.
Distribution: Sindh, Baluchistan, Punjab (Stewart, 1972).

2. Launaea procumbens (Roxb) Ramayya & Rajagopal., Kew Bull. 23(3): 463 (1969); Stewart in Ali and Nasir, Ann. Cat. Vasc. Pl. W. Pak. & Kashm.,762 (1972). **Figure 5B**

Syn. *Chondrilla nudicaulis* L., Mantissa, 273 (1767); *L. fallax* (J. & S.) O. Ketze., Rev. Gen. 350 (1891); *L. procumbens* (Roxb.) Amin in Tackholm, Students Flora Egypt 84 (1956); *L. nudicaulis* (L.) Hk.f., Fl. Brit. Ind. 3: 416 (1881); *Microrhynchus fallax* (J. & S.) Ill., Pl. Or. 3: 106 (1847-50); *Prenanthes procumbens* Roxb.Fl. Ind. 3:404 (1832); *Paramicrorhynchus procumbens* (Roxb.) Khir., Fl. U.R.S.S. 29:237 (1964).

A prostrate perennial herb upto 50 cm high. Stem branched from the base, slender, glabrous, greenish with yellow latex. Leaves ca. 5-15 x 2-3 cm in size, mostly rosette; lower leaves obovate-oblong, pinnatifid with rounded or obtuse segments, spinulose on the margins with white cartilaginous teeth; cauline leaves distant, few sessile, narrowly oblong, pinnatifid. Flowering stem decumbent or spreading, irregularly branched. Heads ca. 1 cm long, cylindric, remotely subracemose along the branches. Involucral bracts 3-seriate, ca. 0.5-1.2 x 0.2-2.5 cm long, all with membranous margins; the outer very short, ovate, apex acute, with strong midrib, subcordate; the inner most thrice as long as the outer, linear-lanceolate, apex subacute, longer than pappus, the midrib thickened in front. Flowers yellow, all ligulate. Pappus-hairs subequal, ca. 10 cm, soft, white, copious. Achenes ca. 2-4 mm long, polymorphous; inner sometimes as if composed of 4 thick ribs; outer slightly curved and flattened with a thick ventral and several thick dorsal ribs, smooth or obscurely uneven.
Flowering period: January-May.
Herbarium specimens examined: NARA DESERT: *Rahmatullah Qureshi*, 110, 970, 1000 (SALUBGH).
Distribution: India, W. Pakistan, westward to the Atlantic (Jafri, 1966).

7. Sonchus

L., Sp. Pl. 794 (1753); Jafri in Fl. Kar., 346 (1966); Matthew, Fl. Tamil Nadu Carnatic, 2: 814 (1983).

Annual or perennial herbs with milky juice. Leaves radical or alternate, the cauline often amplexicaule and auricled, entire, toothed or pinnatifid, the margins sometimes rigidly ciliolate or subspinous. Heads terminal, irregularly subcorymbose, umbellate or paniculate, rarely solitary, yellow, homogamous. Involucre ovoid, campanulate, or cylindric, often dilatem, thickened or conic at the base; bracts ∞-seriate, imbricate, herbaceous, the outer smaller. Receptacle flat, naked. Corollas all ligulate, truncate and 5-toothed at the apex. Anther bases sagittate, the auricles shortly setaceo-acuminate. Style branches slender. Achenes ovoid, obovoid, or ellipsoid, more or less compressed, ribbed, slightly contracted at the base and apex, glabrous, smooth or transversely rugose. Pappus copious; hairs ∝-seriate, very slender, simple, usually pure white, united at the base into a deciduous ring.

Key to species:
1a. Leaf-auricles acute, spreading. Achenes transversely tuberculate..............**1. S. oleraceus**
1b. Leaf-auricles rounded, appressed to the stem. Achenes not tuberculate............**2. S. asper**

1. Sonchus oleraceus L.; Sp. Pl. 794 (1753); Jafri in Fl. Kar., 347 (1966); Stewart in Nasir & Ali., Ann. Cat. Vasc. Pl. W. Pak. & Kashm., 785 (1972). **Figure 5D**

Annual, upto 180 cm tall. Stem erect, branched. Leaves alternate, crowded at the base, upper leaves mostly unlobed, acute-apiculate, amplexicaule, clasping the stem, ca. 6 x 2.5 cm, lower deeply cut into 1-3 lobes each side, obtuse or rounded, ca. 8 x 4 cm. Heads peduncled, peduncle up to 6 cm long; involucral bracts 3-seriate, linear, linear, acute, margin aculeate. Florets irregular, bisexual, all ligulater, homogamous. Corolla surrounded by a pappus of long hairs; corolla tube ca. 3 mm long, having unicellular hairs. Corolla extended on one side into a long strap like structure called ligulate, which surrounds stamens and pistil, ca. 4 x 1 mm. Filaments 5, borne on corolla tube; anthers ca. 1.8 x 0.6 mm, united to form a tube round style. Ovary ca. 1 x 0.5 mm; style with setose hairs, ca. 1.5 mm long, branches into two stigmas, stigmas covered with protuberances (papilae), ca. 1mm across. Achene ca. 2.3 x o.6 mm in size. Pappus longer than cypsela.
Flowering Period: March-April.
Herbarium specimens examined: NARA DESERT: *Rahmatullah Qureshi*, 1424 (SALUBGH).
Field remarks: A common herb found in moist habitat.
Distribution: Europe, S.W. Asia, N. Africa (Matthew, 1983).

2. Sonchus asper (L.) Hill. Herb. Brit. 1: 47. t. 34 (1769); Stewart in Nasir & Ali., Ann. Cat. Vasc. Pl.W.Pak. & Kashm., 784 (1972).
Syn: *S. oleraceus* L. var. *asper* L., Sp. Pl. 794 (1753).

Erect, glabrous, annual herbs up to 2 ft high. Basal leaves petioled in rosette, spathulate-oblong, entire or pinnatipartite, spinously dentate; cauline leaves semi-amplexicaule with rounded basal auricles. Heads in terminal, sub-umbellate corymbs, ca. 1-1.2 cm long, yellow. Involucral bracts 2-3 seriate, lanceolate. Achenes obovoid-oblong, 3-ribbed on lateral faces. Pappus hairs white.
Flowering & Fruiting period: February-May.
Herbarium specimens examined: NARA DESERT: *Rahmatullah Qureshi*, 1423 (SALUBGH).
Field remarks: A common herb found in moist habitat.
Distribution: Wide spread in cooler climates and many tropical countries (Jafri, 1966).

8. Eclipta

(L.) L., Mant. 2: 286. (1771); Jafri in Fl. Kar., 340 (1966).

Annual or perennial herbs, diffuse or erect, branched, more or less strigose. Leaves opposite. Heads small, axillary or terminal peduncled, heterogamous, rayed; ray-flowers ♀, sub-2-seriate, fertile or sterile, white or yellow; disc-flowers bisexual, fertile. Involucre hemispheric or broadly campanulate; bracts sub-2-seriate, herbaceous, subequal or the inner shorter. Receptacle flat or somewhat convex, the paleae narrow, enclosing several flowers. Corollas of ♀ flowers ligulate, spreading, ligules small, narrow, entire or 2-toothed; the corollas of bisexual flowers regular, tubular, the limb shortly 4-5-fid. Anther-bases obtuse, subentire. Style arms flattened, terminated by short or triangular obtuse appendages. Achenes of the ray narrow, triquetrous, often empty, those of the disc laterally subcompressed, top entire, toothed or 2- aristate. Pappus absent.

Eclipta prostrata (L.) L., Mantiss. 286 (1771); Jafri in Fl. Kar., 340 (1966); Stewart in Nasir & Ali., Ann. Cat. Vasc. Pl.W.Pak. & Kashm., 743 (1972); Jafri in Fl. Kar., 340 (1966). **Figure 4D**

Syn: *Verbesina alba* L., Sp. Pl., 902 (1753); *V. prostrata* L. Sp. Pl. 902 (1753); *Eclipta erecta* L. Mantissa, 216 (1771); *E. alba* (L.) Hassk., Pl. Jav. Rar., 528 (1848); F. B. I., 304 (1881).

Annual erect or prostrate herb, up to 60 cm high. Stem much branched, strigose, often rooting at the lower nodes. Leaves ca. 2-7 x 1-2.5 cm in size, sessile, elliptic, oblong or lanceolate, subentire or undulate, with strigose hairs, apex obtuse to subacute. Heads ca. 0.5-1 cm in diam, white, solitary or two together on unequal axillary peduncles. Involucre bracts ca. 9, ovate, acute, with strigose hairs. Ray flowers ca. 3.4 mm long, ♀, ligulate, disk flower ca. 1.5 mm long, bisexual, tubular Corolla tube ca. 1.5x1 mm, 4-lobed, pappus 0. Stamens 4, ca. 0.8 mm long, anther-bases obtuse, subentire; style arms flattened, terminated by short or triangular obtuse appendages. Pistil ca. 1.2 mm. Achenes cuneate, compressed, tuberculate with thickened margin. Pappus absent or of minute hairs.

Flowewring Period: Almost throughout the year.

Herbarium specimens examined: NARA DESERT: *Rahmatullah Qureshi*, 1437 (SALUBGH).

Field remarks: A less common herb found in wetland habitat.

Distribution: Fairly distributed in most warm countries (Jafri, 1966).

9. Iphiona

Cass., Bull. Sci. Soc. Philom. Paris 1817: 153 (1817).

Inula L., Diss. Dass. 25 (1747).

Herbs, rarely shrubs. Leaves radical and alternate. Heads at the end of the branches, solitary, corymbose or panicled, heterogamous, rayed (rarely disciform); ray-flowers females, fertile, α-seriate; disk-flowers bisexual, tubular. Involucre hemispheric or campanulate (rarely ovoid); bracts α-seriate, imbricate, the inner usually rigid and narrow, the outer herbaceous, the outer most often foliaceous. Receptacle flat or tumid, pitted or areolate. Corollas of female flowers ligulate, the ligules long, short or minute, yellow or white, 3-toothed; corollas of the bisexual flowers regular, tubular, yellow, the limb elongate, slightly enlarged, 5-toothed. Anther-bases sagittate, with rather small auricles; tails long, simple or branched. Style-arms of bisexual flowers linear, slightly flattened, broader upwards, obtuse. Achenes subterete, ribbed. Pappus-hairs few or many, 1-2-seriate, rather short, smooth, scabrid or bearded.

A genus of about The 11 species are distributed in Egypt, Israel, Jordan, Sudan, Ethiopia, Somalia, Saudi Arabia, Peoples Democratic Republic Yemen, Oman, Iran and Pakistan (Anderberg, 1985).

Iphiona grantioides (Boiss.) Anderb. Nordic J. Bot. 5(2): 184 (1985). **Figure 5A**

Syn: *Inula grantioides* Boiss. Daig., Ser.II(iii):14 (1856).

Shrubby perennial upto 60 cm in height. Stem stout, woody, much branched at the base, alternately arranged, angular; branches leafy, pale green. Leaves ca. 1-4.5 cm long, variously lobed; the lower leaves petioled, ca. 0.5-1.5 cm long, cuneate, shortly 3-lobed; the upper leaves sessile, linear, entire or 2-3-toothed at the apex, fleshy, glandular hairy, nerves obscure. Heads numerous, ca. 1.5-2.5 cm across, solitary, often with 1 or 2 leafy bracts; peduncle stout, glandular-pubescent. Involucral bracts narrowly linear-lanceolate, apex acute-obtuse, base cuneate-rounded, margin entire, ca. 6-16 x 2-4 mm in size, green, hairy and glandular. Disk florets ca. 10 x 1 mm in size, bisexual, yellow, tubular, acutely toothed. Corolla tube ca. 7.5 mm; lobes 5, ca. 1.5 mm, lanceolate, apex acute, margin entire. Stamens 5; filaments ca. 2 mm; anthers ca. 6 x 0.5 mm in size, narrowly oblanceolate, apex acute, base cuneate, yellow. Ovary ca. 3 x 0.5 mm in size. Ray florets: ligulate, few, yellow, 1-seriate, ca. 9.5 x 1 mm. Ovary ca. 3 x 0.5, obovate, pubescent; styles slender, ca. 4.5 mm; stigma bifid, ca. 2 mm. Pappus rigid, sacbrid, much longer than the achenes, yellowish at first, then turned in reddish. Achenes ca. 2.5-3 mm long, oblong, strongly ribbed, hispid.

Flowering period: November- December.

Herbarium specimens examined: NARA DESERT: Rahmatullah Qureshi, 898, 894, 904, 1173, 1230 (SALUBGH).

Field remarks: A common shrub found in hilly areas.

Distribution: W. Pakistan and S.E. Arabia (Jafri, 1966).

10. Pluchea

Cass., Bull. Soc. Philom. 31 (1817); Hook. f., Fl. Brit. India 3: 272 (1881); Cooke, Fl. Pers. Bomb. reprint ed 2: 81 (1905); Shetty & Singh, Fl. Ind. in: Fl. Rajasthan, 1: 428 (1987); Jafri in Fl. Kar., 335 (1966).

Perennial herbs or undershrub, often woody below, tomentose or glutinous. Leaves alternate, simple. Heads small, discoid, in terminal leafless corymbose, or large and subsolitary, heterogamous, disciform, white, yellow or lilac. Outer florets female, fertile, (-seriate, corolla filiform; inner florets bisexual, few, sterile. Involucre ovoid or broadly campanulate; bracts ovate or lanceolate, usually broad, dry, rigid. Receptacle flat, naked. Corollas of ray florets filiform, shorter than either styles, 3-fid or minutely toothed at the apex, those of the disk florets regular, tubular, with a slightly enlarged limb, 5-fid at the apex. Anther-bases sagittate, the cells tailed. Style-arms of disk florets filiform, entire or 2-fid.

Pappus-hairs slender, 1-seriate, free, or in the sterile achenes very many, more or less connate at the base.

A genus of about 30 species distributed in Tropical and subtropical countries of the world (Cooke, 1906).

Key to species:

1a. Outer involucral bracts oblong, rounded, obt.................................... **1. P. lanceolata**

1b. Outer involucral bracts ovate, acute..**2. P. wallichiana**

1. Pluchea lanceolata (DC.) Oliv. & Hiern, Fl. Trop. Afr., V:329 (1877); Jafri in Fl. Kar., 335 (1966); Stewart in Nasir & Ali., Ann. Cat. Vasc. Pl. W. Pak. & Kashm., 768 (1972). **Figure 5C**

Syn: *Berthelotia lanceolata* DC. Prodr., V:376 (1836).

An erect undershrub upto 90 cm tall. Stem much branched, woody below, herbaceous above, terete, slender, softly ashy-pubescent. Leaves sessile, ca. 1.5-5 x 0.5-1 cm, oblong or oblanceolate, apex obtuse to apiculate, base cuneate, margin entire, main nerves prominent, finely ashy-pubescent on both sides. Heads in compound corymbs, ca. 6-7 mm long, oblong. Florets campanulate, pinkish purple. Involucral bracts contracted at the mouth; outer bracts 2-5-seriate, ca. 2.5-4 x 1.5-2 mm in size, oblong, apex obtuse, silky-pubescent, tinged with purple outside at the apex; inner ones few, longer, linear, apex subacute, scarious. Receptacle flat, naked. Pappus-hairs connate at the base. Achenes minute, oblong.

Flowering period: December.

Herbarium specimens examined: NARA DESERT: *Rahmatullah Qureshi*, 208, 919, 1199, 1202 (SALUBGH).

Field remarks: A common shrub found in calcarious soils.

Distribution: India, W. Pakistan, Afghanistan and N. Africa (Jafri, 1966).

2. Pluchea wallichiana DC. Prodr. 5: 451 (1836); Stewart in Nasir & Ali., Ann. Cat. Vasc. Pl.W.Pak. & Kashm., 769 (1972).

Much branched, glabrous shrubs, up to 2 m high. Braches glandular pubescent. Leaves alternate, ca. 3-8 x 1.5-3 cm in size, broadly elliptic or obovate to oblong, base semi-amplexicaule, apex mucronate, margin dentate to entire, glandular pubescent. Heads 5-6 x 3-4 mm, in terminal, compound corymbs, white. Outer involucral bracts much shorter, ovate-lanceolate, glandular hairy outside, inner ones narrower, with a pencil of hairs at the tips. Achenes 1 mm long, 5-angled with fine hairs along the angles, brown. Pappus hairs uniseriate, dirty white, antrosely barbellate.

Flowering & Fruiting Period: November-March.

Herbarium specimens examined: NARA DESERT: Chotiari reservoir, *Rahmatullah Qureshi*, 1422 (SALUBGH).

Field remarks: A less common herb found in wetland habitat recorded only from Chotiari wetlands complex.

Distribution: Baluchistan (Pakistan) and India (Bandari, 1978).

IV. Liliopsida

53. Hydrocharitaceae

Annual or perennial, monoecious or dioecious (rarely 2-sexual), aquatic usually submerged herbs in fresh or saline water. Stem corm-like or slender, stoloniferous or not. Leaves exstipulate, sessile or petiolate, radical or cauline and spirally whorled, opposite or alternate, submerged, rarely partly emergent or floating, variously shaped; veins ± parallel, straight or curved. Flowers solitary or clustered, enclosed at first by a bifid sessile or long stalked pair of opposite united and spathe-like or free bracts, bisexual or unisexual, when unisexual then male usually more than 1, females solitary, actinomorphic or occasionally ± zygomorphic, mostly trimerous. Sepals 3, free, often green or petaloid, valvate. Petals 3 or 0, membranous, showy or minute and inconspicuous, imbricate or convolute. Male Flower: stamens (1-) 2 to numerous, uni or multiseriate, inner ones rarely staminodial, filaments more or less slender, free or united in pairs, rarely absent; anthers 2-thecous, dehiscing by parallel longitudinal slits; 1 or more pistillodes. Female Flowers: Ovary inferior, composed of 2-15 fused carpels, linear-lanceolate or ovate, generally protruding into a filiform beak or rostrum at the apex 1-celled with parietal placentation and numerous orthotropous to anatropous, erect to pendulous, bitegmic ovules; styles as many as carpels, entire or 2-lobed. Fruit ripening under water, lainear-lanceolate, ovate to globose, dry or pulpy, many-seeded, dehiscing irregularly. Seeds fusiform, elliptic or ovate or globose, glabrous, warty or spinose, exalbuminose; embryo upright.

A family of ca. 17 genera and 76 species, distributed in marine, saline or fresh waters in tropical, subtropical and temperate regions. In Pakistan, the family is represented by 6 genera and 6 species (Ghafoor, 1985).

Hydrilla

L.C. Rich. Mem. Cl. Sci. Math. Phys. Inst. France (Paris) 12(2): 9, 61, 73, t. 2 (1814); A. Ghafoor in Nasir & Ali., Fl. Pak., 169: 3 (1985).

Perennial, submerged, leaf-fresh water herb. Stem slender, branched. Leaves short, 3-4-nately whorls of 3-8 (-12), cauline, sessile, linear-lanceolate with conspicuous midrib, subentire or serrate. Flowers unisexual, minute, actinomorphic, solitary, witrhout nectaries. Male spathe ± sessile, solitary, axillary, flattened, globose, more or less mucronate. Male flower: shortly pedicellate; sepals , ovate-oblong to obovate, sacrious, green, imbricate; petals 3, linear, white; stamens 3, alternating with the petals; pistillode present. Female spathe sessile, cylindrical, apically 2-fid, membranous, rostrate. Female flowers: sepals as in the male, petals somewhat narrower than sepals; ovary 3-carpellate, shorter than the spathe, cylindrical to narrowly conical with filiform beak; styles 3, entire. Fruit cylindrical to narrowly conical, glabrous or softly muricate. Eeds 2-6, oblong-ellipsoidal.

A monotypic genus distributed to Eurasia and Africa to Australia; introduced into the New World (Ghafoor, 1985).

Hydrilla verticillata (L. f.) Royale, Ill. Bot. Himal. T. 376 (1893); Stewart in Nasir & Ali., Ann. Cat. Vasc. Pl. W. Pak. & Kashm., 95 (1972); A. Ghafoor in Nasir & Ali., Fl.Pak., 169: 3 (1985).

Syn: *Serpicula vertcillata* L. f., Suppl. 416 (1781); *Vallisneria verticillata* Roxb., Hort. Bengal. 71 (1814); *H. ovalifolia* Rich., l.c. 12 (2): 9, 96, t. 2 (upper part) (1814); *H. polysperma* Blatter in J. Proc. As. Soc. Beng. n.s. 5. 26: 336 (1931).

Cualescent, glabrous, profusely branched freshwater herb up to 3 m high. Leaves oblong-linear to lanceolate, (8-) 10-40 (-50) mm long, (1-) 2-3 mm wide, green, margin serrate-dentate, apex acute-apiculate; nodal scales 2, axillary, membranous with orange brown cilia on margin. Male spathe 1.25-1.50 mm long, bursting open to liberate male flower, beset with subulate appendages. Male flower: sepals reflexed, 1.5-3 mm long; petals linear-spathulate, equalling the sepals, patent or reflexed, obtuse; stamens with minute filaments. Female spathe c. 5 mm long, reddish-brown striped, bidentate. Ovary oblong, 3.4 mm long, with 1.5-10 cm long slender rostrum. Fruit ± terete. 5-7 mm long. Seeds dark brown, c. 2.5 mm long.

Flowering Period: October-January.

Herbarium specimens examined: NARA DESERT: *Rahmatullah Qureshi*, 1421 (SALUBGH).

Field remarks: A common herb recorded from Chotairi wetland complex.

Distribution: Gregariously found in ditches, pools, lakes, wet rice fields and slow-running water streams in S.E. Europe, Africa, Asia and Australia (Ghafoor, 1985).

54. Potamogetonaceae

Glabrous, aquatic herbs with fibrous roots, over wintering by winter buds or rhizomes. Leaves 2-ranked, alternate or opposite, alike or dimorphic (submerged leaves sessile or petiolate, usually translucent; floating leaves petiolate, usually opaque, lamina dilated). Stipular sheaths free, axillary, or adnate to the leaf base. Inflorescence an axillary or terminal, peduncled, ovoid or cylindrical spike, lax or with flowers irregularly whorled, submerged or emergent, enclosed by stipular sheaths when young. Flowers ebracteate, sessile, tetramerous, small, bisexual, hypogynous. Perianth 4-segmented (sometimes the segments have been treated as outgrowth from the connectives of the anthers). Stamens 4, basally adnate to the perianth segments; anthers 2-locular, extrorse, sessile. Carpels 4 (rarely 3-1), apocarpous, 1-celled with one campylotropous ovule; stigma sessile or subsessile. Fruitlet wall opening with a lid of the dorsal area. Seeds lacking endosperm; hypocotyl massive; seed coat free from fruitlet wall; embryo uniform or spiral.

A family of 2 closely related genera and over 90 species, mostly of fresh water habitat, chiefly in temperate regions and cooler climates; represented in Pakistan by a single genus.

Potamogeton

L., Sp. P1. 126. 1753. Gen. P1. ed. 5: 61 (1754); Hook. f., Fl. Brit. Ind. 6: 565 (1893); Bhandari, Fl. Ind. Desert, 361 (1978).

Leaves submerged or floating, alternate or imperfectly opposite, toothed or entire. Flowers ebracteate, small, tetramerous, actinomorphic, spicate in dense or interrupted ovoid or cylindric spikes (submerged water pollinated, emergent on a peduncle wind pollinated). Perianth segments concave, smooth, obovate. Stamens 4, sessile or sub-sessile. Fruitlets drupaceous, coriaceous, more or less compressed, one seeded, smooth or ornamented with warts, crests or short spines; embryo uniform or spiral.
About 90 species, mostly of fresh water habitat in temperate regions; about a dozen species are reported from Pakistan.

Potamogeton perfoliatus L., Sp. Pl. 126 (1753); Stewart in Nasir & Ali., Ann. Cat. Vasc. Pl. W. Pak. & Kashm., 30 (1972); A. Ghafoor in Nasir & Ali., Fl. Pak., 169: 3 (1985).

Rhizomatous aquatic. Stem branched or simple. Leaves submerged, sessile, lanceolate to orbicular, cordate-amplexicaul, size and shape variable 2-6 cm long, 1.5-2 (-2.5) cm broad, entire or minutely denticulate, rounded to obtuse at the apex, usually alternate below, opposite above, 2-7 (-13)-veined, thin, translucent. Stipules free, small, length variable, less than 1 cm long, soon evanescent. Spike 4-5 cm long, densely flowered. Fruitlets 2-3.5 mm long, 1.5-2 (-3) mm broad, beaked, ventral area convex, irregular or straight, beak more or less keeled.
Flowering Period: November-March.
Herbarium specimens examined: NARA DESERT: *Rahmatullah Qureshi*, 1420 (SALUBGH).
Field remarks: A common herb recorded from Chotairi wetland complex.
Distribution: Europe, Asia, Africa, N. America and Australia.

55. Arecaceae

Mostly monoecious, more rarely hermaphroditic or dioecious trees, shrubs and vines. Underground perennial rhizome sometimes present or more commonly the trunk above ground, erect or climbing. Stem usually single or bearing side branches, rarely forked or profusely branched, smooth, ringed or covered with remains of petiole bases and dead leaves. Leaves pinnately or palmately ribbed or divided, sometimes very large; blades mostly folded, with numerous longitudinal nerves splitting induplicately or reduplicately, sometimes sharp at the apex and prickly on the margins or midrib; petiole base persisting or not and often spreading about or sheathing the trunk, in some genera much expanded and basally enveloping terminal bud and then forming a "crownshaft". Inflorescence usually paniculate, enclosed by one or more bracts in bud condition. Bracts coriaceous or membranous, sometimes sheathing the lateral branches of the inflorescence; bracteoles sometimes connate below the flowers. Flowers small, actinomorphic, usually sessile, sometimes subsessile or pedicelled. Perianth of 6 segments, biseriate or rarely uniseriate, calyx and corolla usually 3-lobed, free or united; sepals generally imbricate or open in bud; petals generally valvate in the male and imbricate in the female flower. Stamens usually 6, sometimes 3, 9 or numerous; anthers 2-celled, dehiscing by vertical slits; filaments distinct, usually short. Carpels usually (1-)3(-more), distinct or variously united, rudimentary or absent in male flowers; ovary superior, 1-3-loculed or rarely 4-10-loculed, having a single ovule in each locule; ovule basal, axile or pendulous. Fruit usually a berry or drupe, epicarp smooth, tubercled, prickly or scaly, mesocarp often fibrous, endocarp often present, thick and woody. Seeds with a small embryo and a large homogeneous or ruminate endosperm.

A fairly large family of about 200-210 genera with 2,800 species, mainly in the tropical and sub-tropical regions of the world. In Pakistan, this family represented by 16 genera and 18 species of which only two genera and three species are wild.

Phoenix

L., Sp. Pl. 1188 (1753); Gen. Pl. ed. 5, 496 (1754); Boiss., Fl. Or. 5: 46 (1811).
Elate L.; *Palma* Miller.

Solitary or clump forming dioecious feather palms. Trunk stout, uniformly thickened, covered with bases of petiole, erect, rarely branched. Leaves pinnate, pinnae numerous, unequal, elongate, coriaceous, ensiform, lanceolate, induplicate, longitudinally folded and attached with folded base to the petiole, lowest pinnae usually modified into strong sharp spines. Inflorescence simply branched, axillary, several glabrous buds covered with one big coriaceous prophyll. Flowers unisexual, small, sessile on erect or drooping peduncle. Male flowers: oblong or ovoid, calyx cup-shaped, 3-toothed, obliquely oblong or ovate. Petals much longer than the calyx, valvate, oblong. Stamens usually 6; filaments subulate, united at the base; anthers erect. Female flowers: petals 3, rounded, imbricate, usually less than two times longer than the calyx. Carpels 3, free, with sessile stigmas; staminodes 6 or joined into a 6-lobed cup; one ovule in each carpel, only one maturing. Fruit one seeded berry, ovoid or oblong, with a fleshy pericarp and a membranous endocarp. Seeds straight, ventrally grooved, endosperm homogeneous.

A genus of about 17 species, scattered in the countries of the old world, from Canary Island in the southern part of Mediterranean region, all over Africa, S. Arabia, Middle East, Persia, Afghanistan, Pakistan, Southern slopes of Himalayas, Assam, South to Malayan Penninsula, N. Samatra and China. Four species occur in Pakistan.

Phoenix sylvestris (L.) Roxb., Fl. Ind. 3: 787 (1832); Stewart in Ali and Nasir, Ann. Cat. Vasc. Pl. W. Pak. & Kashm., 33 (1972).

Syn: *Elate sylvestris* L.

A moderate sized dioecious tree, 7.5-15 m tall, without root suckers, stem clothed with remains of petiole bases. Leaves 96 cm to 4 m long, greyish green, quite glabrous, pinnately divided into numerous leaflets. Leaflets 15-60 cm long and 8-2.5 cm broad, alternate, opposite or fascicled, in several planes, apex almost sharply pointed; lower leaf-lets modified into hard spines, up to c. 12 cm long, petiole short, glabrous, spiny. Inflorescence and flowers as in the *Phoenix dactylifera*. Fruit drupe, about 2.5 cm long, orange yellow, rounded at the ends, sweet, edible. Seeds woody, longitudinally grooved on one side

Flowering Period: March-April.

Herbarium specimens examined: NARA DESERT: *Rahmatullah Qureshi*, 1419 (SALUBGH).

Field remarks: A rare tree found in flat habitat.

Distribution: According to Aitchison this wild date palm is indigenous in the Indus basin. It is also cultivated in Sind and Punjab.

56. Cyperaceae

Perennials or annual herbs; rhizome creeping often stoloniferous, or caespitose sometimes tuberiferous or with bulbils; roots fibrous. Stem 3-gonous, triquetrous or angular, simple, solid, rarely hollow, septate, glabrous but scabrous along edges, smooth. Leaves usually 3-ranked, generally crowded at the base of the stem (upper fewer), with a closed tubular basal sheath, flat, canaliculate or conduplicate, sometimes ligulate. Inflorescence terminal, rarely pseudolateral simple to compound, paniculate, once, twice or more anthelate, capitate or spicate, a few- many-spikeletted, rarely 1. Flower(s) minute, uni- or bisexual, subtended by glumes, 1 or many, distichously aggregated into flattened or spirally into terete spikelets; perfect flower with or without bristle-like perianth. Involucral bracts generally foliar, unequal; prophylls sometimes sheathing at the base. Rachilla straight or flexuous, persistent or disarticulating at base, winged or not. Glumes membranous or coriaceous, often nerved, keel usually distinct, apiculate, mucronate, sometimes recurved, cuspidate or muticous. Perianth 0 or of 2 or more hypogynous bristles, hairs or scales. Stamens 1-3, sometimes more; filaments flattened, persistent, elongating after anthesis; anthers linear-oblong, basifixed, with crested, smooth or setulose connectives. Ovary superior, 1-celled; ovule solitary, basal, erect, anatropous. Style 2- or 3-fid; style-base persistent or deciduous, confluent or articulate with ovary; stigmas 2-3, rarely more, papilate. Disc rarely present. Nut indehiscent, 3-gonous, triquetrous or lenticular, bi-convex, dorsally or laterally compressed, rarely beaked or buttoned, sometimes surrounded by utricle.

A family of 70 genera and 4000 species, distributed throughout the world. In Pakistan, it represents 22 genera and 179 species, of which many are weeds (Kukkonen, 2001).

Key to genera:

1a. Spikes distichously arranged:
2a. Leaves shorter to much longer than stem..**1. Cyperus**
2b. Leaves mostly reduced to 2-3, soft ephemeral sheaths...........................**2. Schoenoplectus**
1b. Spikes not distichously arranged:
3a. Perianth bristles present..**3. Bolboschoenus**
3b. Perianth bristles absent... **4. Fimbristylis**

1. Cyperus

L., Sp. Pl. 1: 44 (1753); Gen. Pl. ed. 5: 26 (1754); Boiss., Fl. Or. 54: 363 (1882); C.B. Clarke in Hook. f., Fl. Brit. Ind. 6: 597 (1893); Kukkonen in Fl. Pak. 206: 82 (2001).

Perennial or sometimes annual, to 150 cm or more high glabrous herbs. Rhizomes often sturdy and creeping, producing stolons or tillers, sometimes tuberous. Stem caespitose or solitary, simple, leafy, only near the base, 3-gonous or terete, rarely septate. Leaves basal, flat or canaliculate, lower ones often scale-like, sometimes reduced to sheaths; ligule mostly 0. Inflorescence terminal, rarely pseudolateral, simple to decompound, often once, twice or more anthelate (central spike overtopped by lateral rays), rarely capitate; involucral bracts foliar; prophylls keeled, tubular at the base. Spikes oblong to obovoid, spicately or digitately spikeletted. Spikelets usually compressed, 4-angular, rarely subterete, often flowered; rachilla straight or flexuous, persistent, winged or not. Flower(s) 1-bisexual, upper ones sometimes male flowers or empty. Glumes distichous, often imbricating, margins usually hyaline, sides nerved or not; keel distinct, rarely muticous, the lowest empty, the above 2-sexual, all ca. equal, deciduous from upwards, the uppermost 1-3 sterile or empty; hypogynous scales or bristles 0. Stamens 2 or 3(1), linear-oblong; connectives often crested. Ovary compressed; styles 2- or 3-fid, continuous with ovary, base not accrescent; stigmas 2 or 3. Nut 3-gonous or biconvex (plano-convex), laterally or dorsally compound, stipitate, apiculate.

A genus of ca. 600 species distributed in tropical and warm regions. Pakistan represents ca. 40 species (Kukkonen, 2001).

Key to species:

1a. Spikelets digitate on the ultimate rays, not spicate:
2a. Annuals... **1. C. difformis**
2b. Perennials with rhizomes creeping in sand:
3a. Culm terete, spetate.. **2. C. articulatus**
3b. Culm trigonous, at leat in distal part, not or obscurely spetate:
4a. Spikelets in one head.. **3. C. arenarius**
4b. Spikelets in contracted umbels.................................... **4. C. conglomerates**
1a. Spikelets not digitae; either spicate or racemose:
5a. Plants annual with tufted stems.. **5. C. iria**
5b. Plants perennial:
6a. Spikes nearly digitately together, suberect, compressed; stolon short:
7a. Leaves radical.. **6. C. rotundus**
7b. Leaves arising from the stem some distance above the base:
8a. Inflorescence with 2 or more foliose involucral bracts; leaf-blades
1.5-3 mm wide; stolon ending to black bulbil.............................. **7. C. bulbosus**
8b. Inflorescence with 3-4 foliose involucral bracts; leaf-blades
7 mm wide; stolon without bulbil.. **8. C. longus**
6b. Spikes in dinstictly pikately arranged, spreading, subterete to terete,
stolon long, producing rooting tuber:
9a. Inflorescence pseuolateral compact group of spikes...................... **9. C. laevigatus**
9b. Inflorescence globose head.. **10. C. difformis**

1. Cyperus difformis L., Cent. Pl. 2: 6 (1756); Stewart in Nasir & Ali., Ann. Cat. Vasc. Pl. W. Pak. & Kashm., 88 (1972); Kukkonen in Nasir & Ali., Fl.Pak., 206: 133 (2001).

An annual sedge up to 45 cm size. Stem c. 3 mm diam., sharply trigonous, smooth. Leaves shorter than stem; sheaths to c. 10 cm, green or yellow-brown, sometimes with reddish tint, mouth oblique; ligule c. 1 mm, papery, arch of attachment higher than wide; blades up to 50 cm, 3-5 mm wide, keeled, margins smooth, apex trigonous, scabrous. Inflorescence of 1-7 globose partial inflorescences, nearly sessile or peduncles to 30 mm; partial inflorescences 7-15 mm diam., axis tightly digitately branching and rebranching, with c. 100 or more sessile spikes. Bracts 2-3, lowest two foliose, to 20 cm or more, sheath-less. Spikes 2-5 x c. 1 mm, with 5-20 flowers; glumes c. 0.7 mm, deeply cymbiform, blunt, wider than long, midnerve strong, sides brown or dark brown, margins narrowly scarious; rachis

compressed, slightly zigzagging, internodes c. 0.8 mm. Stamens 2. Nut c. 0.5 mm, obovoid, sharply trigonous, papillose, yellowish.
Flowering Period: July-October.
Herbarium specimens examined: NARA DESERT: *Rahmatullah Qureshi*, 1418 (SALUBGH).
Field remarks: A less common sedge found in wetland habitat.
Distribution: Common weed in tropical and subtropical areas of all continents; from S. Europe to Turkey, Iraq, Caucasus, Uzbekistan, Tadjikistan, Pakistan and India (Kukkonen, 2001).

2. Cyperus articulatus L., Sp. Pl. 1: 44 (1753); Stewart in Nasir & Ali., Ann. Cat. Vasc. Pl. W. Pak. & Kashm., 86 (1972); Kukkonen in Nasir & Ali., Fl.Pak., 206: 94 (2001).

Perennial, 1-2 m. Stolons, e.g. 3 mm diam., covered by brown or reddish-brown scales, longer than internodes. Stem 4-8 mm diam., terete or obtusely trigonous, green or greyish green, "noded" by transversal internal tissue plates. Leaves reduced; sheaths to 40 cm, soft, yellowish-brown or often reddish, mouth oblique; blades of uppermost sheaths up to 30 mm. Inflorescence a compound anthelodium, 9-16 cm; bracts to 9-30 mm, upright, rigid; primary branches 7-15, up to 12 cm; secondary anthelodia 15-50 mm, with bracts to 8 mm, secondary branches up to 20 mm; cluster of spikes of 5-14 spreading spikes, globular or slightly elongated, greyish green or brown; spikes 5-25 x c.1.5 mm, slightly compressed, with 8-24 glumes, glume-like bract to 2 mm, acute, glume-like prophyll bi-nerved, base spongy; rachis c. 0.5 mm wide, internodes c. 0.8 mm, winged; glumes c. 3 mm, cymbiform, obtuse, mid-nerve area greenish, later pale yellow, with several nerves ending below apex, sides pale yellow, brown or reddish-brown, margins widely scarious. Nut c. 1.6 x 0.5 mm, narrowly obovoid, trigonous, sides concave, brown or silvery brown, glossy, finely reticulate and papillose.
Flowering Period: July.
Herbarium specimens examined: NARA DESERT: *Rahmatullah Qureshi*, 1417 (SALUBGH).
Field remarks: A rare sedge found in wetland habitat recorded from Chotiari dam.
Distribution: Tropical and subtropical regions from Bangla Desh and Sri Lanka west to Africa and Americas (Kukkonen, 2001).

3. Cyperus arenarius Retz., Observ. Bot.. 4: 9 (1786); Stewart in Nasir & Ali., Ann. Cat. Vasc. Pl. W. Pak. & Kashm., 86 (1972); Kukkonen in Nasir & Ali, Fl. Pak. 206: 120 (2001).

Syn: *C. persicus* Boiss. & Hohen., in Boiss., Diagn. Pl. Or. Nov. 13: 39 (1854); *C. persicus* Steud, & Boiss., in Steud., Syn. Pl. Cyperac.: 28 (1855).

A glaucous-green, pale, creeping sedge. Rhizome wiry, elongate, striate, rigid, appearing dichotomously branched, clothed with withering, chestnut brown, striate, 2-5 cm long sheaths; rootlets thin, glabrous from most of the underground nodes. Stem sparsely tufted, up to 16 cm high and 1-1.5 mm broad, terete, covered with long, withering sheaths till they reach the ground level, attenuate towards the base, veins brown, disintegrating into fibres with age, mouth margin straight. Leaves ca. 6-20 x 0.5 cm in size, convolute, linear, spreading and recurved, rigidly coriaceous, almost nerveless, not keeled; ligule 0.

Inflorescence ca. 1.5-2 cm across, solitary, terminal or pseudo-lateral globose, brown; involucral bracts 2, overtopping, leaf-like, ca. 5-11 cm long, erect, unequal; primary ray(s) 1 or 2, ca. 2 x 0.1 cm in size, zigzag ; internodes ca. 0.5 mm. Spikelets ca. 0.5-1 x 0.2-0.5 cm in size, digitate or radiating, ca. 5-20 or more in clusters, oblong, subterete, fulvous-grey; rachilla stout, winged, with oblique remnants of glumes. Glumes ca. 4-5 x 2.5 mm, ovate-oblong, boat-like, apex apiculate or minutely mucronate, dorsally rounded, pale brown, the sides ca. 6 strong nerved, mostly reddish brown, margin hyaline; keel obscure. Stamens 3; filaments 4-5 mm long, flat, persistent; anthers exserted from glume, ca. 1.5-2 mm. Styles ca. 1.5-2.5 mm long, thick; stigmas 3, capillary, exserted. Achene ca. 1.5-1.8 mm long, obovoid, biconvex to obscurely 3-gonous, grey-brown, glossy, smooth, apiculate.

Flowering Period: September-November.

Herbarium specimens examined: NARA DESERT: *Rahmatullah Qureshi*, 204, 620, 640, 1086 (SALUBGH).

Field remarks: A common sedge found in low elevated dunes.

Distribution: From W. coasts of Persian Gulf to W and E Coasts of India and Sri Lanka; (E.G. Camus in M.H. Lecomte, Fl. generale Indoch. 7: 19-20.1912); Vietnam, "Cochin:, Mekong delta; obviously introduced to S. Australia and Durban, S. Africa (Kukkonen, 2001).

4. Cyperus conglomeratus Rottb., Descrip. Et Icon., 21 : t. 15., Figure 7 (1773); Stewart in Nasir & Ali., Ann. Cat. Vasc. Pl. W. Pak. & Kashm., 87 (1972); Kukkonen in Nasir & Ali, Fl. Pak. 206: 123-124 (2001).

Syn: *C. arundinaceus* Forsk., Fl. Aegypt.-Arab.: CIII, Nr. 75 (1775); *C. arcuatus* Beck. in Linnaea 35: 542 (1868); *C. ensifolius* Nees & Ehrenb. ex Boeck. in Linnaea 35: 545 (1868).

subsp. **conglomeratus** Kukkonen, Fl. & Veget. Mundi IX: 63-73, Figures 1 & 2 (1991).

A perennial glabrous, glaucous-green grass up to 2 ft high; rhizome short or long, woody, creeping; roots of stout wooly fibres. Stem stout, terete below, 3-gonous above, thickened at the base; clothed with brown ovate acuminate scales. Leaves as long as stem or shorter, ca. 2.5-5 mm wide, very stout, tough, coriaceous, margins much inrolled when dry, shortly cuspidate at apex, nerves obscure; sheath ca. 5-8 cm long, pale, mouth oblique. Umbels simple, of a few, densely crowded spikes, often contracted into one head. Bracts 3-5, ca. 7-12 cm long, slightly narrower than leaves, apex shortly cuspidate, base dilated. Spikelets ca. 1-2.5 x 0.3-0.4 cm in size, oblong-lanceolate, compressed, 8-20-flowered, straw-coloured or pale-brown; rachilla stout, scarred, not winged. Glumes ca. 8 x 1.5 mm in size, oblong-lanceolate or ovate with a minute mucro, many-ribbed, somewhate shining and paler on the hyaline margins, tightly imbricating but not placed very closely on rachilla; lower 2-3 rounded at apex. Stamens 3; filaments ca. 2.5 mm long; anthers slightly longer, narrowly linear, muticous. Styles upto 3 mm long, flattened; stigma 3, capillary. Achenes ca. 2.5-3 mm long, obovoid, unequally 3-gonous, pale brown.

Flowering Period: September-November

Herbarium specimens examined: NARA DESERT: *Rahmatullah Qureshi*, 710, 716, 867, 870, 884, 952, 1290 (SALUBGH).

Field remarks: A common sedge found in low elevated dunes.

Distribution: Tropical Africa, Mediterranean region, Ceylon, Pakistan (Sindh), India, Maharashtra, Gujrat and Rajasthan (Kukkonen, 2001).

5. Cyperus iria L., Sp. Pl. 1: 45 (1753); Stewart in Nasir & Ali., Ann. Cat. Vasc. Pl. W. Pak. & Kashm., 90 (1972); Kukkonen in Nasir & Ali, Fl. Pak. 206: 111 (2001).

Annual, solitary or small tussocks forming sedge up to 56 cm tall. Rhizome small. Stem 2-3 mm diam., trigonous, smooth. Leaves up to equalling stem; sheaths to 20 cm, soft, yellow brown, sometimes with reddish tint, finely brown-dotted, mouth margin straight; ligule 0; blades to more than 30 cm, 2-5 mm wide, flat, keeled, green or greyish green, margins slightly revolute, margins and keel towards the apex scabrous, apex acute, scabrous. Inflorescence a compound anthelodium, 5-20 cm; 4-6 bracts foliose, to more than 25 cm, much longer than inflorescence; primary branches up to 15 cm, rigid; secondary anthelodium 20-50 mm; tertiary anthelodium 20-30 mm; cluster of spikes 10-40 mm, spikate, erect, lax, with c. 40 spikes; spikes 3.5-5 x c. 2 mm, compressed, lax, with 5-20 flowers, glume-like bract c. 1 mm, glume-like prophyll scarious, bi-nerved; rachis c. 0.4 mm wide, 4-angled, flat, brown, narrowly winged, internodes c. 0.6 mm; glumes c. 1.5 mm, cymbiform, keeled, mucronate, sides nerveless, yellow-brown, shiny, margins scarious. Stamens 2; stigmas 3. Nut c. 1.2 x 0.6 mm, rather sharply trigonous, ellipsoid, brown or dark brown, finely papillose.
Flowering Period: May - October.
Herbarium specimens examined: NARA DESERT: *Rahmatullah Qureshi*, 1416 (SALUBGH).
Field remarks: A common sedge of wetland habitat.
Distribution: Weed in Old World tropics and subtropics; most common in SE Asia from India to Japan and Malaysia, west to Tadjikistan, Kazakhstan and Uzbekistan, Caucasus, Iraq; Africa, Madagascar; Australia and Pacific Islands, the Americas (Kukkonen, 2001).

6. Cyperus rotundus L., Sp. Pl., 45, (1753); Stewart in Nasir & Ali., Ann. Cat. Vasc. Pl. W. Pak. & Kashm., 91 (1972); Kukkonen in Nasir & Ali, Fl. Pak. 206: 101 (2001).

Syn: *Chlorocyperus rotundus* (L.) Palla, Allg. Bot. Zeitschr. 6: 61 (1900); *Schoenus tuberosus* Burm. f., Fl. Ind: 19 (1768); *C. tuberosus* Rottb. Descr et Icon. 28. t. 7. f. 1 (1773).

Perennial herb up to 20 cm high. Stem ca. 1.5 mm broad, sparsely tufted, rigid, triquetrous. Leaves several, flat, ca. 8-14 x 0.1-0.2 cm in size, nerves prominent, scabrous; sheaths stramineous or brown. Umbels large and open, ca. 4-6 cm; involucral bracts 3, unequal, overtopping or not, upto 10 cm high; primary rays 5-7; secondary ray(s) 1-3, slender. Spikes 2-8-spikeletted. Spikelets ca. 0.5-2.5 x 0.1-0.2 cm in size, linear-oblong, spicate, pale or purplish or stramineous, 10-20-flowered; rachilla persistent, wing ca. 1.5 mm, pale. Glumes closely imbricating, in fruits obliquely spreading, ovate, 3-4 mm, membranous, apex mucronulate, sides 5-nerved. Stamens 3; filaments ca, 4 mm long; anthers ca. 3 mm long, red-crested. Stigmas 3. Nut ca. 1 mm long, oblong, 3-gonous, stramineous, stipitate, apex apiculate.
Flowering period: June-September.
Herbarium specimens examined: NARA DESERT: *Rahmatullah Qureshi*, 48, 1025, 1026 (SALUBGH).

Field remarks: A very common sedge of moisted areas.
Distribution: Australia, Mauritius, Ceylon, Indo-Pak subcontinent (Kukkonen, 2001). Commonly found in wet places between dunes.

7. Cyperus bulbosus Vahl, Enum. Pl. 2: 342 (1805); Stewart in Nasir & Ali., Ann. Cat. Vasc. Pl. W. Pak. & Kashm., 87 (1972); Kukkonen in Nasir & Ali., Fl.Pak., 206: 98 (2001).
Syn: *Cyperus jeminicus* Retz., Obs. Bot. 4: 11 (1786), non Rottb. (1773).

Perennial, 10-30 cm, often forming small tufts; stolons thin, ephemeral. Roots fibrous. Stem 0.5-1.5 mm, sharply trigonous, smooth; base swollen, often with remains of bulbil, emitting stolons 0.5 mm diam. ending into a bulbil, 8-12 mm, with dark black-brown scales. Leaves numerous, shorter to much longer than the stem; sheaths to 80 mm, yellowish, grey or brownish, soft, mouth margin slightly concave; blades to more than 20 cm, 1.5-3 mm wide, keeled, revolute, often curved, green or greyish-green, margins smooth, apex trigonous or flat, very long attenuate, acute, scabrous. Inflorescence a much reduced anthelodium or multiple spike, 30-90 mm, primary branches few and short, each ending with compact cluster of 3-10 spirally arranged spikes, or basal part often two spikes close together (on a short branch) and supported by thin foliose bract longer than the spike; at base of branches tubular prophyll up to 7 mm; bracts up to 12 cm, several foliose, thin, clearly spaced. Spike 12-15 x 1.5 mm, slightly compressed, brown; rachis winged; glumes c. 3 mm, cymbiform, back rounded, mid-nerve reddish-brown and ending into mucro just above rounded glume tip, sides with 3-4 prominent nerves, margins scarious. Nut 1.3-1.5 x c. 0.7 mm, trigonous, plano-convex, obovoid, grey, finely reticulate.
Flowering Period: August.
Herbarium specimens examined: NARA DESERT: *Rahmatullah Qureshi*, 1415 (SALUBGH).
Field remarks: A less common sedge recorded from sandy areas of Achhro Thar.
Distribution: Cap Verde Islands, W Africa, Sahara and Africa N of Sahara, E Africa from Ethiopia to Egypt, Iran, Pakistan, India, Sri Lanka, and to Vietnam, Malaysia and Australia (Kukkonen, 2001).

8. Cyperus longus L., Sp. Pl. 1:45 (1753); Stewart in Nasir & Ali., Ann. Cat. Vasc. Pl. W. Pak. & Kashm., 90 (1972); Kukkonen in Nasir & Ali., Fl.Pak., 206: 104 (2001).

A perennial sedge with about 85 cm size. Rhizome strong, with short stolons. Stem 2-3 mm diam., sharply trigonous, smooth. Leaves shorter than stem; sheaths to 20-30 cm, brown or reddish-brown,lowest bladeless, upper yellowish or with reddish tint, mouth margin straight; blades to c. 50 cm, 7 mm wide, flat or keeled, grey-green, margins and midnerve below scabrous, margins slightly revolute, apex acute, trigonous, scabrous. Inflorescence up to c. 30 cm long, rather narrow anthelodium, primary branches 5-15, up to c.20 cm, longest often with secondary anthelodia up to 10 cm, with scarious, brown, tubular prophyll; lowest 3-4 bracts foliose, to more than 50 cm; cluster of spikes by 3-15 spikes almost digitately arranged, 10-20 mm in diam., tight. Spikes 5-15 x c. 1.5 mm, with 10-15 flowers, glume-like bract c. 1.5 mm, glume-like prophyll c. 1.5 mm, two-nerved; rachis compressed, quadrangular, slightly zigzagging, not winged; glumes c. 3 mm, cymbiform, blunt, mid-vein

green, sides brown or reddish brown, margins scarious. Anthers c. 1.5 mm. Nut c. 1 x 0.5 mm, obovoid, trigonous, grey-brown, finely reticulate.
Flowering Period: June -October.
Herbarium specimens examined: NARA DESERT: *Rahmatullah Qureshi,* 1414 (SALUBGH).
Field remarks: A common sedge recorded from cultivated areas.
Distribution: C and S Europe, N, E and W Africa, in east as far as eastern Kazakhstan, Pakistan and N.W. India.

9. Cyperus laevigatus L., Mantissa. alt. 179 (1771); Stewart in Nasir & Ali., Ann. Cat. Vasc. Pl. W. Pak. & Kashm., 90 (1972); Kukkonen in Nasir & Ali., Fl. Pak., 206: 141 (2001).
Syn: *Juncellus laevigatus* (L.) C.B.Clarke in Hook.f., Fl. Brit. Ind. 6: 596 (1893); *Acorellus laevigatus* (L.) Palla in W.D.J. Koch, Syn. deut. schweiz. Fl. ed.3, 2: 2558 (1905).

Perennial, (5-)20-65 cm, forming small tufts or stems in rows. Rhizome horizontal, creeping, 1.5-2.5 mm in size, covered with dark brown or reddish scales. Stem 0.5-5 mm diam., terete or obtusely trigonous, yellow green or grey green, smooth. Leaves mostly much reduced; sheaths up to 20 cm, brown to dark brown, sometimes reddish, mouth margin concave; ligule 0; blade up to 10 or reduced to a point. Inflorescence capitate, pseudolateral, 7-25 mm, one cluster of (1-)3-10(-more than 30) spikes; bract usually one, as a continuation of stem, 5-50 mm, subterete, obtuse (sometimes another short, second bract can be distinguished); spikes 6-20 x (1.5-)2-3.5 mm, often curved, turgid, with 10-40 closely imbricate glumes, glume-like bract, c. 2 mm, glume-like prophyll c. 2 mm; rachis 4-angled, concave side 1-1.2 mm wide, convex side 0.5-0.8 mm wide, internodes 0.7-1 mm, winged; glumes 1.8-3 mm, membranaceous, widely cymbiform, obtuse or mucronate, mid-nerve at apex slightly keeled, from yellowish or grey to dark reddish brown, glossy, margins narrowly scarious. Flowers on concave side of rachis. Nut 1.3-1.7 x c. 1 mm, ellipsoid or obovoid, grey-brown to black-brown, adaxial side flat, pressed against the rachis, abaxial side rounded, glossy, smooth or more or less clearly reticulate.
Flowering Period: Practically all year round.
Herbarium specimens examined: NARA DESERT: *Rahmatullah Qureshi,* 1413 (SALUBGH).
Field remarks: A common sedge found in wetland habitat recorded from Chotiari dam.
Distribution: From C. Asia, India and Pakistan W to southern N. America and S. America (Kukkonen, 2001)

2. Schoenoplectus

(H.G.L.Reichenb.) Palla, Bot. Jahrb. 10: 298 (1888) "1889"; *nom. cons*; Kukkonen in Rech.f., Fl. Iran. 173: 22 (1998); Kukkonen in Nasir & Ali., Fl. Pak., 206: 12 (2001). *Scirpus* sect. *Schoenoplectus* H.G.L. Reichenb., Icon. Fl. Germ. 8: 40 (1846).

Annuals or perennials. Rhizome creeping, or short, with or without stolons. Stem from 10 cm to more than 2 m, terete or trigonous, smooth. Leaves mostly reduced to 2-3, soft, ephemeral sheaths; ligule probably allways present. Inflorescence pseudolateral, from group of spikes to anthelodium; lowest bract forming continuation to stem, erect or later often spreading, or reflexed. Glumes spirally arranged. Flowers bisexual, with or without perianth bristles, bristles retrorsely barbed or plumose, stamens 2-3, stigmas 2 or 3. Nut trigonous or bi-convex, compressed, surface transversely rugulose or smooth; amphicarpous nuts present in some groups. Resembling Scirpus s. str. and Actinoscirpus, but leaf blades reduced and inflorescence smaller, pseudolateral; perianth bristles present or missing.

A genus with c. 60 species, distributed on all continents; represented in Pakistan by 11 species.

Schoenoplectus litoralis (Schrad.) Palla, Bot. Jahrb. 10: 299 (1889); Kukkonen in Nasir & Ali., Fl. Pak., 206: 20 (2001).
Subsp. **litoralis**
Syn: *Scirpus litoralis* Schrad., Fl. Germanica 1: 142 (1806).

Erect, perennial sedge up to 135 cm tall. Rhizome ca. 2 mm diam., short, stoloniferous. Stem 3-6 mm diam., from sharply trigonous to obtusely trigonous above and terete below, smooth, green. Sheaths up to c. 25 cm, green, smooth, scarious side sometimes reddish, mouth deeply oblique with widely scarious margins; ligule 0.5-1 mm, scarious, margin smooth; blades mostly reduced, some up to 15 cm, sometimes longer, up to 5 mm wide, obtusely pointed, keeled, thick or channelled, margins smooth. Inflorescence a multiple spike or anthelodium, 20-90 mm; lowest bract to more than 12 cm, erect, trigonous in transverse section, edges smooth; primary branches 2-14, to 70 mm, smooth or slightly scabrous; tubular prophyll to 10 mm, mouth oblique or bilobed; secondary multiple spikes 30-40 mm diam., with 3-8 mostly solitary and pedunculate spikes, to 15 mm, with occasional tertiary multiple spikes of 3-4 spikes; tertiary branches up to 10 mm. Spikes as a rule solitary, occasionally 2 together, 7-18 x 3-5 mm, ovoid; glumes 3-4 mm, cymbiform, apex rounded, mucro to 0.7 mm, slightly recurved, sides obscurely nerved, brown, margins widely scarious. Perianth bristles 4, upper half plumose; stamens 3, anther c. 2 mm; stigmas 2. Nut 1.6-2.3 x 1.1-1.7 mm, obovoid, apiculate, bi-convex or almost plano-convex, mat or semi-glossy, very finely reticulate, dark brown.

Herbarium specimens examined: NARA DESERT: *Rahmatullah Qureshi*, 1412 (SALUBGH).

Field remarks: A common sedge found in wetland habitat recorded from Chotiari dam.

Distribution: Mediterranean, Turkey, Caucasus, Kazakhstan, China and W. Mongolia, Iran, Afghanistan and Pakistan (Kukkonen, 2001).

3. Bolboschoenus

(Aschers.) Palla, Synops. Fl. Germ. ed. 3, 3: 2531 (1907); Kukkonen in Rech.f., Fl. Iran. 173: 16 (1998); Kukkonen in Nasir & Ali., Fl. Pak., 206: 9 (2001).
Scirpus subdiv. *Bolboschoenus* Ascherson, Fl. Prov. Brandenb. 1: 753 (1864).

Perennials, up to 1.5 m. Rhizome sturdy, creeping, producing tubers. Stem trigonous, smooth or scabrous along edges. Leaves basal, smaller or equalling stem; sheaths up to 17 cm, soft; ligule 0; blades up to 7 mm wide, keeled, scabrous. Inflorescence a multiple spike,

sometimes anthelodium, or spikes congested to head; bracts much exceeding inflorescence; spikes up to 30 x 8 mm. Flowers bisexual, perianth bristles 4-7, inequal, stamens 3, stigmas 2-3. Nut trigonous or biconvex. Resembling Schoenoplectus, but leaf blades long; inflorescence either with few elongated branches (as in sect. Schoenoplectus) or spikes large and in capitula (Schoenoplectus sect. Actaeogeton); perianth bristles present.
A genus of 16 species distributed in all continents except S. America. Two species reported from Pakistan (Kukkonen, 2001).

Key to species:
1a. Inflorescence often yellow brown sessile spikes; stigmas 2.................... **1. B. affinis**
1b. Inflorescence mostly multiple dark brown to pale yellow spike; stigmas 3... **2. B. glaucus**

1. Bolboschoenus affinis (Roth) Drobov, Trav. Mus. Bot. Acad. Sc. Imp. Pétersbourg. 16: 139 (1916); Kukkonen in Nasir & Ali., Fl. Pak., 206: 9-10 (2001).

Syn: *Scirpus affinis* Roth in Roem. & Schult., Veg. 2: 140 (1817); *S. maritimus* L. subsp. *affinis* (Roth) T. Norl. in Bot. Not. 125: 404 (1972); *B. maritimus* (L.) Palla subsp. *affinis* (Roth) T. Koyama in Brittonia 31: 284 (1979); *B. strobilinus* (Roxb.) V. Krecz, in P.N. Ovchinnikov (ed.), Fl. Tadjikskoi SSR, 2: 47 (1963); *Scirpus strobilinus* Roxb., Hort. Bengal. 6 (1814). nom. nud.; Fl. Ind. 1: 219 (1820); *Scirpus vulpinicolor* T. Koyama in Acta Phytotax. Geobot. 17: 50 (1957); *Scirpus schmidii* Raymond in M. Koeie & Rech.f. in Dan. Biol. Skr. 14:15 (1965).

Perennial, 7-60 cm. Rhizome creeping or with stolons ending in a tuber c. 10 mm diam. Stem sharply trigonous above, edges smooth or scabrous, grey green, basal internodes elongated. Leaves c. equalling stem; sheaths up to 10 cm, green or yellowish brown, soft, margin almost straight; blades up to 30 cm x 1.5-6 mm, flat or folded, keeled, grey green, seldom green, smooth or margins and apex scabrous. Inflorescence compact, up to 30 mm diam., formed by 1 - c. 15 spikes, sometimes with 1-2 branches, c. 10 mm, ending with cluster of 1-3 spikes; bracts up to 15 cm. Spikes 8-20 x 5-8 mm, ovoid, obtuse, turgid, sometimes almost sphaerical, yellow brown, seldom brown; glume-like bract, e.g. 5 mm, glume-like prophyll, e.g. 4 mm; glumes c. 50 in one spike, 5.1-8.5 x 2.2-3.3 mm, cymbiform, arista 1-2 mm, slightly recurved, scabrous, sides nerveless, more or less hairy, or barbed, yellow brown or widely scarious, apex truncate, trifid and often ciliate. Perianth bristles usually 4, easily detached, inequal, c. half of nut; anthers 1.5-3.5 mm; stigmas predominantly 2, often 3. Nut 2.1-3.8 x 1.5-2.3 mm, ovoid or obovoid, lenticular, or trigonous and plano-convex, apex from rounded to conical, surface reticulate, white or yellowish brown, glossy, epidermal cells hexagonal, c. 0.1 mm diam., much deeper than wide, outer wall translucent.
Flowering Period: January-April.
Herbarium specimens examined: NARA DESERT: *Rahmatullah Qureshi*, 1411 (SALUBGH).
Field remarks: A common species found in wetland habitat recorded from Chotiari dam.
Distribution: From Europe to India (Kukkonen, 2001).

2. Bolboschoenus glaucus (Lam.) S.G. Smith, Novon. 5: 101 (1995); Kukkonen in Nasir & Ali., Fl. Pak., 206: 10-11 (2001).

Syn: *Scirpus glaucus* Lam., Tabl. Encycl. 1: 142 (1791).

Perennial, 20-30 cm. Rhizome creeping or with stolons ending with tubers, c. 7 mm diam. Stem 1-1.5 cm diam., sharply trigonous, even narrowly winged above, smooth or slightly scabrous, gryish green. Leaves as long as stem or longer; sheaths up to 10 cm, yellowish brown, soft, margin almost straight; blades to 25 cm long, 1-2.5 mm wide, flat, folded or recurved, keeled, grey green, smooth. Inflorescence up to 25 mm long, small multiple spike of 1-4 spikes; lateral spikes each with peduncles to 12 mm long; lowest bract longer than inflorescence. Spikes 6-12 x 4-6 mm, ovoid or globular, brown; tubular prophyll up to 5 mm, white, scarious; glumes 4-6 x 2-2.5 mm, cymbiform, more or less hairy, brown, sides nerveless, widely scarious, apex truncate, trifid, eroded, scarious, arista to 1.5 mm, straight to slightly recurved. Perianth bristles 4-6, ca. half of the nut, persistent; stigmas 3. Nut 2.5-3 x 1.4-1.7 mm, obovoid, planoconvex, trigonous with rounded dorsal edge, brown to dark brown, apex conical, surface finely reticulate, glossy, epidermal cells small.

Flowering Period: September-October.

Herbarium specimens examined: NARA DESERT: Chotiari dam, *Rahmatullah Qureshi*, 1438 (SALUBGH).

Field remarks: A common sdge found in wetland habitat.

Distribution: From Senegal and S. Africa to Pakistan; introduced to N. America (Kukkonen, 2001).

4. Fimbristylis

M. Vahl, Enum. Pl. 2: 285 (1806), *nom. cons.*; Gen. Pl. ed. 5: 26 (1754); Boiss., Fl. Or. 54: 363 (1882); C.B. Clarke in Hook. f., Fl. Brit. Ind. 6: 597 (1893); Kukkonen in Fl. Pak. 206: 74 (2001).

Anuual or rhizomatous perennials. Stems usually tufted. Leaves basal, flat, canaliculated or involute, cellular-reticulate, rarely reduced to sheaths; ligule membranous or hairy, sometimes absent. Flowers hermaphrodite, basal one empty, upper most tabescent. Inflorescence usually terminal, rarely pseudolateral, simple or compound, sometimes capitate, rarely single spikeletted; involucral bracts foliar, much reduced; prophylls tubular. Spikelet(s) solitary or clustered, terete, angular, usually many-flowered, pedicelled, rarely sessile; rachilla persistent, winged. Glumes usually spiral, exceptionally distichous, acropetally deciduous. Hypogynous bristles 0. stamen(s) 1-3; anthers with crested connectives. Style articulate with ovary, base dilated, deciduous, sometimes (sub) persistent; button 0; stigmas 2 or 3, often papilate. Nut obovoid, exceptionally terete or linear, 3-gonous or biconvex, often with sharp borders, smooth, verrucose or tuberculate, often striate.

Key to species:

1a. Ligule present as a fring of short hairs.. **1. F. bisumbellata**

1b. Ligule absent.. **2. F. cymosa**

1. Fimbristylis bisumbellata (Forssk.) Bubani, Dodecanthia. 30 (1850); Kukkonen in Fl. Pak. 206: 75 (2001).

Syn: *Scirpus bisumbellatus* Forssk., Fl. Aeg.-Arab. 1: 15 (1775); *F. dichotoma* auct. non (L.) Vahl, Enum. Pl. 2: 287 (1805).

Greyish-green, small tufts forming annual, up to 30 cm. Stem terete, 0.5-1 mm diam., glabrous. Leaves c. 1/2 of stem length; sheaths open, villous in upper parts; lowest bladeless, bicarinate, with scarious margins; ligule compact fringe of hairs; blades 0.5-1 mm wide, flat, adaxial side almost glabrous, abaxial side villous esp. in proximal parts, margins towards apex scabrous. Inflorescence to 5 cm diam., with 15-30 spikes, mostly solitary, occasionally a few sessile grouped together; lowest bract usually shorter than inflorescence; peduncles grooved. Spikes 3.5-4.5 x 1.3-1.5 mm, narrowly ellipsoid, more or less angular, acute, light brown to greyish brown; rachis brown to dark brown, with scarious wings; glumes 1.3-1.8 x c. 1 mm, ovate, glabrous or, occasionally, slightly hirsute, keeled, with prominent mid-nerve, arista conspicuous, margins widely scarious, slightly ciliate. Stamens 1, sometimes 0; filaments scarious, compressed; anthers c. 1 mm; style brown, c. 1 mm, flat, ciliate, stylopodium whitish, c. 0.25 x 0.25 mm, stigmas 2. Nut c. 0.8 x 0.7 mm, lenticular, widely obovoid, trabeculate, yellowish to almost white.
Flowering Period: July-October.
Herbarium specimens examined: NARA DESERT: *Rahmatullah Qureshi*, 1409 (SALUBGH).
Field remarks: A common sedge found in wetland habitat.
Distribution: From Western tropical and SE Africa to southern Europe, and eastwards to Madagascar, Pakistan, India, Malaysia, tropical Australia and New Zealand (Kukkonen, 2001).

2. Fimbristylis cymosa R. Br., Prodr. 228 (1810); Stewart in Nasir & Ali., Ann. Cat. Vasc. Pl. W. Pak. & Kashm., 95 (1972); Kukkonen in Nasir & Ali., Fl.Pak., 206: 66 (2001).

Syn: *F. spathacea* Roth, Nov. Pl. Spec.: 24 (1821); *F. glomerata* (Schrad.) Nees, fide Goetgheb. & Goudijzer, Bull. Jard. Bot. Nat. Belg. 54 (1984).

Greyish green perennial, forming tight tufts, 10-20 cm. Stem rigid, terete to triangular, deeply grooved, smooth. Leaves less than half stem length; sheaths 1-2.5 cm, pale brown, lustrous, with wide scarious margins; blades 1-1.5 mm wide, stiff, slightly turgid, frequently falcate, involute or flat, margins barbed towards apex, apex short, frequently mucronate. Inflorescence compact or diffuse with elongate peduncles, of 10 or more solitary spikes, or, in compact inflorescences, spikes sessile and clustered; bracts usually short, margins scabrous; branches compressed or from triangular to terete. Spikes 2-5 x 1-2.5 mm, ovoid, more or less terete, greyish brown to brown; rachis articulate, castaneous brown, with prominent brown or colourless wings; glumes 1.8-2 x 1-1.2 mm, keeled, light brown, with green or pale mid-rib not reaching apex, margins widely scarious. Stamens 2; anthers 1.7-1.8 mm, with small basal lobes; style 1-1.2 mm, brown; stigmas 2 (or 3), ciliate, 0.4-0.5 mm. Nut c. 0.8 x 0.6 mm, ovoid, lenticular, slightly rugulose, dark brown.
Flowering Period: October.
Herbarium specimens examined: NARA DESERT: *Rahmatullah Qureshi*, 1328 (SALUBGH).
Field remarks: A common sedge found in wetland habitat.
Distribution: Probably pantropical (Kukkonen, 2001).

57. Poaceae

Annual or perennial herbs, rarely shrubs or trees, often with rhizomes or stolons. Stem erect, ascending or creeping (often branched at the base in perennials) usually mixed with sterile shoots and flowering stems (culms), the latter cylindrical, rarely flattened, jointed, with hollow internode and solid nodes. Leaf-sheaths surrounding the culm, with free, overlapping or connate margins, often swollen at the base, sometime auriculate at mouth; ligule normally membranous, sometime a row of hairs or rarely absent. Blades solitary at the nodes, sometimes crowded and the base of the stem, alternate and 2-rowed, linear, ovate or lanceolate, usually long and narrow, rarely broad, flat and rolled or terete, parallel-nerved, rarely with transverse connections, usually passing gradually into the sheath, sometimes amplexicaul or with falcate auricles, sometimes narrowed into a false petiole or articulated with the sheath. Inflorescence terminal (rarely terminal and lateral), mad up of variously arranged spikelets, paniculate, racemose, capitate, simply or compoundly spicate (rarely of a single spikelet). Spikelets consisting of bracts distichously arranged along a slender axis (rachilla); the 2 lower bracts (involucral glumes) form an involucre to the spikelet and are empty; the succeeding (floral glumes) 1 to many bracts (lemmas) each enclosing a flower and opposed by a hyaline 2-keeled or 2-nerved dorsal scale (palea); floral glumes differing commonly in structure and size from the involucral glumes, and forming with the palea and the flower proper false flower (florets), which are alike or different in structure and sex; base of spikelet or floret sometimes with a horny prolongation downwards (callus); glumes or lemmas often bearing 1 or more stiff bristles (awns), this basic pattern of spikelet structure consistent throughout of parts. Flowers usually bisexual, sometimes unisexual, small and inconspicuous. Perianth presented by 2, rarely 3, minute, hyaline or fleshy scales (lodicules). Stamens hypogynous, 1-6, rarely more, usually 3; filaments slender, delicate, usually free; anthers versatile, fugacious, 2-thecous, opening by a longitudinally slit or rarely a terminal pore. Ovary entire, superior, 1-locular; ovule 1, anatropous, adnate to the adaxial side of the locule; styles 2, rarely 1 or 3; stigma plumose. Fruit mostly a caryopsis with thin pericarp adnate to the seed, rarely seed free, still more rarely a nut or berry; caryopsis commonly combined with various parts of spikelets, or less often the inflorescence, to form a false fruit. Seed with starchy endosperm, an embryo at the base of abaxial face, and a point or line (hillum) on the base of adaxial face making the connection between pericarp and seed.

About 620 genera and 10,000 species in 60 tribes are reported from the whole world. This family represents ca. 158 genera and 492 species in 26 tribes from Pakistan (Cope, 1982).

Key to Tribes

1a. Spikelets falling entire at maturity from the persistent axis of
spike-like panicles or racemes...I. **ZOYSIEAE**

1b. Spikelets breaking up at maturity above the persistent glumes:

2a. Inflorescence racemes or spikes, solitary, digitate or scattered along an axis,
rarely a dense ovoid, spike-like panicle..II. **CHLORIDEAE**

2b. Inflorescence a panicle, either open or contorted and spike-like:

3a. Lemmas bearing a 3-branched awn...III. **ARISTIDEAE**

3b. Lemmas with an unbranched awn or awnless:

4a. Lemmas awned from the back, rarely awnless; ligule membranous...........IV. **AVENEAE**

4b. Lemmas awned from the sinus of the prominently 2-lobed tip;
ligule a ciliate fringe..V. **ARUNDINEAE**
5a. Lower glume smaller, often minute or suppressed; spikelets
usually all fertile; fertile lemma and palea firm in texture,
awnless or nearly so..VI. **PANICEAE**
5b. Lower glumes longer than the second or lemma; spikelets normally in pairs;
one sessile and the other pedicelled; fertile lemma hyaline,
usually awned...VII. **ANDROPOGONEAE**
6a. Lemma 1-3-nerved, if 5-nerved then the two lowermost lemmas empty
and resembling the glumes...VIII. **ERAGROSTIDEAE**
4b. Lemma 9-11-nerved...IX. **AELUROPODEAE**

Tribe I. ZOYSIEAE

1. Leptothrium

Kunth, Rev. Gram. 1: 156 (1829); Cope in Nasir & Ali., Fl. Pak., 143: 151-153 (1982).
Latipes Kunth, Rév. Gram. 1: 156 (1829).

Perennial. Leaf-blades flat, narrow; ligule a line of hairs. Inflorescence a loose many-sided false raceme, the spikelets solitary or paired and falling together with the enlarged common peduncle. Spikelets linear to lanceolate, awnless; glumes well developed, dissimilar, smooth or tuberculate-spinulose, the lower (but only one of them when the spikelets are paired) modified into a long flat recurved tail, the upper tightly enfolding the floret; lemma broadly ovate, shorter than the glumes, hyaline, 1-nerved; palea minute, up to half as long as the lemma, linear, hyaline; stamens 3. Caryopsis subterete to ellipsoid, slightly laterally compressed.

A genus of 2 species in Pakistan, tropical Africa and the Caribbean region; 1 species occurs in Pakistan (Cope, 1982).

Leptothrium senegalense (Kunth) W.D. Clayton in Kew Bull. 27: 151 (1972); Cope in Nasir & Ali., Fl. Pak., 143: 151-153 (1982).

Syn: *Latipes senegalensis* Kunth, Rèv. Gram. 1: 261 (1830); *Lappago latipes* Steud., Syn. Pl. Glum. 1: 112 (1854), *nom. Superfl.*, based on *Latipes senegalensis* Kunth; *L. inermis* Chiov., Pl. Nov. Mint. Aethiop. 22 (1928).

Short-lived perennial forming tough bunches; culms 10-60 cm high, thin, wily, ascending. Leaf-blades narrow, flat or often rolled, 2-10 cm long, 1-3 mm wide. Inflorescence 2-17 cm long, the clusters more or less distant upon its wavy main axis, brightly coloured with purple and light green; peduncle cuneate, flattened, 1-5 mm long, ciliate with hooked hairs on the margins, the spikelets paired (sometimes solitary) and side by side on its truncate tip. First spikelet: lower (outer) glume narrowly lanceolate, 2.5-8 mm long, coriaceous, flattened, ± caudate-acuminate, sharply recurved, the margins ciliate with hooked hairs above; upper glume oblong to narrowly oblong in side-view, 3-5.5 mm long, densely tuberculate and often spinulose, pectinate with hooked spines near one margin, rostrate at the tip: lemma 1.6 mm long. Second spikelet (when present): lower (inward) glume narrowly lanceolate, 2.5-4 mm long, loosely clasping the upper glume, tuberculate-spinulose; upper glume lanceolate in side-view, 3-5 mm long, tuberculate-spinulose, rostrate above with a crest of hooked prickles; lemma 1.5 mm long.

Flowering & Fruiting Period: January-March.

Herbarium specimens examined: NARA DESERT: *Rahmatullah Qureshi*, 1408 (SALUBGH).

Field remarks: A less common grass found in Achhro Thar.

Distribution: Pakistan (Sindh); Tropical East Africa and Senegal (Cope, 1982).

Tribe II. CHLORIDEAE

2a. Inflorescence 1-sided spikes with 1-flowered spikelet; palea present..............**2. Cynodon**

2b. Inflorescence many digitate spikes; spikelets 2-4-flowered; palea absent.............**3. Chloris**

2. Cynodon

Rich. in. Pers., Syn. Pl. 1: 85 (1805); Boiss., Fl. Or. 5: 553 (1884); Hook. f., Fl. Brit. Ind. 7: 288 (1896); Sultan & Stewart, Grasses W. Pak. 2: 273 (1959); Bor, Grasses Burma Ceyl. Ind. Pak. 468 (1960); Bor in Rech. f., Fl. Iran. 70: 450 (1970); Cope in Fl. Pak. 143: 116 (1982).

Rhizomatous or stoloniferous perennials, usually low-growing, often sward forming. Leaf-blades linear and flat, sometimes filiform or lanceolate; ligule membranous, often ciliate on the upper edge. Inflorescence of digitate 1-sided spikes which are sometimes borne in 2 or more closely spaced whorls. Spikelets narrowly ovate, 1-flowered, with or without a rachilla extension (very rarely bearing a tiny vestigial floret), strongly laterally compressed, imbricate in 2 rows, awnless. Glumes 2, narrow, subequal, shorter than the floret, 1-nerved (rarely the upper 3-nerved); lemma narrowly ovate in side view, more or less ciliate on the keel and often also on the side nerves, usually glabrous on the flanks, its callus glabrous. Caryopsis ellipsoid, laterally compressed.

A genus of 8 species in tropical and warm temperate regions. In Pakistan it is represented by 2 species (Cope, 1982).

Cynodon dactylon (L.) Pers., Syn. Pl.,1:85 (1805); Stewart in Nasir & Ali., Ann.Cat.Vasc.Pl.W.Pak. & Kashm., 142 (1972); Cope in Nasir & Ali., Fl.Pak., 143:117 (1982). **Figure 21C**

Syn:*Panicum dactylon* L., Sp.Pl., 1:85 (1753).

Perennial grass extensively creeping by rhizome or by strong flat stolen and rooting at nodes. Culms ca. 5-22 cm high, decumbent, finally erect, slender. Leaf-blades ca. 2-10 x 0.1-0.2 cm in size, narrowly linear or lanceolate, finely acute, more or less glaucous, soft, smooth, conspicuously distichous in the barren shoots and at the base of the culm; sheaths light, glabrous or hairy, bearded at the mouth. Spikelets 2-6, radiating from the top of a slender peduncle, ca. 2.5-5 cm long; rachis slender, compressed or angled, scaberulous. Spikelets ca. 1.7-2.5 mm long, 1-flowered; rachilla produced, very slender, equalling half the length of spikelets. Involucral glumes laceolate, acute to subulate, apex mucronate, the lower ca. 1-1.6 mm long, the upper slightly longer; floral glumes obliquely oblong, ca. 2 mm long. Grain ca. 1 mm long, oblong.

Flowering period: throughout the year.

Herbarium specimens examined: NARA DESERT: *Rahmatullah Qureshi*, 1197 (SALUBGH).

Field remarks: A common grass found in flat plains.

Distribution: Pakistan (Sindh, Baluchistan, Punjab, N.W.F.P. & Kashmir); tropical and warm temperate regions throughout the world (Cope, 1982).

3. Chloris

Sw., Prod. Veg. Ind. Occ. 25 (1788); Boiss., Fl. Or. 5: 554 (1884); Hook.f., Fl. Brit. Ind. 7: 289 (1896); Sultan & Stewart, Grasses W. Pak. 2: 268 (1959); Cope in Nasir & Ali., Fl. Pak., 143: 118 (1982).

Tufted, rhizomatous or stoloniferous perennials or annuals. Leaf-blades linear, flat or folded, acute to acuminate or obtuse; ligule membranous, short, ciliolate; sheaths rounded or keeled and flabellate. Inflorescence of few to many slender, typically digitate spikes. Spikelets 24-flowered, shortly pedicelled or subsessile, alternate in 2 rows on a tough rhachis; glumes unequal, membranous, usually lanceolate in side-view, 1-nerved, acute, acuminate or with an awn-point, usually persistent; lowest floret bisexual; lemma cartilaginous or coriaceous, mostly laterally compressed and keeled, usually ciliate on the margins and keel, entire or 2-lobed at the apex, with a subapical awn; callus rounded or pungent, ciliate; 2nd-6th florets sterile, sometimes male, rarely bisexual, embraced by or projecting from the side of the lowest floret; lemmas more or less reduced or vestigial, awned or awnless; paleas usually absent. Caryopsis ellipsoid to linear, usually trigonous, occasionally dorsally compressed.

A genus of about 40 species in tropical and warm temperate regions of both hemispheres. In Pakistan, 5 species reported.

Chloris barbata Sw., Fl. Ind. Occ., 1: 200 (1797); Stewart in Nasir & Ali., Ann.Cat.Vasc.Pl.W.Pak. & Kashm., 142 (1972); Cope in Nasir & Ali., Fl. Pak., 143: 121 (1982).

Syn: *Andropogon barbata* L., Mant. 2: 302 (1771); *C. inflata* Link., Enum. Pl. 1: 105 (1821); *C. refuscens* Steud., Syn. Pl. Glum. 1: 206 (1854), *non* Lag. (1805).

A loosely tufted, stoloniferous, perennial grass up to 1 m size, mostly glabrous below the inflorescence. Leaf-blades flat, rarely involute, up to 40 cm long, 2-3 mm wide, tapering towards the apex. Inflorescence of 5-12(22) digitate, loose to spreading, purplish, feathery spikes 4-8 cm long. Spikelets 3-flowered, 3-awned; lower glume 1.2-1.5 mm long; upper glume 1.7-25 mm long; lowest lemma narrowly ovate-elliptic in side-view, 2-2.5 mm long, pallid, sparsely to densely ciliate on the margins and keel, the awn 4.5-7 mm long; callus rounded, ciliate; 2nd lemma a glabrous clavate scale 1-1.5 mm long, projecting from the side of the lowest lemma, different in shape, its awn 2.5-7 mm long; 3rd lemma also reduced to a glabrous clavate scale 1 mm long, its awn 3-5 mm long.

Flowering period: April-May.

Herbarium specimens examined: NARA DESERT: *Rahmatullah Qureshi,* 1396 (SALUBGH).
Field remarks: A common grass of plain areas.
Distribution: Pakistan (Sindh); widespread throughout the tropics (Cope, 1982).

Tribe III. ARISTIDEAE

1a. Central awn-branch glabrous..**4. Aristida**
1b. Central awn-branch plumose..**5. Stipagrostis**

4. Aristida

L., Sp. Pl. 1: 82 (1753);Gen. Pl., ed. 5, 35 (1754); Cooke in Fl. Presidency Bombay, iii: 528 (1906); Bor, Grasses Burma Ceyl. Ind. Pak. 404 (1960); Cope in Fl. Pak., 143: 40 (1982); Matthew, Fl. Tamilnadu Carn., 3: 1801 (1983).

Densely tufted annuals or perennials, usually without a rootstock; node glabrous. Leaf-sheaths compact; blades linear, usually in a basal tuft, but in annual species borne mainly on the culms; ligule short, with a ciliate rim. Panicles effuse, contracted or spike-like, narrow or spreading and broad. Spikelets 1-flowered; rachilla short, disarticulating above the upper glume. Glumes oblong, convolute, often aristate; lemma with a bearded callus, terete, glabrous or scabrous, with a 3-branched awn (rarely the lateral branches obsolete); awn with or without column, branches glabrous or scabrid; callus well developed, obtuse, 2-fid or acuminate; palea very small. Anthers 3. Caryopsis narrowly ellipsoid.

A genus of ca. 260 species in areas of poor dry soils with low rainfall (but scarcely true deserts into which Stipagrostis extends) throughout the tropics and subtropics especially along the Tropics of Cancer and Capricorn. Only seven species recorded in Pakistan (Cope, 1982). In the Nara Desert, only 2 species recorded.

Key to species:

1a. Lower glume shorter than the upper, not awned. Awn less than 2 cm long, tri-fid, not articulated.. **1. A. adscensionis**
1b. Lower glume always longer than the upper, awned. Awn usually more than 2 cm long:
2a. Articulation at the top of the column, just below the awns....................**2. A. mutabilis**
2b. Articulation at the base of the column:
3a. Geniculately ascending plant up to 2 feet high; lower glume always longer than the upper; base of the callus pointed...**3. A. funiculata**
3b. A dwarf plant up to 15 cm high with branching from the base; lower glume shorter than the upper; base of the callus forked..................................**4. A. hystricula**

1. Aristida adscensionis L., Sp. Pl., 82 91753); Stewart in Ali and Nasir, Ann. Cat. Vasc. Pl. W. Pak. & Kashm., 131 (1972); Cope in Ali and Nasir, Fl. Pak., 143: 41 (1982).

Syn: *A. depressua* Retz., Obs. Bot. 4: 22 (1786); *A. caerulescens* Desf., Fl. Atlant. 1: 109, t.21/2 (1798); *A. pumila* Dcne. in Ann. Sci. Nat., Ser. 2, 4: 85 (1835); *A. ehrenbbergii* Trin. & Rupr. in Mem. Acad. Sci. Petersb., ser. 6, 7(1): 136 (1842); *A. adscensiionis* L. var. *pumila* (Dcne.) Coss. & Dur., Fl. Alger. Glum. 84 (1855); *A.*

adscensionis L. var. *ehrenbergii* (Trin. & Rupr.) Henr. in Meded. Rijks-Herb. No. 54: 158 (1926).

An annual grass up to 15-60 cm long. Culms tufted, slender, simple or branched, smooth, erect or ascending from a shortly decumbent base. Leaf-blades 10-30 cm x 1-2 mm in size, convolute, filiform, smooth or scabridulous; sheaths smooth with rounded auricles; ligule a ridge of short and fine hairs. Panicle up to 30 cm. Long interrupted, subsecund, sometimes oblong, more or less lax on one side of the rachis; branches single or 2-nate, unequally branched from the base, rachis filiform, smooth; pedicel capillary. Spikelets erect, ca. 6-9 mm long, 1-flowered, often purplish. Lower involucral glume ca. 5-8 mm long, oblong-lanceolate, acute, membranous, often purple with a scabrous keel. Upper involucral glume lanceolate, ca. 6-9 mm long, 2-toothed, apex apiculate; keel smooth. Floral glume ca. 8 mm long, 3-nerved, linear, laterally compressed, as long as upper glume or slightly longer, rarely shorter, scabrid along keel, otherwise usually smooth. Callus ca. 0.5 mm long, pointed, base hairy, awn 3-fid, not articulate; central branch up to 15 mm long, lateral branches diverging, slightly shorter than the central branch.
Flowering & fruiting period: March-December.
Herbarium specimens examined: NARA DESERT: *Rahmatullah Qureshi*, 97, 642, 682, 948, 960, 1093, 1099, 1106, 1108 (SALUBGH).
Field remarks: A very common grass found in the whole desert.
Distribution: Pakistan (Sindh, Baluchistan, Punjab, N.W.F.P., Gilgit & Kashmir); throughout the tropics (Cope, 1982).

2. Aristida mutabilis Trin. & Rupr., Sp. Gram. Stip. 150 (1842); Stewart in Nasir & Ali., Ann. Cat.Vasc. Pl. W. Pak. & Kashm., 132 (1972); Cope in Nasir & Ali., Fl.Pak., 143: 47 (1982).

Syn: *A. articulata* Edgew. in J. L. Soc., Bot. 6: 209 (1862).

Loosely tufted annual; culms (10)30-70 cm high. Leaf-blades linear, 6-14 cm long, narrow or convolute. Panicle narrowly ovate, 12-20 cm long, loose or ± contracted, the spikelets in narrow spikes at the tips of the main branches. Spikelets grey or pallid; glumes typically subequal, lanceolate, scaberulous on the keel and sometimes also on the flanks, the lower 4-5 mm long, the upper 6-7 mm long, acute or apiculate, less often with a short awn-point; lemma terete, 4-5 mm long, scabrid upwards; callus linear, 1-1.5 mm long, finely obtuse at the tip; column 2-5 mm tong, articulated just below the branching point of the awn; awn-branches 10-30 mm long, slender, parallel or divergent.
Flowering & fruiting period: September-December.
Herbarium specimens examined: NARA DESERT: *Rahmatullah Qureshi*, 1382 (SALUBGH).
Field remarks: A common grass found in interdunal valleys.
Distribution: Pakistan (Sind, Baluchistan, Punjab N.W.F.P. & Kashmir); Mauritania to India; perhaps also in tropical East Africa (Cope, 1982).

3. Aristida funiculata Trin. & Pupr., Sp. Gram. Stip. 159 (1842) and in Mem. Acad. Sci. petersb., ser. 6, 7(2): 159 (1843); Stewart in Ali and Nasir, Ann. Cat. Vasc. Pl. W. Pak. & Kashm., 132 (1972); Cope in Ali and Nasir, Fl. Pak., 143: 45 (1982).

Syn: *Aristida royleana* Trin. & Rupr., l.c. 160; *A. macrathera* A. Rich., Tent. Fl. Abyss. 2: 393 (1851); *A. kotschyi* Hochst. ex Steud., Syn. Pl. Glum. 1: 142 (1854); *A. mallica* Edgew. in J. L. Soc., Bot. 6: 209 (1862); *A. funiculata* Trin. & Rupr. var. *royleana* (Trin. & Rupr.) Hook. f., Fl. Brit. Ind. 7: 227 (1896); *A. funiculata* Trin. & Rupr. var. *mallica* (Edgew.) Henr. in Meded. Rijiks-Herb. No. 54A: 388 (1927).

Annual grass up to 40 cm high. Culms slender, tufted, wiry, branched, geniculately ascending. Leaf-blades ca. 5-15 0.1-0.2 cm in size, flat or convolute, ciliate at the base with long, white hairs in axil; sheaths glabrous; ligule a small, ciliate membrane. Panicle ca. 10-15 cm long, lax, narrow; rachis angular, branches short, erect, capillary. Spikelets ca. 17-27 mm long. Lower involucral glume ca. 1.5-2.7 cm long, linear-lanceolate, acute, 1-nerved, tapering into a slender awn; upper involucral glume ca. 1.5-2.3 cm long; floral glume small, smooth, awned; awn articulate on glume; column ca. 2-4.5 cm long, with 3, capillary branches of which central awn ca. 3-6 cm long, lateral awn up to 5 cm long.
Flowering period: June-September.
Herbarium specimens examined: NARA DESERT: *Rahmatullah Qureshi*, 110, 666, 826, 828, 709, 7071, 851, 962, 1094, 1097, 1193 (SALUBGH).
Field remarks: A fairly common species found growing on slopes.
Distribution: Pakistan (Sindh, Baluchistan, Punjab, N.W.F.P., Gilgit & Kashmir); tropical East Africa, through to the Middle East and India (Cope, 1982).

4. Aristida hystricula Edgew., in L. Soc., Bot. 6: 208 (1862); Stewart in Nasir & Ali., Ann. Cat. Vasc. Pl. W. Pak. & Kashm., 132 (1972); Cope in Nasir & Ali., Fl. Pak., 143: 46 (1982).

Tufted annual; culms erect, 5-15 cm high. Leaf-blades tightly inrolled, curved, up to 4 cm long and 1 mm wide. Panicle up to 4 cm long, narrow, few-flowered, the branches erect. Spikelets green or tinged with purple; glumes very unequal; lower glume lanceolate, 4.5-5.5 mm long, bifid at the apex with an awn up to 1 mm long in the sinus; upper glume narrowly lanceolate, 95-11 mm long, bifid at the apex with an awn up to 1.5 mm long in the sinus; lemma 2.5-3.5 mm long, terete, convolute, hispidulous, articulated at the summit; callus 0.6 mm long, oblong, 2-toothed; column up to 20 mm long, strongly twisted; central branch of the awn up to 6 cm long; lateral branches about two-thirds the length of the central.
Flowering Period: June-September.
Herbarium specimens examined: NARA DESERT: *Rahmatullah Qureshi*, 1383 (SALUBGH).
Field remarks: A less common grass found in plains.
Distribution: Known only from Pakistan (Sind) and North West India (Cope, 1982).

5. Stipagrostis

Nees in Linnaea 7: 290 (1832); Bor in Rech. f., Fl. Iran. 70: 368 (1970); Cope in Fl. Pak., 143: 48 (1982).

Densely tufted perennials, suffrutice with knotty rhizomatous base or delicate annuals. Leaf-blades narrow or terete, sometimes early deciduous leaving sheaths and chlorophyll-bearing culms as the main photosynthetic organs. Panicles open or contracted. Spikelets 1-flowered with scarious glumes; lemma cylindrical, indurated, produced into a 3-

branched awn; awn with a column, articulated at the base of the column or near the middle of the lemma, the branches or at least the central one plumose; callus well developed, pungent or minutely 2-fid, usually oblique; palea usually less than half as long as the lemma.

A genus of ca. 50 species in dry or desert regions of the Africa and the Middle East to Northwest India. Seven species are present in Pakistan (Cope, 1982).

Stipagrostis plumosa (L.) Munro ex T. Anders. in J. L. Soc., Bot. 5, suppl. 1: 40 (1860); Stewart in Ali and Nasir, Ann. Cat. Vasc. Pl. W. Pak. & Kashm., 133 (1972); Cope in Ali and Nasir, Fl. Pak., 143: 50-51 (1982). **Figure 23D**

Syn: *Aristida plumosa* L., Sp. Pl., ed. 2, 2: 1666 (1763); *A. brachypoda* Tausch in Flora, Jena 19(2): 506 (1836); *A. plumosa* L. var. brachypoda (Tausch) Trin. & Rupr., Sp. Gram. Stip. 166 (1842) and in Mem. Acad, Sci. Petersb., ser. 6, 7(2): 166 (1843); *Stipagrostis brachypoda* (Tausch) De Winter in Kirkia 3: 133 (1963); *S. plumosa* (L.) Munro ex T. Anderss. var. brachypoda (Tausch) Bor in Taxon 16: 467 (1967).

Densely caespitose perennial grass up to 35 cm high; lowest sheaths and internodes thick flocculent fugacious wooly. Leaf-blades ca. 4-12 cm long, filiform, rolled, smooth, pungent, curved, sometimes forming a semi-circle or complete circle. Panicles ca. 5-18 cm long, more or less contracted. Spikelets pallid; glumes glabrous, unequal, lower one ca. 10-14 mm long, the upper ca. 11-15 mm long; lemma ca. 2.5-4 mm long, smooth, gradually into the awn; callus ca. 1-1.7 mm long, acute, bearded, the hairs increasing in length upward; column of awn ca. 5-10 mm long, glabrous; central branch of the awn ca. 2.5-6 cm long, plumose in the upper 2/3, but with an excurrent naked tip; lateral branches ca. 2.5 cm long, naked.

Flowering period: About round the year.

Herbarium specimens examined: NARA DESERT: *Rahmatullah Qureshi*, 37, 215, 643, 687, 845, 936, 949, 999, 1054, 1095 (SALUBGH).

Field remarks: A very common grass found in the whole desert.

Distribution: Pakistan (Sindh, Baluchistan, Punjab, N.W.F.P., Gilgit & Kashmir); Mediterranean region and the Middle East (Cope, 1982).

Tribe IV. AVENEAE

1a. Panicle silky; dense..**6. Polypogon**

1b. Not so:

2a. Lemma awned from the back..**7. Avena**

2b. Lemma awnless...**8. Phalaris**

6. Polypogon

Desf., Fl. Atland. 1: 66 (1798); Boiss., Fl. Or. 5: 520 (1884); Sultan & Stewart, Grasses W. Pak. 2: 314 1959); Bor, Grasses Burma Ceyl. Ind. Pak. 403 (1960).

Annuals or perennials. Leaf-blades flat, linear. Panicle contracted or spike-like. Spikelets narrowly oblong to oblong, small, falling entire together with the pedicel or part of it; rachilla not prolonged, eventually disarticulating below the floret. Glumes narrow, subequal, 1-nerved, membranous, rough, keeled above, or between the lobes; lemma mostly broadly oblong, shorter than the glumes, truncate, thin, finely 5-nerved, usually awned from near the apex, with the very fine awn often deciduous; palea as long as the lemma or shorter.

A genus of 15 species in subtropical and warm temperate regions. Two species are recorded from Pakistan (Cope, 1982).

Polypogon fugax Nees ex Steud., Syn. Pl. Glum. 1: 184 (1854); Stewart in Nasir & Ali., Ann. Cat. Vasc. Pl. W. Pak. & Kashm., 131 (1972); Cope in Nasir & Ali., Fl. Pak., 143: 468 (1982). **Figure 23B**

Syn: *Polypogon littoralis* auct. n*on* J. E. Sm.; P. lutosus auct. n*on* (Poir) Hitchc.; *P. interruptus* auct. n*on* H.B.K.

Annual herb, culms up to 70 cm high, decumbent at the base and rooting from the lower nodes. Leaf-blades up to ca. 14x1 cm, flat, linear-narrowly lanceolate, acute, margin ciliate; ligule membranuous, ca. 4x3 mm. inflorescence panicle, narrowly ovate or oblong, ca. 10 cm long, pale green. Floret ca. 4 mm long; glumes slightly notched at the apex, hairy on margin, with a fine straight awn, ca. 2x0.5 mm, awn ca. 2.8 mm long; lemma awnless or with an awn.

Flowering Period: February- March.

Herbarium specimens examined: NARA DESERT: *Rahmatullah Qureshi*, 1395 (SALUBGH).

Field remarks: A common grass of wasteland places.

Distribution: Pakistan (Baluchistan, Punjab, N.W.F.P., & Kashmir); Iraq eastwards to Burma, mainly in the Himalayas (Cope, 1982).

7. Avena

L., Sp. Pl. 1: 79 (1753); Sultan & Stewart, Grasses W. Pak. 2: 280 (1959); Bor, Grasses Burma Ceyl. Ind. Pak. 432 (1960).

Annuals. Leaf-blades flat. Inflorescence an open or sometimes contracted panicle with large pendulous spikelets. Spikelets 2-6-flowered (rarely more), the lower 1-3 florets bisexual, the others ± reduced, the terminal floret barren or vestigial; rachilla disarticulating above the glumes and sometimes between the florets, or not at all in cultivated races; glumes lanceolate-ovate or elliptic, equal almost so, rarely very unequal, rounded on the back, scarious-membranuous or herbaceous-membranuous, with thin margins, 7-11-nerved; lemmas lanceolate-ovate or oblong, shorter than or equaling the glumes, rarely exserted, rounded on the back, herbaceous, becoming hardened, 5-9-nerved, with a geniculate (or rarely straight) awn from the back, or awnless in some cultivated races; paleas shorter than the lemmas, 2-keeled; stamens 3; stigmas 2.

A genus of about 10-15 spontaneous species found as weeds of cultivation and 10-12 cultivated races, indigenous to Mediterranean region, Northeast Africa and Central Asia. Four species occur in Pakistan (Cope, 1982).

Avena fatua L., Sp. Pl. 1: 80 (1753); Cope in Nasir & Ali., Fl. Pak., 143: 508-509 (1982). **Figure 20B**

Syn: *Avena sativa* L. Var. *sericea* Hook. f., Fl. Brit. Ind. 7:275. 1896.

Annual plant, up to 160 cm high.Culms erect or geniculately ascending, stout, simple. Leaf-blades up to ca. 30 cm long. Ligule ca. 2x3 mm Penicle with pendulous spikelets. Spikelets 2-3 flowered, lanceolate, finely acute, glabrous; lower glume ca. 20x4 mm; upper glume ca. 21x4.5 mm; lemmas densely hairy, green-brown; upper lemma ca. 12x2 mm, its awn ca. 16 mm; upper palea ca. 10x2 mm, ciliate, Stamens 3, ca. 6 mm long; anthers ca. 2.5 mm long. Ovary ca. 1x0.7 mm, densely covered with long hairs; stigmas 2, ca. 3.5 mm long.

Flowering Period: February- March.

Herbarium specimens examined: NARA DESERT: *Rahmatullah Qureshi*, 1394 (SALUBGH).

Field remarks: A common weed of plain areas.

Distribution: Pakistan (Sindh, Baluchistan, Punjab, N.W.F.P., Gilgit & Kashmir); Europe, western and Central Asia (Cope, 1982).

8. Phalaris

L., Sp. Pl. 1: 54 (1753); Hook.f., Fl. Brit. Ind. 7: 220 (1896); Sultan & Stewart, Grasses W. Pak. 2: 366 (1959); Cope in Nasir & Ali., Fl. Pak., 143: 494 (1982).

Annuals or perennials. Leaf-blades linear; ligule membranous. Inflorescence a contracted or spike-like panicle. Spikelets all alike, bisexual, 3-flowered with the 2 lower florets reduced, disarticulating above the glumes at maturity; glumes subequal, usually glabrous, papery, boat-shaped, keeled, the keel often broadened into a wing, enclosing the florets; sterile florets up to about half as long as the fertile lemma, subulate or chaffy, usually pubescent, sometimes one or both reduced to a fleshy scale at the base of the fertile floret; fertile lemma glabrous or pubescent, often shiny, laterally compressed, indurated, awnless; stamens 3.

Phalaris minor Retz., Obs. Bot. 3: 8 (1783); Cope in Nasir & Ali., Fl. Pak., 143: 494 (1982). **Figure 22D**

Annual herb, up to 100 cm tall, with erect or decumbent-ascending culms. Leaf-blades linear, ca. 20 cm long; ligule ca. 5 mm long, membranuous. Panicle up to 6 cm long, ovate-oblong. Glumes keeled, keel broadened into a wing, wing margin usually erose-denticulate. Sterile floret 1, glabrous; fertile floret broadly lanceolate-ovate, ca. 3.8x1.5 mm, shiny, pubescent. Lower glume ca. 5x1.5 mm; upper glume 5.5x1.8 mm. Stamens 3.

Flowering Period: January- March.

Herbarium specimens examined: NARA DESERT: *Rahmatullah Qureshi*, 1407 (SALUBGH).

Field remarks: A common weed of plain areas.

Distribution: Pakistan (Baluchistan, Punjab, N.W.F.P., & Kashmir); throughout the world, but apparently native only in Mediterranean region and eastward to Baluchistan and the Northwest Himalayas (Cope, 1982).

Tribe V. ARUNDINEAE

9. Phragmites

Adans., Fam. Pl. 2: 34, 59 (1763); Boiss., Fl. Or. 5: 563 (1884); Hook. f., Fl. Brit. Ind. 7: 303 (1896); Sultan & Stewart, Grasses W.Pak. 2: 222 (1959); Cope in Fl. Pak. 143: 23 (1982).

Trichoon Roth in Archiv. Bot. Roemer 1(3): 37 (1798); *Czemya* Presl, Cyp. & Gram. Sicul. 22 (1820); *Oxyanthe* Steud., Syn. Pl. glu. 1: 197 (1854).

Aquatic or semiaquatic tall perennials with a creeping rhizomes. Stem stout, hollow, leafy upwards. Leaf-blades linear-narrowly lanceolat, flat, deciduous; ligule a ring of long or short hairs. Panicles lax, usually very large and profusely branched. Spikelets conspicuously silky from the long hairs on the callus, 3-11-flowered, the lowest floret male or barren, the following bisexual, the uppermost reduced; rachilla disarticulating above the lower and between the floral glumes, slender, penicillate with long hairs, not produced beyond the flowering glumes. Glumes glabrous; invlucral glumes unequal, oblong-lanceolate, acute, 3-nerved, membranous; floral glumes bisexual, the lowest linear-lanceolate, much exceeding the involucral glumes, the following very thin, caudate-acuminate, hyaline, 3-nerved; callus long, slender, densely clothed with very long silky hairs. Lemmas narrow, rounded or slightly keeled on the back, glabrous; lowest lemma much longer than the glumes, 3-7-nerved, persistent; fertile lemmas acuminate, 1-3-nerved; palea linear-oblong, ca. ½ as long as their glumes, 2-keeled. Lodicules 2(3). Stamens 3 or 2 in the lowest floret. Styles 2, distinct, rather shorter; stigmas laterally exserted, densely plumose. Grain oblong, semiterete.

A cosmopolitan genus of 4 species, 2 species are reported from Pakistan (Cope, 1982).

Key to species:

1a. Leaf-blades smooth beneath, the tips filiform and flexuous;
panicle ca. 20-30 x 6-10 cm..**1. P. australis**

1b. Leaf-blades scabrid beneath, the tips attenuate, stiff;
panicle ca. 30-60 x 8-12 cm..**2. P. karka**

1. Phragmites australis (Cav.) Trin. ex Steud., Nom. Bot., ed. 2, 2: 324 (1841); Cope in Nasir & Ali., Fl. Pak., 143: 25-26 (1982). **Figure 23A**

Syn: Arundo phragmites L., Sp., Pl. 1: 81 (1753); *A. australis* Cav. in Anales Hist. nat. Madrid 1: 100 (1799); *Phragmites communis* Trin., Fund. Agrost. 134 (1820), based on *Arundo phragmtes* L.; *P. communis* Trin. var. *stenophylla* Boiss., Fl. Or. 5: 563 (1884); *Trichoon phragmites* (L.) Rendle, Cat. Afr. Pl. Welw. 2: 218 (1899); *Phragmites maxima* (Forssk.) Blatter & McCann, Bombay Grasses 202 (1935), *pro parte*, based on *Arundo maxima* Forssk., Fl. Aegypt.Arab. 24 (1775); *P. australis* (Cav.) Trin. ex Steud. var. *stenophylla* (Boiss.) Bor in Rech. f., Fl. Iran. 70: 353 (1970).

A semi-aquatic perennial, with creeping rhizomes. Culms erect, up to 1.5-2 m high. Leaf-blades ca. 50 cm, glabrous, smooth beneath with filiform tips. Ligule ca. 1.7 mm long.

Panicles ca. 20-30 x 6-10 cm; spikelets ca. 16 mm long; rachilla bearded with long silky hairs, copious. Upper glume ca. 4.5x0.5 mm, lanceolate, sharply acute, mostly apiculte; lowest lemma linear-lanceolate; upper lemma 5.5x0.6 mm long. Stamens 3.
Flowering Period: July- October.
Herbarium specimens examined: NARA DESERT: *Rahmatullah Qureshi*, 1384 (SALUBGH).
Field remarks: A common grass found near water bodies.
Distribution: Pakistan (Punjab & Kashmir); temperate regions of both hemispheres in the Old World and the New (Cope, 1982).

2. Phragmites karka (Retz.) Trin. ex Steud., Nom. Bot., ed. 2, 2: 324 (1841); Stewart in Ali and Nasir, Ann. Cat. Vasc. Pl. W. Pak. & Kashm., 134 (1972); Cope in Ali and Nasir, Fl. Pak., 143: 26 (1982).

Syn: *Arundo karka* Retz., Obs. Bot. 4: 21 (1786); *A. roxburghii* Kunth, Rev. Gram. 1: 79 (1829); *Phragmites roxburghii* (Kunth) Steud., Nom. Bot., ed. 2, 2: 324 (1841); *P. nepalensis* Nees ex Steud., Syn. Pl. Glum. 1: 196 (1854); *P. maxima* (Forssk.) Blatter & McCann, Bombay Grasses 202 (1935), pro parte, based on *Arundo maxima* Forssk., Fl. Aegypt-Arab. 24 (1775).

A perennial reed with creeping rhizomes. Culm erect, ca. 2-10 m high, slender, smooth, shiny, glabrous. Leaf-blades ca. 20-75 x 1-3.5 cm in size, linear-narrowly lanceolate, flat, apex stiff and attenuate, glabrous above, scabrid below; sheaths ciliate; ligule shortly hairy. Panicles ca. 30-60 x 8-12 cm in size; basal branches in whorls, silky hairy. Spikelets ca. 1 cm long, 4-10-flowered; rachilla jointed, internodes ca. 1 mm long, with hairs ca. 4-8 mm long; pedicel slender, ca. 0.4-1 cm long. Glumes persistent; lower glume 1(3)-nerved, ca. 3.5 mm long; upper glume 3-nerved, ca. 6.5 mm long, very narrowly elliptic; lowest lemma ca. 0.4-0.8 cm long, narrowly elliptic; fertile lemmas lanceolate, ca. 1-1.5 cm long, 1-3-nerved, slightly keeled, apex acuminate. Stamens 3, ca. 1.5 mm long. Ovary linear.
Flowering period: April-October.
Herbarium specimens examined: NARA DESERT: *Rahmatullah Qureshi*, 998 (SALUBGH).
Field remarks: A common grass of wetland habitat.
Distribution: Pakistan (Sindh, Baluchistan, Punjab, N.W.F.P., Gilgit & Kashmir); tropical Africa, Polynesia, northern Australia and tropical Asia (Cope, 1982).

Tribe VI. PANICEAE

1a. Upper floret coriaceous to crustaceous;
lower lemma entire at the tip, awnless.. **10. Panicum**
1b. Upper floret thinly cartilaginous; lower lemma often bilobed at the tip or awnd:
2a. Lower glume present:
3a. Glumes acuminate or awned, rarely only acute;
upper lemma not mucronate... **11. Echinochloa**
3b. Glumes awnless, if acuminate then with the upper
lemma not mucronate.. **12. Paspalidium**
4a. Lemma of the upper floret thinly cartilaginous,

usually with flat, hyaline margin..**13. Digitaria**
4b. Lemma of the upper floret more or less crustaceous or coriaceous, usually with narrow, inrolled margins exposing much of the palea:
5a. Lower glume turned towards the rachis, the back of upper lemma turned away from it. i.e. spikelets adaxial..............................**14. Brachiaria**
5b. Lower glume turned away from the rachis of the racemes of spikes, the back of the upper lemma facing it i.e. soikelets abaxial:
6a. Inflorescence of spiciform racemes. Involucre of bristles thickened into a hard cup..**15. Cenchrus**
6b. Inflorescence of effuse or spiciform panicles. Involucre of bristles not thickened as above..**16. Setaria**
2b. Lower glume absent...**17. Paspalum**

10. Panicum

L.., Sp. Pl. 1: 55 (1753); Sultan & Stewart, Grasses W. Pak. 1: 22 (1958); Bor, Grasses Burma Ceyl. Ind. Pak. 319 (1960); Bor in Rech. f., Fl. Iran. 70: 468 (1970); Cope in Fl. Pak. 143: 161 (1982); Matthew, Fl. Tamilnadu Carn., 3: 1879 (1983).

Annuals or perennials. Culms often densely clustered; nodes bearded or not. Leaf-sheaths hispid or tuberculate-pubescent; ligule membranous, fimbriate or with a tuft of elongate pilose hairs; blades rolled or flat, linear-ovate. Inflorescence a panicle, open with ascending or spreading branches. Spikelets lanceolate or ellipsoid, ovoid or oblong, glabrous or not somewhat oblique, falling entire or nearly so, 2-flowered. Glumes unequal, dissimilar, hyaline to membranous; lower glume shorter than the upper and clasping the base of the spikelets; upper glume as long as the spikelets; lower floret male or sterile; lemma similar to upper glume; palea often reduced or absent. Upper floret bisexual, lemma as long as spikelets, crustaceous, smooth or indurated-rugulose, clasping the margins of the palea. Anthers 3. Caryopsis ellipsoid, dorsally compressed.

A genus of ca. 600 species in tropical and warm temperate regions; 15 species are recorded from Pakistan of which 7 are probably native, 3 are established aliens and 5 are found only in cultivation (Cope, 1982). Only 1 species represented by the Nara Desert.

Panicum turgidum Forsk., Fl. Aegypt.-Arab. 18 (1775); Stewart in Ali and Nasir, Ann. Cat. Vasc. Pl. W. Pak. & Kashm., 122 (1972); Cope in Ali and Nasir, Fl. Pak., 143: 167 (1982).

Glaucous suffruticose perennial grass forming bushes up to 1 m high. Roots fibres thick, felty. Culms ca. 2.5-3 mm across in middle, terete, woody, solid, smooth, polished; branches in fascicles from nodes, swollen at the base. Leaf-blades few, basal, ca. 2.5-7.5 cm long, those of the upper nodes often reduced to open, chartaceous sheaths with a setiform blade which are flat, coriaceous, linear-lanceolate, apex acuminate, smooth, with distant spinules along margins; sheaths ca. 2-7 cm long, firm to coriaceous, the upper very light, glabrous, smooth, finely striate; ligule a dense, ciliate rim. Panicles terminal, ca. 4-10 cm long, subpyramidal, branches at first erect, later more or less spreading; rachis angular, grooved, glabrous. Spikelets ca. 3-4 mm long, solitary or rarely 2-nate, glabrous-glaucous, subsecund, ovoid, turgid, white; pedicels subterete, scabridulous, tip cupular. Glumes 4,

firmly membranous, subequal, subcoriaceous; lower involucral glume ca. 2.5 mm long, lanceolate, concave, 5-7-nerved; upper floral glume much smaller than lower, ovate-oblong, apex acute-rotundate or apiculate, convex on both sides.
Flowering period: April-May.
Herbarium specimens examined: NARA DESERT: *Rahmatullah Qureshi*, 606, 700, 942, 1089, 1091, 1105 (SALUBGH).
Field remarks: A less common grass found growing on dunal areas.
Distribution: Pakistan (Sindh & Baluchistan); from Somalia westward through Ethiopia and the Sudan to Morocco, and eastwards from Egypt through Arabia and southern Jordan to Pakistan (Cope, 1982).

11. Echinochloa

P. Beauv., Ess. Agrost. 53 (1812); Sultan & Stewart, Grasses W. Pak. 1: 43 (1958); Cope in Ali and Nasir, Fl. Pak., 143: 192 (1982).

Annuals or perennials. Leaf-blades linear, flat; ligule absent or represented by a line of hairs. Inflorescence composed of racemes arranged on a central axis; rhachis triquetrous, scabrid, often setose, the spikelets paired (very rarely single) or on short side branchlets and forming (2-)4 or more rows. Spikelets narrowly elliptic to sub-rotund, convex on the back, flat on the front, ± hispid, cuspidate or awned, rarely obtuse and then the spikelet borne upon a short stipe; lower glume ovate, ± one-third the length of the spikelet, typically acute to acuminate; upper glume as long as the spikelet, acute to acuminate, rarely with a short awn-point or obtuse; lower floret male or barren, with or without a palea, its lemma resembling the upper glume but often awned; upper lemma crustaceous, smooth and shining, acute (very rarely obtuse), clasping only the margins of the palea; upper palea acute (rarely obtuse), with the tip briefly reflexed and ± protuberant between the lemma margins. Caryopsis broadly elliptic, dorsally flattened.

A genus of about 20 species in tropical and warm temperate regions through-out the world; 5 species occur in Pakistan.

Echinochloa colona (L.) Link, Hort. Berol. 2: 209 (1833); Cope in Ali and Nasir, Fl. Pak., 143: 196 (1982).

Syn: *Panicum colonum* L., Syst. Nat., ed. 10, 2: 870 (1759); *P. brizoides* L.,Mant. 2: 184 (1771); *Oplismenus colonum* (L.) H.B K., Nov. Gen. Sp. 1: 108 (1816); *Echinochloa crusgalli* subsp. *colona* (L.) Honda in Bot. Mag., Tokyo 37:122 (1923).

Annual; culms 10-65 cm high, erect or ascending. Leaf-blades 5-30 cm long, 2-8 mm broad, occasionally marked with purple bars; ligule absent; sheaths glabrous. Inflorescence typically linear, 1-15 cm long, the racemes neatly 4-rowed, seldom over 3 cm long, simple, commonly ± half their length apart and appressed to the axis but sometimes sub-verticillate and spreading. Spikelets ovate-elliptic to subglobose, 1.5-3 mm long, pubescent; lower lemma acute to cuspidate (rarely with a subulate point up to 1 mm long); upper lemma 2-3 mm long.
Flowering & Fruiting Period: May -September

Herbarium specimens examined: NARA DESERT: *Rahmatullah Qureshi*, 1406 (SALUBGH).

Field remarks: A less common weed found in flat habitat.

Distribution: Pakistan (Sindh, Baluchistan, Punjab, N.W.F.P. & Kashmir); through-out the tropics and subtropics (Cope, 1982).

12. Paspalidium

Stapf in Prain, Fl. Trop. Afr. 9: 582 (1920); Sultan; & Stewart, Grasses W. Pak. 1: 39 (1958); Cope in Ali and Nasir, Fl. Pak., 143: 188 (1982).

Annuals or perennials, often aquatic. Leaf-blades linear; ligule a ciliate rim Inflorescence composed of several to many short straight racemes spaced along a common axis and ± appressed to shallow hollows in it; rhachis triquetrous or winged, usually ending in a point, bearing single imbricate spikelets, their lower glume abaxial. Spikelets ovate, dorsally compressed, glabrous; lower glume much shorter than the spikelet; upper glume mostly almost as long as the spikelet; lower floret male or barren, its lemma resembling the upper glume; upper lemma crustaceous, acute, clasping only the edges of the palea; upper palea acute, the tip briefly reflexed. Caryopsis elliptic, dorsally compressed.

A genus of about 27 species distributed in the tropics, 19 of them confined to Australia. Three species occur in Pakistan (Cope, 1982).

Paspalidium geminatum (Forssk.) Stapf in Prain, Fl. Trop. Afr. 9: 583 (1920); Cope in Ali and Nasir, Fl. Pak., 143: 189 (1982).

Syn: *Panicum geminatum* Forssk., Fl. Aegypt.-Arab. 18 (1775); *P. fluitans* Retz., Obs. Bot. 3: 8 (1783); *P. paspaloides* Pers., Syn. Pl. 1: 81 (1805).

Perennial, with creeping or floating ± spongy rhizomes; culms 10-60 cm high, prostrate and rooting at the lower nodes. Leaf-blades 5-30 cm long, 2-10 mm wide, flat or folded, setaceously acuminate. Inflorescence 5-30 cm long; racemes 0.54 cm long, usually less than their own length apart, their rhachis narrowly winged from a triquetrous midrib, 0.5-1 mm wide, glabrous or sparsely ciliate. Spikelets ovate, 1.6-2.6(2.9) mm long; lower glume truncate, a quarter to a third the length of the spikelet; upper glume two-thirds to four-fifths the length of the spikelet; lower floret with a palea, its lemma as long as the spikelet, the nerves not raised; upper lemma granulose.

Flowering & Fruiting Perioid: mostly October-January, but sometimes as early as August or as late as March.

Herbarium specimens examined: NARA DESERT: *Rahmatullah Qureshi*, 1405 (SALUBGH).

Field remarks: A less common grass of wetland habitat.

Distribution: Pakistan (Sindh): Old World tropics (Cope, 1982).

13. Digitaria

Haller, Hist. Stirp. 2: 244 (1768); Sultan & Stewart, Grasses W. Pak. 1: 5 (1958); Cope in Nasir & Ali., Fl. Pak., 143: 216 (1982).

Syntherisma Walt., F. Carol. 76 (1788); *Trichachne* Nees, Agrost. Bras. 85 (1829); *Digitariopsis* C.E. Hubbard in Hook., Icon. Pl. 35: t. 3420 (1940); *Digitariella* de Winter in Bothalia 7: 467 (1961).

Annuals or perennials. Leaf-blades mostly linear and flat; ligule short, scarious or membranous, glabrous or ciliate. Inflorescence composed of racemes, these digitate or borne upon an elongated central axis, rarely solitary, sometimes with secondary branchlets; rhachis flat or triquetrous, bearing the spikelets in appressed groups of 1-5 or more, their lower glume abaxial; pedicels terete or triquetrous, usually smooth or scaberulous. Spikelets lanceolate to oblong-elliptic, flattened on the front, convex on the back; lower glume small or suppressed; upper glume membranous, as long as the spikelet, or much shorter and exposing the upper lemma; lower floret barren, represented by a prominently nerved membranous lemma as long as the spikelet (rarely much reduced), usually hairy, typically with the hairs forming stripes between the 1st and 2nd lateral nerves and along the margin; upper floret bisexual, the lemma chartaceous to cartilaginous, finely longitudinally striate, with its hyaline margins enfolding and concealing most of the plea, ± dorsally compressed, usually rounded on the back, subacute to acuminate, rarely rostrate; caryopsis oblong, plano-convex in section, mostly acute to subacute.

A genus of about 200 species in tropical and warm temperate regions, particularly in the Old World; 14 species occur in Pakistan, 2 of them in cultivation only, 1 an alien weed.

Digitaria sanguinalis (L.) scop., Fl. Cam., ed. 2, 1: 52 (1772); Cope in Nasir & Ali., Fl. Pak., 143: 231 (1982).

Syn: *Panicum sanguinale* L., Sp. Pl. 1: 57 (1753); *Paspalum sanguinale* (L.) Lam., Tab. Encycl. 1: 176 (1791); *Digitaria sanguinalis* subsp. *vulgaris* var. *rottleriana* Henr., Monogr. Genus Digitaria 986 (1950); *D. sanguinalis* subsp. *aegyptiaca* var. *frumentacea* Henr., l.c. 985; *D. sanguinalis* subsp. *pectiniformis* Henr., l.c. 984; *D. cruciata* auct. *non* Nees ex Steud. (1854).

Annual; culms 20.60 cm high or sometimes more, decumbent at the base, geniculately ascending. Leaf-blades broadly linear to narrowly lanceolate, 3-30 cm long, 3-17 mm wide. Inflorescence digitate or subdigitate (rarely with an axis up to 15 cm long in robust specimens), composed of 2-16 racemes; racemes stiff, 3-20 cm long, the spikelets binate and overlapping by about two-thirds their length on a winged rhachis with triquetrous midrib; pedicels triquetrous, scabrid, obtuse or slightly expanded at the tip. Spikelets narrowly elliptic to ovate elliptic, 2.3-3.5 mm long, sharply acute; lower glume an ovate scale 0.2 mm long; upper glume a third to half as long as the spikelet, 3-nerved; lower lemma as long as the spikelet, 7-nerved, the nerves evenly spaced (or occasionally with a wide interspace flanking the midrib), scaberulous on the nerves with minute siliceous spines, obscurely and appressedly pubescent, rarely with brownish glassy bristles very rarely with a ciliate frill; fruit lanceolate, grey or light brown.

Flowering & Fruiting Period: July-September.

Herbarium specimens examined: NARA DESERT: *Rahmatullah Qureshi*, 1397 (SALUBGH).

Field remarks: A common weed of plain areas.

Distribution: Pakistan (Baluchistan, Punjab, N.W.F.P., Gilgit & Kashmir); warn temperate regions throughout the World, penetrating into the tropics (Cope, 1982).

14. Brachiaria

(Trin.) Griseb. in Ledeb., Fl. Ross. 4: 49 (1853); Sultan & Stewart, Grasses W. Pak. 1:30 (1958); Cope in Nasir & Ali., Fl. Pak., 143: 201 (1982).

Annual or perennials. Culms erect, creeping or floating. Leaf-blades linear to lanceolate; ligule a fringed membrane or a fringe of hairs. Inflorescence composed of racemes borne upon a common axis; rachis triquetrous to ribbon-like, bearing single or sometimes paired spikelets. Spikelets 2-flowerd, awnless or mucronate, terete or abaxially flattened, ovate to oblong, plump, obtuse to acute, sometimes the lowest internode elongated, then often acrescent to the sheathing base of the lower glume and forming a short stipe; lower floret male or sterile; upper floret bisexual; glumes 2, unequal; lower glume mostly shorter than the spikelet; upper glume as long as the spikelet, membranous or herbaceous; lemma of lower floret as long as the spikelet, shorter or absent; lemma of upper floret turned away from the lower glume; upper palea obtuse to subacute. Caryopsis elliptic, dorsally compressed.

Key to species:

1a. Upper lemma smooth, shining, obtuse; spikelets imbricate................**1. B. eruciformis**
1b. Upper lemma granulose to rugose, subacute to mucronulate:
2a. Upper glume and lower lemma thinly cartilaginous,
usually smooth and slightly shiny, sometimes papilose pubescent;
spikelets 2.5-4.5 mm long...**2. B. ovalis**
2b. Upper glume and lower lemma membranous, glabrous or pubescent;
spikelets 2.5-3.5 mm long...**3. B. ramosa**

1. Brachiaria eruciformis (J. E. Smith) Griseb. in Ledeb., Fl. Ross. 4: 469 (1853); Cope in Nasir & Ali., Fl. Pak., 143: 203-205 (1982). Figure 20C

Syn: *Panicum eruciforme* J. E. Smith in Sibth. & sm., Fl. Graeca 1: 44- 59 (1806); *Braciaria isachne* (Roth ex Roem. & Schult.) Stapf in prain, Fl. Trop. Afr. 9: 552 (1919).

Annual herb, erect up to 50 cm tall. Culms erect, geniculately ascending. Leaf-blades linear-lanceolate, up to 12x0.8 cm, pubescent, margin ciliate with setose hairs (unicellular), apex acuminate. Ligule a ring of hairs, ca. 1 mm wide. Inflorescence composed of racemes up to 7.5 cm tall, oblong, raceme branches 7-10, up to 2 cm, second, imbricate on a triquetrous rachis. Spkelets elliptic, 2 x 1 mm, densely pubescent, subacute; lower glume 0.5 mm across; upper glume 2 x 1mm; lower lemma pubescent, 2x0.8 mm, its palea 1.5 x 0.5 mm; upper lemma smooth, shiny 2 x 0.6 mm; palea 1.8 x 0.5 mm. Stamens 3; anthers ca. 0.5 mm long, basifixed. Stigma ca. 0.5 mm long, plumose.

Flowering Period: July-September.

Herbarium specimens examined: NARA DESERT: *Rahmatullah Qureshi*, 1391 (SALUBGH).

Field remarks: A common weed of plain areas.

Distribution: Pakistan (Sind, Baluchistan, Punjab, N.W.F.P. & Kashmir); South Africa to Mediterranean; eastward to India (Cope, 1982).

2. Brachiaria ramosa (L.) Stapf in Prain, Fl. Trop. Afr. 9: 542 (1919); Stewart in Nasir & Ali., Ann. Cat. Vasc. Pl. W. Pak. & Kashm., 116 (1972); Cope in Nasir & Ali., Fl. Pak., 143: 207 (1982).

Syn: *Panicum petiveri* Trin., Gram. Pan. 144 (1826); *P. ramosum* L., Mant. 1: 29 (1767); *Urochloa ramosa* (L.) Nguyen in Nov. Sist. Vyssh. Rast. 13 (1966).

Loosely tufted annual; culms 10-70 cm high. Leaf-blades broadly linear, 2-25 cm long, 4-14 mm wide. Inflorescence of 3 racemes borne on an axis 3-10 cm long; racemes 1-8 cm long, simple or the longest with branchlets at the base, bearing mostly paired loosely contiguous spikelets appressed to the triquetrous rhachis; pedicels shorter than the spikelets, 1-2 mm long. Spikelets elliptic to broadly elliptic, 2.5-3.5 mm long, glabrous or pubescent, acute to cuspidate, with or without a stipe up to 0.5 mm long; lower glume a third to half the length of the spikelet; upper glume and lower lemma membranous, or rarely the latter coriaceous; upper lemma rugose, subacute to acute.

Flowering & Fruiting Period: July-October.

Herbarium specimens examined: NARA DESERT: *Rahmatullah Qureshi*, 1392 (SALUBGH).

Field remarks: A common weed of plain areas.

Distribution: Pakistan (Sind, Baluchistan, Punjab, N.W.F.P. & Kashmir); Senegal to Yemen and southwards to Malawi, Rhodesia and South Africa; tropical Asia (Cope, 1982).

3. Brachiaria ovalis (L.) Stapf in Prain, Fl. Trop. Afr. 9: 546 (1919); Stewart in Nasir & Ali., Ann. Cat. Vasc. Pl. W. Pak. & Kashm., 115 (1972); Cope in Nasir & Ali., Fl.Pak., 143: 207 (1982).

Tufted annual; culms 10-50(150) cm high. Leaf-blades broadly linear to narrowly lanceolate, 3-25 cm long, 2-12 mm wide. Inflorescence of 3-15 spreading or appressed racemes borne on an axis 4-15 cm long; racemes 3-7 cm long, bearing single or paired spikelets loosely spaced on a triquetrous rhachis; pedicels shorter than the spikelets. Spikelets elliptic, 2.5-4.5 mm long, glabrous or papillose pubescent, obtuse to acute, with a stipe 0.3-0.7 mm long; lower glume a quarter to half the length of the spikelet; upper glume and lower lemma thinly cartilaginous, smooth, slightly shiny; upper lemma granulose to rugulose, as long as or up to 1 mm shorter than the spikelet, subacute to mucronulate.

Flowering & Fruiting Period: September-October.

Herbarium specimens examined: NARA DESERT: *Rahmatullah Qureshi*, 1393 (SALUBGH).

Field remarks: A less common weed of plain areas.

Distribution: Pakistan (Sind); Kenya and Sudan to southern Arabia (Cope, 1985).

15. Cenchrus

L., Sp. Pl. 2: 1049 (1753); Gen. Pl. ed. 5: 470 (1754); Hook. f., Fl. Brit. Ind. 7: 89 (1896); Sultan & Stewart, Grasses W. Pak. 1: 57 (1958); Cope in Fl. Pak. 143: 240 (1982); Matthew, Fl. Tamilnadu Carn., 3: 1816 (1983).

Annuals or perennials; culms ascending, glabrous at nodes. Leaf-sheaths flabellate; ligule membranous, with an apical tuft of elongate pilose hairs; blades flat or folded. Inflorescence a cylindrical spike-like panicle with an angular rachis, pubescent. Spikelets 2 or 3, lanceolate to ovate, dorsally compressed, acute to acuminate, enclosed by a deciduous involucre of bristles in 1 or more whorls; bristles barbellate, flattened and often spiny, basally connate, unequal . Glumes unequal; over glume nearly half as long as spikelets, sometimes suppressed; upper glume slightly shorter; lower lemma male or sterile, as long as spikelet, membranous; upper lemma rigid, the margins covering up to two-thirds of the palea. Anthers 3. Caryopsis ellipsoid or obovoid, dorsally compressed.

A genus of 22 species in tropical and warm temperate regions, mainly in Africa and America. Five species are recorded form Pakistan (Cope, 19882).

Key to species

1a. Bristles of the invloucres retrosely scabrid..**1. C. biflorus**
1b. Bristles of the invloucres antrosely scabrid:
2a. Bristles connate at the base only:
3a. Annuals; involucres with wide, naked, connate base;
inner bristles stouter at base, subequal, ciliate on both margins,
outer longer, flexuous, scabrid upto 2 cm long....................................**2. C. prieurii**
3b. Perennials; involucres with small base, elliptic, naked;
inner bristles not very stout at base, ciliate on margins,
not more than 1.5 c long..**3. C. ciliaris**
2b. Bristles connate into a cup, ca. 1-3 mm long:
4a. Inner bristles very slender, widened into a cup at the base,
tapering to a setiform tip, the outer bristles slender numerous........**4. C. pennisetiformis**
4b. Inner bristles rigid, flattened and subulate, acute at the tip,
not drawn out; the outer few or absent..**5. C. setigerus**

1. Cenchrus biflorus Roxb. Fl. Ind. I:238, (1820); Stewart in Nasir & Ali., Ann.Cat.Vasc.Pl.W.Pak. & Kashm., 116 (1972); Cope in Nasir & Ali., Fl. Pak., 143: 243-44 (1982). **Figure 20D**

Syn.: *C. barbatus* Schum. Baskr. Guin. Pl. 43. (1827); *C. catharticus* Delie, Cat. Hort. Monsp. 1838:4, (1839).

Annual grass upto 1 m high. Culm geniculately ascending, rarely rooting at lower nodes, slender, slightly compressed or terete, simple or branched, scabridulous towards the panicle, otherwise glabrous and smooth. Leaf-blades ca. 2.5-12 x 0.3-0.6 cm in size, linear-lanceolate, finely scabrid on both side, apex finely acuminate; sheath glabrous; ligule a densely ciliate rim. Racemes dense, ca. 2.5-15 x 1-1.2 cm in size, more or less enclosed in upper most leaf-sheath, straw-coloured, rarely purplish; rachis angular, scabrid on the angles;

involucres subsessile or minutely stalked, ca. 5-6 mm across; bristles numerous, united at base into a shallow crescent-shaped or elliptic disc, the outer row of shorter, equal-lengthed, rigid, retrosely scabrid, at length horizontally spreading or reflexed, hook-like; inner bristles 7-14, subequal, ca. 10 mm long, rigid, subulate, flattened, erect, grooved on the outside, with 1-3 green nerves in the grooves, densely ciliolate on the margins inside, retrosely scabrid above, tips usually pungent and slightly spreading. Spikelets shorter than the inner bristles. Glumes 4, thinly membranous, apex obtuse or acute; lower involucral glume ca. 2 mm long, lanceolate, apex acute, hyaline, the upper ca. 4 mm long, ovate or broadly ovate, 1-5-nerved, hyaline; lower floral glume ovate, apex acuminate, 3-5-nerved, often mucronate barren; upper floral glume as long as the lowers one, cuspidately acuminate, 5-nerved, membranous, chartaceous. Grains ca. 2.5 mm long, elliptic or elliptic-oblong, dorsally compressed, rugulose.

Flowering period: January-April & again September-November.

Herbarium specimens examined: NARA DESERT: *Rahmatullah Qureshi*, 626, 623, 837, 45, 1052, 1089, 110, 41 (a), 652, 1148, 1180, 1243 (SALUBGH).

Field remarks: A very common grass found in flat area.

Distribution: Pakistan (Sindh, Baluchistan, Punjab, N.W.F.P., Gilgit & Kashmir); tropical Africa, extending through Arabia to India (Cope, 1982).

2. Cenchrus prieurii (Kunth) Maire in Bull, Mus. Hist. Nat. Paris, ser.2, 3: 523 (1931); Stewart in Nasir & Ali., Ann. Cat.Vasc. Pl. W. Pak. & Kashm., 117 (1972); Cope in Nasir & Ali., Fl.Pak.,143: 244-45 (1982)

Syn: *Cenchrus macrostachyus* Hochst. ex Steud., Syn Pl. Glum. 1: 109 (1855); *Pennisetum brevifolium* Steud. l.c. 107; *Pennisetum prieurii* Kunth, Rév. Gram. 2: 411, t. 119 (1831).

Annual; culms 12-35 cm high. Leaf-blades 5-7 cm long, 3.5-5 mm wide. Panicle cylindrical, 5-9 cm long, the rhachis flexuous and scabrid; involucre elongate, 15-27 mm long; inner bristles greatly exceeding the spikelets, flattened and connate only at the base, forming a shallow oval or diamond-shaped disc, densely ciliate below, the outer surface with one or two shallow grooves, filiform and flexuous above, antrorsely scaberulous; outer bristles filiform. Spikelets 1 or 2 per burr, about 4.3 mm long, acutely lanceolate; lower glume short, up to 0.8 mm long, or absent; upper glume a little shorter than the spikelet.

Flowering period: September- October.

Herbarium specimens examined: NARA DESERT: *Rahmatullah Qureshi*, 1378 (SALUBGH).

Field remarks: A less common grass found in interdunal areas.

Distribution: Pakistan (Sind); northern Africa, Arabia and India (Cope, 1982).

3. Cenchrus ciliaris L., Mant. 2: 302 (1771); Stewart in Nasir & Ali., Ann. Cat.Vasc. Pl.W. Pak. & Kashm., 116 (1972); Cope in Nasir & Ali., Fl.Pak.,143:241-42 (1982). **Figure 21A**

Syn: *Pennisetum cenchroides* Rich. in Pers., Syn. Pl. 1: 72 (1805), nom. Superfl., based on *C. ciliaris* L.; *P. ciliare* (L.) Link, Hort. Berol. 1: 213 (1827); *Cenchrus longifolius* Hochst. ex. Steud., Syn. Pl. Glum. 1: 109 (1854).

Perennial grass, forming mats or tussocks. Culm up to 120 cm high, ascending from a decumbent base, rarely erect, compressed, rigid, tufted; branches solitary or fascicled; lower internodes glabrous or scanty hairy, smooth or scabrid towards the inflorescence. Leaf-blades ca. 10-25 x 0.3-0.6 cm in size, linear, tapering into a fine point, glabrous or loosely beset with fine, stiff, tubercle-based hairs, flat or convolute when dry; sheaths glabrous or scabridulous, shorter or longer than internodes; ligule a densely ciliolate rim. Racemes ca. 3-15 x 0.8-1.3 cm in size, dense, slender or stout, pale or purplish; rachis angular, scabridulous. Involucels subsessile, spreading; bristles numerous, the outer very slender, scabridulous, shorter or slightly longer than the spikelets, the inner thickened and flattened, connate towards base into a ring, ciliate around with fine, spreading, tubercle-based hairs, ca. 7-12 cm long. Spikelets 1-3 in each involucels, ca. 4-5 mm long, oblong-lanceolate. Glumes 4, ovate to broadly ovate-oblong, acute or acuminate, membranous; lower involucral glume ca. 2.5 mm long, ovate-lanceolate, acute, faintly 1-nerved, hyaline; upper involucral glume ca. 3 mm long, ovate, acuminate, 1-3-nerved, hyaline; lower floral glume male or barren, rarely bisexual, ca. 4-5 mm long, ovate-lanceolate; upper as the lower, 5-nerved, membranous. Grains ca. 2 mm long, broadly oblong, truncate, dorsally compressed.

Flowering period: February- March.

Herbarium specimens examined: NARA DESERT: *Rahmatullah Qureshi*, 255, 933, 930, 1098, 1092, 1088, 41 (b), 1297 (SALUBGH).

Field remarks: A very common grass found in the whole area.

Distribution: Pakistan (Sindh, Baluchistan, Punjab, N.W.F.P., Gilgit & Kashmir); tropical Africa, extending through Arabia and the Middle East to India; it is widely introduced elsewhere in the Old World (Cope, 1982).

4. Cenchrus pennisetiformis Hochst. & Steud. ex Steud., Syn. Pl. Glum. 1:109 (1854); Stewart in Nasir & Ali., Ann. Cat.Vasc. Pl. W. Pak. & Kashm., 116 (1972); Cope in Nasir & Ali., Fl.Pak.,143: 242 (1982).

Syn: *C. echinoides* Wight ex Steud., Nom. Bot., ed 2, 1: 317 (1840), *nom. nud.*; *Pennisetum cenchroides* var. *echinoides* Hook. f., Fl. Brit. Ind. 7: 88 (1896).

Annual or short-lived perennial; culms 10-40 cm high, ascending. Leaf-blades 2-20 cm long, 2-5 mm wide. Panicle 2-6 cm long; involucre elongate, 6-16 mm long; inner bristles greatly exceeding the spikelets, one of them longer and stouter than the rest, flattened at the base, connate for 1-2.5 mm above the basal disc to form a cup, almost glabrous to sparsely ciliate below, grooved on the face or not, filiform above, flexuous, often wavy, antrorsely scaberulous; outer bristles filiform. Spikelets 1-3 per burr, 3-5 mm long.

Flowering & Fruiting Period: February-April and again August-October.

Herbarium specimens examined: NARA DESERT: *Rahmatullah Qureshi*, 1379 (SALUBGH).

Field remarks: A less common grass found in Achhro Thar.

Distribution: Pakistan (Sind & Baluchistan); tropical East Africa through Arabia to India; introduced to Australia (Cope, 1982).

5. Cenchrus setigerus Vahl, Enum. Pl. 2: 395 (1806); Stewart in Nasir & Ali., Ann. Cat.Vasc. Pl. W. Pak. & Kashm., 117 (1972); Cope in Nasir & Ali., Fl.Pak., 143: 245-46 (1982)

Syn: *C. biflorus* of Hook. f., Fl. Brit. Ind. 7: 89 (1896); *C. bulbifer* Hochst. ex Boiss., Fl. Or. 5: 448 (1884); *C. montanus* Nees ex Royle, Ill. Bot. Him. 416 (1840), nom. nud.; *C. uniflorus* Ehr. ex Boiss.. l.c.; *Pennisetum vahlii* Kunth, Rév. Gram. 1: 49 (1829).

Perennial, forming clumps from a ± bulbous base; culms 5-80 cm high, geniculately ascending. Leaf-blades 2-20 cm long, 2-7 mm wide. Panicle 2-12 cm long; involucres cup-shaped, 3-7 mm long; inner spines short, flattened, connate for a quarter to half their length to form a cup, glabrous or obscurely puberulous and grooved on the face, occasionally sparsely ciliate on the margins, green to dark purple, antrorsely scaberulous, narrowly triangular at the tip; outer spines few, short, often suppressed, rarely almost as long as the inner. Spikelets 1-3 per burr, 3-5 mm long.

Flowering & Fruiting Period: August-January and again in April.

Herbarium specimens examined: NARA DESERT: *Rahmatullah Qureshi*, 1380 (SALUBGH).

Field remarks: A less common grass found in Achhro Thar.

Distribution: Pakistan (Sind, Punjab & N.W.F.P); tropical East Africa, through Arabia to India; introduced in several tropical countries (Cope, 1982).

16. Setaria

P. Beauv., Ess. Agrost. 51 (1812); Hook. f., Fl. Brit. Ind. 7: 77 (1896); Sultan & Stewart, Grasses W. Pak. 1: 52 (1958); Cope in Fl. Pak. 143: 177 (1982); Matthew, Fl. Tamilnadu Carn., 3: 1899 (1983).

Cymbosetaria chweik. in. Hook., Icon. Pl. 34: t.3320 (1936).

Annuals or perennials. Culms tufted or not, rarely stout. Leaf-sheaths compact or loose, glabrous; ligule a ciliate to fimbriate membrane; blades flat or folded, rarely plicate. Inflorescence a panicle, open or spike-like. Spikelets (or only the terminal) subtended by 1 or more persisting bristles, oblong to ovoid, plano-convex, awnless. Glumes unequal, membranous; lower glume much smaller, abaxial, clasping at base; upper glume as long or nearly so; lower floret sterile or male; lower lemma thin; upper floret bisexual, the upper lemma crustaceous, convex on the back, rugose, clasping margins of palea. Stamens 3. Caryopsis oblong or ellipsoid.

A genus of ca. 100 species in the tropics and subtropics. Six species are recorded from Pakistan (Cope, 19882).

Setaria pumila (Pouir.) Roem. & Schult., Syst. Veg. 2: 891 (1817); Cope in Ali and Nasir, Fl. Pak., 143: 181 (1982). **Figure 23C**

Syn: *Panicum pumilum* Poir., Encycl. Meth. Bot., Suppl. 4, 273 (1816); *P. pallide-fuscum* K. schum., Beskr. Guin. Pl. 58 (1827); *P. rubiginosum* Steud., Syn. Pl. Glum. 1: 50 (1854); *Setaria rubiginosa* (steud.) Miq., Fl. Ned. Ind. 3: 467 (1857); *S. pallide-fusca* (K. Schum.) Stapf & Hubbard in Kew Bull. 1930: 259 (1930); *S. glauca* auct. Non (L.) P. Beauv.

A loosely tufted annual grass. Culms ca. 50 cm long, geniculate, glabrous; nodes glabrescent. Leaf-blades ca. 4.5-12 x 0.4-0.7 cm in size, lanceolate, flaccid, flat; sheaths ca. 2.5-4.5 cm long, glabrous; ligule ca. 1.5 mm. Inflorescence of spikes or speciform panicles, compact, ca. 2-4.5 x 1 cm in size; rachis pilose; involucral bristles ca. 1-8 mm, 9-11 per whorl, antrosely barbed. Spikelet(s) 1 or 3 in a cluster, ca. 2 mm, ovoid. Lower glume ca. 1 mm long, ovate, 1-3-nerved; upper glume ca. 1.5 mm, 5-7-nerved, anastomosing; lower lemma ca. 2 mm long, 5-nerved; palea ca. 1.5 mm long; upper lemma boat-shaped, ca. 2 mm long, indurated, transversely rugose. Anthers ca. 1.5 mm. Caryopsis ca. 1.5 mm long, ellipsoid.

Flowering Period: June-August.

Herbarium specimens examined: NARA DESERT: *Rahmatullah Qureshi*, 1265, 1288 (SALUBGH).

Field remarks: A common grass found in moist habitat.

Distribution: Pakistan (Sindh, Baluchistan, Punjab, N.W.F.P., Gilgit & Kashmir); tropical and warm temperate regions of the Old World; introduced to North America (Cope, 1982).

17. Paspalum

L., Syst. Nat., ed. 10, 2: 855 (1759); Sultan & Stewart, Grasses W. Pak. 1: 36 (1958); Cope in Ali and Nasir, Fl. Pak., 143: 210 (1982).

Annuals or perennials. Leaf-blades flat, linear to narrowly lanceolate; ligule short, membranous. Inflorescence composed of racemes, these digitate or borne upon an elongated central axis, rarely solitary; rhachis flat, narrowly or broadly winged, bearing solitary or paired spikelets, their lower glume abaxial. Spikelets orbicular to oblong, mostly plano-convex; lower glume suppressed, rarely represented by a minute scale; upper glume membranous, as long as the spikelet, rarely shorter or absent; lower floret barren, without a palea, its lemma resembling the upper glume; upper lemma ± crustaceous, usually obtuse, clasping only the edges of the palea. Caryopsis plano-convex.

A genus of about 250 species, mostly distributed throughout the tropics but mainly in the New World. The genus represented 3 species in Pakistan (Cope, 1982).

Key

1a. Spikelets suborbicular to ovoid; upper glume 5-7-nerved.................**1. P. scrobiculatum**

1b. Spikelets ovate; upper glume 3-nerved..**2. P. paspalodes**

1. Paspalum scrobiculatum L., Mant. 1: 29 (1767); Cope in Ali and Nasir, Fl. Pak., 143: 181 (1982).

Syn: *Paspalum orbiculare* Forst., Fl. Ins. Austr. Prodr. 7 (1786); *P. commersonii* Lam., Tab. Encycl. 1: 175, t.43/1 (1791); *P. kora* Willd., Sp. Pl. 1 (1): 332 (1798); *P. polystachyum* R. Br., Prodr. 188 (1810); *P. cartilagineum* Presl, Rel Haenk. 1: 216 (1830); *P. scrobiculatum* var. *commersonii* (Lam.) Stapf in Prain, Fl. Trop. Afr. 9: 573 (1919).

Perennial grass; culms 10-150 cm high, 1-6 mm in diameter, 2-17-noded, the nodes commonly exposed, erect, or ascending from a procumbent base and rooting at the lower

nodes. Leaf-blades linear, 5-40 cm long, 3-15 mm wide, tapering to a filiform tip. Inflorescence composed of 1-20 racemes, these digitate or borne on an axis up to 8 cm long, the lowest raceme 4-15 cm long, with spikelets borne singly on a ribbon-like rhachis 1-2.5 mm wide. Spikelets broadly elliptic, obovate or suborbicular, 1.4-3 mm long, green, becoming brown; lower glume absent; upper glume papery; lower lemma similar or rarely coriaceous, 3-5-nerved (in the latter case the nerves evenly spaced or the laterals close together); upper lemma finely striate, brown at maturity.
Flowering & Fruiting Period: April-May and again October-December.
Herbarium specimens examined: NARA DESERT: *Rahmatullah Qureshi*, 1404 (SALUBGH).
Field remarks: A less common grass found in Achhro Thar.
Distribution: Pakistan (Sindh); throughout the Old World tropics (Cope, 1982).

2. Paspalum paspalodes (Michx.) Scribner in Mem. Torrey Bot. Club. 5: 29 (1894); Cope in Ali and Nasir, Fl. Pak., 143: 211 (1982).

Syn: *Paspalum distichum* auct. non L., Syst. Nat., ed. 10, 2: 855 (1759); *Digitaria paspalodes* Michx., Fl. Bor. Amer. 1: 46 (1803).

A creeping stoloniferous perennial grass up to 35 cm high. Leaf-blades 5-20 cm long, 2.5-5 mm wide. Inflorescence composed of 2(4) conjugate racemes, each raceme 1.5-7 cm long, the spikelets single in 2 rows on a flattened rhachis 1-2 mm wide. Spikelets ovate, 2.5-3.5 mm long, plano-convex, relatively plump, pale-green; lower glume often present as a small scale; upper glume appressed-puberulent, herbaceous, with a distinct mid-vein; lower lemma similar but glabrous: upper lemma smooth, pallid at maturity.
Flowering Fruiting: April-May and again August-September.
Herbarium specimens examined: NARA DESERT: *Rahmatullah Qureshi*, 1403 (SALUBGH).
Field remarks: A less common grass found in Achhro Thar.
Distribution: Pakistan (Sindh, Punjab, N.W.F.P. & Kashmir); tropics and sub-tropics throughout the world (Cope, 1982).

18. Saccharum

L., Sp. Pl. 1: 54 (1753); Gen. Pl., ed. 5: 28 (1754); Sultan & Stewart, Grasses W. Pak. 1: 74 (1958); Bor, Grasses Burma Ceyl. Ind. Pak. 208 (1960); Bor in Rech. f., Fl. Iran. 70: 514 (1970); Cope in Fl. Pak. 143: 259 (1982); Matthew, Fl. Tamilnadu Carn., 3: 1894 (1983).

Erianthus Michx., Fl. Bor. Amer. 1: 54 (1803); Bois., Fl. Or. 5: 454 (1884); Bor, Grasses Burma Ceyl. Ind. Pak. 149 (1960); *Narenga* Bor in Indian Forester 66: 267 (1940); Bor, Grasses Burma Ceyl. Ind. Pak. 197 (1960).

Perennial tall grass, culms tout, solid. Leaf-blades linear, narrow, flat, sometime involute; ligule scarious or a line of hairs. Inflorescence of terminal, plumose panicles, often silvery; racemes usually flexuous, the rachis fragile and bearing; internodes linear, slender, usually more than half the length of the sessile spikelet. Spikelets lanceolate, binate, one sessile, the other pedicelled, surrounded by long silky hairs from callus, falling with joints of rachis; callus very short, truncate. Glumes equal, membranous or coriaceous, the lower flat or rounded on the back; lower floret reduced to lemma, lanceolate, shorter than spikelets, hyaline; upper floret bisexual; lemma lanceolate, hyaline, awnless or sometimes a straight awn, entire or rarely 2-lobed, sometimes reduced to a subulate vestige. Palea small or suppressed. Stamens 2-3. Caryopsis subglobose or oblong.

A genus of 35-40 species throughout the tropics, extending to warm temperate regions. Representing in Pakistan by 9 species (Cope, 1982). Two species are reported from the Nara Desert.

Key to species:

1a. Leaf-blades flat or markedly channelled, the midrib white,
occupying the greater part of the width.......................................**1. S. bengalensis**

1b. Leaf-blades becoming petiolate towards the base, the lamina gradually
reduced to narrow wing on either side of the midrib:

2a. Leaf-blades up to 7.5 mm wide; callus bearded with silky
white hairs 2-3 times longer than spikelets; glumes glabrous..............**2. S. spontaneum**

2b. Leaf-blades up to 2 mm wide on the either side of midrib;
callus bearded with dense cream or yellowish hairs up to 4 mm long;
glumes hairy above..**3. S. griffithii**

1. Saccharum bengalense Retz., obs. Bot. 5: 16 (1789); Stewart in Ali and Nasir, Ann. Cat. Vasc. Pl. W. Pak. & Kashm., 113 (1972); Cope in Ali and Nasir, Fl. Pak., 143: 264 (1982).

Syn: *Sacchrum munja* Roxb., Fl. Ind. 1: 250 (1820); *S. sara* Roxb., 1.c.249; *S. ciliare* Anderss. in Ofvers. K. Vetensk Akad. Forth. 12: 155 (1855); *S. arundinaceum* Hook. f., Fl. Brit. Ind. 7: `119 (1896), non Retz. (1786); *Erianthus munja* (Roxb.) Jeswiet in Archf Suik. Ind. Ned.-Indie (1925), No.12:399 (1925); *E. ciliaris* (Anderss.) Jeswiet., 1.c.; E. sara (Roxb.) Rumke in Archf Suik. Ind.Ned.-Indii (1934), No. 7: 223, t. 2 (1934); *Ripidium bengalense* (Retz.) Grrassl in Proc. 14th Ong., ISSCT: 244 (1972).

A tall perennial, erect up to 6 m high. Culms up to 1.5 cm across, solid above, fistular below, polished, silky below panicles and minutely silky below upper leaf-insertion. Leaf-sheaths up to 20 cm, glabrous; ligule ovate or deltoid, membranous, often fimbriate when old, up to 5 mm high, pilose; blades erect, glaucous, ca. 30-60 cm long, 3-12 mm broad, flat or markedly channelled, the midrib white, occupying the greater part of the width, margins scabrid, often villous above the base. Panicle 20-75 x 4-5 cm across, conical-oblong; branches ascending, reddish or purplish; peduncle glabrous; racemes ca. 2-4.5 cm long, relatively shorter than the supporting branches; internodes and pedicels hirsute with ca. 3-6 mm long hairs. Spikelets slightly heteromorphous, ca. 3.8-6 mm long, the callus bearded with whitish or greyish hairs up to 2 mm long; glumes equal, membranous, lower glume of sessile spikelet hairy on the back, the upper glume glabrous, both glume of pedicelled spikelet hairy; hairs ca. 4-10 mm long; lower lemma oblong-elliptic, hairy on the back; upper lemma ovate-lanceolate, ciliate on the margins, apex acute or very shortly awned, the awn not visible beyond the glumes.

Flowering & fruiting period: October- January.

Herbarium specimens examined: NARA DESERT: *Rahmatullah Qureshi*, 105, 195, 622, 624, 1245 (SALUBGH).

Field remarks: A very common grass found around the water bodies.
Distribution: Pakistan (Sindh, Punjab & N.W.F.P.); North and Northwest India;? Afghanistan (Cope, 1982).

2. Saccharum spontaneum L., Mant. 2: 183 (1771); Stewart in Ali and Nasir, Ann. Cat. Vasc. Pl. W. Pak. & Kashm., 113 (1972); Cope in Ali and Nasir, Fl. Pak., 143: 263 (1982).

Syn: *Imperata spontanea* (L.) P. Beauv., Ess. Agrost. 8 (1812); *Saccharum semidecumbens* Roxb., Fl. Ind. 1: 241 (1820); *S. canaliculatum* Roxb., l.c. 251; *S. propinquum* Steud., Syn.Pl. Glum. 1: 406 (1855).

A rhizomatous perennial, erect grass upto 7 m high. Culms 1.5 cm broad, solid above, fistular below, polished, silky below panicles and minutely silky below upper leaf-insertions. Leaf-blades ca. 0.5-1.5 m long, up to 7.5 mm broad, erect, gradually tapering towards the base into a linear wing on either side of the petiole, glaucous, midrib white, margins scabrid, often villous above the base, immediately above the ligules; sheath longer than internodes, villous at mouth, minutely pubescent at the base, otherwise glabrous with scattered, adpressed hairs, sulcate; ligule ovate or deltoid, membranous, fimbriate when old, upto 6 mm high, minutely ciliate dorsally. Panicles 15-35 x 3-4.5 cm in size, conical-oblong; branches ascending, reddish or purplish in colour, callus hair adpressed to branches of panicle; primary rachis silky with long, white hairs. Spikelets in pairs, awnless, lanceolate, ca. 2.5 mm long, sessile and pedicelled similar, each 1-flowered, bisexual; callus minute, bearded with spreading, silky hairs, upto 2 cm long. Glumes 4; lower involucral glume lanceolate, subulate, apex acuminate; upper involucral glume as much long, lanceolate, obscurely keeled, 1-nerved; lower floral glume ovate-lanceolate, subacute, ciliate, hyaline, nerveless; upper floral glume very slender, ciliate; palea minute, ciliate.
Flowering period: July-September.
Herbarium specimens examined: NARA DESERT: *Rahmatullah Qureshi*, 46, 992, 1273 (SALUBGH).
Field remarks: A common grass found in saline soils.
Distribution: Pakistan (Sindh, Punjab, N.W.F.P., Gilgit & Kashmir); widely distributed in the warmer regions of the Old World (Cope, 1982).

Saccharum griffithii Munro ex Boiss., Fl. Or. 5: 453 (1884); Stewart in Ali and Nasir, Ann. Cat. Vasc. Pl. W. Pak. & Kashm., 113 (1972); Cope in Ali and Nasir, Fl. Pak., 143: 265 (1982).

Syn: *Erianthus griffithii* (Munro ex Boiss.) Hook. f. Fl. Brit. Ind. 7: 122 (1896); *Saccharum ciliare* var. *griffithii* (Munro ex Boiss.) Hack. In DC., Monogr. Phan. 6: 119 (1889).

Caespitose perennial; culms up to 2 m high. Leaf-blades up to 60 cm long, markedly channelled or strongly involute, almost petiolate at the base, the lamina restricted to a narrow wing up to 2 mm wide on either side of the thick white midrib, glaucous. Panicle 20-35 cm long, the peduncle glabrous; racemes 1.5-2.5 cm long, shorter than the supporting branches, the internodes and pedicels hirsute with hairs 4-6 mm long. Spikelets all alike, 4-5(-8) mm long, the callus bearded with dense cream or yellowish hairs up to 4 mm long; glumes equal, membranous, sparsely or densely hairy on the back; lower lemma narrowly lanceolate, acute or awned with an awn up to 4 mm long, rarely as long as 10 mm, often projecting beyond the tips of the glumes.

Flowering & Fruiting Period: June-September.

Herbarium specimens examined: NARA DESERT: *Rahmatullah Qureshi,* 1381 (SALUBGH).

Field remarks: A less common grass found in flat habitat.

Distribution: Pakistan (Sind, Baluchistan, Punjab & N.W.FY.); Afghanistan (Cope, 1982).

19. Imperata

Grasses W. Pak. 1: 70 (1958); Cope in Ali and Nasir, Fl. Pak., 143: 252 (1982).

Rhizomatous perennials with erect, few-noded culms. Leaves mostly basal, the blades narrow and erect; ligule scarious. Inflorescence a narrow, often spiciform panicle, its numerous primary branches bearing very short secondary racemes; raceme rhachis not fracturing at maturity, bearing paired similar spikelets each supported on a slender pedicel. Spikelets lanceolate to oblong, ± terete, enveloped in long silky hairs from callus and glumes; callus very short, truncate; glumes equal, as long as the spikelet, lanceolate to oblong, membranous; lower floret reduced to a lemma, this lanceolate to oblong, shorter than the spikelet, hyaline; upper floret bisexual; lemma lanceolate to oblong, shorter than the spikelet, hyaline, awnless; palea broad, hyaline, enfolding the flower; stamens 1-2, caryopsis ellipsoid.

A genus of about 8 species, occurring throughout the tropics and extending to warm temperate regions; 1 species occurs in Pakistan.

Imperata cylindrica (Linn) Raeuschel., Nom. Bot., ed. 3, 10 (1797); Cope in Ali and Nasir, Fl. Pak., 143: 252 (1982).

Syn: *Lagurus cylindricus* L., Syst. Nat., ed. 10, 2: 878 (1759); *Saccharum cylindricum* (L.) Lam., Encycl. Méth. Bot. 1: 594 (1785); *S. koenigii* Retz., Obs. Bot. 5: 16 (1789); *S. thunbergii* Retz., l.c. 17; *Imperata arundinacea* Cyr., Pl. Rar. Neap. 2: 27, t. 11 (1792); *I. allang* Jungh. in Tijdschr. Natuural. Gesch. Physiol. 7: 295 (1840); *I. koenigii* var. *major* Nees, Fl. Afr. Austr. 90 (1841); *I. arundinacea* var. *africana* Anderss. In Öfvers.K. Vetensk.Akad. Förh. Stockh. 12(3): 159 (1855); *I. arundinacea* var. *europaea* Anderss., lc., based on *I. cylindrical* (L.) Raeuschel. var. *thunbergii* (Retz.) Stapf in Thiselt.-Dyer, Fl. Cap. 7: 320 (1898).

An aggressive rhizomatous perennial, forming tufts of leaves from a scaly rhizome. Culms up to 85 cm high, erect. Leaf-blades basal, flat (in Southeast Asia), 3-100 cm or more long, 2-20 mm wide, stiffly erect. Panicle spiciform, cylindrical, sometimes with the lowermost branches loose, 3-22 cm long, obscured in copious silky white hairs. Spikelets 2.2-6 mm long; stamens 2.

Flowering & Fruiting Period: April-June.

Herbarium specimens examined: NARA DESERT: *Rahmatullah Qureshi,* 1402 (SALUBGH).

Field remarks: A less common grass found in Achhro Thar near water bodies.

Distribution: Pakistan (Sindh, Baluchistan, Punjab, N.W.F.P., Gilgit & Kashmir); throughout the Old World tropics, extending to the Mediterranean and the Middle East; also in Chile (Cope, 1982).

20. Cymbopogon

Spreng., Pl. Pugill. 2: 14 (1815); Sultan & Stewart, Grasses W. Pak. 1: 110 (1959); Bor, Grasses Burma Ceyl. Ind. Pak. 121 (1960); Bor in Rech. f., Fl. Iran. 70: 541 (1970); Cope in Fl. Pak. 143: 326 (1982); Matthew, Fl. Tamilnadu Carn., 3: 1826 (1983).

Robust perennials, tufted, often aromatic. Culms erect, sometimes basally decumbent, glabrous; nodes swollen, bearded or not. Leaf-sheaths glabrous, sometimes tomentose; ligule membranous or scarious; blades linear. Inflorescence a large complex panicles, borne on a short common peduncle and enclosed by boat-shaped spatheole, raceme-bases equal or unequal, usually deflexed at maturity, often the lower raceme-bases and/or lowermost pedicel swollen; rachis and pedicels terete or flat, hollow. Spikelets binate, lowest pair of one or both racemes homogamous, male or empty. Sessile spikelets: Dorsally compressed; callus obtuse, inserted in the concave or cupular tip of the internode; lower floret empty, upper one bisexual. Lower glume chartaceous, often streaked with oil glands, equal to spikelets, 2-keeled, narrowly or broadly winged along the keels, flat, concave or with a V-shaped, 1-keeled at back, with or without intercarinal-nerves; lower floret reduced to a hyaline lemma, glabrous, often ciliate along margins, epaleate; upper lemma hyaline, deeply 2-fid, lobes filiform or linear, sinus with a geniculate awn. Anthers 3. Caryopsis oblong. Pedicelled spikelets: Lower glume as long as spikelets, flat at back, 7- or more-nerved; upper glume boat-shaped, 1-keeled; lower lemma hyaline, epaleate. Anthers 3

A genus of about 40 species in the Old World tropics and subtropics, elsewhere introduced or in the cultivation. Six species are occurring in Pakistan (Cope, 1982).

Cymbopogon jwarancusa (Jones) Schult. Syst. Veg. Mant. 2:458, (1824); Stewart in Nasir & Ali., Ann.Cat.Vasc.Pl.W.Pak. & Kashm., 108 (1972); Cope in Nasir & Ali., Fl.Pak., 143: 328 (1982). **Figure 21B**

Syn: *Andropogon jawarancusa* Jones in Asiat. Res. 4:109, (1795); *A. himalayensis* Gand. in Bull. Soc. Bot. Fr. 46: 421 (1899), non Steud. (1854).

subsp. **Jwarancua**

A densely tufted, tall, perennial grass with aromatic roots; culms erect or geniculately ascending up to 135 cm long, ca. 2 mm broad, sparsely branched; nodes densely ciliate. Leaf-sheath ca. 3-5 cm long, ca. 2-4 mm broad; blades linear, whitish, narrowed at the base, attenuate to a filiform tip, ca. 3-22 cm long, ca. 2-4 mm broad, glaucouscent within, pubescent above; ligule ca. 2-3 mm, hyaline, glabrous, flat or coiled. Panicles ca. 10-35 cm, exserted; spathes 2.5 cm, spatheoles ca. 1.5 x 0.4 cm in size, broadly lanceolate, narrowly elliptic, apex acute, base cuneate, margin hyaline, glabrescent above. Spikelets lanceolate; lower glume lanceolate, apex acute, shallowly concave on the back, enveloping other parts, ca. 3 x 2 mm, green, hairy on margin and above. Lemma deeply bifid, ca. 4-5 x 1 mm, elliptic, hyaline, hairy above; awn ca. 0.7 mm, filiform. Anthers elliptic-oblong, creamy, ca. 3

x 1 mm in size, base amplexicaul. Ovary ca. 0.5 x 0.3 mm, ovoid, greenish, glabrous; styles 2. Caryopsis ca. 1.5 x 0.7 mm, elliptic, apex mucronate, base slightly reddish, creamy, glabrous.
Flowering period: March-May & again in September.
Herbarium specimens examined: NARA DESERT: *Rahmatullah Qureshi*, 627, 684, 885, 1090, 1213, 1227 (SALUBGH).
Field remarks: A common species found in low-lying area between dunes along with calcareous soil.
Distribution: Pakistan (Sindh, Baluchistan & Punjab); India (mainly in the west) and Nepal (Cope, 19882).

21. Dichanthium

Willemet in Usteri, Ann. Bot. 18: 11 (1796); Sultan & Stewart, Grasses W.Pak. 1: 140 (1958); Bor, Grasses Burma Ceyl. Ind. Pak. 133 (1960); Bor in Rech. f., Fl. Iran. 70: 540 (1970); Cope in Fl. Pak. 143: 279 (1982); Matthew, Fl. Tamilnadu Carn., 3: 1833 (1983).

Eremopogon Stapf in Prain, Fl. Trop. Afr. 9: 182 (1917); Sultan & Stewart, l.c. 120; Bor, Grasses Burma Ceyl. Ind. Pak. 148 (1960); Bor in Rech. f., Fl. Iran. 70: 536 (1970).

Rhizomatous, Annuals or perennials; culms tufted, sometimes densely branched at the base. Ligules membranous. Leaf-blades narrow, flat or rolled. Inflorescence of single or subdigitate racemes, spike-like; joints and pedicels slender, solid with 1-3 homogamous spikelets pairs at base. Sessile spikelets dorsally compressed; callus very short, rounded. Glumes equal; lower glume broadly convex to concave on back, abruptly rounded at apex, with or without a circular pit, 2-keeled; upper glume boat-shaped, 1-keeled. Lemmas usually epaleate; lower lemma hyaline, upper one stipiform, apex entire or 2-dentate, with a scabrid awn. Stamens 3. Caryopsis oblong, dorsally compressed. Pedicelled spikelets awnless, similar to sessile.

A genus of about 20 species in the Old World tropics. Only 3 species are reported from Pakistan (Cope, 1982).

Dichanthium annulatum (Forssk.) Stapf. in Oliver, Fl. trop. Africa 9: 178 (1917); Stewart in Nasir & Ali., Ann. Cat. Vasc. Pl. W.Pak. & Kashm., 108 (1972); Cope in Nasir & Ali., Fl.Pak., 143: 280 (1982).

Syn: *Andropogon annulatus* Forsskal, Fl. Aegypt.-Arab. 173 (1775); *A. scandens* Roxb., Fl. Ind.. 1: 262 (1820); *Dichanthium nodosum* Willemet in Usteri, Ann. Bot. 18: 11 (1796), nom. superfl., based on *A. annulatus* Forssk., but the description is that of *D. aristatum* (Poir.) C.E. Hubbard; *Andropogon nodosus* (Willemet) Nash, N. mer. Fl. 17: 122 (1912).

A densely tufted perennial grass. Culms up to 85 cm, erect, rarely ascending, slender, terete, smooth; nodes bearded. Leaf-sheath ca. 5-10 cm long, terete, striate, densely ciliate at tip; ligule ca. 1-3 mm long, oblong, obtuse, glabrous; blades ca. 5-30 x 0.3-0.7 cm in size, linear, finely acuminate, glabrous or sparsely hairy above; hairs small, bulbous based; margin scabrid, slightly contracted at base. Racemes 2-9, ca. 2.5-6 cm long, subdigitately erect,

pinkish or nearly white spikes; common rachis filiform, ca. 4-12 mm long. Sessile spikelets ca. 3.5-4 mm long; elliptic-oblong, closely imbricating; callus shortly bearded at base; lower involucral glume thinly ca. 3 mm long, chartaceous, tips purplish, sharply ciliolate from middle upwards to the truncate apex, margins narrowly incurved, the keel ciliate; back hairy, 3-9-nerved, nerves not reaching the tip; upper involucral glume more or less equalling the lower but narrower, lanceolate, subacute, glabrous or ciliolate, 3-nerved; lower floral glume linear-oblong, obtuse, nerveless, glabrous; upper floral glume represented by a flattened, white base of a scabrid, slender awn, the latter up to 2 cm long; lower lemmas 2 mm; upper lemmas stipiform; awn ca. 2 cm, scaberulous. Pedicelled spikelets 3-5 mm long, male or neuter, usually darker; lower involucral glume ca. 3 mm long, elliptic-oblong, obtuse, 7-13-nerved; keel bristly; upper involucral glume ciliate; upper floral glume small or obsolete.
Flowering period: March-April.
Herbarium specimens examined: NARA DESERT: *Rahmatullah Qureshi*, 253, 831, 1003, 1087, 1294, 1297 (SALUBGH).
Field remarks: A common grass found in sheltered hedges and on roadsides.
Distribution: Pakistan (Sindh, Baluchistan, Punjab, N.W.F.P., Gilgit & Kashmir); Kenya, Tanzania and Senegal, through the Middle East to Indonesia; introduced to southern Africa, Tropical America and Australia (Cope, 1982).

22. Lasiurus

Boiss., Diagn., Ser. 2, 4: 145 (1859); Sultan & Stewart, Grasses W.Pak. 1: 143 (1958); Bor, Grasses Burma Ceyl. Ind. Pak. 189 (1960); Jafri in Fl. Kar. 39 (1966); Bor in Rech. f., Fl. Iran. 70: 526 (1970); Cope in Fl. Pak. 143: 342 (1982).

Perennial. Leaf-blades linear; ligule hyaline of hairs. Inflorescence a single racemes, terminal on the culm or from auxillary branches; racemes fragile, flat, with 2 sessile and 1 pedicelled spikelet at each node (rarely 1 sessile and 2 pedicelled) or 1 sessile and 1 pedicelled; internodes linear, usually widening upwards, the pedicels free. Sessile spikelet 2-flowered, flat across the back, callus truncate; lower glume subcoriaceous, 2-keeled, densely ciliate on the back and kneels; lower floret male with hyaline lemma and palea; upper floret bisexual, the lemma entire and awnless. Caryopsis oblong, terete. Pedicelled spikelets similar to the sessile or shorter, both florets male.

A monotypic genus in Southwest Asia and northern Africa (Cope, 19882).

Lasiurus sindicus Henr.in Blumea 4: 514 ((1941); Stewart in Ali and Nasir, Ann. Cat. Vasc. Pl. W. Pak. & Kashm., 111 (1972); Cope in Ali and Nasir, Fl. Pak., 143: 342-33 (1982). **Figure 22B**

Syn: *Saccharum hirsutum* Forssk., Fl. Aegypt.-Arab. 16 (1775), non *Rottboellia hirsuta* Vahl, nec *Lasiurus hirsutus* (Vahl) Boiss.; *Lasiurus ecaudatus* Satyan. & Shankar. in. Bombay nat. Hist. Soc. 60(3): 763 (1963).

A perennial, tufted; rootstock rhizomatous, more or less oblique and woody, with many extra and intravaginal shoots from the base, often appearing shrubby. Culms up to 1 m high, simple or suffruticose, erect from a branching base, terete, smooth or the lower internodes slightly rough, more or less hairy, rigid. Leaf-blades ca. 7-22 x 0.2-0.5 cm in size, linear, apex finely acuminate, convolute or flat; sheaths terete, glabrous, the upper sheaths shorter than the internodes, glaucous or mealy white, the lower crowded, long, persistent, glabrous or sparingly ciliate upwards; ligule a fringe of hairs. Racemes ca. 5-11 cm long, solitary, erect, densely silky villous and more or less fragile from internodes; joints 4-5 x 2 mm, shorter than sessile spikelets, subclavate, densely villous on the back, dorsally much compressed, rather ciliolate along the angles; pedicels keeled along the inner sides, ca. 4.5 x 1-1.5 mm in size. Sessile spikelets ca. 5-12 mm long, 1 or 2 at each node of the rachis, imbricate, pale green. Glumes 4; lower involucral glume ca. 1 cm long, coriaceous, 7-9-nerved, acuminate, unequally 2-toothed at the apex, densely silky villous; upper involucral glume 6 mm long, ovate, apex aristately acuminate, 3-5-nerved, apex acuminate; lower floral glume hyaline, ovate-oblong, acute, 3-nerved, paleate; upper floral glume ovate, acute, glabrous, with incurved margins. Pedicellate spikelets similar to the sessile one but shorter, with both flowers male or reduced.

Flowering period: March-May and September-October.

Herbarium specimens examined: NARA DESERT: *Rahmatullah Qureshi*, 712, 1111, 1123, 1145 (SALUBGH).

Field remarks: A rare species found on slopes of the desert.

Distribution: Pakistan (Sindh, Baluchistan, Punjab, N.W.F.P.); Northwest India; westwards to Iraq; tropical Arabia; Egypt; Ethiopia, Somalia and Mali (Cope, 1982).

Tribe VIII. ERAGROSTIDEAE

23. Dactyloctenium

Willd., Enum. Hort. Berol. 1029 (1809); Boiss., Fl. Or. 5: 556 (1884); Sultan & Stewart, Grasses W. Pak. 2: 250 (1959); Bor, Grasses Burma Ceyl. Ind. Pak. 488 (1960); Bor in Rech. f., Fl. Iran. 70: 439 (1970); Cope in Fl. Pak. 143: 105 (1982). *Cynosurus* L., Sp. Pl. 72 (1753).

Tufted or stoloniferous annuals or perennials. Leaf-blades linear, flat or loosely folded; ligule membranous, truncate, often ciliolate. Inflorescence digitate, composed of several linear to narrowly oblong secund spikes which disarticulate at maturity from the top of the culm (very tardily in some species); uppermost spikelets abortive, the spike terminating in a pointed extension of the flattened rachis. Spikelets several-flowered, elliptic-ovate, laterally compressed, 2-seriate, closely overlapping, disarticulating above the glumes but not between the florets; glumes 1-nerved, keeled, subequal, shorter than the lemmas, persistent, the lower sharply acute, the upper with a stout spreading awn usually arising just below the broadly rounded emarginate tip; lemmas 3-nerved, with the lateral nerves shorter than the lemmas and obscure, keeled, membranous, glabrous, the tip entire, acute to shortly awned; palea subequalling the lemmas, the keels sometimes winged. Grain angular, ornamented, enclosed within a free hyaline pericarp, which ruptures to release the grain.

A genus of 13 species distributed mainly Africa to India but one species is widespread in warm countries of the Old World and introduced in America. Three species are reported from Pakistan (Cope, 1982).

Key to species:

1a. Annual grass; spikes up to 5 cm long:

2a. Inflorescence open, the spikes 1.2-5 cm long, linear-narrowly oblong, ascending or radiating from the culm tip..**1. D. aegyptium**

2b. Inflorescence compact, the spikes 0.8-1.8 cm long, oblong to broadly oblong, clustered in a dense head..**2. D. aristatum**

1b. Stoloniferous, perennial, spikes 0.5-2 cm long....................................**3. D. scindicum**

1. Dactyloctenium aegyptium (L.) P. Beauv., Ess. Agrost. Expl. 15 (1812); Stewart in Ali and Nasir, Ann. Cat. Vasc. Pl. W. Pak. & Kashm., 144 (1972); Cope in Ali and Nasir, Fl. Pak., 143: 106 (1982). **Figure 21D**

Syn: *Cynosurus aegyptius* L., Species. Pl. 1: 72 (1753); *Eleusine aegyptia* (L.) Desf., Fl. Atlant. 1: 85 (1789); *Chloris mucronata* Michx., Fl. Bor.-Amer. 1: 59 (1803); *Dactyloctenium mucronatum* (Michx.) Willd., Enum. Hort. Berol. 1029 (1809).

An annual grass usually with geniculately ascending and rooting at the lower nodes, frequently stoloniferous and mat-forming, less often erects. Culms ca. 15-40 cm high. Leaf-blades ca. 3-12 x 0.2-0.4 cm in size, linear, tapering to a fine point, flat, ciliate with a bulbous-based hairs; ligule a small, ciliolate, line. Inflorescence composed of 2-6 linear to narrowly oblong spikes, ca. 1.2-5 cm long, ascending or radiating from the culm tip; rachis 3-gonous, rigid, excurrent into a pungent mucro. Spikelets many, 3-5-flowered, spreading at right angles to rachis, upto 3.2 mm long. Glumes divaricate; lower involucral glume ca. 2 mm long, ovate, acute; upper equally long, suborbicular, midrib produced into a recurved awn as much long; floral glume ca. 3.2 mm long, gibbously ovate, apex mucronate or awned. Grain ca. 1 mm across, subglobose, reddish.

Flowering period: July-October.

Herbarium specimens examined: NARA DESERT: *Rahmatullah Qureshi*, 846, 848, 950, 1106 (SALUBGH).

Field remarks: A common species found in low-lying area.

Distribution: Pakistan (Sindh, Baluchistan, Punjab, N.W.F.P., Gilgit & Kashmir); widely distributed in tropical and warm temperate regions of the Old World; introduced in America (Cope, 1982).

2. Dactyloctenium aristatum Link, Hort. Berol. 1: 59 (1827); Stewart in Nasir & Ali., Ann. Cat.Vasc. Pl. W. Pak. & Kashm., 114 (1972); Cope in Nasir & Ali., Fl.Pak., 143: 107-8 (1982).

Syn: *D. aegyptium* var. *aristatum* (Link) A. Chev. in Revue Bot. Appl. Agric. Trop. 14: 130 (1934); *D. glaucophyllum* var. *villosum* Mattei in Boll. R. Orto. Bot. Palermo 7: 168 (1908); *D. semipunctatum* Courb. in Annls. Sci. nat. sér. 4, 18: 135 (1862).

Sprawling tufted annual; culms slender, 4-38 cm high, geniculately ascending from a decumbent base, often rooting at the lower nodes. Leaf-blades flat, 1-13 cm long, 1.5-7 mm wide, usually conspicuously papillose-hispid, sharply acuminate. Inflorescence compact, composed of (2)4-7(-11) oblong to broadly oblong, spikes ca. 0.8-1.8 cm long clustered together in a dense, often subglobose head. Spikelets 3-5-flowered, broadly ovate, 4.1-5.2 mm long; glumes subequal, 1.7-2.3 mm long, the lower narrowly lanceolate-oblong in profile, the keel thick, scabrid and often narrowly winged, the upper narrowly elliptic-oblong in profile, the keel extended into a stout awn shorter than or equalling the glume; lemmas lanceolate to narrowly ovate in profile (3-)3.34.3 mm long, acuminate, the keel concave, thick and scabrid above the middle and extended into a stout mucro up to 1 mm long; palea-keels winged; anthers 0.3-0.5 mm long. Grain 0.75-1.1 mm long, broadly elliptic to obovate in profile, dome-shaped across the top, finely granular or granular-striate.

Flowering period: July-October.

Herbarium specimens examined: NARA DESERT: *Rahmatullah Qureshi,* 1385 (SALUBGH).

Field remarks: A less common species found in low-lying areas.

Distribution: Pakistan (Sind, Baluchistan & Punjab); Kenya north to Sudan and eastwards through Arabia to India.

3. Dactyloctenium scindicum Boiss., Diagn. ser. 2, 4:131 (1859); Stewart in Nasir & Ali., Ann. Cat.Vasc. Pl. W. Pak. & Kashm., 114 (1972); Cope in Nasir & Ali., Fl.Pak., 143: 108 (1982).

Syn: *D. glaucophyllum* Courb. in Annls. Sci. nat. sér. 4, 18: 133 (1862); *D. glaucophyllum* var. *elongatior* Courb, 134; *D. glaucophyllum* var. *robustior* Courb., i.c. 134; *Eleusine glaucophylla* (Courb.) Benth. in J. L. Soc. Bot. 19: 107 (1881); *E. scindica* (Boiss.) Duthie, Fodder Grasses 58 (1888).

Stoloniferous perennial forming extensive spreading mats; culms slender with swollen bases, 7-45 cm high, erect. Leaf-blades flat or loosely folded, tough and rather glaucous, 1-11 cm long, 1.5-3 mm wide, scattered papillose-hispid especially along the margins. Inflorescence composed of 3-4(5) slightly falcate oblong spikes ca. 0.8-2 cm long forming a compact head, the spikes readily disarticulating at maturity from the top of the culm. Spikelets 3-9-flowered, broadly lanceolate to ovate, 4-8 mm long; glumes broadly elliptic in profile, the lower 1.7-2.5 mm long, the keel narrowly winged, the upper 1.5-2.3 mm long, the prominent scabrid keel extended into an awn half to as long as the glume; lemmas lanceolate in profile, 3-3.8 mm long, obtuse to acute with the scabrid keel extended into a mucro up to 0.8 mm long; palea-keels finely scabrid, unwinged; anthers 1.1-2 mm long. Grain 0.7-1 mm long, transversely rugose.

Flowering & Fruiting Period: July-September.

Herbarium specimens examined: NARA DESERT: *Rahmatullah Qureshi,* 1386 (SALUBGH).

Field remarks: A less common species found in low-lying areas.

Distribution: Pakistan (Sind, Baluchistan, Punjab & N.W.F.P.); Kenya north to the Sudan and eastward to Northwest India (Cope, 1982).

24. Dinebra

Jacq., Fragm. 77, t. 121/1 (1809); Boiss., Fl. Or. 5: 557 (1884); Hook. f., Fl. Brit. Ind. 7: 296 (1896); Sultan & Stewart, Grasses W. Pak. 2: 224 (1959); Bor, Grasses Burma Ceyl. Ind. Pak. 491 (1960); Cope in Fl. Pak. 143: 79 (1982).

Annuals. Leaf-blades linear, flat; ligule membranous, truncate, lacerate or ciliate. Inflorescence composed of several-many-flowered secund spikes irregularly arranged along a central axis; spikes varying from short, broad and crowded to linear and distant, deciduous from the main rachis at maturity or persistent but with the lower spikelets on each side often replaced by small deciduous branchlets. Spikelets 1-several-flowered, wedge-shaped, laterally compressed, sessile, 2-seriate, closely overlapping, disarticulating between the florets; glumes subequal, acuminate-aristate, much longer than the florets, strongly keeled, coriaceous or membranous; lemmas 3-nerved, lightly keeled, thinly membranous, pilose on the nerves, the tip acute to 2-lobed. Caryopsis elliptic-oblong, shallowly concave on the hilar side, 3-gonous on cross-section.

A genus of 3 species in the tropics and subtropics from Africa to India. One species occurs in Pakistan (Cope, 1982).

Dinebra retroflexa (Vahl.) Panzer in Denkschr. Acad. Muench. 1813: 270, t.12 (1814); Stewart in Ali and Nasir, Ann. Cat. Vasc. Pl. W. Pak. & Kashm., 145 (1972); Cope in Ali and Nasir, Fl. Pak., 143: 79 (1982).

Syn: *Cynosurus retroflexa* Vahl., Symb. Bot. 2: 20 (1791); *Dinebra arabica* Jacq., Fragm. 77, t.121/1 (1809); *Dcatylis paspaloides* Willd., Enum. Hort. Berol. 111 (1809), nom. superfl., based on *Cynosurus retroflexa* Vahl; *Leptochloa arabica* (Jacq.) Kunth., Rev. Gram. 1: 91 (1829).

An annual loosely tufted grass up to 100 cm high, usually straggling and ascending from a decumbent base, often rooting at the lower nodes; nodes glabrous. Leaf-sheaths ca. 2.5-7 cm long, glandular; blades ca. 10-20 x 0.3-0.6 cm in size, flat, base cordate; ligule ca. 3 mm long, membranous. Panicles pyramidal, spicate, ca. 20 x 4 cm in size; racemes ca. 1-2.5 x 0.3-0.4 cm in size, stiff, ascending at first, oblong. Spikelets ca. 5 mm, secund, 2-seriate, sessile, 2- or 3-flowered; rachilla slender, jointed at the base. Upper glumes strongly keeled, apex caudate, persistent; lower glume ca. 5 mm, asymmetric, lanceolate, glandular along the keel; lemmas narrowly ovate, ca. 2 mm long, 3-nerved, appressed-pilose along nerves; paleas 3-lobed, lobes erose, 2-keeled, ciliate. Stamens 3. Caryopsis ellipsoid-oblong.

Flowering period: May-September.

Herbarium specimens examined: NARA DESERT: *Rahmatullah Qureshi*, 1260, 1308 (SALUBGH).

Field remarks: A less common species found in low-lying areas.

Distribution: Pakistan (N.W.F.P.); South Africa, through Africa to Egypt; Iraq to India (Cope, 1982).

25. Desmostachya

Stapf in Thiselt.-Dyer, Fl. Cap. 7: 632 (1900); Sultan & Stewart, Grasses W. Pak. 2: 238 (1959); Bor, Grasses Burma Ceyl. Ind. Pak. 488 (1960); Bor in Rech. f., Fl. Iran. 70: 491 (1960); Cope in Fl. Pak. 143: 109 (1982).

Stapfiola O. Ktze. in Post & O. Ktze., Lexic. Gen. Phan. 532 (1903).

Perennials. Leaf-blades linear, flat or inrolled; ligule a short lacerate membrane. Inflorescence a narrow erect linear panicle composed of short densely clustered or spaced ascending or spreading spikes. Spikelets several-many-flowered, laterally compressed, sessile or subsessile, crowded and secund on the branches, falling entire; glumes unequal, lanceolate,

1-nerved, membranous; lemmas keeled, ovate-acute, muticous, membranous or chartaceous to coriaceous, glabrous, faintly 3-nerved the side-nerves shorter than the lemmas; palea 2-keeled, shorter than the lemmas. Caryopsis obliquely ovoid, triquetrous in cross-section, pericarp adnate.

A single species in the Old World tropics (Cope, 1982).

Desmostachya bippinata (L.) Stapf in Dyer, Fl.Cap. 7:632 (1900); Stewart in Nasir & Ali., Ann. Cat. Vasc. Pl. W.Pak. & Kashm., 145 (1972); Cope in Nasir & Ali., Fl. Pak., 143:109-110 (1982). **Figure 22A**

Syn: *Briza bipinnata* L., Syst. Nat., ed. 10, 2:875 (1759); *Poa cynosuroides* Retz., Obs.Bot. 4:20 (1786); *Eragrostis cynosuroides* (Retz.) P.Beauv., Ess. Agrost. 162. (1812); *Desmostachya cynosuroides* Stapf in Haines, Bot. Bih. & Or. 962 (1924).

A perennial harsh tussocky rihozomatous grass. Culms upto 1 m high, creeping, much branched at the base, tufted, smooth, covered with shining sheaths. Leaf-blades ca. 40-50 x 1 cm in size; basal fascicled, rigid, acuminate, tips filiform, margins hispid; sheath glabrous; ligule a hairy line. Panicles ca. 15-45 x 1.5-4 cm in size, erect, narrowly pyramidal, often interrupted; rachis puberulent; branches many, short, scarcely ca. 2.5 cm long, crowded, clothed from the base with sessile imbricating spikelets. Spikelets sessile, 2-seriate, crowded, deflexed, pale-brown, ca. 13 mm long, shining, 20-30-flowered. Involucral glumes very unequal; lower ca. 0.5 mm long; upper ca. 1.6 mm long, obtuse; floral glumes ca. 1.6-2 mm long, ovate, acute, coriaceous; palea shorter than its glumes with minutely scabrid keels. Grains ca. 0.5-0.6 mm long, obliquely ovoid, laterally compressed, obliquely 3-gonous.

Flowering period: July-November.

Herbarium specimens examined: NARA DESERT: *Rahmatullah Qureshi*, 1259, 1310 (SALUBGH).

Field remarks: A common species found in low-lying areas.

Distribution: Pakistan (Sindh, Baluchistan, Punjab, N.W.F.P., Gilgit & Kashmir); throughout the Middle East to Indo-China; North and tropical Africa (Cope, 1982).

26. Ochthochloa

Edgew. in. Asiat. Soc. Beng. 11: 26 (1842); Cope in Fl. Pak. 143: 81 (1982).

Tufted stoloniferous perennials. Leaf-blades usually folded; ligule a short membrane with a long ciliate fringe. Inflorescence composed of (2-)3-5 short secund digitate racemes, these deciduous at maturity. Spikelets several-flowered, ovate-elliptic, strongly laterally compressed, disarticulating above the glumes but not between the florets. Glumes shorter than the adjacent lemmas, the upper with a thickened 3-nerved keel; lemmas 3-nerved, keeled, elliptic-lanceolate, entire, acute or with a short awn-point. Caryopsis enclosed within a free hyaline pericarp.

A monotypic genus from Northwest India through Arabia to Northeast Africa (Cope, 1982).

Ochthochloa compressa (Forssk.) Hilu. in Kew Bull. 36:000 (1981); Cope in Ali and Nasir, Fl. Pak., 143: 81 (1982). **Figure 22C**

Syn: *Panicum compressum* Forssk., Fl. Aegypt.-Arab. 18 (1775); *Eleusine flagellifera* Nees in Linnaea 16: 220 (1842); *Ochthochloa dactyloides* Edgew. in J. Asiat. Soc. Beng. 11: 27 (1842); *Eleusine caespitosa* A. Rich., Tent. Fl. Abyssss. 2: 412 (1850); *E. arabica* Hochst. ex Steud., Syn.Pl.Glum. 1: 211 (1854), in syn.; *E. compressa* (Forssk.) Asch. & Schweinf. ex C. Christ. in. Dansk. bot. Ark. 4(3): 12 (1922).

A perennial sprawling stoloniferous grass. Stem smooth, rooting at distant, thickened, leafy nodes; internodes ca. 10-20 cm long; culms prostrate or geniculately ascending slender, forming clumps up to 75 cm high. Leaf-blades ca. 1-3.5 x 0.5-3 cm in size, linear-lanceolate, apex acuminate, rigid, distant, flat, recurved; sheath glabrous; ligule a hairy line. Spikes 3-6, digitate, ca. 2-3.8 x 0.40.8 cm in size; rachis slender, pubescent. Spikelets ca. 3-8.5 mm long, imbricate, sessile or shortly pedicellate, 4-6-flowered. Lower involucral glume ca. 3 mm long, ovate, apex acute, membranous. Upper involucral glume ca. 4 mm long, lanceolate, apex aristate, coriaceous, the margins broad, membranous; floral glumes ca. 5 mm long, ovate, apex acute-apiculate, chartaceous, 3-nerved; midnerve compound, margins membranous, hairy below; palea oblong with ciliate keels.

Flowering period: March-October.

Herbarium specimens examined: NARA DESERT: *Rahmatullah Qureshi*, 100, 101, 102, 216, 280, 621, 724, 833, 937, 1082, 1267 (SALUBGH).

Field remarks: A very common species found in the whole area.

Distribution: Pakistan (Sindh, Baluchistan, Punjab & N.W.F.P.); Northwest India to North Africa (Cope, 1982).

27. Eragrostis

Wolf., Gen. Pl. Vocab. Char. Def. 23 (1776); Boiss., Fl. Or. 5: 580 (1884); Hook. f., Fl. Brit. Ind. 7: 312 (1896); Sultan & Stewart, Grasses W. Pak. 2: 229 (1959); Bor, Grasses Burma Ceyl. Ind. Pak. 495 (1960); Cope in Fl. Pak. 143: 82 (1982); Matthew, Fl. Tamilnadu Carn., 3: 1852 (1983).

Diandrochloa De Winter in Bothalia 7: 387 (1960); *Roshevitzia* Tzvelev in Bot. Zh. SSSR 53: 311 (1968).

Annuals or perennials; culms sometimes glandular, nodes glabrescent to sparsely bearded. Leaf-sheaths glandular (as a crateriform warts or band-like patches), pilose or not at mouth; liglue a fimbriate rim, membranous; blades linear-oblong, flat or inrolled, often glandular. Inflorescence a terminal, open, contracted or spike-like panicle; branches ascending, whorled or not; rachis often glandular. Spikelets laterally compressed, variously disarticulating, awnless; florets shedding from base, or apex, at maturity; rachilla persistent or not. Glumes subequal, shorter than lowest lemma, 1-nerved, asperulous on keel. Lemmas imbricate or not, glabrous, 3-nerved, acuminate-emarginate, rarely mucronulate, scabrid to asperulous on back, sometimes ciliate on margins; palea slightly shorter than the lemma, deciduous with their lemmas or persistent, 2-keeled, sometime long-ciliate, often scabrid, rarely winged. Stamens 2-3. Caryopsis globose-ellipsoid, smooth or finely striate to reticulate, the pericarp adherent (sometimes slightly loose but never free).

A genus of ca. 300 species, commonly distributed in tropics and subtropics, particularly as ruderal and pioneer species. In Pakistan, it is represented by 14 species (Cope, 1988). One species recorded from the Nara Desert.

Key to species:

1a. Spikelets breaking up from above downward; rachis fragile:
2a. Panicles speciform, often ± lobed or interrupted,
the spikelets densely clustered... **1. E. ciliaris**
2b. Panicles effuse, not lobed or interrupted, the spikelets not in clusters............**2. E. tenalla**
1b. Spikelets breaking up from below upwards; rachis tough:
3a. Grain subglobose, ca. 0.5 mm...**3. E. cilianensis**
3b. Grain oblong or elliptic, ca. 0.6-1 mm long:
4a. Leaves with wart-like glands on the margin;
grain broadly oblong, ca. 0.7-0.8 mm long..**4. E. minor**
4b. Leaves eglandular,
grain elliptic-oblong, 0.6-1 mm long..**5 E. barrelieri**

1. Eragrostis ciliaris (L.) R.Br. in Tuckey, Narr. Exp. Congo App. 478 (1818); Stewart in Nasir & Ali., Ann. Cat. Vasc. Pl. W. Pak. & Kashm., 146 (1972); Cope in Nasir & Ali., Fl. Pak., 143: 86 (1982).

Syn: *Poa ciliaris* L., Syst. Nat., ed. 10, 2: 875 (1759); *Megastachya ciliaris* (L.) P. Beauv., Ess. Agrost. 74, 167 (1812); *Eragrostis pulchella* Parl. in Atti Riun. Scient. Ital. 8: 586 (1847); *E.ciliaris* var. *brachystachya* Boiss., Fl. Or. 5: 582 (1884).

Tufted annual; culms 5-60 cm high, erect. Leaf-blades flat, up to 12 cm long and 5 mm wide. Panicle spike-like, but often ± lobed or interrupted, 1-20 cm long, woolly, the spikelets densely clustered. Spikelets 6-12-flowered, ovate, 2-4.5 mm long, fluffy, often purplish, breaking up from the apex, the rhachilla fragile; glumes lanceolate, subequal, 0.7-1.2 mm long, acute; lemmas oblong-elliptic, 0.9-1.5 mm long, the keels (at least in the upper lemmas) bearing a few short stiff hairs, otherwise smooth, broadly obtuse, more or less obscurely mucronate; paleakeels tuberculate-ciliate, the hairs 0.6-0.7 mm long, exceeding the width of the adjacent floret; anthers 2, 0.2 mm long. Caryopsis ellipsoid, 0.3-0.5 mm long.
Flowering & Fruiting Period: Almost throughout the year.
Herbarium specimens examined: NARA DESERT: *Rahmatullah Qureshi,* 1389 (SALUBGH).
Field remarks: A less common species found in low-lying area.
Distribution: Pakistan (Sind, Baluchistan & Punjab); tropical and South Africa, extending through Arabia and the Mascarene Islands to India; tropical America (Cope, 1982).

2. Eragrostis tenella (L.) P. Beauv. ex Roem. & Schult., Syst. Veg. 2: 576 (1817); Stewart in Nasir & Ali., Ann. Cat. Vasc. Pl. W. Pak. & Kashm., 147 (1972); Cope in Nasir & Ali., Fl. Pak., 143: 88 (1982).

Syn: *Poa tenella* L., Sp. Pl. 1: 69 (1753); *P. amabilis* L., Sp. Pl. 1: 68 (1753); *P. plumosa* Retz., Obs. Bot. 4: 20 (1786); *Eragrostis plumosa* (Retz.) Link, Hort. Berol. 1: 192 (1827); *E.amabilis* Wight & Arn. ex Nees in Hook. & Arn., Bot. Beechy Voy.

251 (1838); *E. tenella* var. *plumosa* (Retz.) Stapf in Hook.f., Fl. Brit. Ind. 7: 315 (1896).

Delicate tufted annual; culms 6-50 cm high, erect or geniculately ascending. Leaf-blades flat, up to 9 cm long and 4 mm wide. Panicle elliptic or pyramidal, 2-14 cm long, open, hairy in the axils or not, the branches spreading and bearing oblong yellowish glands but not sticky. Spikelets 4-8-flowered, oblong-ovate, 1.5-2.5 mm long, breaking up from the apex, the rhachilla fragile; glumes ovate to narrowly ovate, subequal, 0.5-1 mm long, acute; lemmas ovate-oblong, 0.7-1 mm long, smooth or scaberulous, broadly obtuse; palea-keels ciliate with hairs 0.1-0.3 mm long, normally shorter than the width of the adjacent floret; anthers 3, 0.2 mm long. Caryopsis ellipsoid, 0.5 mm long.
Flowering & Fruiting Period: March-September.
Herbarium specimens examined: NARA DESERT: *Rahmatullah Qureshi*, 1390 (SALUBGH).
Field remarks: A rare species found in low-lying areas.
Distribution: Pakistan (Sind, Punjab & Kashmir); throughout the tropics (Cope, 1982).

3. Eragrostis cilianensis (All.) Lut. ex F.T. Hubbard in Philipp. J. Sci. C, Bot. 8: 159 (1913); Stewart in Nasir & Ali., Ann. Cat. Vasc. Pl. W. Pak. & Kashm., 146 (1972); Cope in Nasir & Ali., Fl.Pak., 143: 93-95 (1982).

Syn: *Briza eragrostis* L., Sp. Pl. 1: 70 (1753), *non Poa eragrostis* L.; *Poa cilianesis* All., Fl. Pedem. 2: 246, t.91/2 (1785); *P. megastachya* Koel., Descr. Gram. 181 (1802); *Eragrostis major* Host, Gram. 181 (1802), based on *Briza eragrostis* L.; *E. megastachya* (Koel.) Link, Hort. Berol. 1: 187 (1827).

Loosely tufted annual; culms 10-90 cm high, erect or ascending. Leaf-blades flat, up to 15 cm long and 8 mm wide, mostly glabrous and usually with a row of warty glands along the margin. Panicle ovate, 4-30 cm long, fairly dense, contracted, stiffly branched, usually with glands on pedicels and branchlets. Spikelets 5-60-flowered, narrowly ovate to narrowly oblong, rarely linear, 3-20 min long, 2-4 mm wide, pallid yellowish green to leaden grey, breaking up from the base, the rhachilla often breaking off above the glumes before all the lemmas have fallen; glumes subequal, oval, boat-shaped, 1.5-2.2 mm long, 1-3-nerved, often glandular on the keel, acute; lemmas broadly ovate to subrotund, (1.7)2-2.8 mm long, chartaceous, often glandular on the keel, the lateral nerves distinct, emarginate or obtuse (occasionally bluntly acute) when flattened, usually obtuse in side-view; palea ± scabrid on the keels, persistent; anthers 3, 0.3 mm long. Caryopsis typically subglobose, 0.5(0.7) mm long, rarely oblong, dark reddish brown.
Flowering & Fruiting Period: May-October.
Herbarium specimens examined: NARA DESERT: *Rahmatullah Qureshi*, 1388 (SALUBGH).
Field remarks: A common species found in low-lying areas.
Distribution: Pakistan (Sind, Baluchistan, Punjab, N.WF.P. & Kashmir); tropical and warm temperate regions of the Old World; introduced to the New World (Cope, 1982).

4. Eragrostis minor Host, Gram. Austr. 4: 15 (1809); Cope in Nasir & Ali., Fl. Pak., 143: 95 (1982).

Syn: *Poa erogrostis* L., Sp. Pl. 1: 68 (1753); *Eragrostis poaeoides* P. Beauv., Ess. Agrost. 162 (1812), based on *Poa eragrostis* L.; *E. multiflora* (Forssk.) Aschers. var. *pappiana* Chiov. in Annur. R. Ist. bot. Roma 8: 65 (1903); *E. pappiana* (Chiov.) Chiov., l. c. 371 (1908).

An erect annual densely tufted branched grass. Culms up to 45 cm high, slender. Lea-blades ca. 3-13 x 0.2-0.5 cm in size, linear-narrowly lanceolate, apex acute, flat, usually glandular along the margins; sheath shortly bearded at mouth; ligule a pubescent ridge. Panicles ca. 5-15 cm long, oblong-ovate, stiff; rachis glabrous; branches spreading when mature. Spikelets ca. 6 x 1.5-2 mm in size, upto 12-flowered, linear-ovate, pale-green to olive-grey; pedicels ca. 3-4 mm long; rachilla tough. Involucral glumes ca. 1.2 mm long, upper slightly longer or subequal, ovate, apex acute, 1-nerved; floral glume ca. 1.5-2 mm long, ovate, apex acute, strongly nerved; palea 1 mm long, oblong, keels scabrid or slightly ciliolate. Caryopsis ca. 0.5 mm long, broadly oblong, slightly flattened, dark brown in colour.
Flowering period: June-September.
Herbarium specimens examined: NARA DESERT: *Rahmatullah Qureshi*, 644, 671, 1017, 1088, 1102, 1182, 1199, 1231 (SALUBGH).
Field remarks: A very common species found in low-lying areas.
Distribution: Pakistan (Sindh, Baluchistan, Punjab, N.W.F.P., Gilgit & Kashmir); warm temperate and subtropical regions of the Old World; occasionally found as an intoduced in the tropics and the New World (Cope, 1982).

5. Eragrostis barrelieri Dav. in J. Bot., Paris 8: 289 (1894); Stewart in Nasir & Ali., Ann. Cat. Vasc. Pl. W. Pak. & Kashm., 146 (1972); Cope in Nasir & Ali., Fl.Pak., 143: 97 (1982).

Loosely tufted annual; culms 5-50(80) cm high, ascending, Leaf-blades flat or inrolled, up to 10 cm long and 5.5 mm wide, mostly glabrous, devoid of warty crateriform glands, somewhat glaucous. Panicle oblong-lanceolate, 3-20 cm long, usually open, stiffly branched, with glands on branchlets and pedicels. Spikelets 5-30-flowered, narrowly oblong, 5-20 mm long, 1.3-2 mm wide, yellowish green or tinged with grey or purple, breaking up from the base, the rhachilla persistent; glumes unequal, lanceolate, the lower 0.9-1.3 mm long, the upper 1.5-2 mm long, acute; lemmas broadly ovate-oblong, 1.7-2.6 mm long, the lateral nerves distinct, obtuse; palea scabrid on the keels, persistent; anthers 3, 0.2 mm long. Caryopsis elliptic-oblong, 0.65-1 mm long, dark brown.
Flowering & Fruiting Period: June-September.
Herbarium specimens examined: NARA DESERT: *Rahmatullah Qureshi*, 1387 (SALUBGH).
Field remarks: A rare species found in low-lying areas.
Distribution: Pakistan (Baluchistan & N.W.F.P); Mediterranean region to the Middle East; tropical Africa; introduced in South Africa.

Tribe IX. AELUROPODEAE

28. Aeluropus

Trin., Fund. Agrost. 143 (1820); Hook.f., Fl. Brit. Ind. 7: 333 (18 6); Cooke, Fl. Pres. Bombay, III: 567 (1906); Sultan & Stewart, Grasses W.Pak. 2: 200 (1959); Bor, Grasses Burma Ceyl. Ind. Pak. 380 (1960); Bor in Rech. f., Fl. Iran. 70: 419 (1970); Cope in Fl. Pak. 143: 62 (1982).

Low much branched very rigid, perennial grasses. Leaves subdistichous, short, strict, usually convolute, coriaceous, pungent. Inflorescence a panicle in which the subessile spikelets are closely packed or rachides arranged on one side of the axis. Spikelets 4-18-flowered, minute, sessile, densely crowded in terminal villous heads, laterally compressed, not articulate at the base; rachilla obscurely jointed at the base, not produced beyond the upper glumes; internodes very short. Glumes many, oblong, apiculate or mucronate, chartaceous, unequal, keeled, the margins and tips broadly hyaline; lemmas similar in texture, 9-11-nerved, keeled, mucronate or subaristate; palea very larger, broadly cuneate, 3-lobed, lobes erose, flaps broad, 2-keeled, ciliolate on the keels. Styles short, free; stigmas short, plumose. Grain oblong-elliptic or oblong-obovoid, free within the glumes, flat adaxially, rounded abaxially.

A genus of 6 species from the Mediterranean region to India. Three species are reported from Pakistan (Cope, 1982).

Aeluropus lagopoides (L.) Trin. ex. Thw., Enum. pl., Zeyl. 374 (1864); Stewart in Nasir & Ali., Ann. Cat. Vasc. Pl. W.Pak. & Kashm.,125 (1972); Cope in Ali and Nasir, Fl. Pak., 143: 63 (1982). **Figure 20A**

Syn: *Dactylis lagopoides* L.,Mant. 1:3,(1767); *D. repens* Desf., Fl. Atlant. 1: 79, t.15 (1798); *D. brevifolia* Koen. ex Willd.,Sp. Pl. ed. 4, 1(1): 410 (1798); *Aeluropous laevis* Trin., Fund. Agrost. 143 (1820); *A. brevifolia* (Koen. ex Willd.) Nees ex Steud., Nom. Bot., ed. 2, 1: 30 (1840); *A. repens* (Desf.) Parl., Fl. Ital. 1: 462 (1848); *A. villosus* Hook. f., Fl. Brit. Ind. 7: 334 (1896).

A tufted, rigid, perennial grass. Culms ca. 10-15 cm high, erect to creeping, crowded, resembling stolons, giving off branchlets at nodes but not the roots, arising from a woody root stock with stout root fibres; nodes glabrous. Leaf-sheaths compressed, ca. 3-6 mm long; ligule a ciliate rim; blades ca. 0.5-2.5 x 0.2-0.3 cm in size, narrowly lanceolate, acuminate, flat or convulate, glabrous or sparsely ciliate. Panicle ca. 1-1.5 x 0.8-1 cm in size, oblong-subglobose, pinkish. Spikelets ca. 2.5-3.5 mm, 4-8-flowered, crowded, sessile, villous with soft hairs pale-green o white. Glumes 3, elliptic; lower involucral glume ca. 2.5 mm long, ovate-oblong, subacute, hairy, margins ciliate; upper involucral glume ca. 3.5 mm long, or like the lower; lemma ca. 2.5-3 mm long, ovate, subobtuse, apex apiculate, 9-nerved, hairy and ciliate; palea ca. 2 mm, softly ciliate, 2-lobed. Anthers 2, ca. 1.5 mm long. Ovary ellipsoid. Grain ca. 0.5-0.6 mm long, ovoid-oblong, compressed.

Flowering period: April-November.

Herbarium specimens examined: NARA DESERT: *Rahmatullah Qureshi*, 853, 934, 1074, 1091, 1103 (SALUBGH).

Field remarks: Found only on salt marshy grounds.

Distribution: Pakistan (Sindh, Baluchistan & Punjab); northern Africa from Morocco to Somalia; Sicily and Cyprus through the Middle East to Central Asia and India (Cope, 1982).

58. Typhaceae

Usually tall, perennial of aquatic or marshy, erect, with or without nodes, monoecoius herbs. Rhizomes creeping, submerged, sacly. Leaves radical or cauline, elongated, linear, parallel-veined, sessile. Inflorescence a terminal, cylindrical, superposed spike; male above and female below, parts separated or sometimes united. Flowers numerous, densely crowded. Male flowers 1-3 stamens and surrounded by one or more hairs, some representing the perianth; anthers basifixed, linear. Female flowers usually intermixed with clavate pistillodes, covered with hairs and bracts or bracts absent; ovary superior, 1-locular and 1-ovuled, stipitate; style long; stigma linear or lanceolate. Fruit a follicle, dehiscing longitudinally by a slit with persistent style. Seed striate.
Unigeneric family, with ca. 10 species, distributed to marshy and aquatic habitat of both the hemispheres (Omer & Hashmi, 1987).

Typha

L., Sp. Pl. 971(1753); Gen. Pl. ed. 5: 924 (1754); Boiss., Fl. Or. 5: 49 (1884); Hook. f., Fl. Brit. Ind. 6: 448 (1849); Riedl in Rech. f., Fl. Iran. 71: 1 (1970); Saldanha & Nicolson in Fl. Hassan Dist. Karn. Ind., 793 (1976); Jafri, in Fl. Kar., 82 (1966); Omer & Hashmi, Fl. Pak., 177: 1-8 (1987).

Characters of the genus that of the family.
In Pakistan, this family represents 5 species, distributed from plains to mountainous regions up to ± 3300 m (Omer & Hashmi, 1987).

Key to species:

1a. Leaves upto 12 mm broad, semicylindric above the sheath, flat dorsally; stigma linear.. **1. T. domingensis**
1b. Leaves 25-40 mm broad, trigonous above the sheath, angularly keeled dorsally; stigma lanceolate....................................... **2. T. elephantina**

1. Typha domingensis Pers., Syn. Pl. 2: 532 (1807); Omer & Hashmi in Nasir & Ali., Fl.Pak.,177: 5 (1987).

Syn: *T. australis* Schum. & Thonn., Danske Vidensk. Afhandl. 4, Ser. 4: 175 (1829); *T. angustata* Bory & Chaub., Exp., Sc. Moree 111: 388 (1832); Stewart in Nasir & Ali., Ann.Cat.Vasc.Pl.W.Pak.& Kashm., 28 (1972); *T. angustifolia* auct. non Linn: Roxb. Fl. Ind. 3: 567 (1832).

A robust shrub ca. 2-3 m tall, stem somewhat stout, terete. Leaves distichous, linear, erect, ca. 2 m long and 3 cm broad, apex tapering, acute, base sheathing, usually open at throat, margin free; lamina ca. 5-12 mm broad. Male spike(s) 1(2), 10-20 cm long; rachis (separating male from female) 3 cm long. Stamens 2 or 3; anthers linear, ca. 2 mm long; connectives shortly produced, apex rounded. Hairs around stamens linear, spathulate. Female spike(s) 1(2), ca. 20 cm long and ca. 1-1.5 broad, cushion-like; bracteoles numerous.

Gynophore white, ca. 1.8 mm long. Ovary linear, ca. 1 mm long; style ca. 3 mm long; stigma linear and usually equal to the hairs and bracts, flattened. Fruit ca. 2 mm long, fusiform.
Flowering period: March-April.
Herbarium specimens examined: NARA DESERT: *Rahmatullah Qureshi*, 1401 (SALUBGH).
Field remarks: Found in wetlands.
Distribution: Pantropical (Omer & Hashmi, 1987).

2. **Typha elephantina** Roxb., Fl. Ind., 3:566 (1832); Stewart in Nasir & Ali., Ann.Cat.Vasc.Pl.W.Pak.& Kashm., 28 (1972); Omer & Hashmi in Nasir & Ali., Fl.Pak.,177: 5 (1987).

A tall aquatic perennial herb up 4 m high. Leaves linear or broadly linear; trigonous above the sheath, angularly keeled dorsally, 3-angled; lamina 25-40 mm broad. Upper part flat, usually as long as inflorescence; male and femal parts separate; axis of the male spike covered with hairs; female spike cylindric, blackish brown or brown; pistillodes absent; female flowers bracteate; bracts spathulate and longer than the hairs; stigma lanceolate.
Flowering period: March-August.
Herbarium specimens examined: NARA DESERT: *Rahmatullah Qureshi*, 1400 (SALUBGH).
Field remarks: Found in wetlands.
Distribution: Africa, Asia (Bangladesh, India, Nepal, Pakistan, Iran and Turcomania) (Omer & Hashmi, 1987).

59. Liliaceae

Annual or mostly perennial herbs (rarely shrubs or small trees), generally with underground stems (corms, bulbs, tubers or rhizomes). Stem erect or climbing, leafy or scapose. Leaves simple, normally well developed, sometimes scaly, radical and basal or cauline, usually parallel-veined, rarely reticulate, cladodes occasional. Inflorescence terminal and/or axillary, of simple or branched racemes, umbels, sometimes in corymbs, rarely flower solitary. Flowers bisexual, regular. Bracts usually small, scarious, sometimes when the flowers are umbellate, spathe-like. Perianth tubular, herbaceous or petaloid, shallowly or deeply 6-lobed, imbricate (rarely valvate in bud); lobes often spreading. Stamens 6 (rarely 3 or more), usually included, usually hypogynous, often adnate to the base of perianth-tube; filaments free, rarely connate. Ovary commonly superior, sometimes inferior, 3-celled; ovule(s) 1-∞ per cell on axile placentae; style(s) 1 or 2, simple or branched; stigma(s) simple or 3-lobed, sometimes lobes recurved. Fruit a loculicidal capsule or berry, angular, winged. Seed(s) 1 or more per cell, globose or compressed; albumen horney or fleshy; embryo small, terete.

About 240 genera and nearly 4000 species, distributed throughout the world (Jafri, 1966).

2. Asphodelus

L., Gen. Pl., ed. 5: n. 379 (1754); Jafri, Fl. Kar., 84 (1966).

Annual or perennial herbs with fibrous roots, fastigiate and united in the upper part. Leaves radical, sheathing at the base, linear, terete, fistular. Flowers bracteate, racemose on leafless scapes. Perianth white, withering without falling down, 6-partite, segments connate below. Stamens 6. Ovary 3-celled; ovules 2 per cell. Capsule globose, loculicidal. Seeds 1 per cell, black; endosperm cartilaginous.

A genus of 6-7 species mostly distributed in S. Europe to India (Jafri, 1966).

Asphodelus tenuifolius Cav. in An. Cienc. Nat. 3: 64. t. 27 f. 2 (1801); Jafri in Fl. Kar., 85 (1966); Stewart in Nasir & Ali., Ann. Cat. Vasc. Pl. W. Pak. & Kashm., 50 (1972).

Syn: *A. fistulosus* var. *tenuifolius* Baker in J.L.S., xv: 272 (1875).

An annual, leafy herb. Stem much branched, reduced. Leaves many, in a dense, basal, erect rosette, ca. 6-15 cm long and ca. 2.5-8 mm broad, semiterete, fistular, tapering, sheathing at base, striate and papilose-scabridulous. Flowers on ca. 42 cm high scapes, much branched, panicled scape; racemes laxly flowered. Flowers solitary in the axil of each bract, white; pedicel ca. 3-6 mm long, jointed above the middle; bracts broadly ovate, cymbiform, acute, ca. 2-3 mm long, scarious with a strong, brownish keel. Perianth segments ca. 5 x 2.5-3.2 mm in size, oblong, obtuse, white with a strong, conspicuous, brown keel. Stamens 3 mm long; filaments swollen near apex, epiphilous; anthers brown. Capsule ca. 4-5 mm across, erect, transversely wrinkled. Seeds ca. 3 mm long, black, sharply 3-gonous, acute.

Flowering Period: November-February.

Herbarium specimens examined: NARA DESERT: *Rahmatullah Qureshi*, 192, 193, 628, 854 (SALUBGH).

Field remarks: A common species of plain areas.

Distribution: Canary Island, Tropical Africa and Indo-Pak subcontinent (Jafri, 1966).

Figures

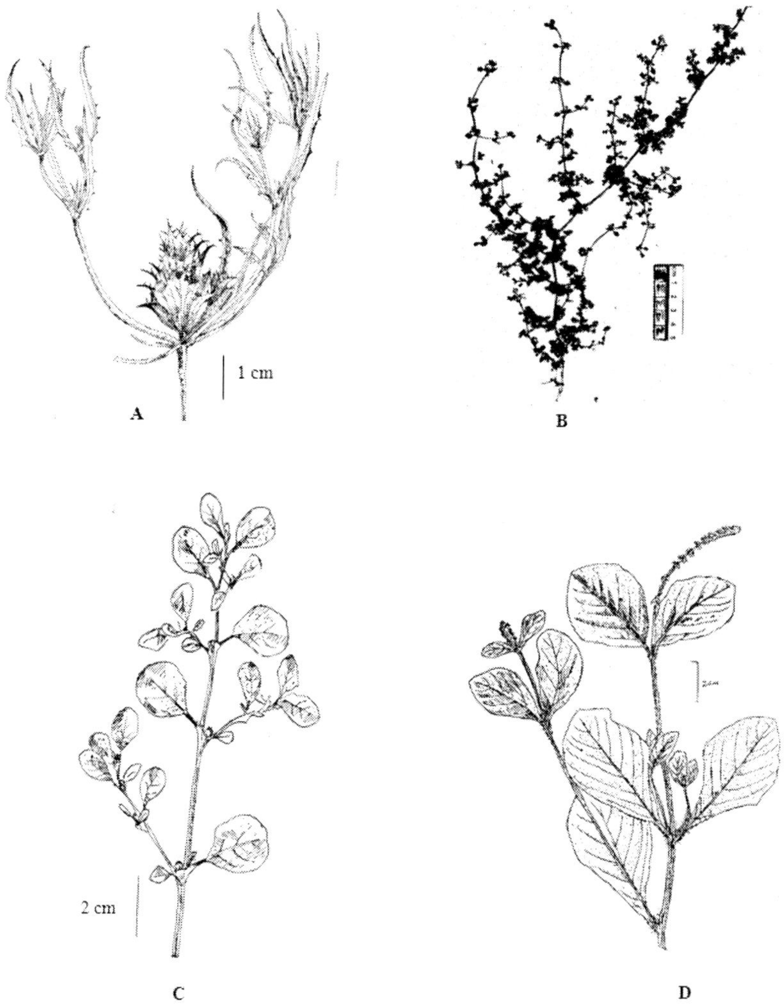

Figure 1. Blepharis scindica (Acanthaceae) [A]; Limeum indicum (Aizoaceae) [B]; Trianthema portulacastrum (Aizoaceae) [C]; Achyranthus aspera (Amaranthaceae) [D].

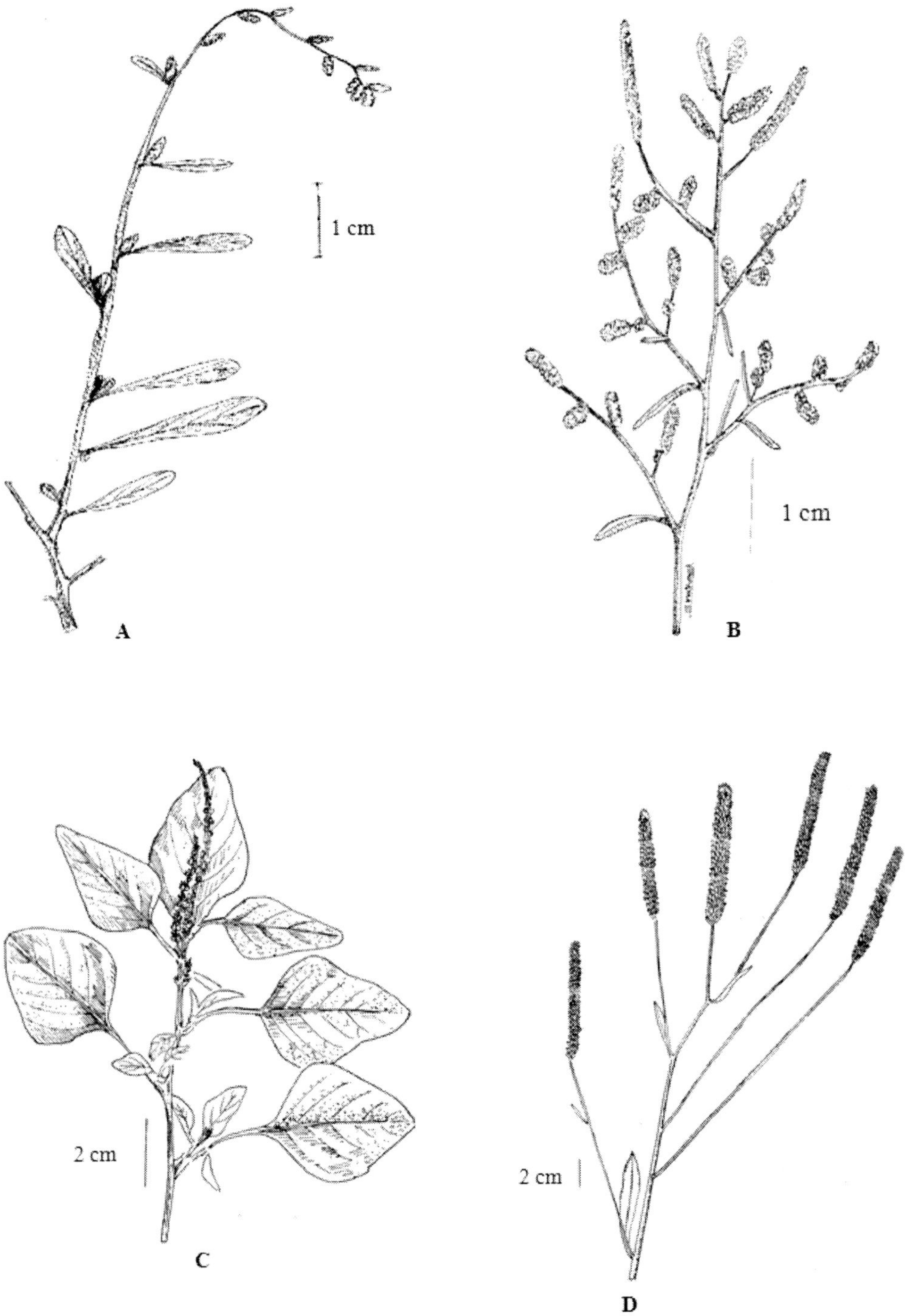

Figure 2. Aerva javanica var. bovei (Amaranthaceae) [A]; A. javanica var. javanica (Amaranthaceae) [B]; Amaranthus viridis (Amaranthaceae) [C]; Celosia argentea (Amaranthaceae) [D].

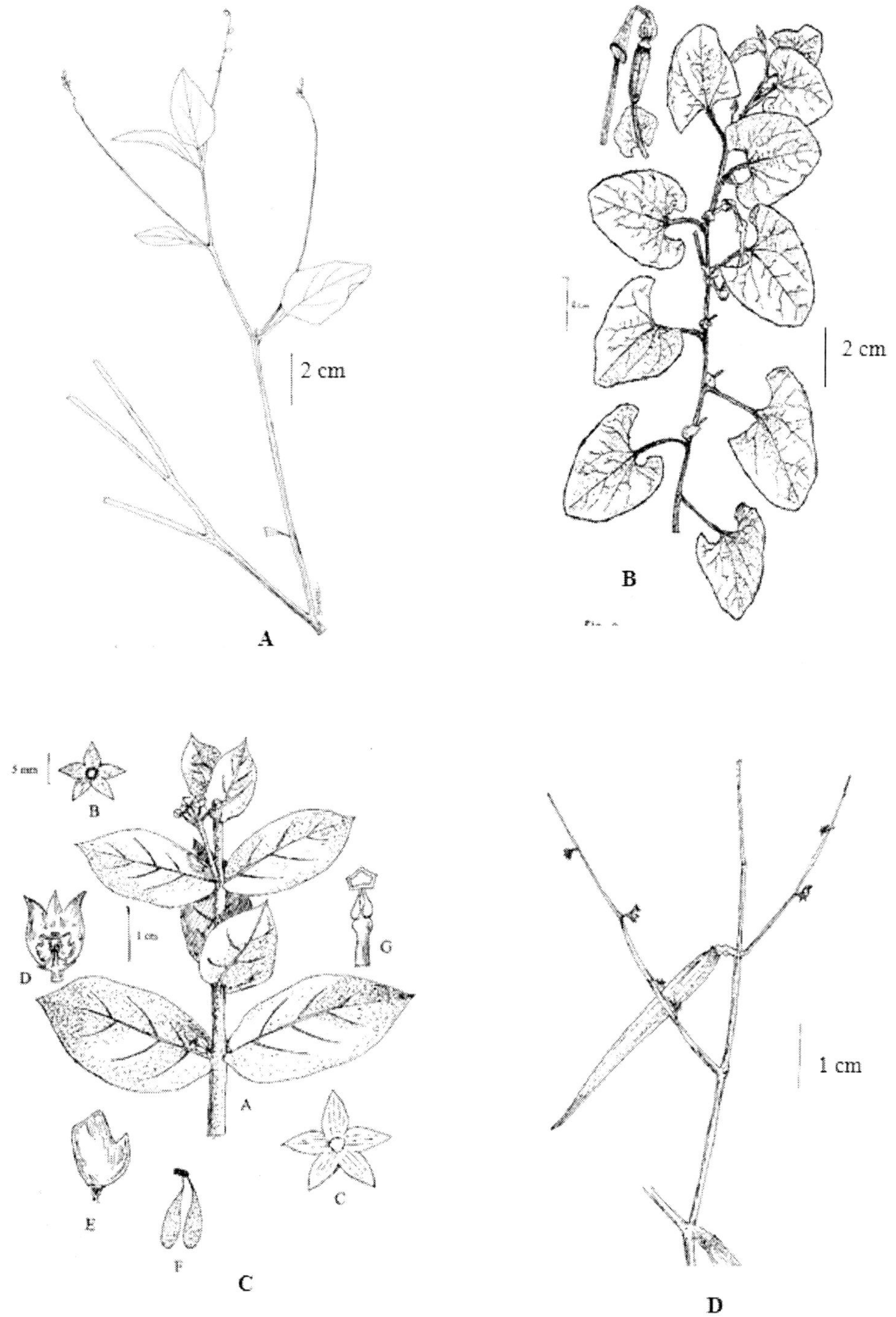

Figure 3. Digera muricata (Amaranthaceae) [A]; Aristolochia bracteolata (Aristolochiaceae) [B]; Calotropis procera (Asclepiadaceae) [C]; Leptadenia pyrotechnica (Asclepiadaceae) [D].

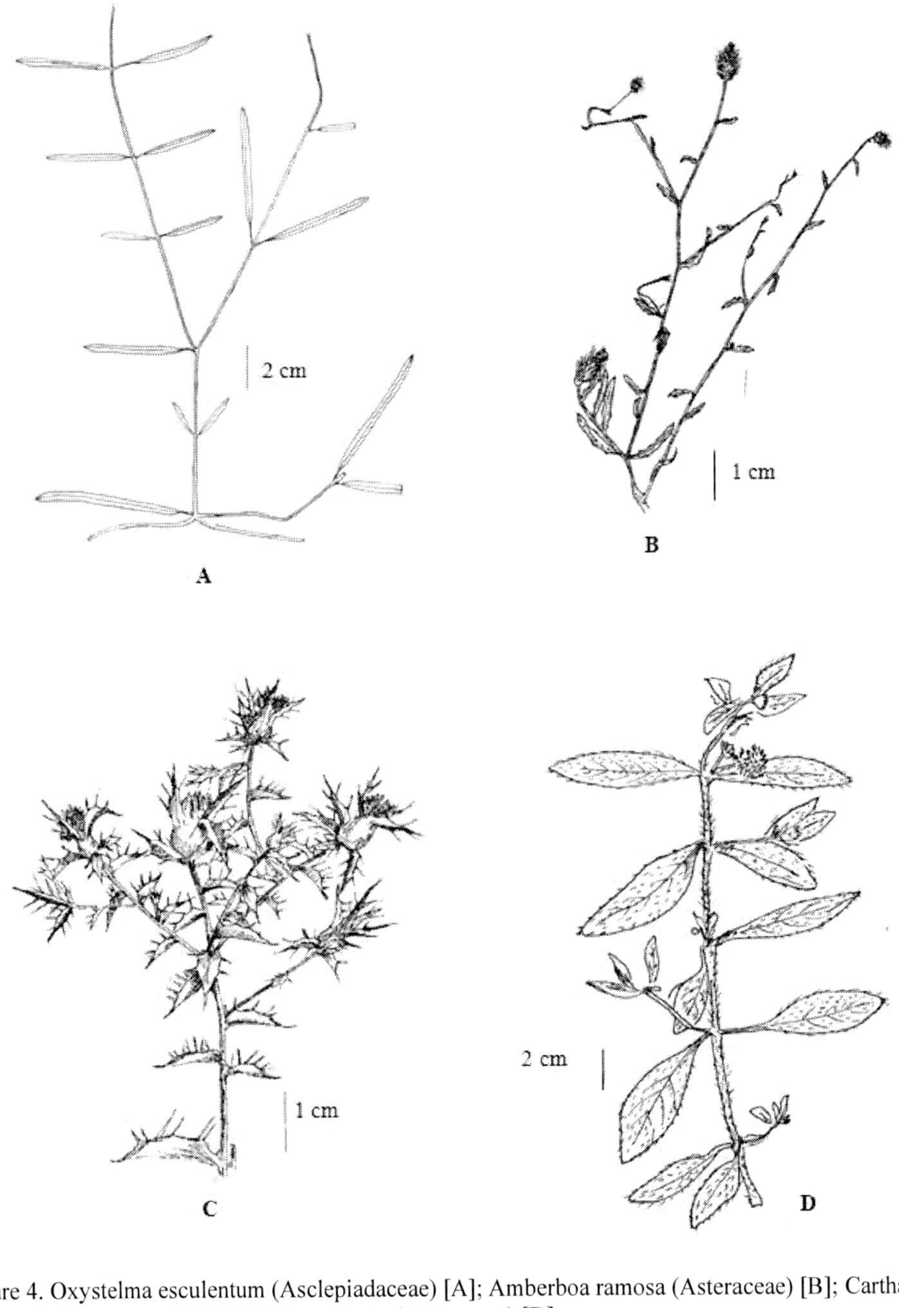

Figure 4. Oxystelma esculentum (Asclepiadaceae) [A]; Amberboa ramosa (Asteraceae) [B]; Carthamus oxycantha (Asteraceae) [C]; Eclipta prostrata (Asteraceae) [D].

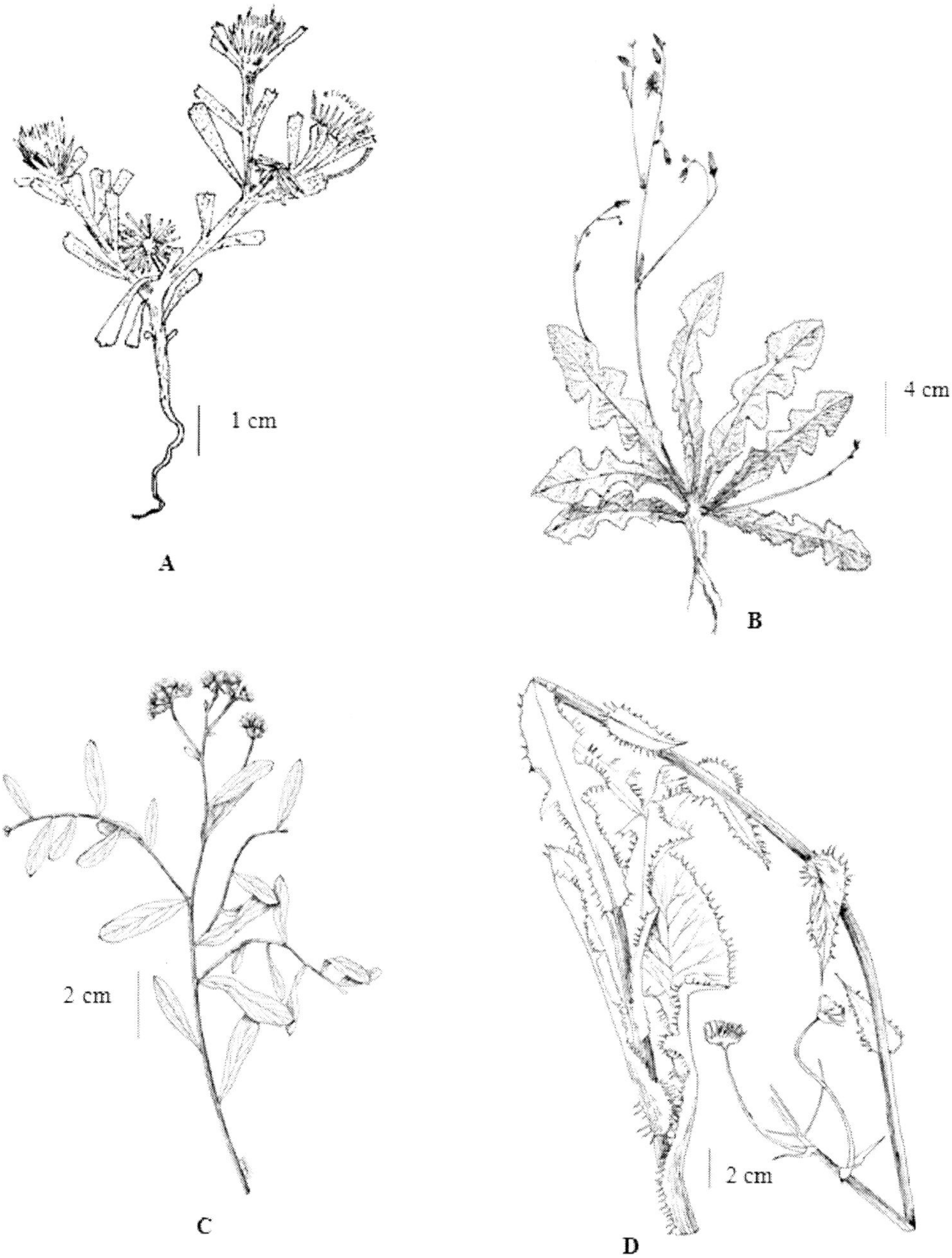

Figure 5. Iphiona grantioides (Asteraceae) [A]; Launaea procumbens (Asteraceae) [B]; Pluchea lanceolata (Asteraceae) [C]; Sonchus oleraceus(Asteraceae) [D].

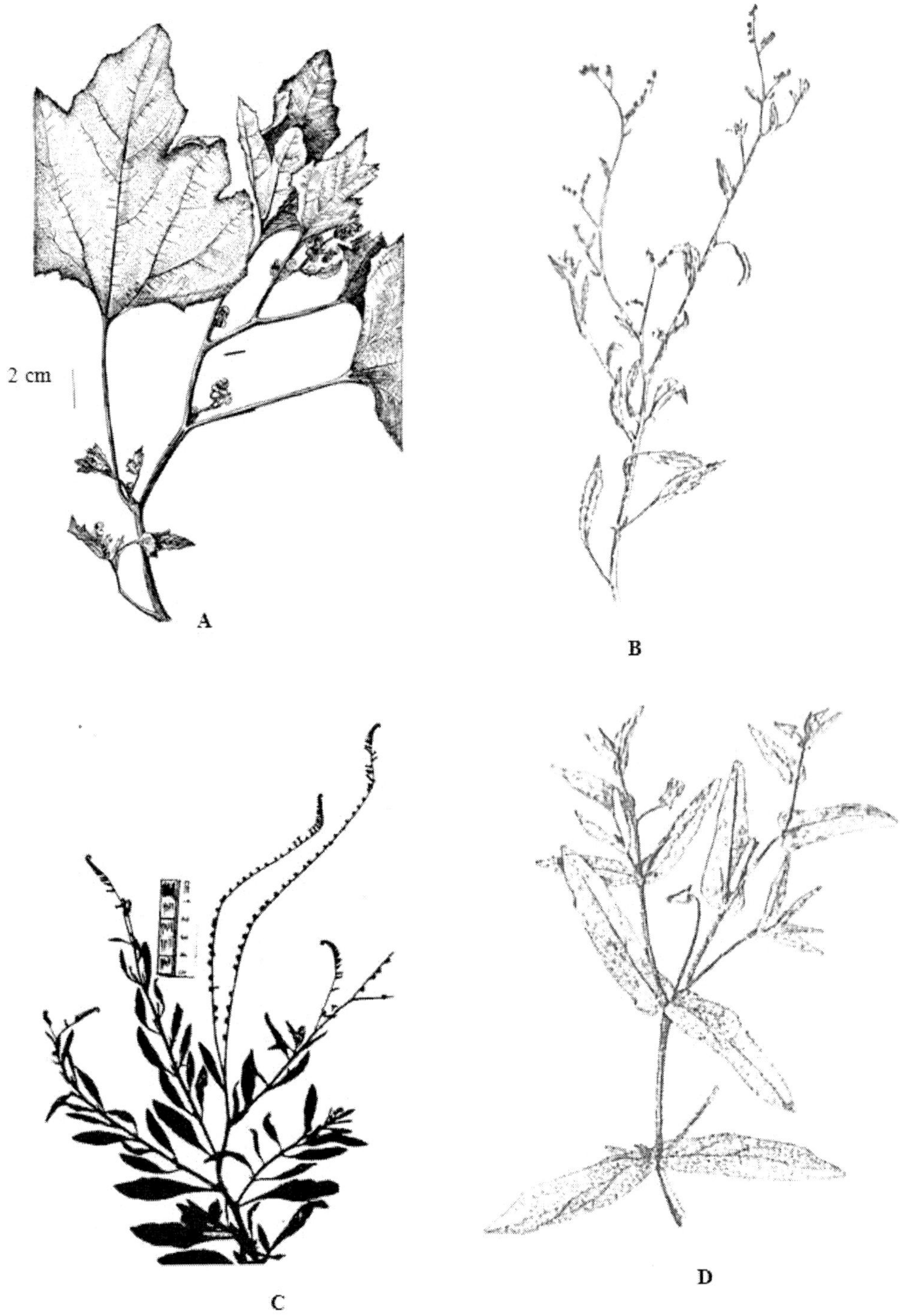

Figure 6. Xanthium indicum (Asteraceae) [A]; Heliotropium crispum (Boraginceae) [B]; Heliotropium subulatum (Boraginceae) [C]; Trichodesma indicum (Boraginceae) [D].

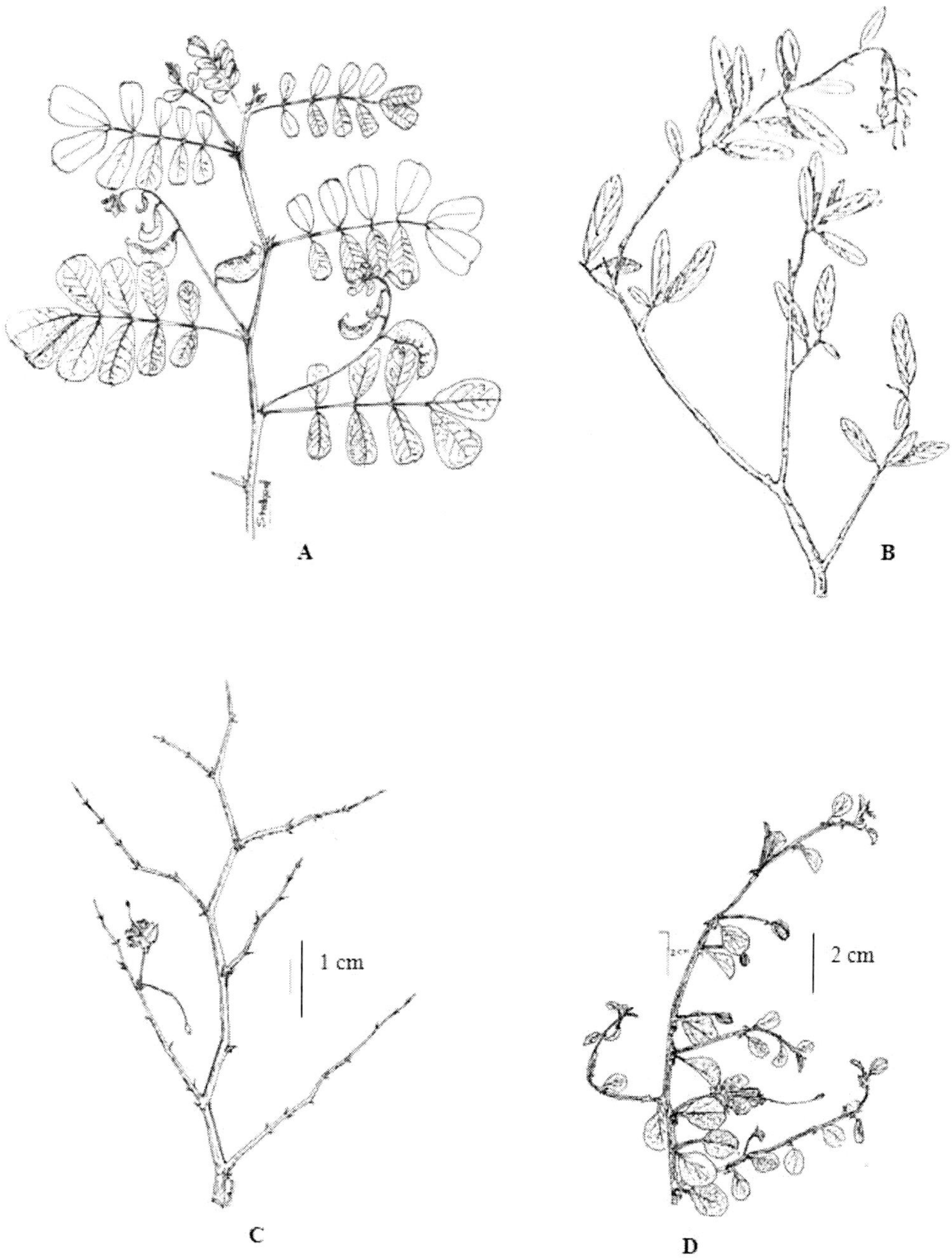

Figure 7. Senna italica (Caesalpiniaceae) [A]; Cadaba farinosa (Capparidaceae) [B]; Capparis decidua (Capparidaceae) [C]; C. spinosa (Capparidaceae) [D].

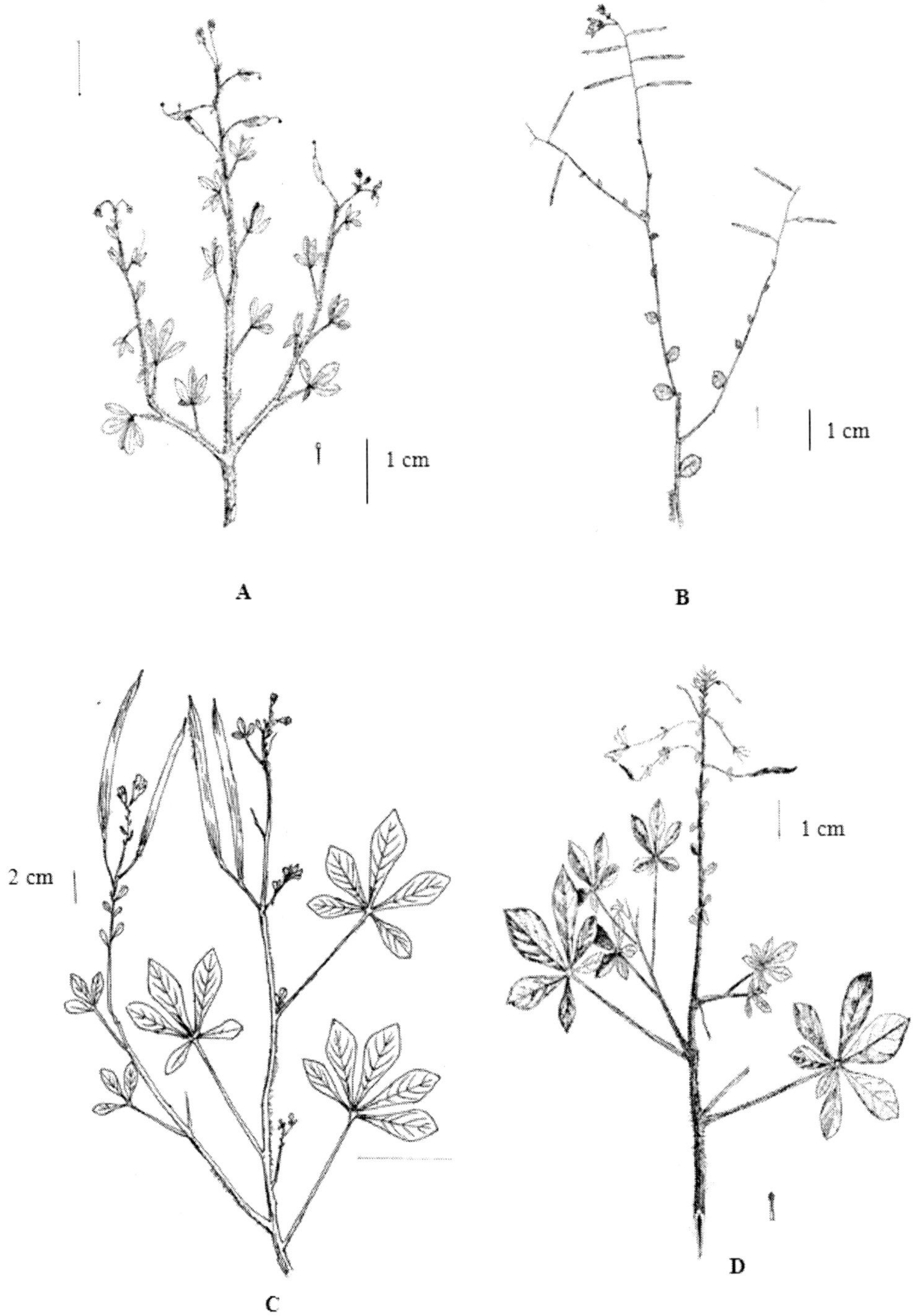

Figure 8. Cleome brachycarpa (Capparidaceae) [A]; C. scaposa (Capparidaceae) [B]; C. viscosa (Capparidaceae) [C]; Gynandropsis gynandra (Capparidaceae) [D].

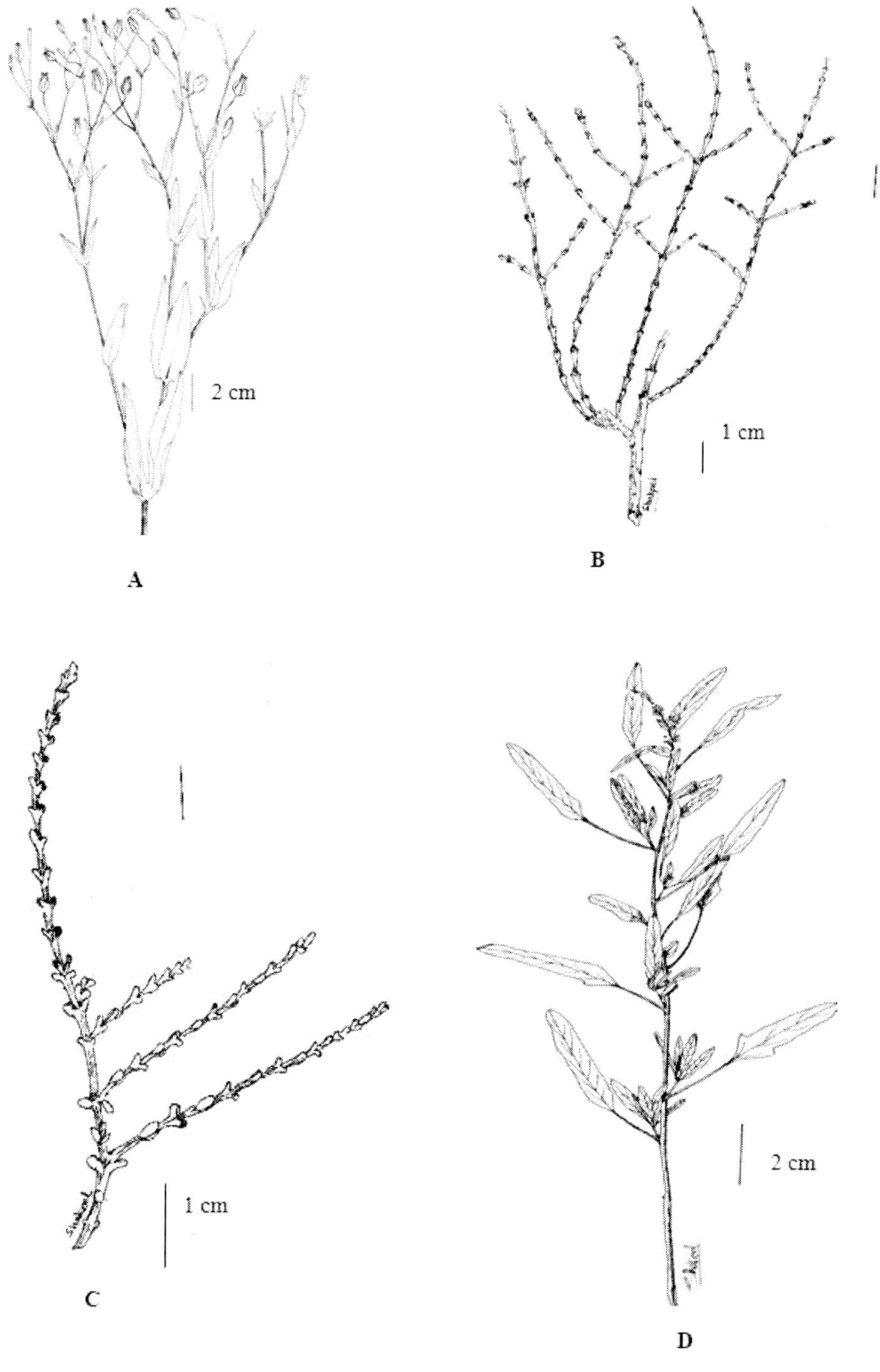

Figure 9. Vaccaria hispanica (Caryophyllaceae) [A]; Haloxylon salicornicum (Chenopodiaceae) [B]; H. stocksii (Chenopodiaceae) [C]; Chenopodium album (Chenopodiaceae) [D].

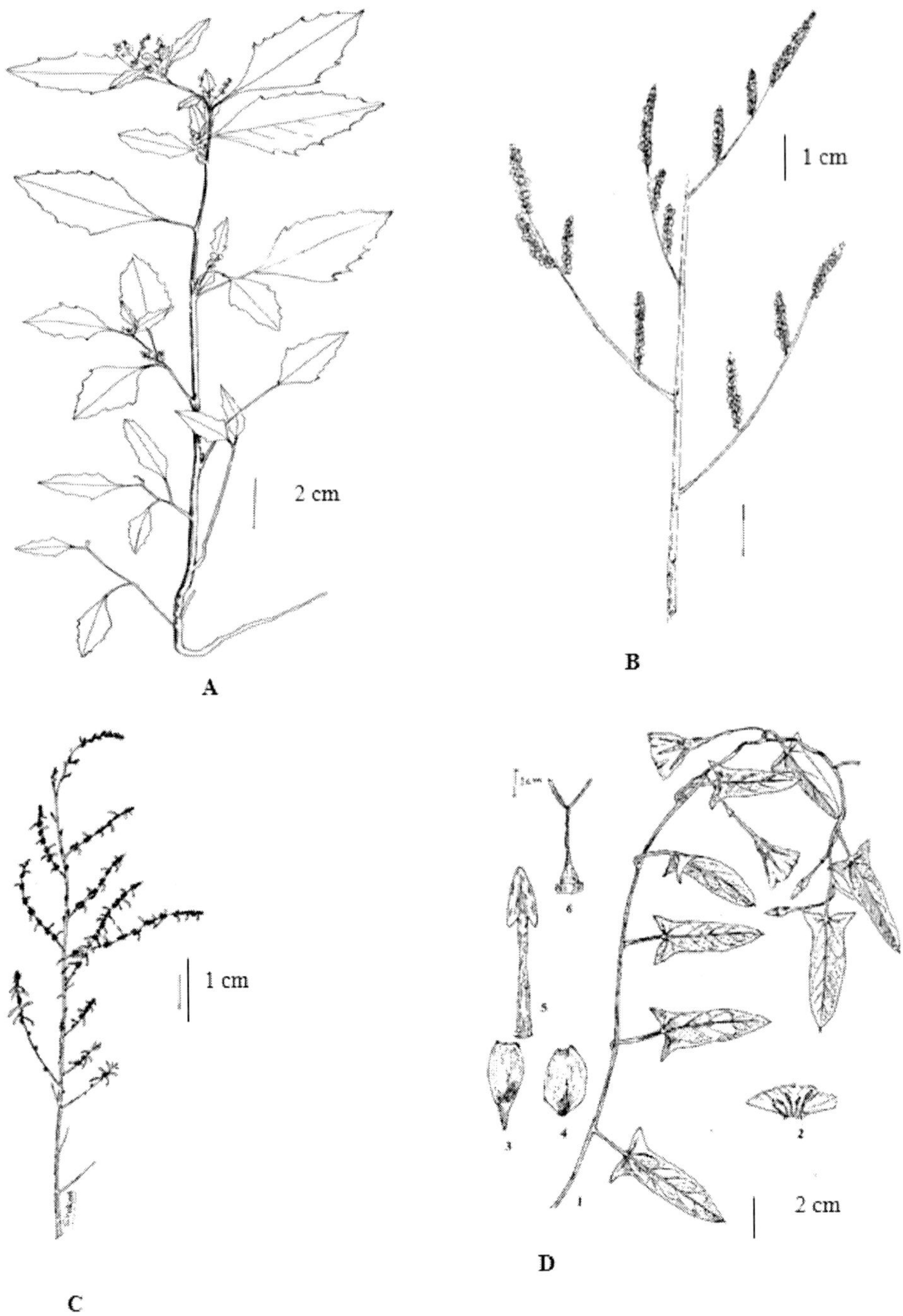

Figure 10. Chenopodium murale (Chenopodiaceae) [A]; Salsola imbricata (Chenopodiaceae) [B]; Saueda fruticosa (Chenopodiaceae) [C]; Convolvulus arvensis (Convolvulaceae) [D].

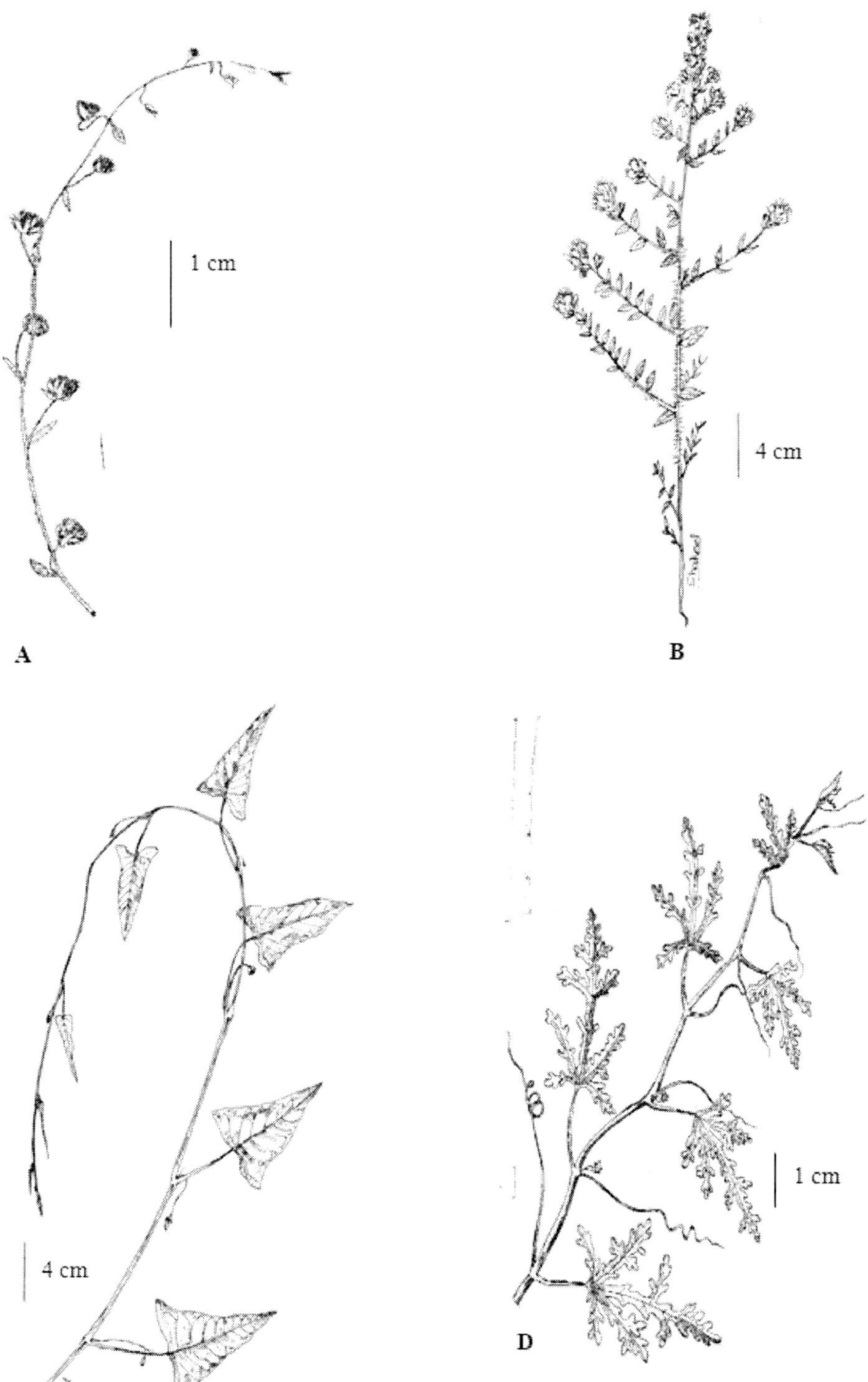

Figure 11. Convolvulus prostratus (Convolvulaceae) [A]; Cressa cretica (Convolvulaceae) [B]; Ipomoea aquatica (Convolvulaceae) [C]; Citrullus colocynthis (Cucurbitaceae) [D].

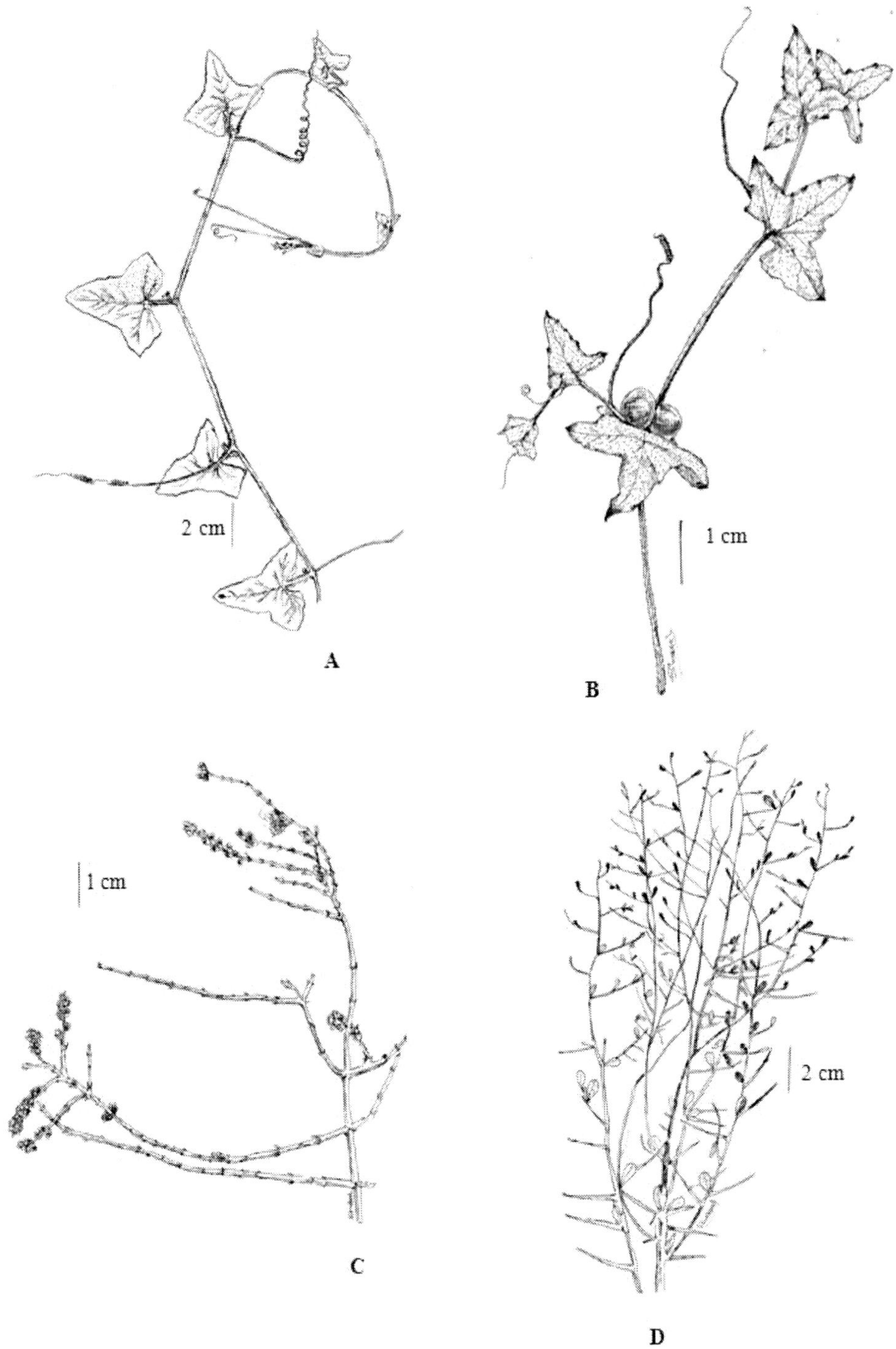

Figure 12. Cucumis melo var. agrestis (Cucurbitaceae) [A]; Mukia maderaspatana (Cucurbitaceae) [B]; Bergia aestivosa (Elatinaceae) [C]; Alhagi maurorum (Fabaceae) [D].

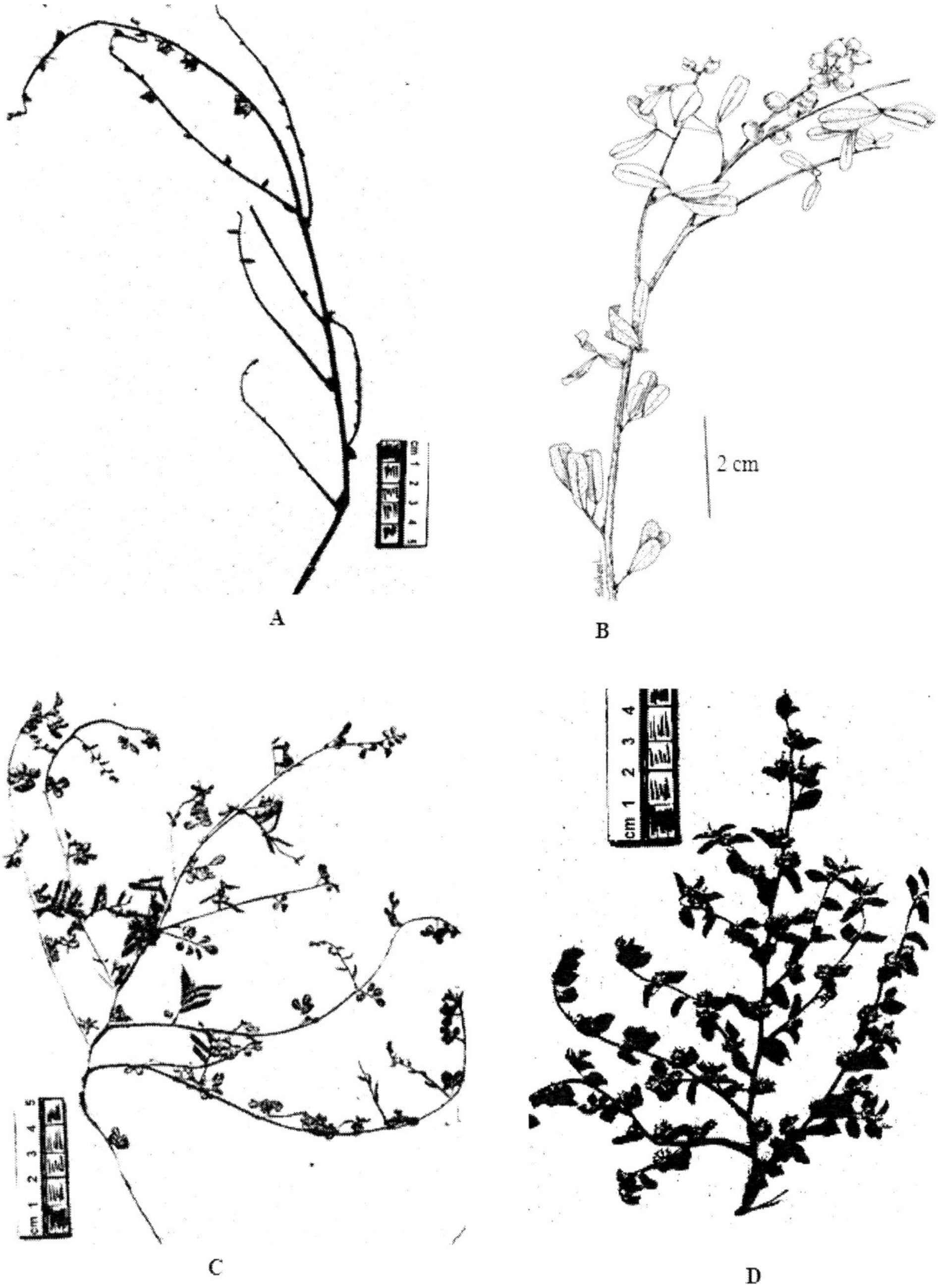

Figure 13. Crotalaria burhia (Fabaceae) [A]; Crotalaria medicagnea (Fabaceae) [B]; Indigofera argentea (Fabaceae) [C]; I. cordifolia (Fabaceae) [D].

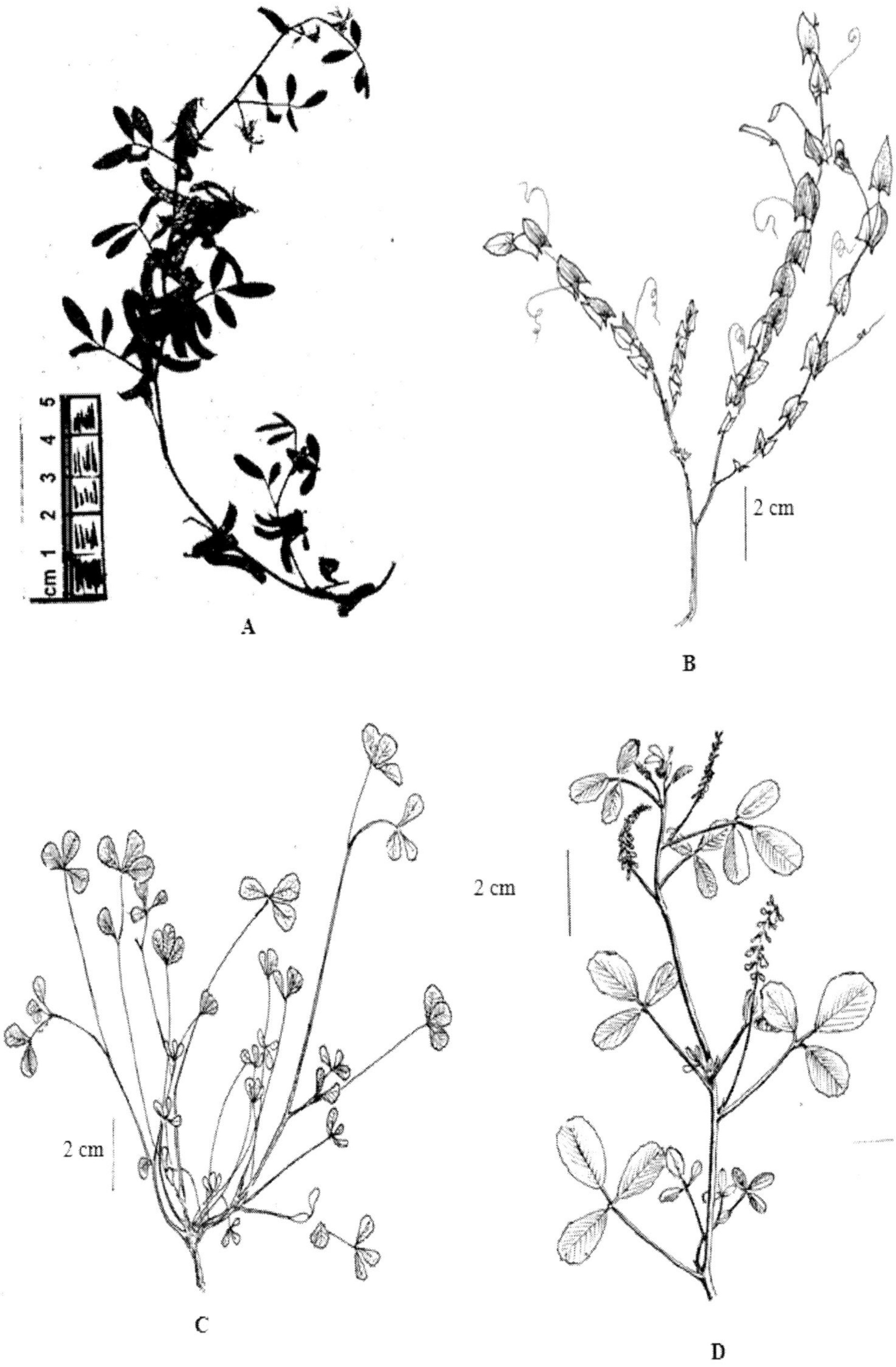

Figure 14. Indigofera hochstetteri (Fabaceae) [A]; Lathyrus aphaca (Fabaceae) [B]; Medicago polymorpha (Fabaceae) [C]; Melilotus alba (Fabaceae) [D].

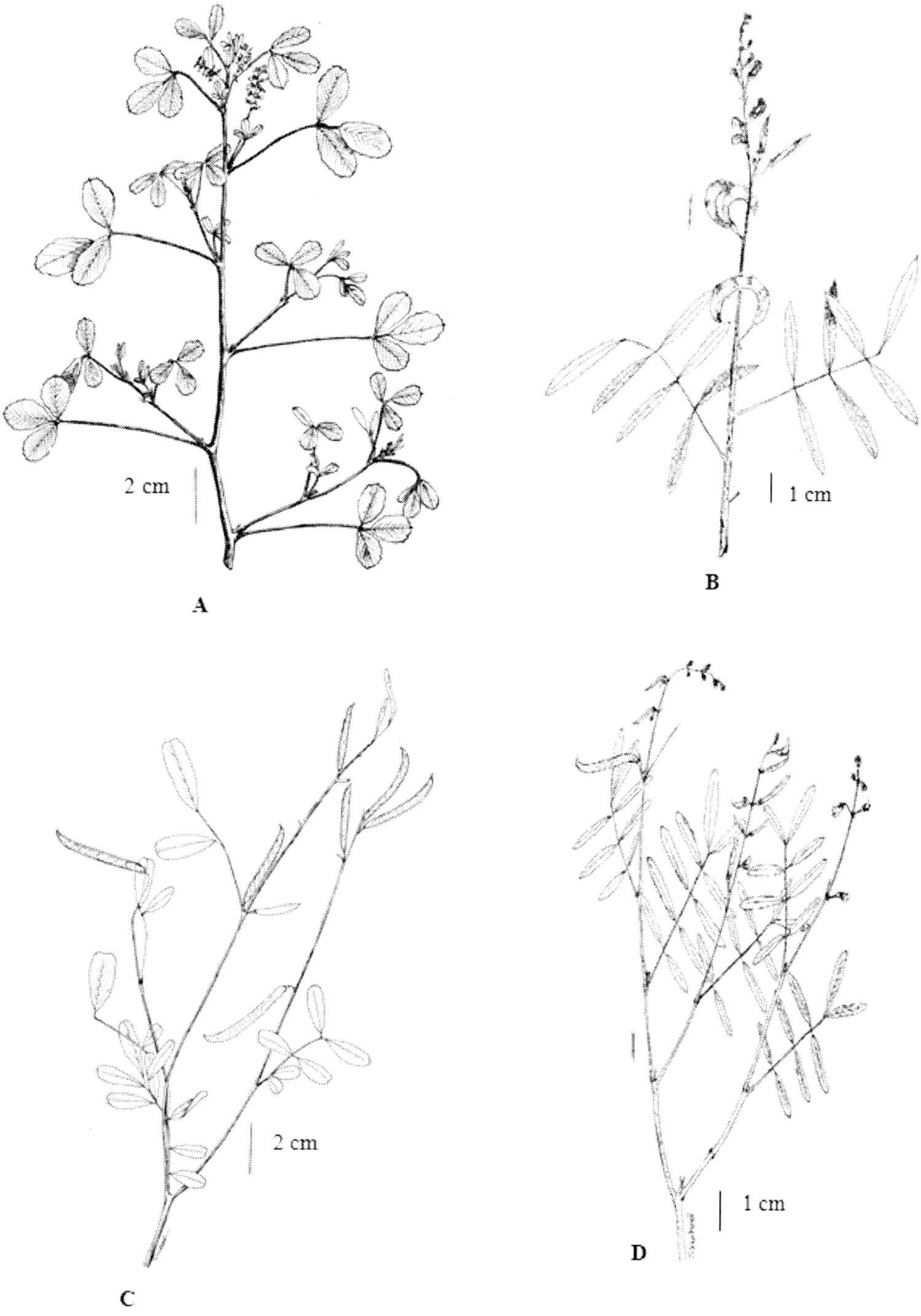

Figure 15. Melilotus indica (Fabaceae) [A]; Tephrosia falciformis (Fabaceae) [B]; T. uniflora var. petrosa (Fabaceae) [C]; T. villosa (Fabaceae) [D].

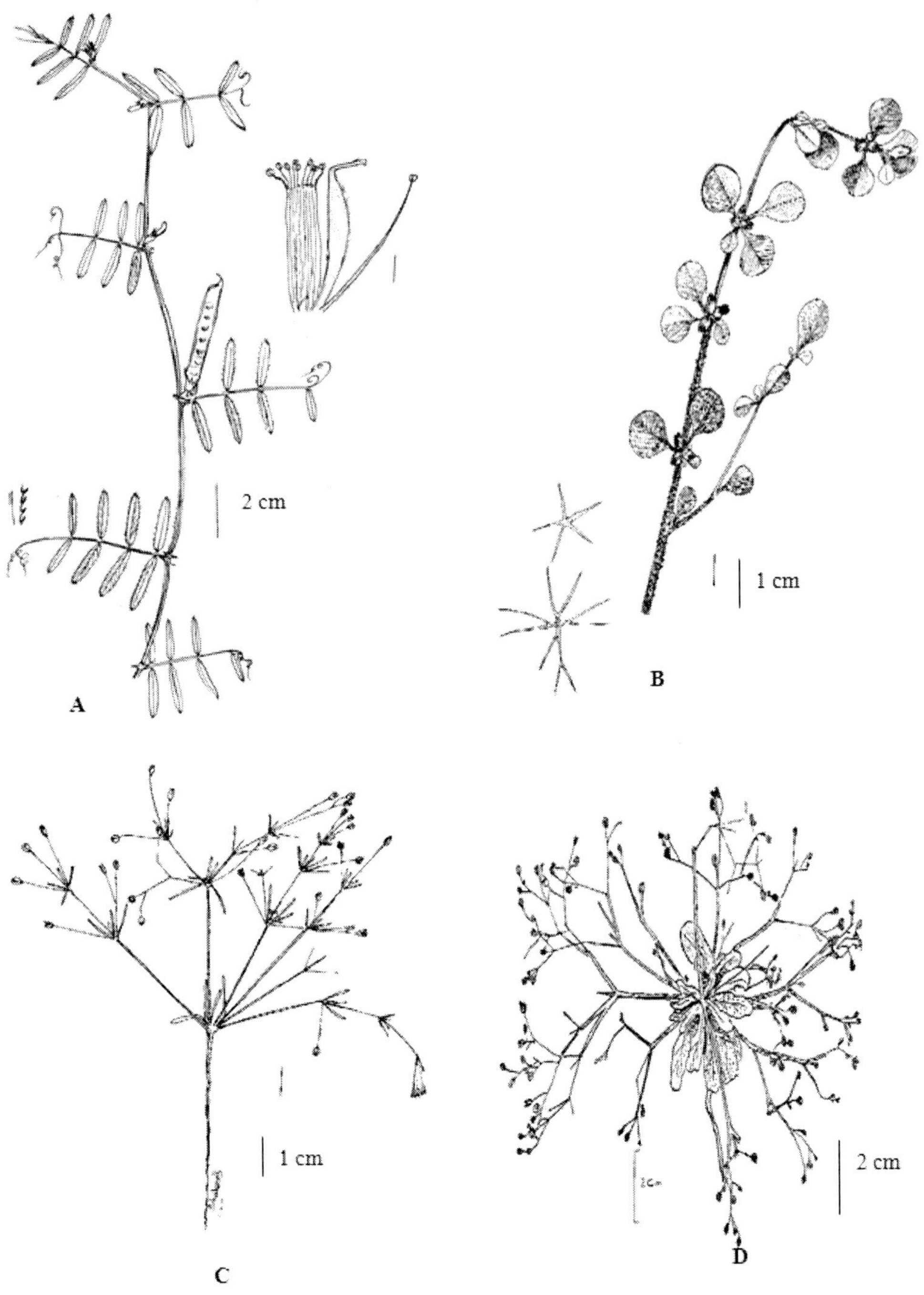

Figure 16. Vicia sativa (Fabaceae) [A]; Glinus lotoides (Molluginaceae) [B]; Mollugo cerviana (Molluginaceae) [C]; M. nudicaulis (Molluginaceae) [D].

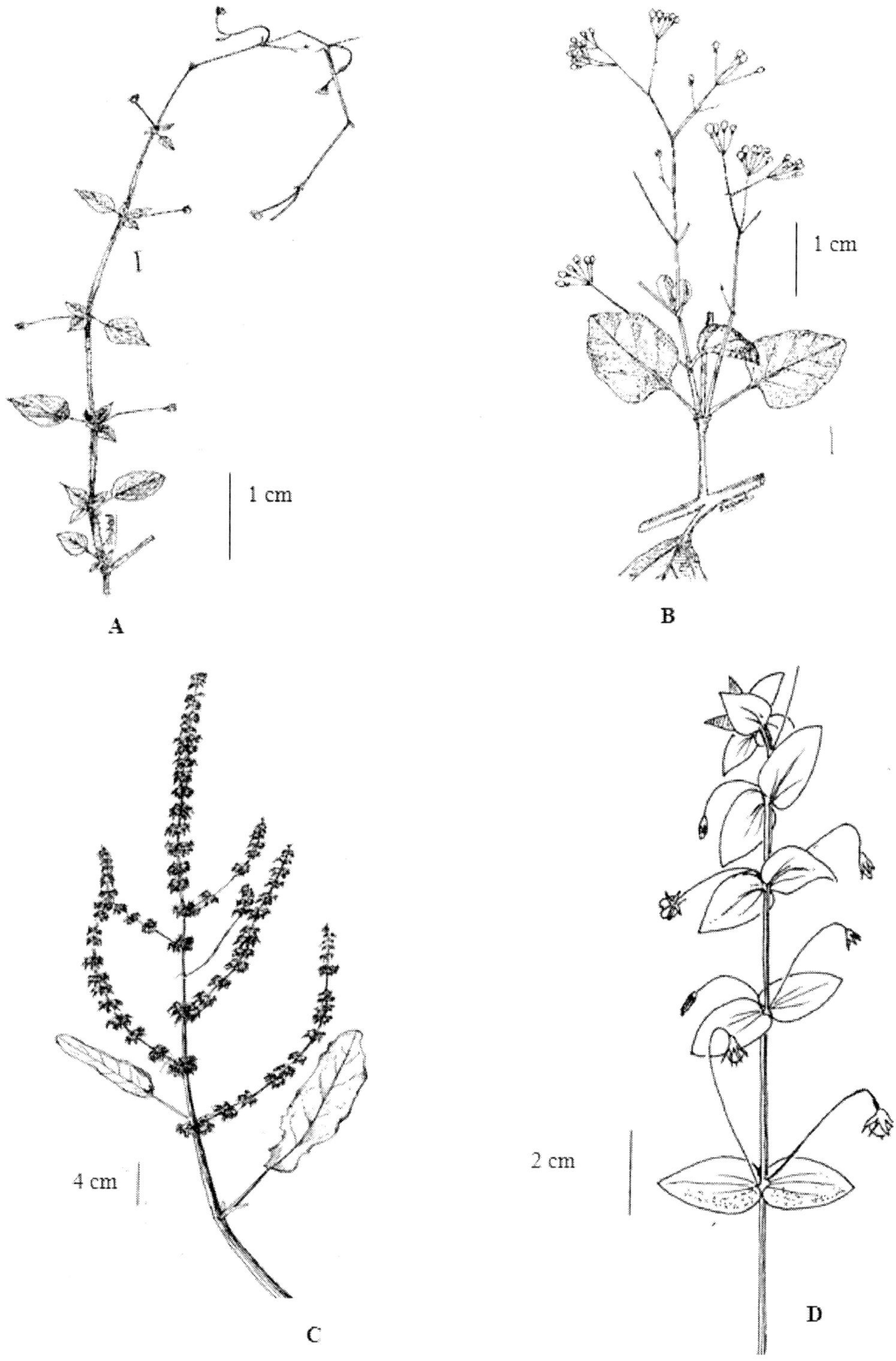

Figure 17. Boerhavia procumbens (Nyctaginaceae) [A]; Commicarpus boisseiri (Nyctaginaceae) [B]; Rumex dentatus (Polygonaceae) [C]; Anagallis arvensis var. coerulea (Primulaceae) [D].

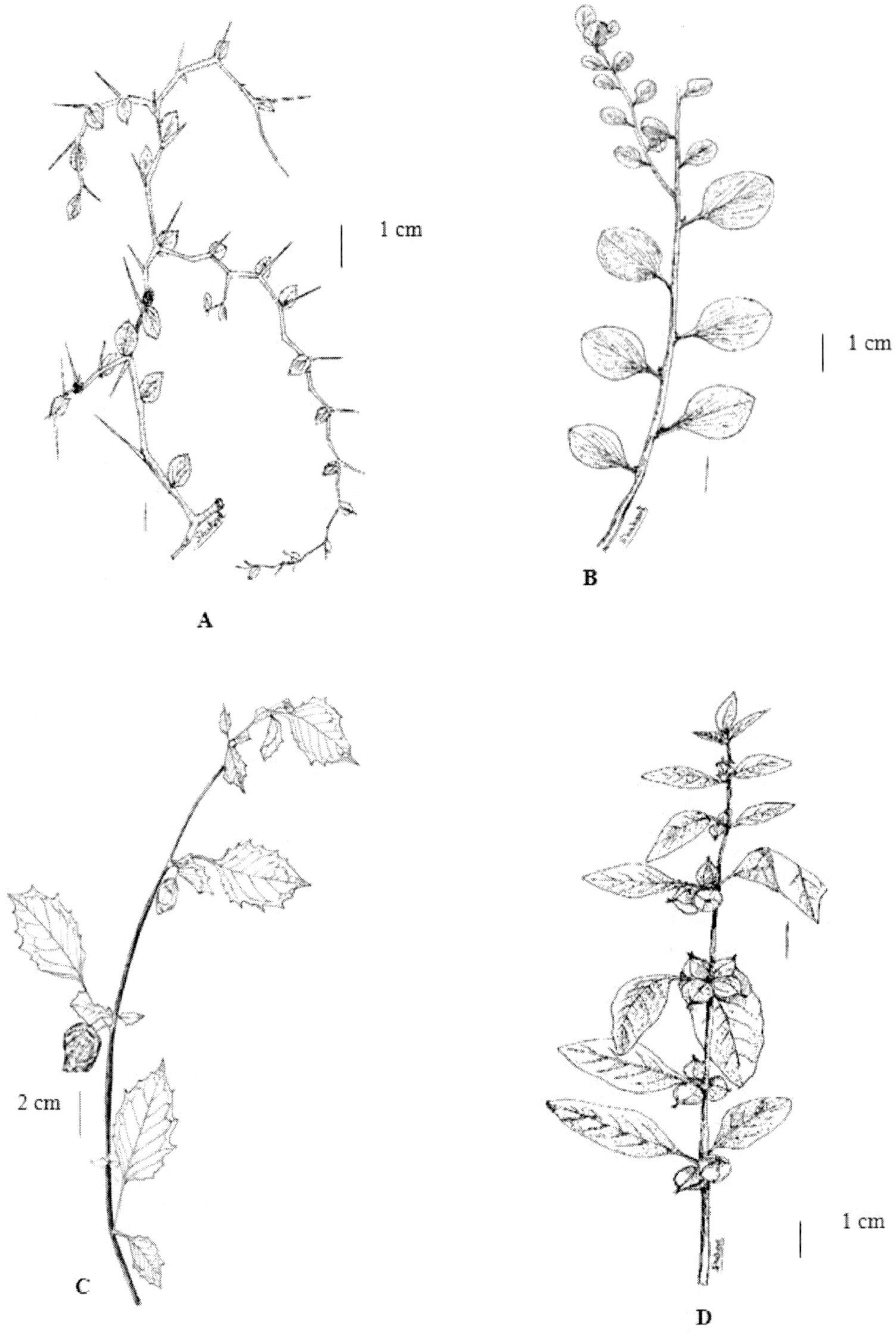

Figure 18. Zizyphus nummularia (Rhamnaceae) [A]; Schweinfurthia papilionacea (Scrophulariaceae) [B]; Physalis minima (Solanaceae) [C]; Withania somnifera (Solanaceae) [D].

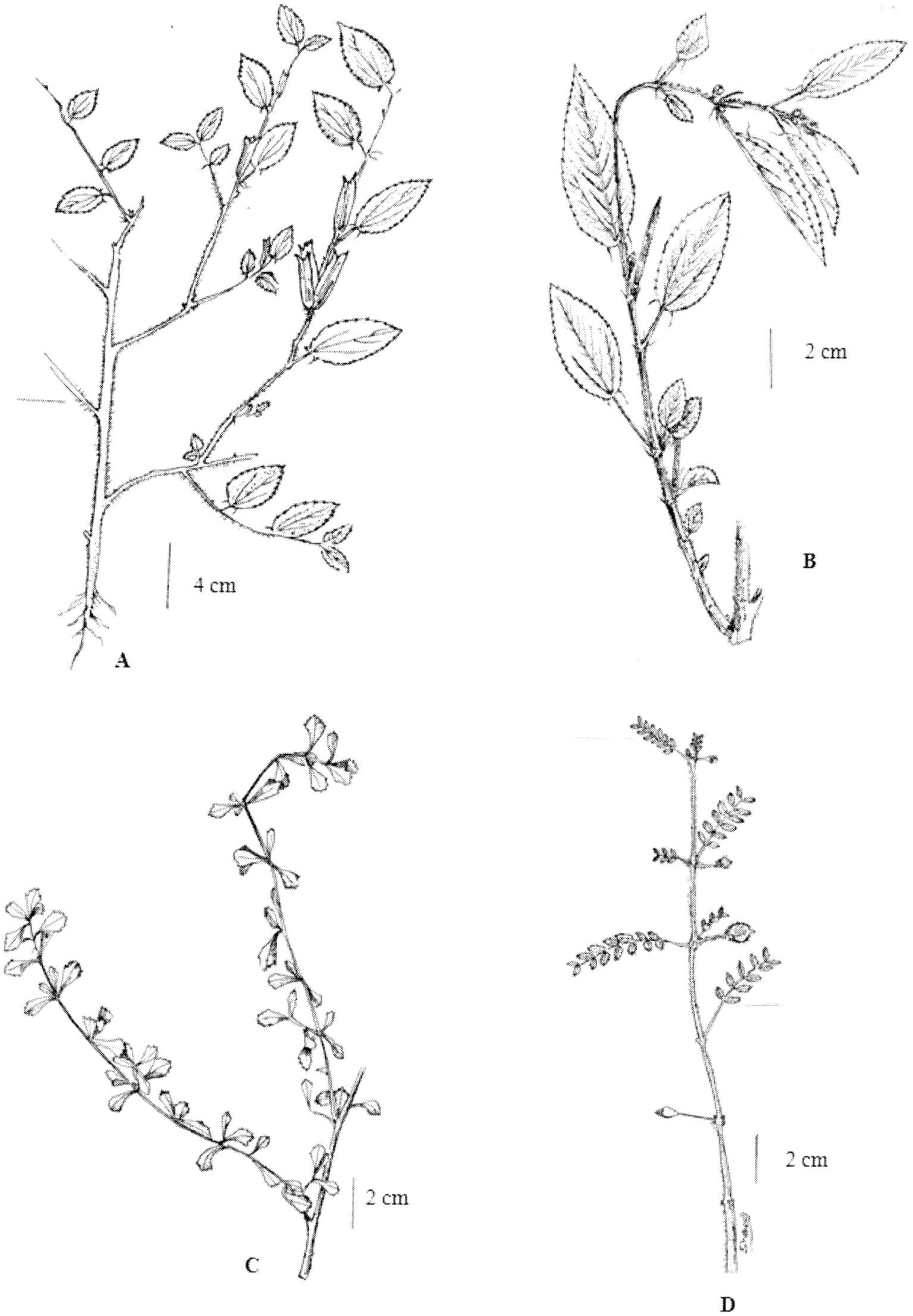

Figure 19. Corchorus aestuans (Tiliaceae) [A]; C. tridens (Tiliaceae) [B]; Phyla nodiflora (Verbenaceae) [C]; Tribulus terrestris (Zygophyllaceae) [D].

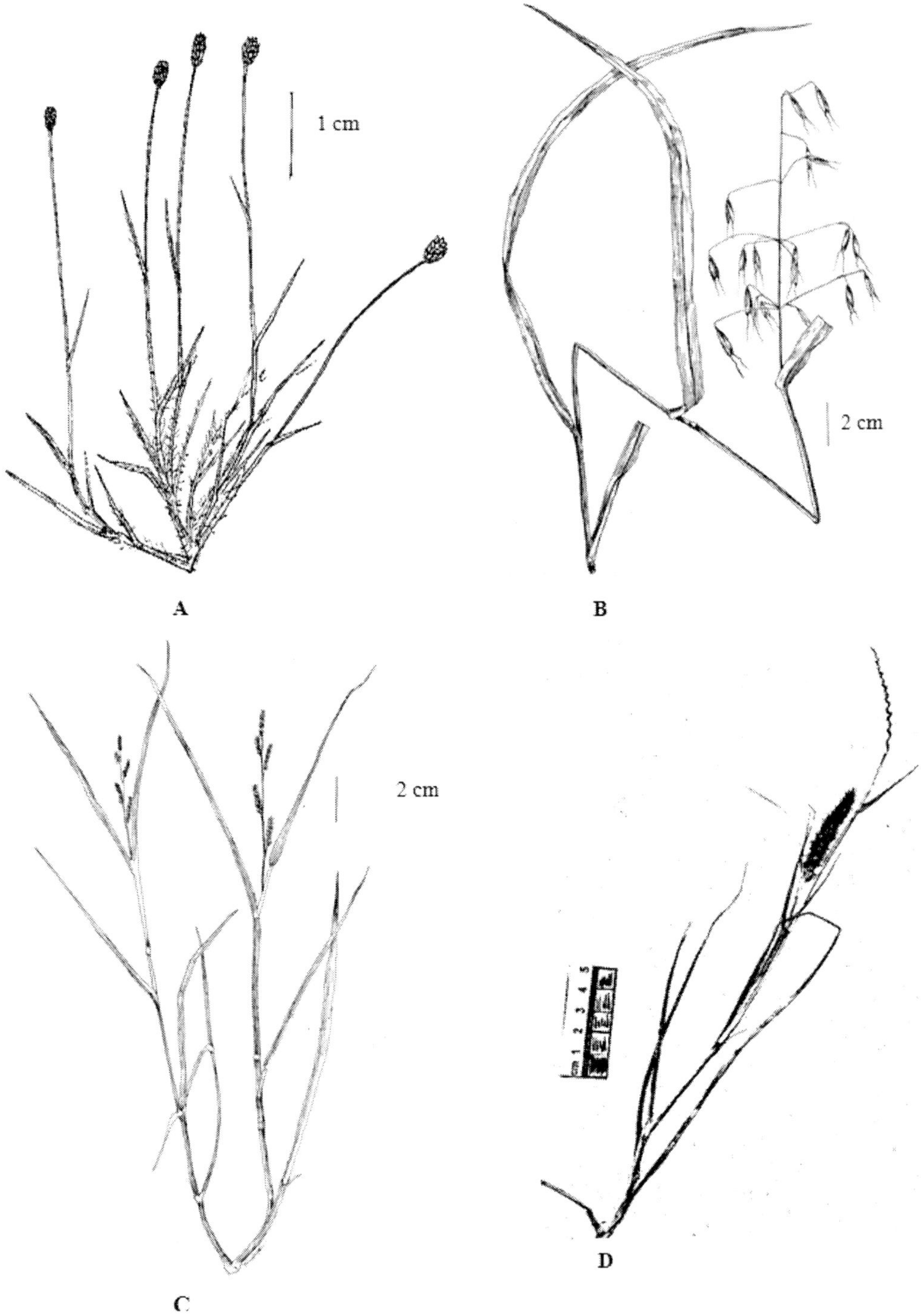

Figure 20. Aeluropus lagopoides (Poaceae) [A]; Avena fatua (Poaceae) [B]; Brachiaria eruciformis (Poaceae) [C]; Cenchrus bifloris (Poaceae) [D].

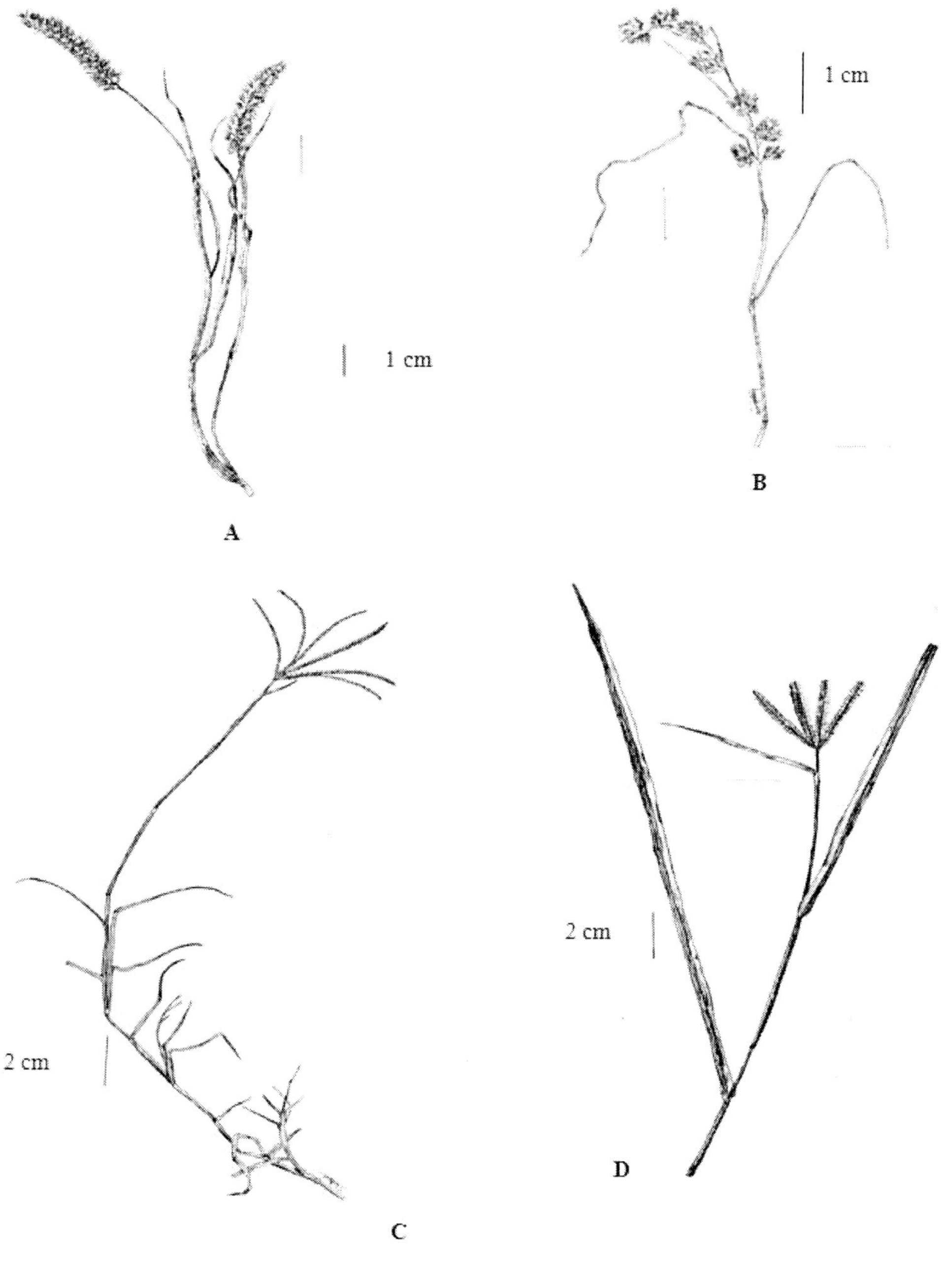

Figure 21. Cenchrus ciliaris (Poaceae) [A]; Cymbopogon jwarancusa (Poaceae) [B]; Cynodon dactylon (Poaceae) [C]; Dactyloctenium aegyptium (Poaceae) [D].

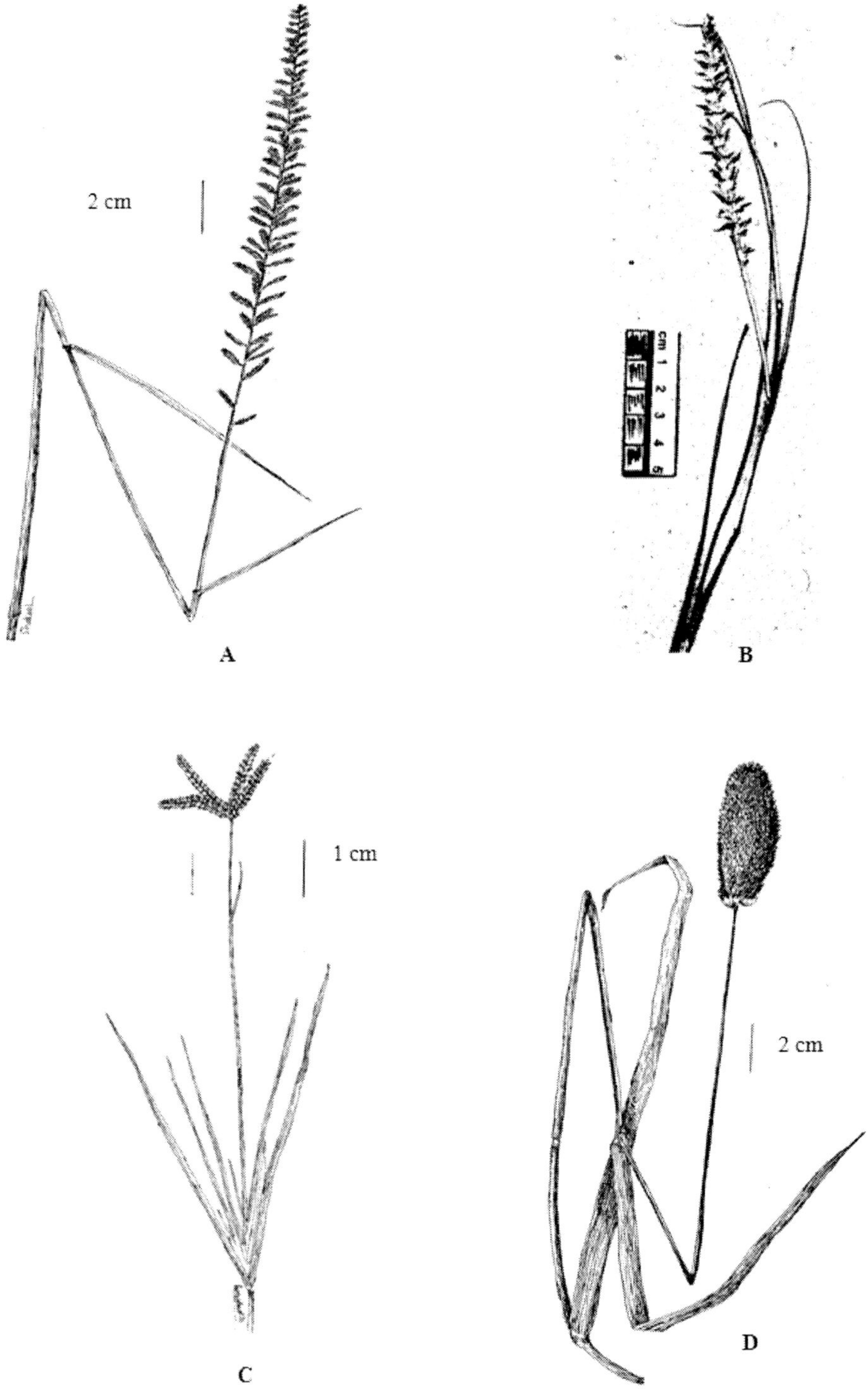

Figure 22. Desmostachya bipinnata (Poaceae) [A]; Lasiurus scindicus (Poaceae) [B]; Ochthochloa compressa (Poaceae) [C]; Phalaris minor (Poaceae) [D].

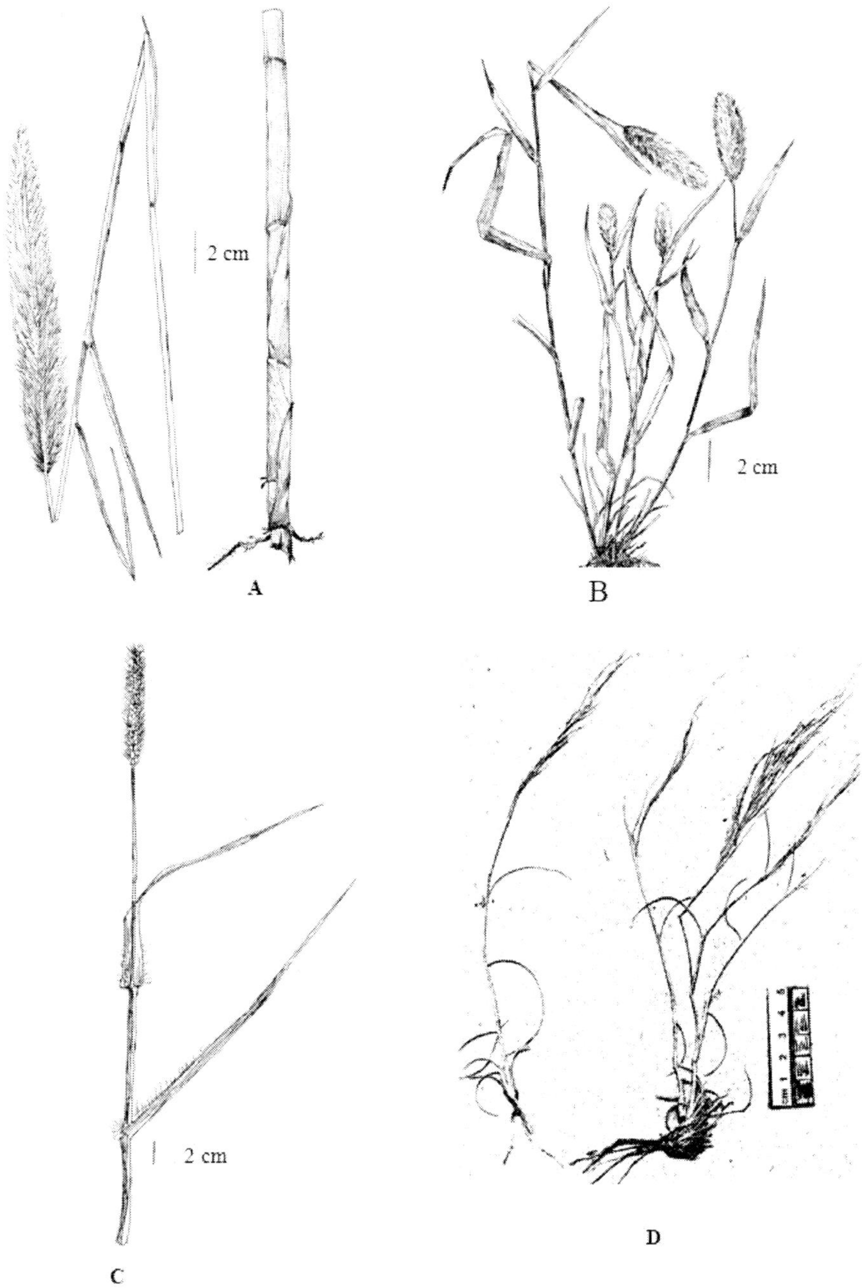

Figure 23. Phragmites australis (Poaceae) [A]; Polypogon fugax (Poaceae) [B]; Setaria pumila (Poaceae) [C]; Stipagrostis plumosa (Poaceae) [D].

References

[1] Abedin, S. and Ali, S.I. in Nasir, E. and Ali, S.I. (1974). *Flora of Pakistan*, No. 71 (Burseraceae), Islamabad, Karachi.

[2] Abedin, S. in Nasir, E. & Ali, S.I. (1979). *Flora of Pakistan*, No. 130 (*Malvaceae*). Islamabad, Karachi.

[3] Ali, S.I. and M. Qaiser (Eds.), (1993-2007). *Flora of Pakistan*, Islamabad, Karachi.

[4] Ali, S.I. and Y.J. Nasir (Eds.), (1989-1991). *Flora of Pakistan*, Islamabad, Karachi.

[5] Ali, S.I. in Ali, S.I. in Nasir, E. and Ali, S.I. (1977). *Flora of Pakistan*, No. 130 (Fabaceae), Islamabad, Karachi.

[6] Ali, S.I. in Nasir, E. & Ali, S.I. (1973). *Flora of Pakistan*, No. 36 (Mimosaceae), Islamabad, Karachi.

[7] Ali, S.I. in Nasir, E. & Ali, S.I. (1973). *Flora of Pakistan*, No. 54 (Caesalpiniaceae), Islamabad, Karachi.

[8] Ali, S.I. in Nasir, E. & Ali, S.I. (1977). *Flora of Pakistan*, No. 100 (Papilionaceae), Islamabad, Karachi.

[9] Ali, S.I. in Nasir, E. & Ali, S.I. (1983). *Flora of Pakistan*, No. 150 (Asclepiadaceae), Islamabad, Karachi.

[10] Anderberg, A. (1985). The genus Iphiona (Compositae-Inuleae). *Nordic J. Bot.*, 5(2): 169-194.

[11] Anonymous. (1992). *Pak. Swiss Range and Livestock Improvement Project.*

[12] Anonymous. 2002. *Nara game reserve baseline environmental study.* Halcrow Pakistan, Islamabad.

[13] Austin, D. F. and Ghazanfar, S. in Nasir, E. & Ali, S.I. (1979). *Flora of Pakistan*, No. 126 (Convolvulaceae), Islamabad, Karachi.

[14] Baqri, S. R. H., Hasan, S. A., Awan, M. R. & Khatoon, S. (1992). The preliminary investigations about the changing environment of the Cholistan Desert, *Nat. Hist. Bull.*, 1(1): 1-4.

[15] Batanouny, K.H. (1981). *Ecology and Flora of Qatar.* Centre for scientific and applied Research, University of Qatar, P.O. Box 2713, Doha.

[16] Bhandhari, M.M. (1978). *Flora of Indian Desert.* Scientific Publishers, Jodhpur.

[17] Bhatti, G.R., Muqarrab, S. and Qureshi, R. (1998-2001). *Floristic study of arid zone (Desert Nara, Region), Sindh, Pakistan,* Final Technical report, PSF Project. No. S-SALU/ENVR (45).

[18] Boulos, L. (1991). *Flora of Egypt.* Al Hadara Publishing Cairo, Egypt, Vol. 1.
[19] Chaudhri, I.I. and Chuttar, M.S. (1966). *The vegetation and range Flora of Thar desert,* W. Pak. For. dept., Hyderabad.
[20] Cooke, T.C.I.E. (1903-1906). *The Flora of the presidency of* Bombay, Botanical Survey of India, Calcutta. Vol. I-III.
[21] Cope, T.A. in Nasir, E. & Ali, S.I. (1982). *Flora of Pakistan,* No. 143 (Poaceae), Islamabad, Karachi.
[22] Cronquist, A. 1981. *An Integrated System of Classification of Flowering Plants.* Columbia University Press, New York, pp. 1-1262.
[23] Dar, M.I. in Nasir, E. & Ali, S.I. (1972). *Flora of Pakistan,* No. 52 (Polygalaceae), Islamabad, Karachi.
[24] Ghafoor, A. in Nasir, E. & Ali, S.I. (1972). *Flora of Pakistan* 19 (Elatinaceae), Department of Botany, University of Karachi, Pakistan
[25] Ghafoor, A. in Nasir, E. & Ali, S.I. (1974). *Flora of Pakistan,* No. 75 (Tiliaceae), Islamabad, Karachi.
[26] Ghafoor, A. in Nasir, E. & Ali, S.I. (1974). *Flora of Pakistan,* No. 76 (Zygophyllaceae), Islamabad, Karachi.
[27] Ghafoor, A. in Nasir, E. & Ali, S.I. (1985). *Flora of Pakistan,* No. 51 (Portulacaceae), Islamabad, Karachi.
[28] Ghafoor, A. in Nasir, E. & Ali, S.I. (1985). *Flora of Pakistan,* No. 79 (Potamogetonaceae), Islamabad, Karachi.
[29] Ghafoor, A. in Nasir, E. & Ali, S.I. (1985). *Flora of Pakistan,* No. 171 (Moraceae), Islamabad, Karachi.
[30] Ghafoor, A. in Nasir, E. & Ali, S.I. (1985). *Flora of Pakistan,* No.169 (Hydrocharitaceae), Islamabad, Karachi.
[31] Ghazanfar, S.A. & Nasir, Y.J. in Nasir, E. & Ali, S.I. (1986). *Flora of Pakistan,* No. 175 (Caryophyllaceae), Islamabad, Karachi.
[32] Hedge, I.C. in Nasir, E. & Ali, S.I. (1990). *Flora of Pakistan,* No. 192 (Labiatae), Islamabad, Karachi.
[33] Jafri, S.M.H. (1966). The Flora of Karachi. The Book Corporation, Karachi, Pakistan.
[34] Jafri, S.M.H. and Ghafoor, A. in Nasir, E. & Ali, S.I. (1974). *Flora of Pakistan,* Vol. 77 (Verbenaceae).
[35] Jafri, S.M.H. in Nasir, E. & Ali, S.I. (1973). *Flora of Pakistan,* No. 34 (Capparidaceae), Islamabad, Karachi.
[36] Jafri, S.M.H. in Nasir, E. & Ali, S.I. (1973). *Flora of Pakistan*, No. 55 (Brassicaceae), Islamabad, Karachi.
[37] Jafri, S.M.H. in Nasir, E. & Ali, S.I. (1974). *Flora of Pakistan*, No. 73 (Fumariaceae), Islamabad, Karachi.
[38] Jafri, S.M.H. in Nasir, E. & Ali, S.I. (1976). *Flora of Pakistan,* No. 98 (Orobanchaceae), Islamabad, Karachi.
[39] Kassel, F.H., Edinburgh, I.C.H., Jafri, S.M.H., Groβ-Zimmern, G.H., Aylmer, S.O. and Helsinki, P.U. in Nasir, E. & Ali, S.I. (2001). *Flora of Pakistan,* No. 204 (Chenopodiaceae), Islamabad, Karachi.
[40] Kukkonen in Nasir, E. & Ali, S.I. (2001). *Flora of Pakistan*, No. 206 (Cyperaceae), Islamabad, Karachi.

[41] Malik, K.A. in Nasir, E. & Ali, S.I. (1988). *Flora of Pakistan,* No. 188 (Acanthaceae), Islamabad, Karachi.

[42] Marwat, Q. and Siddiui, M.A. in Nasir, E. & Ali, S.I. (1997). *Flora of Pakistan,* No. 198 (Neuradaceae), Islamabad, Karachi.

[43] Matthew, K.M. (1981-3). *Flora of Tamilnadu Carnatic*, The Rapinat Herbarium, St. Joseph's College, Tiruchirapalli 620002, India, 1-3.

[44] Nasir in Nasir, E. & Ali, S.I. (1984). *Flora of Pakistan,* No. 157 (Primulaceae), Islamabad, Karachi.

[45] Nasir, E. and Nasir, Y.J. in Nasir, E. & Ali, S.I. (1987). *Flora of Pakistan,* No. 186 (Ephederaceae), Islamabad, Karachi.

[46] Nasir, E. in Nasir, E. & Ali, S.I. (1983). *Flora of Pakistan,* No. 149 (Geraniaceae), Islamabad, Karachi.

[47] Nasir, E., Ali, S.I., (eds.) 1970-1989. *Flora of Pakistan* (*Fascicle series*), Department of Botany, University of Karachi, Pakistan

[48] Nasir, E., Siddiqi, M.A. & Ali, Z. in Nasir, E. & Ali, S.I. (1969). *Flora of Pakistan,* No. 10 (Gymnosperms of West Pak), Islamabad, Karachi.

[49] Nasir, N. in Nasir, E. & Ali, S.I. (1975). *Flora of Pakistan,* No. 41 (Aizoaceae), Islamabad, Karachi.

[50] Nasir, Y. in Nasir, E. & Ali, S.I. (1971). *Flora of Pakistan,* No. 7 (Frankeniaceae), Islamabad, Karachi.

[51] Nasir, Y.J. in Nasir, E. & Ali, S.I. (1971). *Flora of Pakistan,* No. 4 (Oxalidaceae), Islamabad, Karachi.

[52] Nasir, Y.J. in Nasir, E. & Ali, S.I. (1973). *Flora of Pakistan,* No. 40 (Molluginaceae), Islamabad, Karachi.

[53] Nasir, Y.J. in Nasir, E. & Ali, S.I. (1977). *Flora of Pakistan,* No. 115 (Nyctaginaceae), Islamabad, Karachi.

[54] Nasir, Y.J. in Nasir, E. & Ali, S.I. (1979). *Flora of Pakistan,* No. 131 (Bignoniaceae), Islamabad, Karachi.

[55] Nasir, Y.J. in Nasir, E. & Ali, S.I. (1983). *Flora of Pakistan,* No. 149 (Geraniaceae), Islamabad, Karachi.

[56] Nasir, Y.J. in Nasir, E. & Ali, S.I. (1985). *Flora of Pakistan,* No.168 (Solanaceae), Islamabad, Karachi.

[57] Nasir, Y.J. in Nasir, E. & Ali, S.I. (1989). *Flora of Pakistan,* No. 191 (Boraginaceae), Islamabad, Karachi.

[58] Nizamuddin S. and Naqvi, S.H. in Nasir, E. & Ali, S.I. (1984). *Flora of Pakistan, No. 154 (Cucurbitaceae)*, Islamabad, Karachi.

[59] Omer, S. and Hashmi R.Y. in Nasir, E. & Ali, S.I. (1987). *Flora of Pakistan, No. 177 (Typhaceae)*, Islamabad, Karachi.

[60] Panhwar, M.N. (1964). *Ground water in Hyderabad & Khairpur Division.* G. sheriff at Capital Press, Hyderabad.

[61] Polhill, R. M. y J. E. Vidal. (1981). Caesalpinieae. Pp. 81–95 en: R. M. Polhill y P. H. Raven (eds.), *Advances in Legume Systematics I.* Royal Botanic Gardens, Kew, Londres.

[62] Qaiser, M. and Nazimuddin, S. in Nasir, E. & Ali, S.I. (1981). *Flora of Pakistan,* No. 140 (Rhamnaceae), Islamabad, Karachi.

[63] Qaiser, M. in Nasir, E. & Ali, S.I. (1973). *Flora of Pakistan,* No. 38 (Moringaceae), Islamabad, Karachi.
[64] Qaiser, M. in Nasir, E. & Ali, S.I. (1977). *Flora of Pakistan,* No. 110 (Aristolochiaceae), Islamabad, Karachi.
[65] Qaiser, M. in Nasir, E. & Ali, S.I. (1982). *Flora of Pakistan,* No. 141 (Tamaricaceae), Islamabad, Karachi.
[66] Qaiser, M. in Nasir, E. & Ali, S.I. (2001). *Flora of Pakistan,* No. 205 (Polygonaceae), Islamabad, Karachi.
[67] Qureshi, R. (2004). *Floristic and Ethnobotanical Study of Desert Nara Region, Sindh.* Department of Botany, Shah Abdul Latif University, Khairpur, Sindh, Pakistan. Ph.D. Thesis. Vol. I: 1-300
[68] Qureshi, R. (2008). Preliminary floristic list of chotiari wetland complex, Nawab Shah, Sindh, Pakistan. *Pak.J.Bot.*, 40(6): 2281-2288.
[69] Qureshi, S. in Nasir, E. & Ali, S.I. (1972). *Flora of Pakistan*, No. 29 (Salvadoraceae), Islamabad, Karachi.
[70] Rajput, M.T.M. and Tahir, S.S. in Nasir, E. & Ali, S.I. (1988). *Flora of Pakistan, No. 189 (Cuscutaceae)*, Islamabad, Karachi.
[71] Rani, N. and Matthew in Matthew, K.M. (1983). *Flora of Tamilnadu Carnatic*, The Rapinat Herbarium, St. Joseph's College, Tiruchirapalli 620002, India, 1:747 (1981).
[72] Rani, N. and Matthew in Matthew, K.M. (1983). *Flora of Tamilnadu Carnatic*, The Rapinat Herbarium, St. Joseph's College, Tiruchirapalli 620002, India, 2: 1067-1107 (1982).
[73] Shetty, B.V. & Singh, V. (1987 & 1991). *Flora of Rajasthan*, Botanical Survey of India. Old Connaught Place Dehra Dun. Vol. I & II.
[74] Siddiqi, M.A. in Nasir, E. & Ali, S.I. (1974). *Flora of Pakistan*, No. 74 (Menispermaceae), Islamabad, Karachi.
[75] Smith, A. R. in Nasir, E. & Ali, S.I. (1986). *Flora of Pakistan, No.* 172 (Euphorbiaceae), Islamabad, Karachi.
[76] Townsend, C.C. in Nasir, E. & Ali, S.I. (1974). *Flora of Pakistan,* No. 71 (Amaranthaceae), Islamabad, Karachi.

Index

A

Abutilon fruticosum Guill. & Perr., 31
Abutilon indicum (L.) Sweet, 93
Abutilon muticum (Del.ex DC.) Sweet, 93
Acacia jacquemontii Benth., 31, 128
Acacia nilotica ssp. *cupressiformis* (J.L. Stewart) Ali & Faruqi, 127
Acacia nilotica subsp. *indica* (Benth.) Brenan, 127
Acacia senegal (L.) Willd., 127
Achyranthes aspera L., 27
Adiantum capillus-veneris L., 25, 40
Aeluropus lagopoides (L.) Trin. ex. Thw, 26, 281
Aerva javanica var. *bovei* Webb., 27, 67
Ak, 28
Akri, 29, 32
Alhagi maurorum Medic., 30, 134
Alysicarpus monilifer (L.) DC., 30, 135
Alysicarpus scariosus Grah. ex Thwaites, 136
Amaranthus graecizans subsp. *thllungianus* (Nevski) Gusev., 27
Amberboa ramosa (Roxb.) Jafri., 28
Anagallis arvensis var. *coerulea* (L.) Gouan, 32, 122
Andhri, 30
Anticharis glandulosa Aschers., 32, 207
Argyrolobium roseum ssp. *ornithopodioides* (Jaub. & Spach) Jafri & Ali, 30, 150
Aristida adscensionis L., 26, 246
Aristida funiculata Trin. & Pupr., 247
Aristida hystricula Edgew., 248
Aristida mutabilis Trin. & Rupr., 247
Aristolochia bracteolata Lamk., 27, 43
Arnebia hispidissima (Sieber) DC., 28, 201
Arthrocnemum indicum (Willd.) Moq., 29, 59
Asphodelus tenuifolius Cavan, 26
Avena fatua L., 26, 250

B

Bacopa monnieri (L.) Wettestein in Engl. & Prantl., 32, 209
Bahan, 32
Bakhro, 33
Banak, 27
Banwar, 31
Barleria acanthoides Vahl, 27, 214
Barr, 31
Basri, 26
Bepari, 30
Bergia aestivosa Wight & Arn., 30
Bhangrio, 28
Bhattar, 28
Bheekar, 30
Bhunjro, 31
Bhurt, 26, 28, 33
Bili buti, 32
Blepharis sindica Stocks ex. Anders, 27, 213
Boerhavia diffusa L., 31, 51
Boerhavia rubicunda Steud., 50
Bolboschoenus affinis (Roth.) Drobov, 25
Bolboschoenus glaucus (Lam.) S.G. Smith, 240
Boobak, 27
Booh, 27
Booro, 27
Brachiaria eruciformis (J. E. Smith) Griseb., 26, 258
Brachiaria ovalis (L.) Stapf, 259
Brachiaria ramosa (L.) Stapf, 259
Bukkan, 33

C

Cadaba farinosa Forssk., 29, 116
Caesalpinia bonduc (L.) Roxb., 28, 129
Calligonum polygonoides L., 32, 79

Calotropis procera (Willd.) R. Br., 28, 176
Capparis decidua (Forssk.) Edgew., 29, 114
Capparis spinosa L., 115
Carthamus oxycantha Bieb., 28, 220
Celosia argentea L., 27, 71
Cenchrus biflorus Roxb., 26, 260
Cenchrus ciliaris L., 261
Cenchrus pennisetiformis Hochst. & Steud. ex Steud., 262
Cenchrus prieurii (Kunth) Maire, 261
Chagg, 30
Charyo Dhatooro, 32
Chenopodium album L., 29, 61
Chenopodium murale L., 62
Chhabar, 26
Chhipri, 31
Chibhri Wal, 29
Chil, 29
Chloris barbata Sw., 26, 245
Chrozophora plicata (Vahl) A. Juss. ex Spreng., 30, 158
Cirsium arvense (L.) Scop., 28, 220
Cistanche tubulosa (Schrenk) Hook. f., 31, 211
Citrullus colocynthis (L.) Schrad., 29, 106
Cleome brachycarpa Vahl., 29
Cleome scaposa D.C., 111
Cleome viscosa L., 112
Cocculus hirsutus (L.) Diels, 31, 44
Cocculus pendulus (J.R. & Forst) Diels, 45
Commicarpus boisseiri (Heimer) Cufod., 31, 51
Commiphora stocksiana (Engl.) Engl., 28, 166
Commiphora wightii (Arn.) Bhandari, 166
Convolvulus arvensis L., 29, 189
Convolvulus prostratus Forssk., 189
Corchorus aestuans L., 33, 88
Corchorus tridens L., 88
Corchorus trilocularis L., 87
Cordia gharaf (Forssk.) Ehren. ex Asch., 28, 195
Cordia myxa L., 195
Coronopus didymus (L.) Smith, 28, 120
Cressa cretica L., 29, 187
Crotolaria burhia Ham. ex Benth., 30
Ctenolapis cerasiformis (Stocks.) Hooker.f., 29, 103
Cucumis melo var. *agrestis* Naudin., 29
Cucumis prophaterum L., 105
Cuscuta chinensis Lam., 30, 193
Cymbopogon jwarancusa (Jones) Schult., 26, 269
Cynodon dactylon (L.) Pers., 26, 244
Cyperus arenarius Retz., 26, 233
Cyperus articulatus L., 233
Cyperus bulbosus Vahl, 236
Cyperus conglomeratus Rottb., 234
Cyperus difformis L., 232
Cyperus iria L., 235
Cyperus laevigatus L., 237
Cyperus longus L., 236
Cyperus rotundus L., 235

D

Dactyloctenium aegyptium (L.) P. Beauv., 26, 273
Dactyloctenium aristatum Link, 273
Dactyloctenium scindicum Boiss., 274
Dakhri, 31
Datehri, 33
Datura fastuosa L., 32, 181
Deer, 27
Desmostachya bippinata (L.) Stapf., 26
Devi, 31
Dhanaar Khathoori, 29
Dhoor Chhapri, 27
Dichanthium annulatum (Forsskal) Stapf., 26
Digera muricata (L.) Mart., 27, 70
Digitaria sanguinalis (L.) scop., 26, 257
Dinebra retroflexa (Vahl.) Panzer., 26
Dinohi, 26
Dipterygium glaucum Decne., 29, 117
Doonhi Buti, 31
Drabh, 26, 27
Dramaaho, 33
Drebarr, 30

E

Echinochloa colona (L.) Link, 26, 255
Echinops echinatus DC., 28, 217
Eclipta prostrata (L.) L., 28, 224
Ephedra ciliata Fisch. & Mey. ex C.A.Meyer., 25
Equisetum arvense L., 25, 39
Eragrostis barrelieri Dav., 26, 280
Eragrostis cilianensis (All.) Lut. ex F.T. Hubbard, 279
Eragrostis ciliaris (L.) R.Br., 278
Eragrostis tenella (L.) P. Beauv. ex Roem. & Schult., 278
Euphorbia caducifolia Haines, 30, 156
Euphorbia clarkeana Hook. f., 157
Euphorbia hirta L., 156
Euphorbia thymifolia L., 157

F

Fagonia bruguieri DC. var. *rechingeri* Had, 33
Fareed Buti, 31
Farsetia hamiltonii Royale., 28

Farsetia jacquemontii Hook. f. & Thoms., 119
Ficus bengalensis L., 48
Ficus religiosa L., 31, 47
Fimbristylis bisumbellata (Forssk.) Bubani, 26, 240
Fimbristylis cymosa R. Br., 241
Frankenia pulveulenta L., 30, 100
Fumaria indica (Hausskn.) H.N. Pugsley, 31, 46

G

Gaamesh, 32
Gaaozaban, 28
Gandheer Gaah, 26
Gandraf, 28
Ghora Wal, 29
Gisekia pharancoides L., 27, 55
Glinus lotoides L., 31, 74
Golaro, 29
Goomi Buti, 31
Grewia tenax (Forssk.) Aschers & Schweinf., 33
Gugur, 28
Gynandropsis gynandra (L.) Briq., 29, 110

H

Haloxylon salicornicum (Moq.) Bunge ex Boiss., 29, 60
Haloxylon stocksii (Boiss.) Benth. & Hook., 60
Harnub, 28
Hazar dani, 30
Heliotropium crispum Desf., 28, 198
Heliotropium curassavicum L., 196
Heliotropium europeum L., 197
Heliotropium subulatum (Hochst. ex DC.) Vatke., 199
Hibiscus lobatus (J.A Murray) O. Ketze., 31, 95
Hiran Buti, 32
Hydrilla verticillata (L. f.) Royale, 31, 228

I

Imperata cylindrica (Linn) Raeuschel., 27, 268
Indigofera argentea Burm.f., 30, 145
Indigofera cordifolia Heyne ex Roth, 144
Indigofera hochstetteri Baker, 142
Indigofera linifolia (L.) Retz., 144
Indigofera sessiliflora DC., 143
Iphiona grantioides (Boiss.) Anderb., 28
Ipomoea aquatica Forssk., 29, 191

J

Jand Laani, 33
Jangli Jai, 26
Jantur, 30
Jar, 32
Jhangli halon, 28
Jhangli khaji, 27
Jhangli matri, 30
Jhangli palak, 32
Jhangli toori, 29
Jhangoori Ber, 32
Jhil, 30

K

Kabah, 26
Kal, 26
Kalar Gaah, 26
Kaluro, 25, 26
Kamho, 30
Kanderi Bhattar, 28
Kanderi Wal, 32
Kandero, 30
Kandi, 31
Kando Mitero, 29
Kanh, 27
Kanwal Buti, 32
Karash, 26
Kattan, 26
Khabar, 29
Khabri Wal, 29
Kharsan, 28
Khathoori, 29
Khatti buti, 32
Khaunr, 31
Kheera Wal, 30
Kher Wal, 30
Khip, 28
Kickxia ramosissima (Wall.) Janchen, 32, 208
Kinni Buti, 29
Kinro, 29
Kirar, 29
Kirhanj, 29
Kotak, 31
Kuari lonak, 32

L

Lai, 33
Lani, 29
Lano, 29

Lasiurus sindicus Henr., 27, 271
Lathio, 28, 30
Lathyrus aphaca L., 30, 137
Launaea procumbens (Roxb) Ramayya & Rajagopal., 28, 222
Launaea resedifolia (L.) Kuntze, 221
Lawo, 33
Leh, 28
Leptadenia pyrotechnica (Forssk.) Decne., 28
Leptothrium senegalense (Kunth) W.D. Clayton, 27, 243
Lesuro, 28
Leucaus aspera (Willd.) Link., 31
Limeum indicum Stocks ex T. Anderson., 27
Lohiro, 28
Lonak, 27, 32
Luffa echinata Roxb., 29, 102
Lulur, 27
Lumb Gaah, 26, 27
Lusani, 25
Lycium edgeworthii Dunal., 32

M

Machhi Gaah, 31, 32
Maerua crassifolia Forssk., 29, 113
Makhni Gaah, 26
Malvastrum coromendelianum (L.) Garcke, 31, 91
Mandhano, 27
Mariro, 27
Marsilea minuta L., 25, 40
Matri, 30
Medicago polymorpha L., 30, 139
Melhania denhamii R. Br., 32, 89
Melilotus alba Desr., 30, 151
Melilotus indica (L.) All, 152
Mena, 30
Mitero, 29
Mohabbat buti, 26
Mollugo cerviana (L.)Ser., 31
Mollugo nudicaulis Lamk., 75
Mollugo pentaphylla L., 76
Moniah Gaah, 26
Monsonia heliotropoides (Cav.) Boiss., 31
Moringa oleifera Lamk., 31, 121
Mukia maderaspatana (L.) M. J. Roem., 29
Mundheri, 33

N

Nar, 27
Naro, 27, 29
Neurada procumbens L., 31, 123
Nil Kanthi, 30
Nor puchh, 27

O

Ochthochloa compressa (Forssk..) Hilu., 27
Oin, 29
Oxalis corniculata L., 32, 174
Oxystelma esculentum (L.f.) R. Br., 28, 178

P

Pahar Wal, 28
Pan, 27, 259
Paneer, 32
Paneer Wal, 32
Panicum turgidum Forsk., 27, 254
Paspalidium geminatum (Forssk.) Stapf in Prain, 27, 256
Paspalum paspalodes (Michx.) Scribner, 27, 265
Pat Teer, 31
Pavonia glechomaefolia S. Abedin, 31
Pentatropis nivalis (J.F.Gmel.) Field & J.R.I.Wood, 28, 179
Peroon, 32
Persicaria barbata (L.) Hara, 32, 81
Persicaria glabra (Willd.) M. Gómes, 81
Phaar Buti, 28
Phair, 29
Phalaris minor Retz., 27, 251
Phoenix sylvestris (L.) Roxb., 27, 230
Phog, 32
Phragmites australis (Cav.) Trin. ex Steud., 27, 252
Phragmites karka (Retz.) Trin. ex Steud., 253
Phuli, 28
Phyla nodiflora (L.) Greene., 33, 202
Phyllanthus maderaspatensis L., 30, 160
Phyllanthus reticulatus Pair., 161
Physalis minima L., 32
Pipul, 31
Pluchea lanceolata (DC.) Oliv. & Hiern., 28
Pluchea wallichiana DC., 226
Pohli, 28
Polygala erioptera DC., 32, 164
Polygonum plebejum R.Br., 32, 82
Polypogon fugax Nees ex Steud.,, 27, 250
Populus euphratica Olivier., 32, 108
Potamogeton perfoliatus L., 32, 229
Prosopis cineraria (L.) Druce, 31, 125
Prosopis juliflora (Swartz) DC., 125
Pursha, 25

R

Rhynchosia minima (L.) DC., 30, 140
Rhynchosia pulverulenta Stocks, 141
Ricinus communis L., 30, 159
Rumex dentatus subsp. *klotzschianus* (Meisn.) Rech. f., 32, 80

S

Saccharum bengalense Retz., 27, 266
Saccharum spontaneum L., 267
Sacho Lano, 29
Salsola imbricata Forssk., 29, 63
Salvadora oleoides Decne., 32, 153
Salvadora persica L., 153
Salvia santolinifolia Boiss., 31, 204
Santh, 25
Sanwak, 26
Saueda fruticosa (L.) Forssk., 29
Sawri, 26, 27
Schoenoplectus litoralis (Schrad.) Palla, 26, 238
Schweinfurthia papilionacea (Burm. f.) Boiss., 32, 210
Seddera latifolia Hochst. et Steud., 29, 188
Seetzania lanata (Willd.) Bullock., 33
Senna holosericea (Fresen.) Greuter, 29, 131
Senna italica Mill., 130
Sesbania bispinosa (Jacq.) W. F. Wight, 30, 150
Sesuvium sesuvioides (Fenzl) Verdc., 27, 57
Setaria pumila (Pouir.) Roem. & Schult., 27, 263
Sewan, 27
Shahtaro, 31
Sindhi senna, 29
Sinjhi, 30
Siringh, 30
Solanum surattense Burm.f., 32, 185
Sonchus asper (L.) Hill., 28, 223
Sonchus oleraceus L., 223
Spergula arvensis L., 29, 78
Stipagrostis plumosa (L.) Munro ex T. Anders., 27, 249
Suhandaro, 31
Suhanjro, 31

T

Taklo, 29
Tamarix aphylla (L.) Karst., 33, 98
Tamarix dioica Roxb. ex Roch, 97
Tamarix indica willd., 98
Tamarix pakistanica Qaiser, 97
Tecomella undulata (Sm.) Seem., 28, 215
Tephrosia falciformis Romaswami., 30
Tephrosia purpurea (L.) Pers., 149
Tephrosia strigosa (Dalz.) Sant. & Mahesh., 147
Tephrosia villosa (L.) Pers., 148
Thor, 30
Trianthema portulacastrum L., 27, 54
Tribulus longipetalus Viv., 33
Trichodesma indicum (L.) R.Br., 28, 200
Trooh, 29
Typha domingensis Pers., 27, 282
Typha elephantina Roxb., 283

U

Ubat Kandri, 27
Utangan, 27

V

Vaccaria hispanica (Miller) Rauschert, 29, 77
Vekar, 30
Vicia sativa L., 30, 138

W

Waho, 27
Wan vehri, 30
Wasanh, 27
Withania coagulans (Stocks) Dunal, 32, 183
Withania somnifera (L.) Dunal, 184

X

Xanthium strumarium L., 28, 219

Z

Zeerawund, 27
Zelya petandra (L.) Jeffrey, 27
Zizyphus nummularia (Burm. F.) Wt. & Arn., 32, 162
Zygophyllum simplex L., 33, 170